Separations Division 2019

Held at the 2019 AIChE Annual Meeting

Orlando, Florida, USA
10 - 15 November 2019

ISBN: 978-1-7138-0526-7

Printed from e-media with permission by:

Curran Associates, Inc.
57 Morehouse Lane
Red Hook, NY 12571

AIChE
120 Wall Street, FL 23
New York, NY 10005-4020

Phone: (800) 242-4363
Fax: (203) 775-5177

www.aiche.org

Additional copies of this publication are available from:

Curran Associates, Inc.
57 Morehouse Lane
Red Hook, NY 12571 USA
Phone: 845-758-0400
Fax: 845-758-2633
Email: curran@proceedings.com
Web: www.proceedings.com

TABLE OF CONTENTS

571591 Demystifying Mass Transfer in Electrolyzers through Rapid Prototyping with 3D Printed Parallel Plate Cells

As a result of the global energy transition, a large percentage of the world's power supply will become electric. For electrochemical engineers this transition generates a sea of opportunities. After all, it is electrolyzers that use electricity to produce some of life's essential chemicals. However, fully embracing these opportunities in a sustainable way is challenging: Future electrolyzers must be safer, cleaner, and far more flexible, while simultaneously increasing productivity and cost-effectiveness. Essentially, they will have to achieve more with less.

One key parameter in the performance of electrolyzers is the rate of mass transfer. Most often, electrochemical reactions are mass transfer limited due to the fast kinetics that depend exponentially on the applied overpotential. Numerous correlations in literature exist to describe the rate of mass transfer in a specific reactor.[1] Though, their predictions can vary up to almost an order of magnitude (see figure 1). The reason for this is the large geometric variety between the reactors that were measured. The design of the inlets in particular is a crucial contributor to the rate of mass transfer. [2] Moreover, only a few lab-scale electrolyzers are commercially available, often with sub-optimal designs, which limits further research. As a consequence, it is difficult to establish a generalized description of the mass transfer performance of parallel plate electrolyzers. Furthermore, most existing correlations only describe the average rates of mass transfer and therefore do not consider any local variation despite the importance thereof.[3]

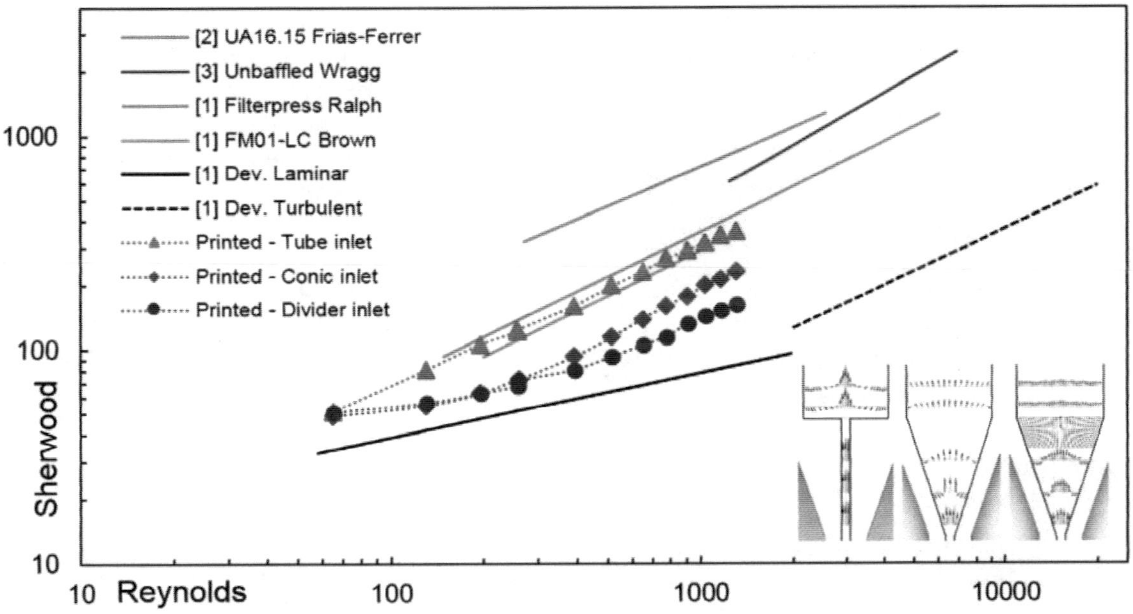

Figure 1: Mass transfer performance in terms of Sherwood vs. Reynolds. Up to an order of magnitude of difference in performance is visible between reactors. The red triangle, square and circle markers denote a selection of inlet manifolds tested with the 3D printed prototype. Small insert: sketches of the inlet manifolds, from left to right: tube inlet, conic inlet, divider-type inlet.

The goal of our study has been to demystify the fundamentals of mass transfer in parallel plate electrolyzers. As this would have required testing a large geometric variety of cells, traditional manufacturing techniques were impractical due to their lead time and cost. Therefore, we have used 3D printing to rapidly construct many different reactor prototypes. Moreover, each prototype could be finetuned to the exact specifications each measurement required. This way, each geometric parameter could be varied and investigated individually.

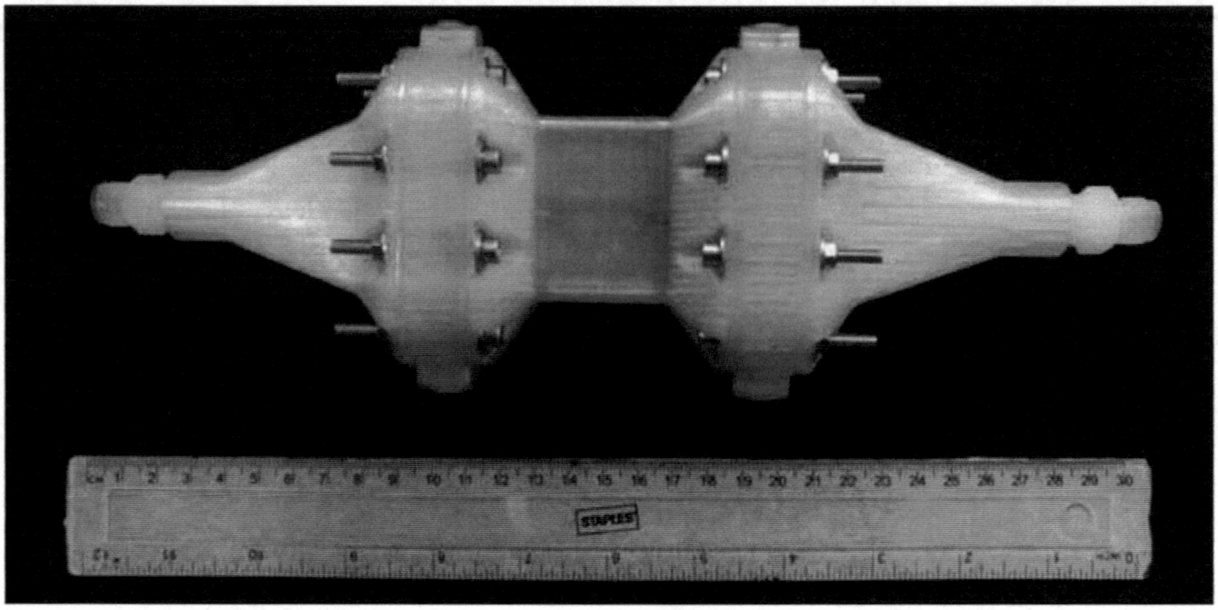

Figure 2: 3D printed electrolyzer prototype for mass transfer studies. The electrode area is 40 cm^2.

Nearly the entire reactor can be 3D printed in less than a workweek's time (see figure 2). Inlet manifolds, the channel and the rubber sealings could all be printed using commercially available equipment. Furthermore, by using 32 independently operated electrode segments we were able to measure the mass transfer performance on a local scale. Sherwood – Reynolds correlations were established for designs with different shapes of turbulence promoters, 3D electrodes and inlet manifolds. The effect of an increased or decreased hydrodynamic entrance length was also investigated.

In the end these prototypes lead to enhanced understanding of electrochemical reactors, allowing us to intensify processes such as alkaline water electrolysis. As a proof of concept, we are developing a 3D printed lab-scale electrolyzer for alkaline water splitting which can be used to test the effect of geometry on the production of hydrogen directly.

References

[1] T.R. Ralph, M.L. Hitchman, J.P. Millington, F.C. Walsh, *Mass Transport in an electrochemical laboratory filterpress reactor and its enhancement by turbulence promoters,* Electrochem. Acta, **1994,** pp. 591-603.

[2] A. Frías-Ferrer, J. Gonzálex-García, V.Sáez, C. Ponce de León, F.C. Walsh, *The effects of manifold flow on mass transport in electrochemical filter-press reactors,* AIChE J., **2008,** pp. 811-823.

[3] A.A. Wragg, A.A. Leontaritis, *Local mass transfer and current distribution in baffled and unbaffled parallel plate electrochemical reactors,* Chem. Eng. J., **1997,** pp. 1-10.

574364 Additive Manufacturing of Catalyst Support Structures with in Operando Adjustable Mass Transport and Flow Characteristics

In conventional randomly packed bed reactors, the flow and mass transport characteristics are predefined by the particle diameter and the tube diameter. Structured reactors offer the opportunity to tailor mass transport and flow field according to the needs of the reaction system. In this regard, additively manufactured so-called Periodic Open Cellular Structures (POCS) as catalyst supports have shown their advantages over conventional randomly packed bed reactors in terms of heat management and pressure drop [1, 2]. However, since the internal flow field, pressure drop and therefore the resulting mass transport characteristics are mainly influenced by the geometrical properties of the unit cell, for POCS the resulting transport characteristics of the structure is already determined and fixed by the choice and design of the unit cell. To achieve more flexibility, we have recently proposed a new type of so-called interpenetrating POCS (interPOCS) which are formed by inserting a diamond unit cell based structure in the cavities of a second diamond cell structure. These interPOCS allow for in operando flow field and mass transport adjustments by shifting the (relative) position of the loose structure [3].

With detailed computational fluid dynamics (CFD) based simulations with an additional in-house particle-tracking algorithm implementation we systematically investigated the flow field and mass transport characteristics within the interPOCS in dependency of the relative positioning of the loose structure. The highly adjustable structure allows for a broad variation of mass transport characteristics in a catalytic reactor without changing the system's periphery. This functionality enables mass transport optimization for a given system and represents a completely new application of additively manufactured catalyst support structures for highly flexible in operando tuning of flow field and mass transport characteristics.

References

[1] C. Busse, H. Freund, and W. Schwieger (2018). *Chem. Eng. Process.* 124, 199-214.

[2] A. Inayat, M. Klumpp, M. Lämmermann, H. Freund and W. Schwieger (2016). *Chem. Eng. J.* 287, 704-719.

[3] G. Do, T. Stiegler, M. Fiegl, L. Adler, C. Körner, A. Bösmann, H. Freund, W. Schwieger and P. Wasserscheid (2017). *Ind. Eng. Chem. Res.* 45, 13402-13410.

561588 Role of 3D Printed and LED-Driven Photostation in Photocatalytic and Photo-Fenton Activity of Iron Oxide Doped Graphitic Carbon Nitride

The degree of photocatalysis or any photolysis reaction depends on the intensity of light. In most cases, a major portion of light is not used as it does not reach the reaction zone thereby wasting a significant portion of light and energy, as well. In order to handle this situation, we have designed and 3D printed a photostation with complex parabolic reflective surface, so that all the incident light beams can reach the reaction zone. The photostation is fitted with two chips-on-board (COB) light emitting diodes (LEDs) reaching the total (visible) light intensity within the reaction zone over 290,000 lux (For comparison, it is commonly reported that the sun emits anywhere from 32,000 to 100,000 lux onto the Earth). Light simulation analysis confirmed that all incident light beams can reach the reaction zone of photostation. In order to test the performance of this photostation, we have used 0.5 to 2 wt.% iron oxide doped graphitic carbon nitride (g-C_3N_4) to degrade a model water pollutant, dicamba (3,6-Dichloro-2-methoxybenzoic acid), which is common herbicide, in presence of traces of hydrogen peroxide and visible light. It has been confirmed experimentally that the rate of degradation of dicamba is directly proportional to the light intensity and geometry of photostation thereby confirming the positive role of photostation design. The rapid degradation of dicamba under this study can be attributed to the four distinct phenomena of photocatalytic activity of carbon nitride, quenching of electron/hole pairs and generation of additional reactive hydroxyl radicals by hydrogen peroxide, Fenton and photo-Fenton activity of iron oxide component of carbon nitride in presence of hydrogen peroxide and photocatalytic activity of iron oxide alone in conjuncture with carbon nitride. Liquid Chromatography Mass spectrometry (LCMS) analysis of the degradation products showed loss of chlorine from the aromatic ring and evidence of free radical addition reactions in the course of photocatalysis.

557053 Solution-Based 3D Printing of Hierarchical Microporous Polymers for Energy-Efficient Adsorption

Current 3D printing methods have strong limitations in the classes of polymers that can be printed. This means that many polymers of technological interest cannot currently be 3D printed. Here, we demonstrate a generalizable method for 3D printing of viscoelastic polymer solutions and apply it to both porous (a polymer of intrinsic microporosity, PIM-1) and non-porous (e.g., cellulose acetate, Matrimid) polymers. Successful ternary ink formulations that result in the creation of three-dimensional structures with hierarchical porosity require balancing of solution thermodynamics (phase separation), mass transfer (solvent evaporation), and rheology. To demonstrate the utility of this approach, a microporous polymer (PIM-1) that cannot be printed with current 3D printing technologies is 3D printed into a high-efficiency respirator exhibiting hierarchical porosity ranging from sub-nanometer to millimeter pores. These short mass transfer contactors (1.27 cm) were challenged with a flow of toluene vapor (10000 ppm) in N_2 gas. Owing to the precisely designed gas distribution channels, the 3D printed respirator were found to create a pure (< 1 ppm toluene) N_2 stream for 1.7 hours, 6 times longer than traditional structures fabricated from the same PIM-1 material, and more than 4,000 times the residence time of gas in the bed. This solution-based additive manufacturing approach significantly extends the material spectrum of 3D printing technologies. By 3D printing adsorptive materials into precisely designed mass transfer contactor, the performance, energy efficiency, and material efficiency can be significantly improved.

559790 Titanium/Titania Electrocatalyst for Hydrogen Evolution Synthesized Via a Combination of 3D Printing and Anodization

A titanium core/titania coating, hierarchically structured catalyst was fabricated via combining 3D printing and in-situ anodization. 3D printing enabled control of the macro-porosity of the titanium core. Anodization was used in order to grow highly-ordered TiO_2 nanotubes (NTs) on the metal core. The dimensions of the TiO_2 nanotubes were controlled via tuning of anodization parameters. The combination of both fabrication techniques allows for hierarchical control of the catalyst structure.

The performance of the reported 3D Ti/TiO_2 electrocatalyst was tested for hydrogen-evolution reaction (HER) via water electrolysis. Titanium dioxide is a known photo-electro-catalysts for water splitting, but competitive electrocatalytic activity for HER in the absence of light has not been reported. Onset potentials for HER as low as -45 mV vs. RHE were obtained for our electrodes in acid medium, making them competitive with current state-of-the-art electrodes aiming to replace platinum. Tafel slopes between 40 and 52 mV dec^{-1} point towards a Volmer-Heyrovsky mechanism, with the Heyrovsky reaction as possible controlling step.

The proposed electrocatalysts feature the following advantages: (1) the fabrication method for the 3D Ti/TiO_2 catalyst is simple, scalable and can be readily incorporated as printed reactor components into different technologies; (2) the 3D Ti/TiO_2 catalyst features a flow-through architecture that not only enables more efficient utilization of the surface area of the catalyst but that could enable large-scale, continuous-flow hydrogen-production reactor technologies; and (3) minimization of electrical resistance between active surface area (TiO_2 NTs) and current collector.

556480 Advanced Manufacturing Enabling Process Intensification for Catalytic Processes

Advanced manufacturing is a transformative and disruptive approach to industrial production that enables the rapid synthesis of virtually any geometry composed of a wide range of materials, such as polymers, ceramics, and metals. Process intensification is a set of novel approaches to process engineering and equipment design that leads to substantially smaller, more selective, and energy-efficient processing plants. The integration of advanced manufacturing with process intensification accelerates process development and deployment through rapid prototyping and enables new hardware that overcomes transport limitations. Advanced manufacturing (ex. 3D printing or additive manufacturing) is the physical backbone of Industry 4.0 that can take advantage of the conclusions of big data and machine learning by allowing design freedom. This expansion of the material design window can be utilized to reduce GHG intensity for novel or existing processes.

Numerous conversion technologies in the energy industry suffer from rapid coke build-up and subsequent deactivation of a catalytic material that is exposed to a carburizing environment. The time scale of deactivation often determines the reactor technology and process configuration employed. Commercial examples of deactivating catalysts range from seconds in a fluid bed catalytic cracking unit to minutes for moving bed propane dehydrogenation and days for continuous catalytic reformers or even years for steam methane reformers. A common step for these processes is the regeneration of the catalytic material via combustion of the coke. However, the cyclic process of coking in a carburizing environment followed by combustion in an oxidizing environment can often be detrimental to catalyst performance.

This work will discuss how advanced manufacturing methods can enable construction of materials capable of rapidly switching between two different environments with enhanced catalytic activity. Such materials capable of withstanding cyclic carburizing and oxidizing steps can be efficiently integrated into novel reactor concepts. The conversion of light gas is an excellent probe reaction for elucidating this behavior. In an effort to search for robust materials, honeycomb structures composed of metal were fabricated by direct metal laser sintering (DMLS), a 3D printing technique that involves the layer-wise successive deposition of micron-sized powder followed by laser induced sintering. The cyclic operation that alternates between carburizing and oxidizing environments leads to compositional modification at the surface of the monolith channels that imparts in-situ catalytic activity.

576800 Next Generation Charged Poly(arylene ether sulfone) Membranes: Integrating Fouling Resistance, Chlorine Tolerance, and Water/Salt Permselectivity

Poly(arylene ether sulfone)s (PAES) have been used for membrane-based separations because of their processability, oxidative stability, and their robust thermomechanical properties. Recently, we have identified a synthetic pathway to introduce a variety of functional groups to the PAES backbone, which can tune the transport characteristics and fouling resistance without degrading the chlorine tolerance. This talk will highlight the adaptable nature of the synthetic pathway and the impact the added functionality has on surface and cross-section characteristics, processability, and thermomechanical performance. In addition, the membrane performance, both the energy efficiency and permselectivity, for the treatment of saline waters with foulants and scalants will be discussed. To date, a variety of source waters with a range of salinities and compositions have been tested, which speaks to the modularity of these promising membranes. In summary, this talk will highlight recent efforts to develop a platform membrane technology suitable to remove a range of contaminants at variable concentrations in order to deliver potable water.

576179 Self-Assembling Random Zwitterionic Copolymers As Charge-Selective Nanofiltration Membranes

The purification of water is vital for many chemical separation processes. However, feedstocks containing foulants and/or chlorine represent harsh conditions in which many membranes are unable to operate. Accordingly, the development of membranes that can function effectively in such conditions is an important research endeavor. A zwitterion is defined as a molecule that contains an equal number of positive and negative charges. Random zwitterionic copolymers (RaZCs) are linear copolymers comprised of randomly distributed zwitterionic repeat units and non-zwitterionic repeat units. RaZCs self-organize to form a network of zwitterionic nanochannels that are surrounded by the non-zwitterionic repeat unit of the RaZC. The zwitterionic channels act as the effective 1 nm pores. Previous work shows that thin film composite membranes featuring these RaZCs as their selective layer exhibit high flux, exceptional fouling resistance, and size-based small molecule selectivity along with low salt rejection. However, for many applications, improved charged solute and divalent salt rejection would be beneficial. This can be achieved by incorporating charged groups into these polymers that segregate into the zwitterionic domains, and exclude ions by hindering their rate of transport through the membrane. In this work, we show that RaZCs that combine zwitterionic, charged, and hydrophobic repeat units are versatile materials for the fabrication of robust membranes that exhibit high flux, divalent ion rejection, fouling resistance, and remain stable upon exposure to chlorine. Well-designed charged RaZCs lead to membranes that exhibit salt rejections comparable to commercial nanofiltration (NF) membranes (e.g. >95% Na_2SO_4 rejection). Charged groups segregated into these nano-scale channels also impart exceptional charge-based selectivity between organic small molecules to the membrane. Furthermore, the pore chemistry of these membranes can be tuned by a simple approach, thus rationally tuning membrane selectivity for desired applications. This approach represents a unique pathway to fabricating industrially relevant membranes with tunable separation properties.

578816 Complex Multicomponent Transport through Polymeric Membranes

Polymeric membranes are utilized for a wide variety of applications: from energy production and desalination to batteries and dialysis. They are used so frequently due to their low energy requirement, low carbon footprint and selectivity in transporting molecules. Multicomponent transport through dense polymeric membranes is a complex topic whose experimental study has traditionally required aliquotic sampling which invokes hydrostatic pressure issues, complicating calculations immensely and has therefore received relatively little attention. Consequently, single component transport experiments have been used to describe membrane permeability and ideal selectivity and ultimately for selecting a membrane for a particular process. In this work, multicomponent, *in-situ* attenuated total reflectance Fourier-transform infrared spectroscopy is utilized to quantify multicomponent concentrations over time in diffusion cell experiments without the need for aliquotic sampling. The permeability and solubility of Nafion® 117 to acetone, ethanol, methanol and n-propanol are determined for single solute and multiple solute solutions to examine the impact of co-permeants through solute-solute interactions and multiple solute-membrane interactions on membrane transport behavior. Ultimately, we find these interactions can impact both the absolute and relative magnitudes of membrane permeabilities to different solutes which should be accounted for when describing membrane transport behavior.

577588 Thermodynamics of Ion and Water Uptake in Phase-Separated Ion-Exchange Membranes

Partitioning of ions and solvent from electrolyte solutions into polymer membranes is an important phenomena for water treatment and numerous electrochemical technologies. Consequently, controlling the partition coefficients of species in these materials provides an avenue to optimize performance of these technologies by tuning selectivity. To understand this process, we present and validate a mathematical model for multicomponent thermodynamic activity in phase-separated cation-exchange membranes (*e.g.*, Nafion). Specifically, this work formulates the membrane and electrolyte solution free energy. The model accounts for long-range electrostatic and short-range solvation and physical interactions between ions. A mechanics-based term accounts for the elastic swelling energy of the membrane's hydrophobic matrix. We also include steric interactions that exclude larger species from the membrane. A thermodynamically consistent treatment of the Donnan potential accounts for species charge. Externally validated solution and membrane properties parameterize the model. The model's calculations agree with measured water and ion uptake into the membrane in dilute and concentrated binary and ternary salt electrolytes. The results show that water and ion uptake is a balance between osmotic and elastic pressures with ion specificity and membrane pretreatment affecting this balance.

576936 Functionalized Nanoporous Ceramic Membranes Towards Low-Cost Electrodialysis

The State of California has experienced annual snow pack depletion and persistent drought that is impacting agriculture, industry and day-to-day life. They are not alone as many parts of the world face a fresh water crisis. Interestingly, California, for example, has 38 times more brackish water than fresh water. Electrodialysis is known to be one of the most efficient technologies for purification of brackish water and yet it is often not considered as a leading solution to the fresh water crisis. This is in part due to the cost and performance of the ion exchange membranes (IEM). IEMs suffer from high cost, high resistance, poor durability and other problems. In this presentation, we discuss the design and scale-up of novel, functionalized ceramic IEMs that are ion selective, experience near-zero swelling and are inexpensive at 30% of the production cost of current IEMs.

In this work, we investigate the use of a surface-modified inorganic membrane created using sol-gel processing of siloxanes without calcination or sintering. Silica is an advantageous material for membranes because of its excellent durability, extremely low cost and lack of swelling in water. Membrane performance can be tuned by size exclusion (*i.e.,* sub-5nm nanometer pores) and surface functionalization (*i.e.,* silane chemistry) to achieve promising anion and cation-selective performance. Permselectivity and water permeance of these membranes are shown to be superior to an industry standard Neosepta, while area specific resistance has room for improvement. Area-normalized ion exchange capacities and accelerated degradation tests show nearly identical performance between the functionalized ceramic and Neosepta membranes. Design strategies for continued optimization will also be discussed. These promising preliminary results highlight the commercial potential for a new, inexpensive IEM with the potential to elevate the status of electrodialysis as a leading solution to the fresh water crisis.

565643 High Temperature Proton Exchange Membrane Based on Polycation – Polybenzimidazole Blends

Phosphoric acid doped polybenzimidazole (PA – PBI) has been the most commonly used membrane for high temperature polymer electrolyte membrane fuel cell and hydrogen pump applications. Some of the limitations of PA – PBI arise from their weak acid – base interaction leading to PA loss under two scenarios: i.) temperatures above 180 °C that causes PA evaporation and ii.) low humidity operations were water supplants PA favorably within the acid-doped membrane matrix. PA doped polycation based membranes exhibit strong cation-anion pair electrostatic interactions that aid in PA retention up to 220 °C and curtail PA loss in the presence of water. However, several PA-polycations variants examined in our lab display lower proton conductivity when compared to PA – PBI because the PBI host features a higher PA loading and in essence a higher effective ion-exchange capacity (i.e., more imidazole groups per tethered cation groups).

This talk will discuss a new class of high temperature anhydrous proton conducting membrane based upon PA – PBI – polycation blends. The polycation is based upon quaternary benzyl pyridinium polysulfone. The blended membranes doped with PA displayed proton conductivity up to 0.28 S cm^{-1} with stability up to 240 °C. Similar proton conductivity results were attained for thin film ionomer variants implying that this material, unlike Nafion, does not lose its proton conductivity under thin film confinement in electrode layers. Finally, the talk will present fuel cell, electrolysis, and supercapacitor device performance at temperatures well above 100 °C demonstrating the versatility and effectiveness of these new membrane and solution processable ionomer binder materials.

561065 Integrating a CO_2-Selective Membrane Process into a Solid Oxide Fuel Cell System to Enhance H_2 Utilization

The recovery of H_2 from the anode exhaust of a high-temperature solid oxide fuel cell (SOFC) system is an emerging method to increase the H_2 utilization, and a polymeric membrane with a high CO_2/H_2 selectivity is the key for this separation task. However, one of the challenges is to maintain the high selectivity at a high temperature of 120 °C. The membrane reported here contained a polyelectrolyte with quarternaryammonium ions as the cations bound to the polymer backbone. Fluoride was chosen as the counterion as it can catalyze the reaction between CO_2 and water for enhancing CO_2 permeance. The composition also included a boric acid to further catalyze the reaction as well as a hydroxide-based mobile carrier for the facilitated transport of CO_2. The membrane showed a CO_2 permeance of 100 GPU and a CO_2/H_2 selectivity of 116 at 120 °C. Moreover, CO_2 partial pressures and membrane hydration were systematically studied to generate membrane performance data for the design of a membrane system.

Next, a high-level techno-economic analysis was used to explore the integration of the membrane process into the SOFC system. Owing to the high selectivity of the membrane, a single stage membrane system was able to achieve a 99% H_2 recovery with a CO_2 removal of 46.9%. Based on the CO_2 permeance of 100 GPU and the CO_2/H_2 selectivity of 116, the CO_2 removal cost was calculated to be $62.8/tonne for a set of optimized parameters. In order to improve the process further, a vacuum has been proposed to be pulled on the permeate side so that the removed CO_2 can be captured, which will decrease the carbon emission of the SOFC process.

Dr. Salim's current affiliation: Membrane Technology and Research, Inc. (MTR), Newark, CA

569011 A Carbon Molecular Sieve Membrane-Based Reactive Separation Process for Pre-Combustion CO_2 Capture

Abstract

Reactive separation processes, such as membrane reactors (MRs) and adsorptive reactors (ARs), are attracting increased interest for industrial applications. We study here a hybrid system combining a MR and an AR in tandem (with the AR following the MR, and the MR's reject stream serving as the AR's feed). We apply this system to the water gas shift (WGS) reaction in the context of its application in the IGCC process for H_2 generation and simultaneous CO_2 capture from coal/biomass. The MR-AR hybrid system under study attains a high WGS reaction conversion exceeding equlibrium, produces a pure H_2 product for power generation, and delivers a high-pressure CO_2 stream ready for sequestration. In our research, specifically, we use carbon molecular sieve (CMS) membranes. Following the field-testing and performance-validation of these CMS membranes under coal/biomass gasification conditions, experiments have been carried-out in the laboratory to determine the membrane characteristics, and the MR performance under the IGCC-relevant conditions, i.e., for temperatures up to 300 °C and pressures up to 25 bar, with different syngas compositions. In our research, the CMS membranes and the commercial sour-shift WGS catalyst ($Co/Mo/Al_2O_3$) we use in the MR have displayed very robust and stable performance during a long-term run (~750 hr of H_2S exposure). Further, we have experimentally evaluated the MR-AR hybrid system in multiple-cycle runs (10-16 cycles, with the AR in the reaction/adsorbent regeneration operation mode), and the system has demonstrated superior performance to that of a conventional PBR with high purities for the hydrogen product which can be directly usable in a hydrogen turbine for power generation. As part of this work, we have also carried-out parametric studies for optimization of the operation of the MR-AR hybrid system by investigating various operating conditions for both the MR and the AR, and a detailed model for the MR-AR system was also developed and validated using the experimental data. Using this model, a technical and economic analysis (TEA) was carried-out for process design/optimization and economic evaluation of the WGS-MR-AR system for a broad range of operating conditions and design parameters. The experimental and model findings manifest the ability of the WGS-MR-AR process to operate under the desired conditions and to improve the efficiency of the WGS reaction. They demonstrate, in addition, the potential of the system to carry out the in-situ separation of H_2 and CO_2 using the CMS membranes. We conclude from our study, that the CMS-membrane-based WGS-MR-AR system is a good candidate technology for incorporation into IGCC power plants for environmentally-benign power generation.

577728 Enhancing CO_2 Permeability of Cross-Linked Poly(ethylene oxide) By Doping with Crown Ethers and Metal-Organic Frameworks for CO_2 Capture

Poly(ethylene oxide) (PEO) based materials are leading membrane materials for CO_2/N_2 separation with a balanced high CO_2 permeability and CO_2/N_2 selectivity. For example, amorphous copolymers prepared from macromonomers of poly(ethylene glycol) diacrylate (PEGDA, 20 wt.%) and poly(ethylene glycol) methyl ether acrylate (PEGMEA, 80 wt%) exhibit high CO_2 permeability (~590 Barrers) and good CO_2/N_2 selectivity (~47) at 35 ^oC. However, for the CO_2 capture from flue gas with enormous volume and low CO_2 partial pressure, membrane materials should have CO_2 permeability as high as possible. We demonstrate that the CO_2 permeability of PEGDA-co-PEGMEA can be further increased by doping with 18-Crown-6 and metal-organic frameworks (MOFs) without sacrificing CO_2/N_2 selectivity. The crown ether can be fully "dissolved" in the PEO matrix and increase the polymer chain flexibility. For example, adding 50 wt% of 18-Crown-6 in the copolymers decreases the glass transition temperature (T_g) from -61 ^oC to -72 ^oC. Consistently, CO_2 permeability increases from 590 Barrers to 1100 Barrers while retaining CO_2/N_2 selectivity at 45 at 35 ^oC. An addition of ~1.0 wt% UiO-66 (functionalized with NH_2) can further increase the CO_2 permeability to 1800 Barrers and retain CO_2/N_2 selectivity of 41 at 35 ^oC. Using a simulated flue gas containing 50% CO_2 and 50% N_2 at 60 ^oC, the doped copolymer shows an impressive mixed-gas CO_2 permeability of ~2200 Barrers and mixed-gas CO_2/N_2 selectivity of 25. These materials show the CO_2/N_2 separation properties above the 2008 Robeson's upper bound, indicating their promise for the practical applications. This presentation will also discuss the structure/property relationship in these interesting materials.

575097 Highly Selective Hollow Fiber Membranes for Carbon Capture Via in-Situ Layer-By-Layer Surface Functionalization

Recent developments in the field of carbon sequestration have generated a renewed effort for efficient separation of CO_2, which can be sent to a aquifers or depleted reservoirs for a long-term storage. However, current CO_2 capture technologies, including absorption and pressure swing adsorption, are expensive and contribute 70-80% to the total cost of overall carbon capture and sequestration process. In the past two decades, cost effective membrane technology for CO_2 separation has gained significant attention because of inherently low energy consumption, small carbon footprint and passive operation.

Homogenous polymeric membranes currently studied for separation of O_2 from air and CO_2 from natural gas possess an intrinsic trade-off between permeability and selectivity [1], while industrial application demands membranes with both high permeability and selectivity. Poly(ethylene oxide) (PEO) is a potential candidate for CO_2 selective membranes owing to the unique interaction of the dipole-quadrupole of polar ether groups. In Our past work, we employed layer-by-layer (LbL) technique to fabricate Polymethacrylic acid (PMAA)/polyethylene oxide (PEO) films on polystyrene which demonstrated high CO_2:N_2 (~ 135) selectivity [2]. However, these membranes had but low CO_2 permeability (0.011 barrer).

This works aims to demonstrate a cost-effective approach of deposition of highly CO_2 selective membranes via LbL route on hollow fibers (HF), provided by our collaborator at NETL. These HFs were first potted inside a set-up and polyelectrolyte solution though the setup to deposit functional thin film. After addition of 20 bi-layers of PEO/PMAA on HF substrate, the selectivity of CO_2 over N_2 increased by 3 orders of magnitude exceeding the Robeson's upper bound for homogenous polymeric films [3]. This work for the first time reports a thin deposition of membranes on HF with high selectivity (~1000) of CO_2 over N_2 as well as significant permeability (~50 barrer). The LbL technique on low-cost and mass-producible HF substrate has a potential use in carbon dioxide capture in power plants and refinery flue gases.

1. Freeman, B.D., *Basis of permeability/selectivity tradeoff relations in polymeric gas separation membranes.* Macromolecules, 1999. **32**(2): p. 375-380.
2. Song, Y., et al., *Highly selective multilayer polymer thin films for CO2/N2 separation.* Journal of Polymer Science Part B: Polymer Physics, 2017. **55**(23): p. 1730-1737.
3. Robeson, L.M., *The upper bound revisited.* Journal of Membrane Science, 2008. **320**(1-2): p. 390-400.

566490 Field Trial of Spiral-Wound Facilitated Transport Membrane Module for CO_2 Capture from Flue Gas

A field trial with 1.4 m^2 spiral-wound (SW) membrane modules fabricated from an amine-containing facilitated transport membrane was conducted with actual flue gas at the National Carbon Capture Center (NCCC) in Wilsonville, AL. Prior to the field test, the membrane modules were systematically evaluated with simulated flue gas to identify the optimal operating conditions. Significant responses to various operating conditions have been noted, particularly with respect to the feed flow rate, feed pressure, operating temperature, and CO_2 concentration. The concentration polarization behavior and feed side pressure drop were significantly affected by the choice of feed flow rate, which retroactively influences the stage cut, or the CO_2 recovery of the module.

With the actual flue gas, the membrane modules showed ca. 1450 GPU CO_2 permeance and 185 CO_2/N_2 selectivity at 67°C with feed and permeate pressures of 4 and 0.3 atm, respectively. The module performance results were essentially identical to those obtained from both membrane modules and flat-sheet membrane samples tested at OSU using simulated flue gas. A 500-h stability was demonstrated despite the interference of system upsets and flue gas outages at NCCC. During the field trial, carbon capture rates of better than 40% were achieved by a single SW module with coal-derived flue gas. Except for a few flue gas upsets, the CO_2 purities in the captured stream were about 95% or better on dry basis. In addition, a short transient time of the membrane system was observed, which benefits the dynamic integration of the carbon capture technology into the power plant. Further analysis of the tested membrane revealed that no detectable amounts of the heavy metals, Cr, As, Se and Hg, were deposited on the membrane. The data obtained from this project validate the membrane material and provide basis for the design and fabrication of the full-scale SW module with a membrane area larger than 50 m^2.

563642 High-Permeance Polymer-Functionalized Graphene Membranes That Surpass the Postcombustion Carbon Capture Target

Postcombustion carbon capture using high-performance CO_2-selective membranes has been identified as one of the most energy-efficient routes for reducing CO_2 emissions [1-3]. However, the capture performance, especially the permeance of the state-of-the-art membranes needs to be significantly increased to further improve the appeal of membranes for postconbustion capture, and importantingly, reduce the needed membrane area for separation because the capital cost of the membrane modules scales inversely with the permeance. Currently, with the exception of Polaris second generation membranes from Membrane Technology Research which yields a CO_2 permeance of 2000 GPU, there is no membrane technology that reaches the CO_2 permeance significantly higher than 1000 GPU, while maintaining a separation factor of 20 or above in the humid flue gas conditions. In this presentation, I will report a new class of sub-20-nm-thick organic-inorganic hybrid membranes, which we call as SPONG, comprising of nanoporous single-layer-graphene with a high porosity (up to 18.5%), functionalized with CO_2-phillic polymeric chains [4]. The hybrid membranes yield CO_2 permeance sixfold larger (6290 GPU) than the performance target while separating the CO_2/N_2 mixture mimicking flue gas streams with a separation factor well within the target. Overall, we report a number of membranes that yield CO_2 permeance up to 11790 GPU and CO_2/N_2 mixture separation factor up to 57.2. The hybrid membrane is conducive for scale-up because of its facile preparation involving simple coating steps.

References

1. Merkel, T. C., Lin, H., Wei, X. & Baker, R. *Memb. Sci.* **359,** 126–139 (2010).
2. Rao, A. B. & Rubin, E. S. *Sci. Technol.* **36,** 4467–4475 (2002).
3. Bara, J. E., Camper, D. E., Gin, D. L. & Noble, R. D. *Chem. Res.* **43,** 152–159 (2009).
4. He, G., Huang, S., Villalobos, L. F., Zhao, J., Mensi, M., Oveisi, E., Rezaei, M., Agrawal, K. V. Submitted, under review (2019).

571515 Facilitated Transport Nanocellulose Membranes Enhanced By Amine Moieties for CO_2 Separation

As a response to the growing emissions of carbon dioxide from anthropogenic sources, more efficient and straightforward technologies are needed, in order to separate it from flue gases. Gas separation membranes, in this concern, do represent a viable and explored alternative and several solutions have reached promising results; regular solution-diffusion based materials, however, tend to suffer from an inherent trade-off between permeability and selectivity [1]. Facilitated transport membranes (FTMs), on the other hand, represent a class of films, where CO_2 can selectively react with aminated functional groups, embedded within the polymeric matrix, and be efficiently transported on the permeate side [2]. Most promising materials for facilitated transport membranes production, however, show low mechanical stability, if not crosslinked or mixed with a supporting polymer.

Recently, another option to stabilize such materials was found in the blending of nanocellulose fibers (NFC), a 1-D nanomaterial with ever growing applications, due to its large availability, renewability and stable mechanical properties. Several works have been indeed carried on in the last years, focusing on exploring its gas separation capabilities, both in pure form and when combined with functionalized molecules [3].

To better understand the potential of nanocellulose in the preparation of facilitated transport membranes, a novel nanocomposite material was developed by coupling carboxymethylated nanocellulose (CMC-NFC) with polyvinylamine (PVAm) and L-Arginine, which represent very well-known fixed and mobile carrier system for CO_2 transport across FTMs.

Membranes were initially fabricated without the aminated polymer via a solvent casting protocol and nanocellulose was combined directly with different amounts of Arginine, an amino acid carrying 3 amine moieties in its molecular structure.

Permeation tests were conducted at low temperature (35 °C, 1 bar a) and variable humidity conditions. These showed a strong improvement of both selectivity and permeability once high loadings of the carrier were reached. Pure CMC-NFC presented a top CO_2 permeability of 29 Barrer and an ideal CO_2/N_2 selectivity of 56, but, when 45 wt% of the amino acid was added, carbon dioxide permeability increased 7-fold, up to 225 Barrer and selectivity up to 186. Both results were obtained in humidity saturated conditions.

To further increase the membrane stability, PVAm was then employed in the blend, which also served as an additional amine source. Once the two polyelectrolytes (carboxymethylated nanocellulose and polyvinylamine) were mixed together, indeed they were able to form a strong ionic interaction and a stable suspension, which could be successfully casted in self-standing films.

Despite being formed by two water soluble materials, the final composite exhibited very good resistance to water, even in liquid form, without losing integrity. Since FTMs must work in large presence of humidity, this significantly improved stability in a highly hydrated environment, respect to pure CMC-NFC and to its mixture with L-Arginine.

Thanks to its improved stability, this new nanocomposite was tested at both low (35 °C) and high (>90 °C) temperature, employing pure gases and mixtures (10 % CO_2, 90 % N_2) at medium-low pressure (1-3 bar a). At 35 °C, self-standing films, containing both PVAm and the amino acid, exhibited good, but not exceptional, performances, topping at 130 Barrer and a CO_2/N_2 selectivity of 58 in saturated conditions.

In an attempt to increase the membrane performances, new permeation tests were carried out in a different system, raising the temperature up to 96°C. Only CMC-NFC/PVAm and CMC-NFC/PVAm/L-Arginine membranes were tested in this condition as CMC-NFC/L-Arginine composite were not able to withstand that operative condition due to intrinsic fragility. Once the temperature was raised, interestingly, a significant increment of membrane properties was observed, which resulted particularly elevated for the ternary system. Membranes loaded with L-Arginine indeed increased their CO_2 permeability up to 600 Barrer at 96 °C, while in similar conditions films made exclusively by fixed carriers could not exceed 200 Barrer. Moreover, selectivity with respect to N_2 was substantially increased with values for the different materials, always higher than 80. This behavior can be considered a strong indication of transport facilitation, since the kinetic of the reactions occurring is heavily dependent on temperature and speeds up significantly with it as the theory suggests [2].

As of now, research is focused in developing supported thin films of the material, to better understand its potential in industrial scale carbon capture applications.

In conclusion, a new polyelectrolyte-based nanocomposite membrane, relying on facilitated transport mechanisms, was successfully fabricated by the combination of carboxymethylated nanocellulose and polyvinylamine. Moreover, the addition of a mobile carrier such as L-Arginine was able to greatly improve the permeation and separation performances of the material, reaching very interesting results, especially at high temperature. Due to its improved stability, tunable performances, high workability and lack of harmful solvents, this material could find its way in many gas separation applications, starting from carbon dioxide capture from flue gas.

[1] L.M. Robeson, The upper bound revisited, J. Memb. Sci. 320 (2008) 390–400. doi:10.1016/j.memsci.2008.04.030.

[2] R. Rea, M.G. De Angelis, M.G. Baschetti, Models for facilitated transport membranes: A review, Membranes (Basel). 9 (2019). doi:10.3390/membranes9020026.

[3] D. Venturi, D. Grupkovic, L. Sisti, M.G. Baschetti, Effect of humidity and nanocellulose content on Polyvinylamine-nanocellulose hybrid membranes for CO2capture, J. Memb. Sci. 548 (2018) 263–274. doi:10.1016/j.memsci.2017.11.021.

574328 Enhanced CO_2 Separation Using Water Swellable Polymer/ Amine Functionalized Zinc Oxide Mixed Matrix Membrane

Enhanced CO_2 Separation Using Water Swellable Polymer/ Amine Functionalized Zinc Oxide Mixed Matrix Membrane

Rajashree Borgohain, Pradip Das, Bishnupada Mandal

Department of Chemical Engineering, Indian Institute of Technology Guwahati, India- 781039

Correspondence to: B. Mandal (E-mail: bpmandal@iitg.ac.in)

The separation and capture of CO_2 is widely believed to be an important route for moderating the problem of global warming. In this regard, membrane separation technology has grabbed the attention of researchers owing to its functional simplicity, cost effectiveness and corrosion free nature. Mixed matrix membranes are gaining popularity as efficient CO_2-selective membranes over polymeric or inorganic membranes, due to their improved thermal and mechanical properties. In the present study, the amine functionalized zinc oxide (AF-ZnO) has been loaded on a water swellable polymer (WSP) i.e. carboxymethyl chitosan (CMC) membrane. The CO_2 separation from CO_2/N_2 mixture has taken place via solution diffusion and facilitated transport mechanism. The synthesized membranes have been characterized employing various microscopic and spectroscopic techniques. The moisture holding behavior of the mixed matrix membrane has also been assessed. The synthesis, characterization and permeation results are shown in Figure 1. The result obtained from the gas permeation study justifies that the proposed membrane can be a budding system for CO_2 separation application.

Figure 1. Overall representation of the synthesis and characterization of CMC/AF-ZnO mixed matrix membrane used for CO_2 separation

578558 Importance of Cocrystal Discovery Workflow for API Purification and Development

Importance of cocrystal discovery workflow for API purification and development

Kulkarni, Samir A., Wood, Geoffrey, Pickard, Frank, Kris Jones and Chekal Brian P.

Pfizer World Research & Development, Groton Laboratories, Groton, CT 06340, United States

In the current work, we describe a systematic and effective method to discover new co-crystals. The method consists of comparison of different *in silico* methods to find the list of co-formers suitable to form co-crystal with the given target compound. The experimental techniques for synthesizing co-crystals are time consuming and expensive, our goal is to develop an *in silico* method to rationalize co-crystal formation. Using the computational and experimental data generated by a researcher at Pfizer, we critically evaluated the predictive prowess of several co-crystal screening tools in use at Pfizer today, including: gFIT[1], multicomponent hydrogen bond propensity[2] with shape matching[3], and quantum chemistry-based clustering analyses. Our computational screens ranked coformers based on their predicted likelihood to form co-crystals. A computational assay was considered "successful" if it able to correctly predict co-crystal synthesis amongst the top ranked coformers. This blind analysis of our current computational methods also allowed us to better understand where they succeed, and more importantly where they fall short. We ultimately used these results to refine our computational workflows, and improve their predictive prowess.

Co-crystal development is important to the pharmaceutical industry because it offers the opportunity to modify the physiochemical properties of crystalline API without altering its covalent structure. While there are many computational assays that promise to rationally predict co-crystals, including: hydrogen bonding propensities, solubility parameters, crystal database screening, thermodynamic characteristics, etc., in reality the most reliable method to discover and prepare new co-crystals is still based on empirical screening.[3-4]

Co-crystal formation is mediated by strong intermolecular interactions, such as: hydrogen bonding, halogen bonding and π-π interactions. This project will study the influence of individual solvents on the formation of co-crystals, specifically what is an API's affinity for self-association relative to its affinity to the coformer. Also, the role of solvent in co-crystal formation remains poorly understood and solvent can be critical in obtaining particular co-crystals from solution. Once we determined the solvents and crystallization conditions, then Crystal16 (a small scale multiple reactor system) was used to determine the change in saturation temperature due to their ability to form stable co-crystals. If the co-crystal is stable, then its solubility will be lower than the solubility of its pure components. We then generated a two-phase diagram by mixing the pure components together in a selected solvent system as well as the solubility of pure components in the given solvent system. This method provides an experimental assay to efficiently screen many different coformers in a high throughput fashion.[3] The solvent drop grinding method will also be used as an initial screening tool for co-crystal screening.

This combined computational and experimental work helped us to evaluate the possible platform to discover new co-crystal systems in various solvent environments, and this high quality experimental reference data will ultimately allow us to critically evaluate available computational methods for predicting co-crystal formation.

References:

1. Abramov, Y. A.; Loschen, C.; Klamt, A., Rational coformer or solvent selection for pharmaceutical cocrystallization or desolvation. *Journal of pharmaceutical sciences* **2012,** *101* (10), 3687-97.
2. Galek, P. T. A.; Allen, F. H.; Fabian, L.; Feeder, N., Knowledge-based H-bond prediction to aid experimental polymorph screening. *CrystEngComm* **2009,** *11* (12), 2634-2639.

3. ter Horst, J. H.; Deij, M. A.; Cains, P. W., Discovering New Co-Crystals. *Crystal Growth & Design* **2009,** *9* (3), 1531-1537.
4. Gadade, D. D.; Pekamwar, S. S., Pharmaceutical Cocrystals: Regulatory and Strategic Aspects, Design and Development. *Advanced Pharmaceutical Bulletin* **2016,** *6* (4), 479-494.
5. Kulkarni, S. A.; McGarrity, E. S.; Meekes, H.; ter Horst, J. H., Isonicotinamide self-association: the link between solvent and polymorph nucleation. *Chemical communications (Cambridge, England)* **2012,** *48* (41), 4983-5.

578183 Study on the Non-Isothermal Dehydration Kinetics of Cefminox Sodium

Drugs containing crystal water account for a large proportion in solid crystalline drugs. The presence of crystal water will affect the internal energy, entropy and enthalpy of the crystal, further affecting its stability, solubility and biological drug efficiency. Therefore, it is of great significance for practical production to study the mechanism of drug crystal dehydration process and establish its dehydration kinetic model Thermal analysis technology is a kind of technology to determine the relationship between physical properties and temperature under the premise of controlling the heating rate. In the field of pharmaceutical research, this technology is mainly applied to the detection of drug polymorphism and the characterization of thermal stability, etc., and its analysis results can often be used as an indicator of process condition evaluation in chemical production. There are seven crystal water in the molecule of Cefminox sodium. The amount of crystal water may change with the change of environment, which may affect the quality of the product. Therefore, in this chapter, the non-isothermal dehydration kinetics model of cefminox sodium was systematically established through the thermal analysis of cefminox sodium, and the kinetic factors of its dehydration process A, E, $f(a)$, were obtained, which provided support for the subsequent process research.

571748 The Role of Crystallizability, Polymorphism and Drug Species in Drug Candidate Selection

In the pharmaceutical industry, for a given intended indication, typically several analogues are transitioned from discovery medicinal chemistry into development. These molecules then undergo careful evaluation to generate data mainly around their potency, stability and toxicity. That information is crucial to make an informed decision about drug candidate selection. In recent years, developability of drugs is becoming increasingly important as growing data indicates that appropriate selection of polymorph and or species can have a profound impact on the development of a viable drug. For example, an ionizable drug with a desired potency and target binding specificity can have stability issues as parent molecule but can form a stable salt that can improve its developability. In this presentation, the increasing importance of drug developability, and in particular, crystallizability, polymorphism and drug species in candidate selection is described. In addition, the rationale behind drug selection with includes contributions of bioavailability, toxicity, stability and manufacturability is explained.

558582 *Spatial Confined Capillary Flow Involved Model for Micro Droplet Crystallization and Particle Screening*

Spatial Confined Capillary Flow Involved Model for Micro Droplet Crystallization and Particle Screening

Jin Li, Mingguang Han, Xiaobin Jiang*, Gaohong He

State Key Laboratory of Fine Chemicals, Research and Development Center of Membrane Science and Technology, School of Chemical Engineering, Dalian University of Technology, Dalian, P. R. China

*Corresponding author: Email: xbjiang@dlut.edu.cn

Abstract

Droplet evaporative crystallization and its relevant crystal phenomena in microscale space have drown great attention recently in fine crystal manufature[1-3], sensor fabrication[4], and biochemistry[5], etc., due to its high precision in mass transfer and heat transfer controlling characteristic. Several fabrication approaches have been reported in micro scaled evaporative crystallization devices, and exhibit tremendous work in advanced application, however, it hasn¡¯t drowned sufficient attention yet especially for the theoretical research during droplet evaporation crystallization. Here, we developed a droplet simplified model with conventional evaporative model and nucleation models to handle the process inside droplet evaporation crystallization, and predict crystal formation and distribution according to the developed droplet model and crystal series characteristics. Firstly, the main idea of droplet evaporation model is shown in the **Figure 1**. A united step based evaporative model is raised to predict potential crystallizing supersaturation by handling concentration and nucleation energy distribution in further crystallization process.

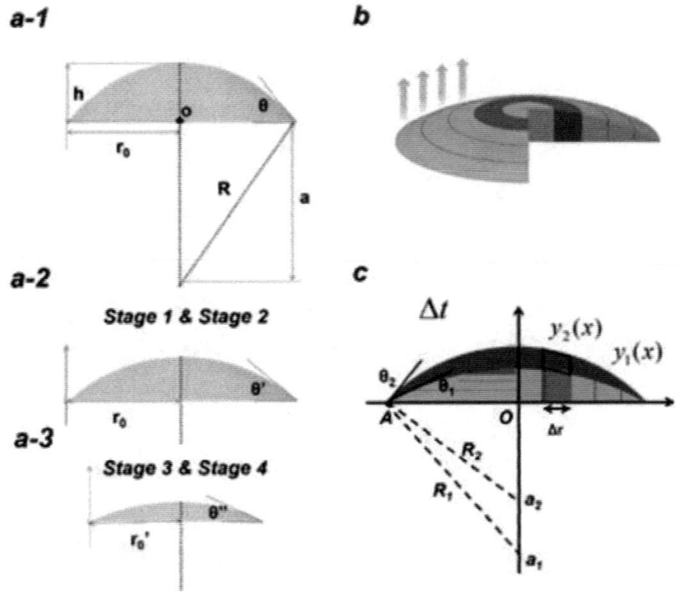

Fig 1: (a) Droplet appearance during 4 evaporated condition. (b) The idea of united step based evaporative model

To describe the superiority of developed models, several inorganic crystal series like sodium urate monohydrate (SUM) is invited in our work with various concentration and conventional nucleation theory [6] (in **Figure 2**). It could be apparently found that, droplet evaporation crystallization model predicted nucleation energy barrier meet the result of crystal distribution observed under higher precision scanning electronic microscopy.

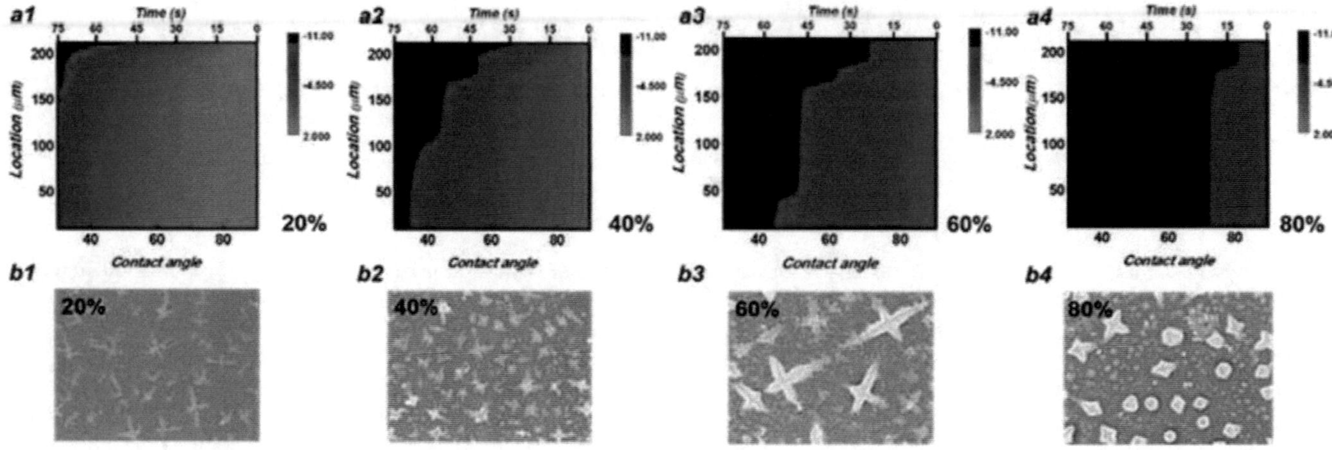

Fig 2: (a1~a4): Droplet nucleation distribution and (b1-b4): SUM crystal distribution and crystal form from various concentration SUM series.

By means of conventional evaporation stages raised by Shananhan[7] and capillary flow ¡°coffee ring¡± raised by Deegan[8], we sketched the crystal map by using NaCl series with various crystal concentration, droplet condition and eventually formed crystal shape and distribution in **Figure 3**, leading to the high precision prediction as long as knowing crystal series, concentration and droplet condition. The controlling mechanism revealed in this work is also contribute for the interfacial based crystallization and paritcle screening process.

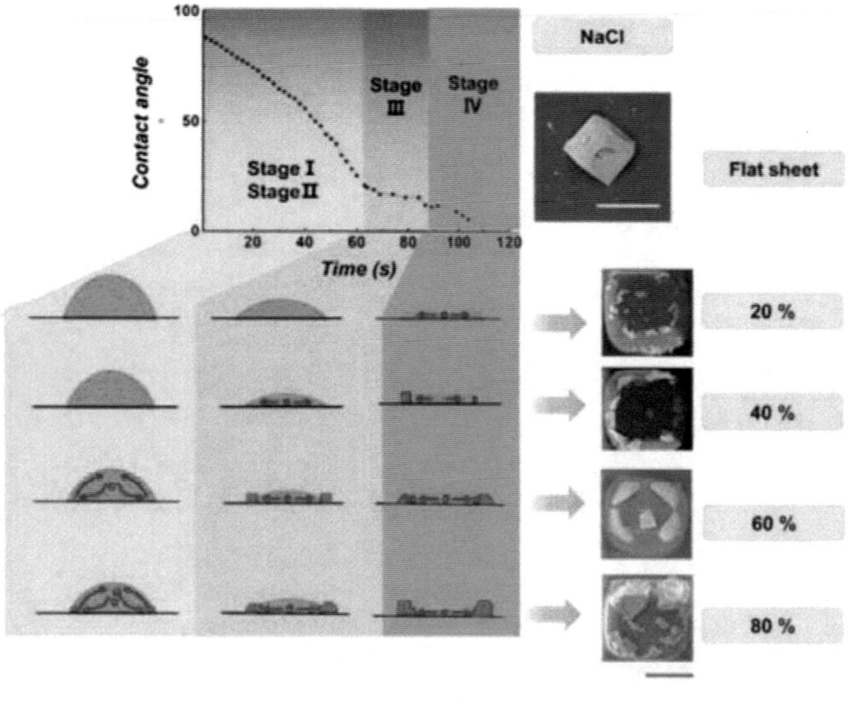

Fig 3: Illustration of NaCl crystal shape, distribution, evaporate stages, capillary flow and droplet condition.

Acknowledgment

We acknowledge the financial contribution from National Natural Science Foundation of China (Grant No. 21527812, 21676043, 21606035), the Fundamental Research Funds for the Central Universities (DUT16TD19, DUT17ZD203) and MOST innovation team in key area (No. 2016RA4053).

References

1. Teh SY, Lin R, Hung LH, Lee AP. Droplet microfluidics. *Lab on a chip.* Feb 2008;8(2):198-220.

2. Eales AD, Routh AF, Dartnell N, Simon G. Evaporation of pinned droplets containing polymer - an examination of the important groups controlling final shape. *AIChE Journal.* 2015;61(5):1759-1767.

3. Devlin NR, Loehr K, Harris MT. The separation of two different sized particles in an evaporating droplet. *AIChE Journal.* 2015;61(10):3547-3556.

4. Wen JT, Ho CM, Lillehoj PB. Coffee ring aptasensor for rapid protein detection. *Langmuir : the ACS journal of surfaces and colloids.* Jul 2 2013;29(26):8440-8446.

5. Liang YR, Zhu LN, Gao J, et al. 3D-Printed High-Density Droplet Array Chip for Miniaturized Protein Crystallization Screening under Vapor Diffusion Mode. *ACS applied materials & interfaces.* Apr 5 2017;9(13):11837-11845.

6. Han M, Li J, He G, et al. Tailored 3D printed micro-crystallization chip for versatile and high-efficiency droplet evaporative crystallization. *Lab on a chip.* Feb 26 2019;19(5):767-777.

7. Shanahan MER, Bourg C. Effects of evaporation on contact angles on polymer surfaces. *INT. J. ADHESION AND ADHESIVES.* 1994;14(3):201-205.

8. Deegan RD, Bakajin O, Dupont TF, Huber G, Nagel SR, Witten TA. Capillary flow as the cause of ring stains from dried liquid drops. *Nature.* 1997;389(6653): 827-829

565767 A Combined Theoretical and Experimental Investigation into the High Throughput Screening of Cocrystal Coformers

The CrystalPredictor[1,2] and CrystalOptimizer[3] codes have been used to explore the space of crystal structures successfully in several crystal structure prediction (CSP) investigations in recent years, including in the series of blind tests organised by the Cambridge Crystallographic Data Centre[4] and in the prediction of the crystal structures of pharmaceutically-relevant molecules[5-7]. Recent advances in CrystalPredictor[8], have allowed for orders of magnitude increases in the efficiency of the global search stage, whilst the capacity of CrystalOptimizer to reuse quantum mechanical calculations for the same molecule using LAM databases, allows for Quantum Mechanical (QM) accuracy in conformation, intramolecular energy and molecular electrostatics, at forcefield cost.

Exploiting these advances, we present a high throughput co-crystallisation study into 4 Active Pharmaceutical Ingredients (API's), combined with 10 coformers, selected from the GRAS list. Having performed a standard, neat, CSP study on each of the molecules, assessing the energy of potential cocrystals of the API and any of the coformers becomes an almost trivial task, through the judicious use of LAM databases. Comparing the energies of the cocrystals, versus the combined single crystal energies, allows the user to make informed decisions on which coformer to crystallise the API with, in order to target specific physical properties.

Combined with the theoretical predictions, experimental co-crystallisation experiments were performed between the API's and each of the coformers; comparisons will be made to assess the accuracy, as well as demonstrate the capacity for the technique to be taken up into standard pharmaceutical development workflows.

[1] M. Habgood, I. J. Sugden, A. V. Kazantsev, C. S. Adjiman, and C. C. Pantelides, J Chem Theory Comput **11**, 1957 (2015).

[2] I. Sugden, C. S. Adjiman, and C. C. Pantelides, Acta Crystallogr B **72**, 864 (2016).

[3] A. V. Kazantsev, P. G. Karamertzanis, C. S. Adjiman, and C. C. Pantelides, J Chem Theory Comput **7**, 1998 (2011).

[4] A. M. Reilly, et al., Acta Crystallogr B **72**, 439 (2016).

[5] D. E. Braun, T. Gelbrich, K. Wurst, and U. J. Griesser, Cryst Growth Des **16**, 3480 (2016).

[6] D. E. Braun, V. Kahlenberg, and U. J. Griesser, Cryst Growth Des **17**, 4347 (2017).

[7] S. L. Price, D. E. Braun, and S. M. Reutzel-Edens, Chem Commun **52**, 7065 (2016).

[8] I. Sugden, C. S. Adjiman, and C. C. Pantelides, Acta Crystallogr B (*Submitted*).

554978 Insight into the Role of Pre-Assembly and Desolvation in Crystal Nucleation: A Case of p-Nitrobenzoic Acid

Crystal nucleation from solution is of central importance but not well understood phenomena in chemistry science. Increasing interest was focus on the relationship between solution chemistry and nucleation dynamics recently [1]. Our current work take p-nitrobenzoic acid (PNBA) as the model compound, studing how pre-assembly species and desolvation process govern the crystal nucleation from solution, and it first consider the case of solvates. The nucleation behaviors and dynamics of PNBA in chloroform, acetonitrile, methanol, DMSO, DMF, NMP and DMA were studied. Vibration spectroscopy, combined with computational method are applied to reveal the pre-assembly species in the solution, which may be retained in the mature crystal structure or undergo rearrangement. Further analysis of density functional theory (DFT) and solvation energy calculation reflect the strength of solute-solvent interactions. Based on above results, the more comprehensive correlation of molecular structure, crystal structure, solution chemistry and nucleation dynamics was established. The finding suggests that the structural similarity between solution chemistry and crystal structure, the specific site interaction and overall solvation strength have a combined influence on the nucleation process, however, none of them should be regarded as the rate-limiting factor.

574371 Prediction of Phase Diagrams for Crystallization Process Development and Controlling Polymorphism

Knowledge of the various phase boundaries (e.g., solid-liquid equilibrium, metastable liquid-liquid equilibrium) exhibited by systems of interest is essential in crystallization process development and polymorph control. A comprehensive methodology that allows a process scientist/engineer to characterize these phase boundaries relevant to crystal form selection is currently lacking. In this work, we present a modeling framework toward predicting the binodal, spinodal, and gelation boundaries exhibited by a system starting from the available solubility data of a solute. We collate the necessary theoretical concepts from the literature and describe a unified approach to model the phase equilibria of solute–solvent systems from first principles. The modeling effort is validated using experimental data reported in the literature.

563381 Surprising Effect of Carbon Chain Length on Inducing Ability of Additives: Elusive Form-II of γ-Aminobutyric Acid (GABA) Induced By Sodium Carboxylate Additives

Surprising Effect of Carbon Chain Length on Inducing Ability of Additives: Elusive Form-II of gamma-Aminobutyric Acid (GABA) Induced by Sodium Carboxylate Additives

Lingyu Wang[1,2], Weiwei Tang[1,2], Junbo Gong[1,2*]

[1] National Engineering Research Center of Industry Crystallization Technology, Tianjin University, Tianjin, China;

[2] The Co-Innovation Center of Chemistry and Chemical Engineering of Tianjin, Tianjin University, Tianjin, China

Keywords: GABA, additives-induced polymorph, carbon chain length, sodium carboxylate

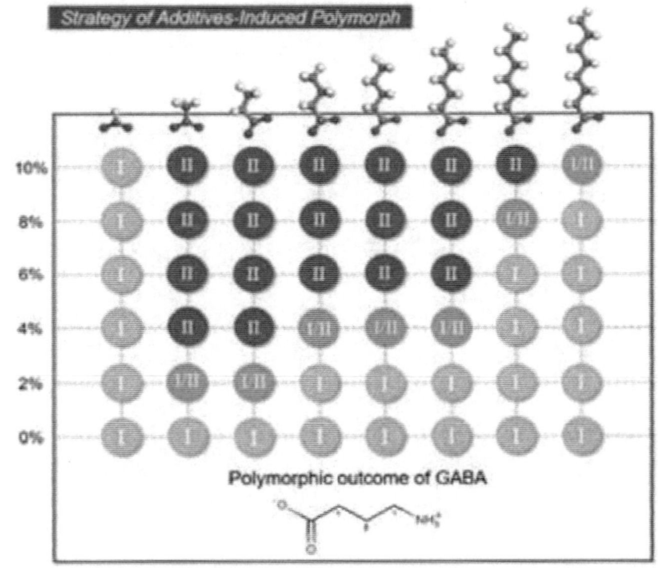

Figure 1. Resulting polymorphs of GABA obtained in presence of different additives with different addition amounts (Green balls: Form-I; Red balls: Form-I or Form-II; Blue balls: Form-II).

Selectively preparing a desired polymorph in a controlled way is an issue of great concern for pharmaceutical industry. The introduction of appropriate additives into crystallization process represents one of important strategies to achieve polymorphic control. Herein, we developed the strategy of additives-induced polymorph, realizing the selective crystallization of elusive Form-II of gamma-aminobutyric acid (GABA), a potent bioactive compound. A series of sodium carboxylate additives with different carbon chain lengths (C2-C8) were identified to induce the crystallization of Form-II. It was found that additive molecules will inhibit the growth of the centrosymmetric Form-I at two opposite ends thereby precluding it, but they will inhibit the growth of the polar Form-II at only one end so that it survives. Notably, these sodium carboxylate additives exhibited distinct inducing ability for Form-II (**Figure 1**), that is, the effective critical addition amount required to induce Form-II increases with the increase of carbon chain length. However, the sodium formate is an exception and it cannot induce the Form-II. The difference on the inducing ability of additives was rationalized based on the competition of attachment and detachment behavior of additives on growth end of Form-I. When the extent to which the attachment is more favorable than the detachment is larger, the corresponding additive exhibits better inhibition effect on Form-I and more remarkable inducing ability for Form-II.

581813 Water Treatment and Desalination:from High Performance Membranes to Flexible Self-Adaptive Systems

Water scarcity and groundwater and surface water quality impairment are on the rise in various regions of the world. In order to increase and diversify the water portfolio, efforts are growing to upgrade and expand the use of otherwise unutilized water resources and lower the cost of water treatment and desalination. Potable water production from seawater, brackish groundwater, and municipal and industrial water reuse, as well as for management and reuse of agricultural drainage water are now becoming increasingly reliant on membrane-based water treatment and desalination technologies. It is noted that of the various water desalination approaches, reverse osmosis (RO) membrane desalination has become the dominant technology accounting for about 65% of the worldwide desalination capacity, followed by followed by multi-stage flash evaporation (21%), with the remaining 14% capacity provided by other thermal desalination methods (e.g., multiple effect distillation and vapor compression) and electrodialysis (ED)/ED reversal (EDR). It is also noted that other membrane-based desalination technologies which may suitable for small-scale specialized applications have been proposed in recent years including capacitive deionization, membrane distillation and forward osmosis. Admittedly, the market growth of RO water purification/desalination is attributed, in part, to its simplicity and technological maturity. However, despite the rapid progress in RO water purification and desalination and associated pre- and post- feed and product treatment processes, respectively, there are various challenges to overcome that include, for example: membrane fouling and mineral scaling, reducing energy consumption, decreasing plant footprint, improving process efficiency and robustness and of course lowering capital and operational costs. Accordingly, a discussion of the above challenges and solutions will be presented focusing on advances through the progression of a multipronged approach with practical considerations of the following elements and their development from fundamental research to field deployment: (a) high performance membranes for filtration and desalination, (b) real time detection of the onset and evolution of fouling/mineral scaling and membrane integrity monitoring, (c) energy-optimal and flexible desalination via process configuration and model-based control, (d) high recovery desalination, and (d) intelligent (self-adaptive) membrane filtration and fouling control with self-learning, and fault detection and isolation.

581812 Manufacturing Fit-for-Purpose Membranes from Nanostructured Polymers

In addition to the improved conservation and responsible management of existing freshwater resources, the use of non-traditional sources of water (e.g., seawater and wastewater) is essential to supporting human life on earth. The success of seawater desalination by reverse osmosis highlights the large role membrane separations will play in using these non-traditional sources to help society meet the demand for water. However, desalination processes are inherently energy-intensive due to the need to overcome the large osmotic pressures generated by saline sources. This energy demand has caused engineers and scientists to consider the design of distributed wastewater treatment systems that incorporate regenerative technologies to provide water at the quality level demanded by its intended application. In this manner, the separation is tailored based on the chemical composition of the source water with the end-use requirements in mind. This talk will discuss a versatile membrane platform fabricated from self-assembled block polymer precursors that can help to enable the fit-for-purpose water paradigm.

Most state-of-the-art membranes utilize a size-selective, steric exclusion mechanism to filter dissolved solutes from solution. In this regard, self-assembled block polymers are a promising platform for producing high-performance membranes because large areas of membrane that contain a high density of well-defined nanoscale pores can be readily produced using the self-assembly and non-solvent induced phase separation (SNIPS) technique. Furthermore, the performance profile of the block polymer membranes can be controlled through clever design of the macromolecular precursors. The performance of these membranes is pushing the limits of size-selective separation mechanisms, which is driving interest in multifunctional membranes with deliberately-assigned pore wall chemistries. As such, block polymer membranes with nanoscale pores lined by moieties that that can be thoughtfully designed to effect selective separations through affinity-based adsorption, electrostatic interactions, and catalytic conversion of target solutes further demonstrate the promise of this platform. In one example, the pore-lining groups are converted to coordinating groups that act as high capacity binding sites for heavy metal ion adsorption. In another example, charge-patterned mosaic membranes, are prepared by functionalizing the copolymer membrane precursors using ink-jet printing devices. The well-defined counter-charged domains that cover the surface of the charge mosaic membranes allow dissolved salts to permeate more rapidly than water¾even though water is three times smaller in size. Through these examples, we will demonstrate that membranes that are based on self-assembled block polymer materials provide a scalable platform that can be tailored to myriad separations for the purification and conservation of fresh water resources.

581811 Control of Polymorphism and Crystal Size Distribution and in Pharmaceutical Crystallization

An overview is provided on the design and operation of pharmaceutical crystallizations to control polymorphism and crystal size distribution. A semi-automated methodology is described for the selective crystallization of metastable and stable polymorphic and solvatamorphic forms based on Attenuated Total Reflection-Fourier Transform Infrared (ATR-FTIR) spectroscopy, laser backscattering, and process data analytics. The methodology enables the design and operation of seeded batch crystallizers to manufacture large crystals of uniform size by suppressing secondary nucleation. A modification of the methodology achieves a target size distribution in a semi-batch crystallization configuration by employing continuously generated seeding. Theoretical, simulation, and experimental results are reported for a variety of pharmaceutical compounds. The presentation ends with a discussion of directions towards control of multiple properties of the crystal product.

581667 Carbon Nanotube Membranes for Separation of Fuels from Water

The application of carbon nanotube (CNT) membranes for the separation of water miscible fuels like alcohols from water is discussed. Previously separating alcohols like ethanol or butanol from water has been achieved through distillation followed by molecular sieves, an energy intensive thermal process typically requiring the combustion of fossil fuels. CNT membranes have recently been shown to achieve the same separation with much lower energy requirements. The CNT membranes are alcohol selective, permeating alcohols while rejecting water transport, in room temperature pervaporation and vapor permeation modes of operation. Very high alcohol concentrations, in some cases anhydrous alcohols, are achieved in single pass extractions using only vacuum and condensation. Experimental data at the pilot scale will be discussed, including data from the direct air capture and conversion of CO2 to transportation fuels.

581666 Manufacturing Advanced Structured Contactors with Integrated Heat Management Capabilities

The rapid increase in global industrialization necessitates technology shifts in energy production, manufacturing, and carbon management techniques. Large energy costs in refineries, power plants, and manufacturing facilities using traditional separation techniques are currently a major opportunity for innovation. Approximately 10-15% of global energy use can be attributed to separation processes, with the vast majority of separations being "thermal" in nature (e.g., distillation). Significant energy and cost savings can be realized using advanced separation techniques such as membranes and sorbents. One of the major barriers to acceptance of these techniques remains linking engineering materials to actual processes that are effective in the presence of aggressive industrial feeds.

The creation of robust materials-enabled advanced separators and their manufacturing into low-cost, energy-efficient devices to meet this global challenge will be the focus of the talk. Engineering novel materials—such as solid-supported amines, metal-organic frameworks, and polymers of intrinsic microporosity—into structured contactor adsorption devices shows promise for emerging separation applications. These include olefin/paraffin separation, carbon capture, and air purification. Structured sorbent contactors can help manage kinetic and engineering factors associated with the separation, including pressure drop, sorption enthalpy effects, and external heat integration (for temperature swing adsorption). Monoliths, fiber sorbents, and 3D printed materials will be discussed as the three main classes of existing or emerging structured sorbent contactors. Recent developments in their manufacture; advantages and disadvantages of each structure relative to each other and to pellet packed beds; recent developments in system modeling; and finally, critical needs in this area of research will all be discussed.

553587 Wind-Powered Ammonia Synthesis

We are developing a small scale ammonia synthesis plant powered by wind energy. This is a major departure, because ammonia is currently made with energy from fossil fuels, especially natural gas. In contrast, the wind energy used here is sustainable but stranded, far from urban population centers but near locations of ammonia fertilizer demand. The wind energy is converted to electricity used for pressure swing absorption of air to make nitrogen, and for electrolysis of water to make hydrogen. Nitrogen and hydrogen are combined in a small-scale Haber process to synthesize ammonia.

To make this wind-based process more competitive, we are making the ammonia with a conventional catalyst but separating the ammonia by absorption using ammines. The rate limiting step can now be not the chemical kinetics or ammonia separation, but the speed of the pump recycling the unreacted gases. Because the operating pressure drops from 150 to 20 bar, the cost for the process is reduced. To further intensify the synthesis, we are testing putting both catalyst and absorbent in the same vessel.

We continue to explore the feasibility of this small process to harvest stranded energy.

574253 Adsorption of CO_2 on High Silica Zeolites: Structural Defects and Differences between Adsorbent Samples

Equilibrium isotherms have been measured for adsorption of CO2 on several different samples of high silica zeolites (silicalite, the aluminum free form of ZSM-5 and DDR, the aluminum free form of ZSM-58), prepared by different synthesis methods and therefore containing different levels of framework hydroxyls. The isotherms for the various samples show significant differences but detailed analysis reveals that the normalized adsorption equilibrium constants are substantially the same for all samples. The differences in the isotherms appear to be due to differences in the saturation capacities. Comparison with theoretical isotherms, derived by molecular simulation, shows that, for both silicalite and DDR, the measured isotherms for the lowest capacity samples are very close to the theoretical values. The theoretical isotherms conform almost exactly to the simple Langmuir model and the heats of adsorption are essentially independent of loading. For the silicalite samples the capacity increases somewhat with hydroxyl content but there is no stoichiometric relationship. The increase in capacity appears to be associated with some mesoporosity arising from structural defects. Only one of the DDR samples showed an adsorptive capacity substantially greater than that of the ideal structure. In contrast to silicalite the increased capacity of DDR appears to be associated with higher energy sites, suggesting access to the smaller cages in some DDR crystals.

Acknowledgement: This study was carried out as part of a GOALI project. The financial support provided by ExxonMobil Corporation and the National Science Foundation is gratefully acknowledged. We are grateful to Wieslaw Roth for zeolite samples and Clarence Chase for the assistance in carrying out the XRD and NMR measurements.

575899 Using Adsorption Microcalorimetry and Manometry to Measure Useful Data for Adsorbent Screening and Process Design

Adsorption microcalorimetry has been proposed in the literature as a useful technique to characterize adsorbent surface and heterogeneities. By coupling a manometric setup to a Tian Calvet calorimeter, for instance, it is possible to simultaneously measure equilibrium adsorption isotherms and differential calorimetric curves. Additionally, it is also possible to assess intraparticle mass transfer kinetics by fitting experimental calorimetric peaks to a convenient mathematical model based on mass and heat balances applied to the sample cell and surge tank. We show different models with increasing degree of complexity and the coefficients of diffusion which may be extracted by fitting experimental data. Therefore, a useful set of parameters may be measured by this single technique: adsorption isotherms, heats of adsorption as a function of loading and intraparticle diffusivity. This is information usually required for adsorbent screening and the design of fixed bed adsorbers. Examples will be shown for different classes of porous solids aiming at CO_2 adsorption. An extension of the technique, originally applied to single gas adsorption, will also be shown for binary gas mixtures.

564417 Using Data Analytics to Find Promising Metal–Organic Frameworks for Adsorption Applications

Data analytics, including methods such as machine learning, are poised to play an important role in the development of new adsorbents, especially for crystalline materials such as zeolites and metal-organic frameworks (MOFs). Recently, we have updated the Computation-Ready, Experimental (CoRE) MOF database with additional analyses and almost 10,000 additional structures, bringing the total to nearly 15,000. We report separate versions with all solvent molecules removed (ASR) and only free solvent molecules removed (FSR). Crystal structures (only for the subset that underwent significant changes during curation), pore analytics, and physical property data are included with the publicly available CoRE MOF 2019 database. A detailed analysis of open metal sites shows that a large fraction of the CoRE MOF structures (34% for FSR and 62% for ASR) contain at least one open metal site. This knowledge can be used to screen the database for selective adsorption of small molecules and catalysis applications. For example, previously, Simon and coworkers carried out computational screening of the CoRE MOF 2014 Database for Xe/Kr separation and found SBMOF-1 (CSD REFCODE: KAXQIL) to be the most selective MOF for Xe/Kr separation. In our recent follow-up work, computational screening of the CoRE MOF 2019 databases discovered structures with similar selectivity as SBMOF-1 but with higher capacity. Based on a molecular fingerprint approach that we developed, we compared the CoRE MOF 2019 Database with a database of hypothetical MOFs. We found that there are at least 16 synthesized MOFs present in the hypothetical MOF database of Wilmer et al. In addition, we report a systematic, machine-readable identification scheme for MOFs. The format, which we call MOFid, includes the MOF topology and SMILES strings for the building blocks. We anticipate that MOFid may be of considerable value for various analyses of MOFs, and we provide a few preliminary examples.

559997 The Role of Adsorption Processes in Lignin Biorefinery to Make Vanillin and Syringaldehyde

Lignin is an undervalued byproduct of the pulp and paper industry, contained in the black liquors. Concentrated black liquors are burned in boilers that usually have a fixed capacity and are very costly to upgrade. The application of lignin for other uses may debottleneck the production of cellulose by allowing larger quantities of black liquor to be burned since it would have a lower heating value after partial lignin precipitation [1]. Lignin is the second most abundant biopolymer and it is produced annually in the order of 1.2 Mt/y for lignin precipitated from different pulping processes, with 10% coming from Kraft processes [2]. There are commercially available technologies to remove lignin from the black liquor, for instance LignoBoost (Valmet) and LignoForce (FPInnovations) [3, 4].

Lignin has also the potential to become an alternative source for aromatic compounds due to its phenylpropane structure . The oxidative depolymerization of lignin produces a range of phenolic compounds that are already highly functionalized but they need a very laborious process of separation. Functional groups can be removed/reduced and the ratio of oxygen/hydrogen decreased all the way to benzene like structures. The functionalized molecules like vanillin and syringaldehyde can be used in the production of pharmaceuticals avoiding several steps of synthesis [5] while vanillic and syringic acids can be directed to the production of polymers [6].

Over the years, an integrated process for the production of monomers compounds including vanillin and syringaldehyde has been put forward (Figure 1). It encompasses several unit operations that include: oxidation of the lignin solution, filtration (ultra- and nanofiltration), adsorption, extraction and crystallization. The integrated process also allows the removal of lignin fractions for other applications from the membrane steps that can be used in the polyurethane resins/foams benefiting from the oxidation step, which increases the number of OH groups in lignin [7].

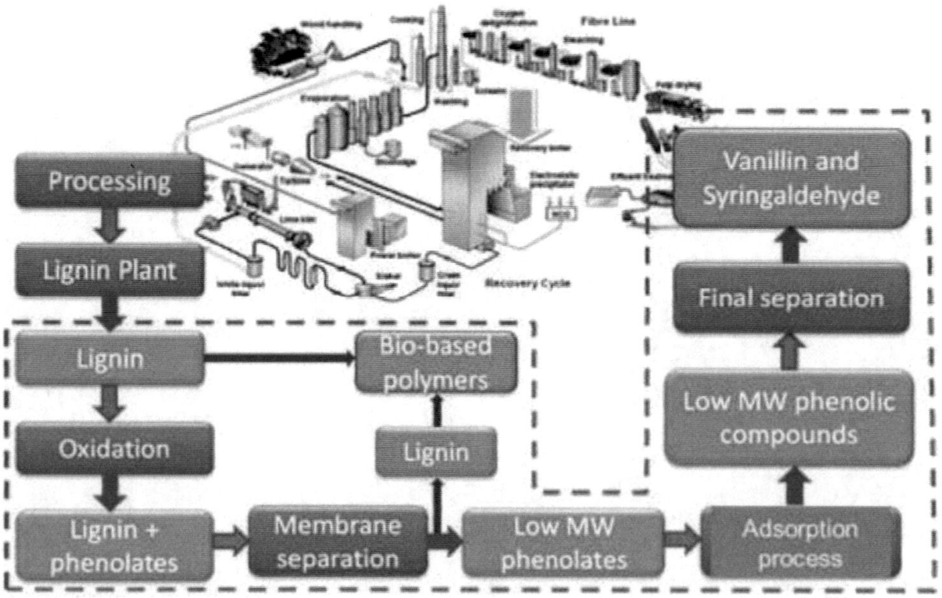

Figure 1 – Integrated process for the production of value added products like vanillin, syringaldehyde and lignin-based polyurethanes from Kraft lignin.

Alkaline wet oxidation is the most widely used method for the depolymerization of lignin, having the best yields. The yields also vary according to the lignin isolation method. The lignin solution is oxidized in a continuous reactor, a structured packed bubble column reactor [8]. The lignin is oxidized by an oxygen rich stream at 10 bar of total pressure, at an initial temperature of 140 C, in the continuous reactor.

The lignin that is not depolymerized is removed by ultra-/ nanofiltration in a flatsheet membrane (140 cm^2) in a SEPA CFII filtration system. After removal of part of the undepolymerized lignin, the permeate is then fed to the chromatographic column. The membranes are able to reduce the total organic solids in solution. The ultrafiltration membrane has a cut-off of 1000 Da and the nanofiltration membrane 600Da. The last membrane plays an important role since the lignin in solution will hinder the chromatographic process and the crystallization step due to contamination.

The chromatographic column is packed with SP700 resin and then the feed is fed at very alkaline pH. Annual plant (Tobbaco) and a softwood (Indulin AT) depolymerized lignin solutions were tested. The chromatographic step is able to fractionate the compounds in the oxidized solution into families of phenolic acids, phenolic aldehydes and phenolic ketones in a single column (Figure 2). The separation is achieved by using a two-eluent desorption scheme where water and ethanol are used [9, 10]. The solution eluting during the feed phase is rich in phenolic acids, (vanillic acid and syringic acid), the solution eluted with water is rich in the aldehydes (vanillin and syringaldehyde) and the solution eluted with ethanol is rich in ketones, (acetovanillone and acetosyringone). The fractionation scheme can be adjusted to collect the most concentrated fractions while reducing the amounts of desorbent used. This opens up great potential, since it is easier to work with solutions rich in vanillin/vanillic acid than with a mixture of several phenolic compounds. The fraction enriched in vanillin is then used to perform crystallization. The remaining contaminants are left in the crystallization broth and can be recycled back to the feed phase of the column since it may still contain phenolic acids and phenolic ketones. The same can be done with the other enriched fractions, achieving higher purities by crystallization.

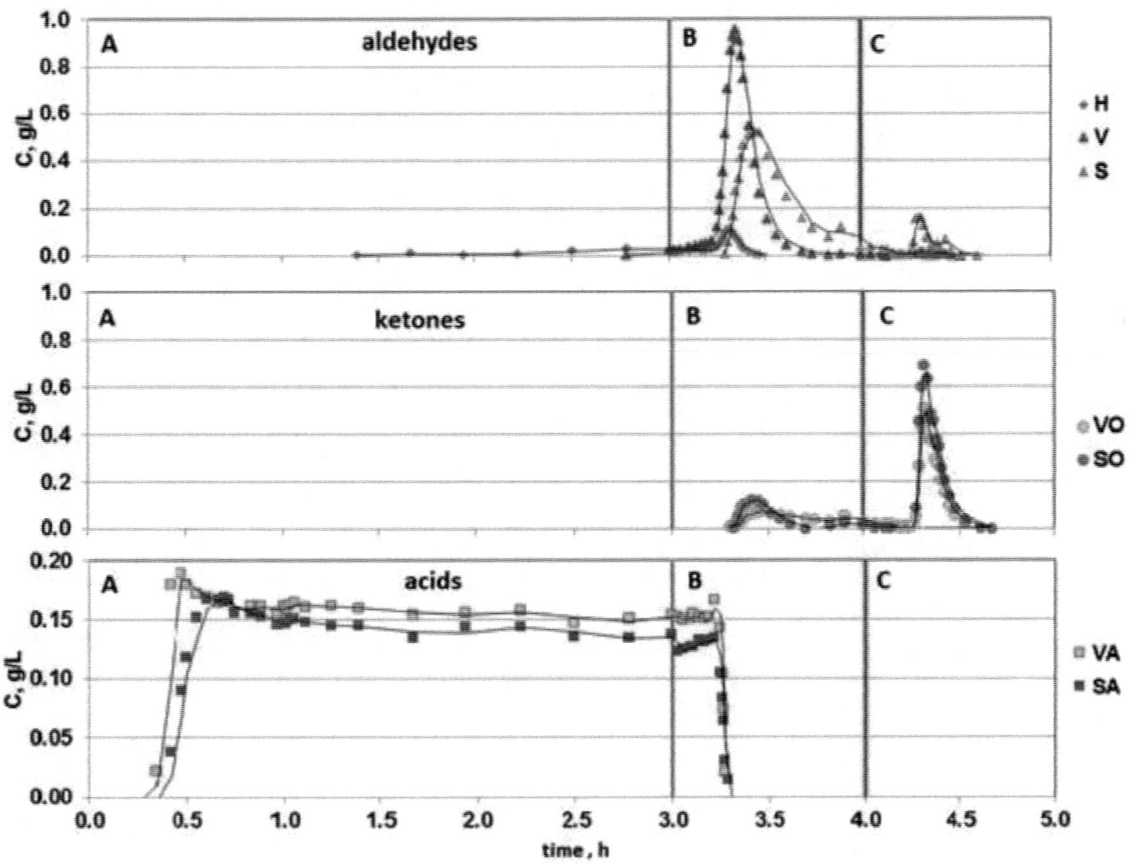

Figure 2 – Fractionation of phenolic aldehydes, phenolic acids and phenolic ketones from a lignin (from tobacco stalks) oxidation mixture in a chromatographic column packed with SP700 resin. H – p-hydroxybenzaldehyde; VA – vanillic acid; V – vanillin; SA – syringic acid; S – syringaldehyde; VO – acetovanillone; SO – acetosyringone. A – feed phase, B – water desorption phase, C – ethanol desorption phase. Feed pH = 10. Feed concentrations: [H]=0.014g/L, [VA]=0.152 g/L, [V]=0.138g/L, [SA]=0.170 g/L, [S]=0.103 g/L, [VO]=0.038 g/L, [SO]=0.037 g/L.

[1] S. Stiefel, C. Marks, T. Schmidt, S. Hanisch, G. Spalding, M. Wessling, Overcoming lignin heterogeneity: reliably characterizing the cleavage of technical lignin, Green Chem. 18 (2016) 531-540.

[2] A. Sørlie, B. Lyngstad, M. Harlem, T.E. Foss-Jacobsen, T. Barnholt, P. Romberg, O.G. Jakobsen, Welcome to Borregaard's Capital Markets Day, in: Borregaard (Ed.)Sarpsborg, Norway, 2016, pp. 71.

[3] L. Kouisni, A. Gagné, K. Maki, P. Holt-Hindle, M. Paleologou, LignoForce System for the Recovery of Lignin from Black Liquor: Feedstock Options, Odor Profile, and Product Characterization, ACS Sustainable Chemistry & Engineering 4 (2016) 5152-5159.

[4] P. Tomani, The Lignoboost Process, Cell Chem Technol 44 (2010) 53-58.

[5] B. Bouguerne, Conception et Synthèse de dérivés phénoliques hautement fonctionnalisés et étude de leur propriétés biologiques vis-à-vis des maladies cardiovasculaires (athérosclérose). Laboratoire de Syntèse et Propriétés physico-chimiques de Molécules d'Intérêt Biologique, cnrs 5068, Institute des Maladies Métaboliques et Cardiovasculaires, INSERM 1048, Université Toulouse III - Paul Sabatier, 2012.

[6] C. Pang, J. Zhang, Q. Zhang, G. Wu, Y. Wang, J. Ma, Novel vanillic acid-based poly(ether–ester)s: from synthesis to properties, Polymer Chemistry 6 (2015) 797-804.

[7] E.A.B. Silva, M. Zabkova, J.D. Araújo, C.A. Cateto, M.F. Barreiro, M.N. Belgacem, A.E. Rodrigues, An integrated process to produce vanillin and lignin-based polyurethanes from Kraft lignin, Chemical Engineering Research and Design 87 (2009) 1276-1292.

[8] J.D.P. Araujo, C.A. Grande, A.E. Rodrigues, Structured packed bubble column reactor for continuous production of vanillin from Kraft lignin oxidation, Catalysis Today 147 (2009) S330-S335.

[9] E.D. Gomes, M.I. Mota, A.E. Rodrigues, Fractionation of acids, ketones and aldehydes from alkaline lignin oxidation solution with SP700 resin, Separation and Purification Technology 194 (2018) 256-264.

[10] E.D. Gomes, A.E. Rodrigues, Lignin biorefinery: Separation of vanillin, vanillic acid and acetovanillone by adsorption, Separation and Purification Technology 216 (2019) 92-101.

574300 Improved Process for Ultra-High Purity Nitrogen Production

Production of UHP Nitrogen is critical for many electronic processing facilities. Previous technology for combined CO/H_2 removal from air utilized a separate reactor vessel and economizer located after the main air compressor for oxidation of CO and H_2 to CO_2 and H_2O. These were then removed in the TSA unit, yielding a purified air stream essentially free of CO, H_2, CO_2, and H_2O. Drawbacks, however, include the requirement of relatively high reactor temperatures (and associated heating equipment), use of a very expensive noble metal-based catalyst, use of an expensive aluminum plate fin economizer and the cost of the separate reactor vessel.

A novel, lower cost approach has been developed for removing both H_2 and CO from air by incorporating a layer of metal oxide catalyst in the TSA unit. It is placed after the alumina and zeolite layers (which are used to remove water and CO_2 from the air). Under the low water, low CO_2 conditions the catalyst can oxidize CO and H_2 and simultaneously adsorb the produced water and CO_2. Development efforts and commercial application opportunities will be highlighted in this talk.

573508 Transport Coefficients in Commercial Beads Measured Using a Thermal Frequency Response Method

Transport coefficients are needed in the design of vacuum swing adsorption processes for air separation. In this contribution we present a novel use of the dual piston pressure swing adsorption (DP-PSA) apparatus, demonstrating its use as a thermal frequency response system.

The method relies on inserting fast response thermocouples (0.25 mm) in commercial beads (1 mm) used in the production of oxygen. With only 3 beads inserted in the DP-PSA unit, the volume modulation produces concentration fluctuations that are not affected by adsorption. Therefore the measured pressure and temperatures can be used to determine the thermal frequency response curves up to 1 Hz.

Operation with different gases allows to determine the mass transfer and heat transfer coefficients at different pressure levels. Helium can be used to determine the relevant heat transfer parameters in the range 0.1 Hz to 1 Hz, where rapid cycling induces a temperature variation in the gas phase due to compression and expansion. Nitrogen is then used with the heat transfer parameters chosen to be consistent with the values found for helium. Finally, air is used to verify the response for a mixture.

571962 MOFs for Sour Gas Treatment: Design Strategies and Stability Considerations

Acid gases are pervasive in applications that span the oil and gas industry, the general chemicals industry, and the energy sector more broadly. Whether the target is natural gas upgrading, clean gas for solid oxide fuel cells, or flue gas treatment, advances in materials and methods for efficient and cost-effective removal of acid gases require a deep understanding of how materials adsorb, react, and behave in complex mixtures. Achieving this understanding necessitates the evaluation of sorbents, membranes, and catalysts in realistic, multicomponent mixtures. Unfortunately, academic research almost exclusively focuses on testing materials in highly idealized process streams, typically ignoring any potential contaminants. Multicomponent adsorption data are difficult to obtain, even for simple gas mixtures, but the inclusion of toxic gases such as H_2S and SO_2 adds several levels of complexity and difficulty to these types of measurements.

Nevertheless, many future large-scale innovations in energy-intensive industries will require development or implementation of new materials that must operate under complex conditions. This presentation will highlight our recent work in understanding the acid gas adsorption in metal-organic frameworks.

571650 HCl Removal from H_2 Gas By Adsorption on Hydrated Ion-Exchanged Zeolites

The removal of hydrogen chloride from hydrogen gas is of great importance for the chemical industry and the environment. In this work, asystematic study is presented on HCl removal from hydrogen gas by adsorption on zeolites, including cation free (i.e. silicalite) and high cation loaded materials (i.e. LTA and 13X zeolites), as well as ion-exchanged zeolites (with alkali, alkaline earth and transition metal ions). The HCl removal performance was studied by breakthrough experimentation with packed beds under high gas velocities, at high pressures (30 bar) and with low HCl concentrations (< 200 ppm). The zeolite screening indicated that the 13X zeolite outperforms the other tested materials. The ion-exchanged X zeolites were extensively characterized via SEM-EDX, XRD, FTIR-ATR, TGA and argon adsorption isotherms. The characterization after HCl adsorption supports the hypothesis that HCl is taken up by the material by forming salt molecules within the zeolite cavities by reaction with the cations and, which is, moreover, enhanced by the presence of pre-adsorbed water. The type of cation present in the zeolite framework structure notably affected the HCl removal adsorption capacity as well as uptake rate The investigation revealed important parameters such as cation radius, atomic mass and electronegativity, which played a noteworthy role in defining the HCl removal performance of ion-exchanged zeolites.

573207 Hollow Fiber Membrane-Based Continuous Crystallization Processes

At NJIT we have been developing hollow fiber membrane-based continuous crystallization processes for some time. This has involved anti-solvent crystallization using porous hollow fiber membranes. We have also developed solid hollow fiber membrane-based process for cooling crystallization. Recently we have developed also continuous processes for polymer coating of particle/drug crystal suspensions for controlled release using hollow fiber membrane-based anti-solvent crystallization as well as cooling crystallization. We have now extended such processes over long lengths of time to demonstrate that such processes can be carried out successfully with same performance for extended periods of time. These processes have now been further extended to producing nanocrystals on a continuous basis. This presentation will describe recent results and provide a perspective on how these membrane-based technologies are shaping up.

565211 Continuous Crystallisation Oscillatory Flow Platform for Protein Purification and Bio-Separation

Continuous Crystallisation Oscillatory Flow Platform for Protein Purification and Bio-Separation

Xiaoyu Li [1], Huaiyu Yang[1], Wenqian Chen [1] and Jerry Y. Y. Heng [1] *

Department of Chemical Engineering, Imperial College London, South Kensington Campus, London SW7 2AZ, UK

* Corresponding Author: Jerry Y. Y. Heng. E-mail: jerry.heng@imperial.ac.uk

Abstract

Progress of advanced biotechnology has enhanced the development of commercial biopharmaceutical products and the advances in biotechnology have led to over 246 approved macromolecular drugs and vaccines with cumulative revenues reaching $140 billion [1]. A cost effective and reliable downstream route becomes a major industrial challenge in therapeutic protein manufacture due to the expanded market remands. Crystallisation has been used for the purification of more than 95% small-molecule pharmaceuticals in various conventional chemical industries [2]. Crystallisation of proteins as downstream separation and isolation steps is considered to be more efficient and cost-effective alternative compared to conventional purification techniques such as chromatography [3-5]. Moreover, crystalline proteins have a higher purity and stability which can benefit formulation, storage, and drug delivery steps [3, 6].

The aim of this work is to develop a continuous biocrystallisation platform for cost-effective downstream manufacturing of therapeutic proteins. Additionally, selective crystallisation of a protein and simultaneous crystallisation of proteins from binary protein mixtures were investigated. Mixtures were prepared using model proteins: lysozyme and thaumatin. $>10^3$ of 2-4 μl hanging drops containing mixture of protein and precipitant solutions were set-up. The crystallisation processes were monitored under microscope for 1-2 weeks. Protein crystals of these proteins were distinguishable from their habits (shapes). Both model proteins can be crystallised out by adding the same precipitant solution. Selective crystallisation of each protein from mixtures was achieved by manipulating crystallisation conditions such as types of precipitants, precipitant concentration and pH of the solution. Diagrams were produced to show the operation windows of selective crystallisation of target protein from the mixtures. This work demonstrates our proof of concept that crystallisation possesses the potential of separating a target protein from a complex mixture environment.

Further, protein crystallisation from mixture is conducted at a laboratory scale of a few μl to hundreds of μl. For future scale-up study of the biocrystallisation process, we designed and built a continuous oscillatory flow crystallisation platform at the ml scale, combining two syringe pumps for feeding protein and precipitation solutions, respectively, and an extra syringe pump to provide the oscillatory flow [7]. Crystallisation using this platform was conducted with lysozyme as model protein. Lysozyme concentration ranging from 30 to 100 mg/mL, oscillatory conditions with amplitude (x_0) from 10 to 25 mm and frequency (f) from 0.05 to 0.25 Hz were investigated in a batch oscillatory flow crystallisation platform and then successfully transferred to a continuous oscillatory flow crystallisation platform. The steady state of continuous crystallisation was reached at resident time 200 minutes, and the final product crystal shape and size were consistent during the continuous process. This study demonstrates a robust method to develop continuous oscillatory flow crystallisation based on batch crystallisation of protein, which can contribute to the future development of downstream bio-separation processes.

Conclusions

We demonstrate that the selective crystallisation for bioseparation of proteins is viable, as demonstrated for the case of lysozyme-thaumatin mixtures. Further, the process for protein crystallisation is successfully transferred from the -μl scale to a -ml scale and finally a continuous oscillatory crystalliser. The findings in this study can be translated to purification of other proteins with high pharmaceutical values using crystallisation as a simpler and cost-effective alternative to conventional multi-step chromatography to benefit the society.

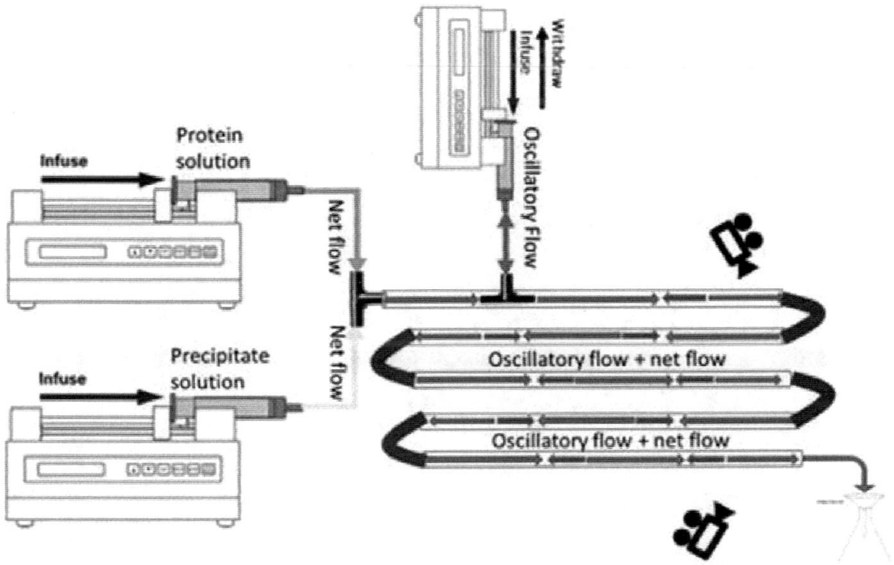

Figure 1: Illustration of continuous oscillatory flow crystalliser [7].

References

1. Walsh, G., Biopharmaceutical benchmarks 2014. Nat. Biotechnol. 2014, 32 (10), 992-1000.

2. Yang, H. and J.H. ter Horst, Crystal nucleation of small organic molecules, in New Perspectives on Mineral Nucleation and Growth. 2017, Springer. p. 317-337.

3. Basu, S.K., Govardhan CP, Jung CW, Margolin AL., Protein crystals for the delivery of biopharmaceuticals. Expert opinion on biological therapy, 2004. 4(3): p. 301-317.

4. Parambil, J.V., Schaepertoens, M., Williams D.R., and Heng, J.Y.Y., Effects of Oscillatory Flow on the Nucleation and Crystallization of Insulin. Crystal Growth & Design, 2011. 11(10): p. 4353-4359.

5. Delmas, T., Roberts, M.M., and Heng, J.Y.Y., Nucleation and crystallization of lysozyme: Role of substrate surface chemistry and topography. Journal of Adhesion Science and Technology, 2011. 25(4-5): p. 357-366.

6. Harrison, R.G., Todd, W.P., Rudge, S.R., and Petrides, D.P., Bioseparations science and engineering. 2015: Topics in Chemical Engineering.

7. Yang, H., Peczulis, P., Inguva, P., Li, X., and Heng, J.Y.Y., Continuous protein crystallisation platform and process: Case of lysozyme. Chemical Engineering Research and Design, 2018. Volume 136: p. 529-535.

568848 Implementation of Continuous Crystallization to Circumvent a Meta-Stable Liquid-Liquid Separation

High concentrations of complex organic solute may lead to meta-stable liquid-liquid separations due to solvent-solute interactions, even in otherwise miscible solvent systems. The formation of such non-ideal phases is generally undesirable in crystallization operations, as they tend to unevenly distribute the solute across the resulting liquids. Some of the potential side effects of this phenomenon are uncontrolled nucleation and growth, inadequate impurity purge, solvent entrainment, and morphological inconsistencies in the solids. Despite the inherent challenges that arise due to the formation of non-ideal phases, they have become increasingly common in pharmaceutical process development due to the current shift in the industry towards higher-complexity molecular structures and greater sustainability. The growing prevalence of these complex systems solicits the design and implementation of novel development tools that can assist process engineers in creating robust crystallization procedures. Traditional batch crystallization relies on modifications to the cooling curves, or overall solvent/anti-solvent composition to avoid meta-stable liquid-liquid separations; however, some chemical systems lack the flexibility required to incorporate these corrective measures, and the potential recourse may lead to unexpected consequences, such as polymorphic changes and slower crystallization kinetics. Using a Mixed Suspension Mixed Product Removal (MSMPR) continuous crystallizer, we captured a steady state equilibrium point in the saturation curve that avoids the formation of the undesired meta-stable phase for the purification of a commercially relevant small-molecule. The resulting continuous crystallization process circumvents original design limitations of the system, providing controlled recovery of the desired material without altering the original solvent/anti-solvent composition. In this talk, we will discuss and compare the different development challenges encountered when implementing the process in batch versus continuous mode. Additionally, we will outline the benefits of implementing continuous crystallization as a tool to obtain crystallization kinetic data, which can be incorporated with a 3D population balance and mass balance model to optimize conditions for yield and productivity.

571578 Comparison of Isolation-Free Continuous Crystallization Operations: Falling Film Solution Layer Crystallization and Confined Suspension Crystallization

Continuous Manufacturing and Flow Synthesis have seen significant interest and uptake within the pharmaceutical industry. This has been driven by a number of distinct advantages over traditional batch chemistry and manufacturing approaches. Compared to batch processing continuous flow provides improved heat and mass transfer, selectivity and process intensity that enables improved safety, capital costs, and accessibility to a significantly wider range of process scale chemistries. In-spite of these advantages continuous synthesis and particularly multiple step integrated flow syntheses, that characterize the relatively complex continuous synthetic active pharmaceutical ingredient (API) routes have limitations in comparison to batch manufacturing of clinical and commercial supplies, requiring significant telescoping, which constrains the ability to provide intermediate purification, relies heavily on liquid extraction and increases complexity in process development. Batch synthetic routes are characterized by steps which typically are demarked by an intermediate crystallization, these steps enable subsequent steps to begin with unconstrained solvent choice, optimal concentration and purified intermediate feeds. This aligns well with the almost unique combination of purity, regulatory requirements, constrained development timescales, uncertainty of product commercialization and need for flexibility that characterize pharmaceutical development.

This study aims to build upon the previous demonstration of isolation free crystallization[1,2] via falling film solution layer crystallization to provide the ability to conduct significant intermediate purification and solvent swapping via crystallization without the need for isolation, thus allowing many of the flexibilities and advantages of batch manufacturing to be retained in fully telescoped multi-step continuous synthesis. Falling film solution layer crystallization [1,2] is conducted in a novel low-cost modular 3D printed configuration in this study. Its performance is compared to confined suspension crystallization an approach outlined here where suspended solids are retained within a crystallization volume rather than confined to a layer. Both operations negate the need to conduct standard continuous crystallisations with isolation and redissolution for intermediate purification in telescoped flow synthesis.

References

1. N Yazdanpanah, ST Ferguson, AS Myerson, BL Trout. Crystal Growth & Design 16 (1), 285-296
2. N Yazdanpanah, AS Myerson, BL Trout. *Ind. Eng. Chem. Res.*, **2016**, *55* (17), pp 5019 –5029

572539 Integrated Continuous Crystallization and Spray Drying of Beclomethasone Dipropionate for Pulmonary Drug Delivery

Pulmonary drug delivery can offer improved bioavailability of active pharmaceutical ingredients (APIs) compared to oral drug delivery for both the treatment of systemic and local diseases due to the large surface area of the lungs, direct delivery to the site of action, and reduced metabolism. Delivery of the API in dry powder form through inhalation is beneficial compared to other aerosol forms, because a dry powder requires no propellant, has superior chemical stability and resistance to bacterial, and is portable and easy to use. However, the manufacture of dry powders for pulmonary drug delivery is challenged by stringent requirements for the solid-state properties.(1) Small particles (< 1 μm) do not deposit sufficiently fast within a single breathing cycle and large particles (> 5 μm) impact in the back of the throat followed by swallowing. Crystalline particles offer advantages related to stability and flowability compared to amorphous particles. Finally, needle-like particles have a better aerosolization behavior compared to spherical particles. Drug particles that do not deposit effectively in the lungs increase the cost of a single dose and may lead to side-effects for the patient. Therefore, the design and operation of pharmaceutical processes that can produce pulmonary drugs with optimal properties reliably is of critical importance to minimize costs and enhance therapeutic efficacy. Micronization such as milling is often needed after conventional batch crystallization processes to reduce the size of the API crystals for pulmonary drug delivery. However, micronization lowers the process yield and typically leads to poorer aerosolization behavior.

Continuous manufacturing offers inherent advantages to improve the manufacture of APIs for dry powder inhalation. A high supersaturation is needed to create a large number of nuclei that can grow into the optimal size range with high attainable yield. Anti-solvent crystallization can create such high supersaturation, which works well in continuous flow due better mixing conditions compared to batch operation. Furthermore, continuous crystallization can be integrated with spray drying, which inherently operates in continuous flow, to minimize work-up steps and transfer time.

An integrated continuous process combining plug-flow crystallization and spray drying for the improved manufacture of pulmonary drugs is presented.(2) A segmented-flow crystallizer(3-5) is integrated with spray drying to produce crystals with optimal properties for pulmonary drug delivery in a single step. The process is illustrated for the case of beclomethasone dipropionate, which is a commercial pulmonary drug used, for example, for the treatment of asthma. First, the continuous anti-solvent crystallization is optimized by screening suitable solvent/anti-solvent pairs and varying common operational variables such as residence time, temperature, and initial supersaturation. Subsequently, spray drying is optimized by varying the inlet temperature and atomizer gas flow rate. Finally, both steps are integrated into a single process, which has been optimized by varying the most critical process parameters.

The dry powder is analyzed in terms of solid-state form, morphology and drug delivery performance using a Next Generation Impactor™ and commercial dry powder inhaler. The results demonstrate that under optimal conditions the integrated process produces a dry powder consisting of needle-like crystalline particles with an emitted fraction (i.e., the fraction of the powder leaving the capsule in the inhaler device) of 75% and a fine particle fraction of 33% (i.e., the fraction of the powder expected to deposited at the intended sites in the lungs) without any size correction steps or addition of excipients, which represents a good performance compared to commercial products. Furthermore, the overall process yield is 65% and the total residence time is only 10 minutes. The presented process is simple and flexible due to continuous operation and the clear separation of process functions. Future work may focus on the extension towards other applications, dedicated equipment design to further improve process yield, process modeling, control, and alignment with quality-by-design principles and recently published regulatory guidelines for continuous pharmaceutical manufacturing to enable commercial implementation.(6)

Acknowledgment

This research is funded by the Hong Kong Innovation and Technology Commission, under ITSP Tier 3

Scheme (Project No. ITS/137/16).

References

1. Hadiwinoto GD, Kwok PCL, Lakerveld R. A review on recent technologies for the manufacture of pulmonary drugs. Ther Delivery. 2018;9(1):47-70.
2. Hadiwinoto GD, Kwok PCL, Tong HHY, Wong SN, Chow SF, Lakerveld R. Integrated Continuous Plug-Flow Crystallization and Spray Drying of Pharmaceuticals for Dry Powder Inhalation. Ind Eng Chem Res. 2019;Submitted.
3. Eder RJP, Schrank S, Besenhard MO, Roblegg E, Gruber-Woelfler H, Khinast JG. Continuous Sonocrystallization of Acetylsalicylic Acid (ASA): Control of Crystal Size. Cryst Growth Des. 2012;12(10):4733-8.
4. Jiang M, Zhu Z, Jimenez E, Papageorgiou CD, Waetzig J, Hardy A, et al. Continuous-Flow Tubular Crystallization in Slugs Spontaneously Induced by Hydrodynamics. Cryst Growth Des. 2014;14(2):851-60.
5. Kudo S, Takiyama H. Production of Fine Organic Crystalline Particles by Using Milli Segmented Flow Crystallizer. J Chem Eng Jpn. 2012;45(4):305-9.
6. Food and Drug Administration. Quality Considerations for Continuous Manufacturing: Draft Guidance. U.S. Department of Health and Human Services: Rockville, MD, 1997.

557167 Continuous Crystallization Development for the End-to-End Manufacturing of Ciprofloxacin HCl in Refrigerator-Sized Modules

Crystallization has recently gained an increased interest for continuous pharmaceutical manufacturing, particularly for the use of Mixed Suspension Mixed Product Removal (MSMPR) crystallizers for intermediate purification and isolation of Active Pharmaceutical Ingredients (APIs). The ability to sustain high suspension densities of crystals grown under known, stable kinetics, provides an advantage for ensuring consistent product quality in continuous operation.

This work was part of the development of an end-to-end process for Ciprofloxacin hydrochloride in compact, reconfigurable modules. The overall process is divided in four steps, namely: synthesis, workup, isolation, and formulation. This presentation will focus on the downstream modules for workup and isolation, including continuous precipitation of Ciprofloxacin base as an intermediate and the final separation of Ciprofloxacin hydrochloride monohydrate via continuous crystallization.

The process development milestones are described first, with particular focus on how the critical crystallization conditions were selected and the role of each MSMPR step on defining the drug substance's critical quality attributes. Then, the workup and isolation parts of the end-to-end manufacturing campaign are described, with emphasis on how the MSMPR processes were integrated with fundamentally batch processes (e.g. filtration, drying), and how the implemented process controls ensured an extended continuous operation and a consistent drug substance quality.

560531 Achieving Particle Size and Impurity Control for a Continuous Crystallization Process Using a Digital Design Approach

The attainable regions for critical quality attributes, such as Particle Size Distribution (PSD) and impurity levels, in the manufacture of Solid Oral Dosage Forms (SODF) can be highly dependent upon the type and operation of the crystallization process. In this work, we outline a step-wise workflow consisting of model validation, model-based technology transfer and process optimisation employed for the digital design of a continuous cascade cooling crystallization and wet milling process for manufacturing an Active Pharmaceutical Ingredient (API). Following this workflow, the mechanistic model was first validated using batch crystallization data and subsequently applied to describe the continuous crystallization and wet milling process, and explore the impact of varying process parameters and process configurations (namely the position of the wet mill unit in the continuous crystallization process – in recycle with the first or the third crystallization stages) on the attainable processing regions for particle size and purity, subject to various process constraints.

The step-wise approach taken for model validation and its subsequent application was as follows:

1. Model validation of crystallization kinetics for the pure API solid phase from batch de-supersaturation experimental runs.
2. Application of the validated crystallization kinetics to describe the continuous crystallization of pure API in a three-stage cascade process, as shown in Figure 1 below (wet milling was not considered at this stage).
3. Refinement of the crystallization kinetic parameters for the pure API solid phase, utilising targeted data sets suggested by the model fitted in 1 using batch data sets.
4. Optimisation of the continuous crystallization process without wet milling to determine its feasibility.
5. Model validation of breakage kinetic parameters using batch data from a wet milling unit operated in a recycle with a batch crystallizer. A rotor-stator wet mill was utilised for this process, with a range of rotor frequencies and milling head generator configurations probed to assess the impact on the PSD over time and to fit the breakage kinetic parameters.
6. Optimisation of the combined crystallization and wet milling process for pure API to achieve a target product PSD. The configuration of the process, in terms of the position of the wet mill unit, in a recycle with the first or third crystallization stages, was also probed.
7. Model validation of crystallization kinetics for an API dimer solid phase from batch de-supersaturation experimental runs. The API dimer forms as a separate solid phase that can crystallize during the process and is considered to be a process impurity.
8. Optimisation of the combined crystallization and wet milling process for the pure API and API dimer solid phases (impurity) to achieve target product PSD and desired level of impurity.

Conclusions

The main conclusions of the work include the following:

- The continuous crystallization process was unable to achieve the desired product PSD without wet milling, due to very low levels of secondary nucleation in the system. Optimal PSD quantiles (d10, d50 and d90) predicted for the product were significantly higher than the target PSD quantiles required to achieve the desired product performance for this compound.
- Addition of a wet milling step in a recycle with the continuous crystallization significantly increased the lower end of the attainable region in terms of PSD quantiles and impurity level, allowing the process to comfortably achieve the desired range for product PSD quantiles.
- In terms of configurations for the continuous process, it was found that having the wet mill placed in a recycle with the first crystallization stage was more effective at reducing product PSD and at decreasing the impurity level of the material produced, compared to having the wet mill placed in a recycle with the third stage.
- Mechanistic modelling approaches can be utilised to significantly reduce the development timelines, material consumption and efficiency of continuous API crystallization production processes.

Figure 1: Simulation flowsheet used model for optimisation of continuous crystallization and wet milling process.

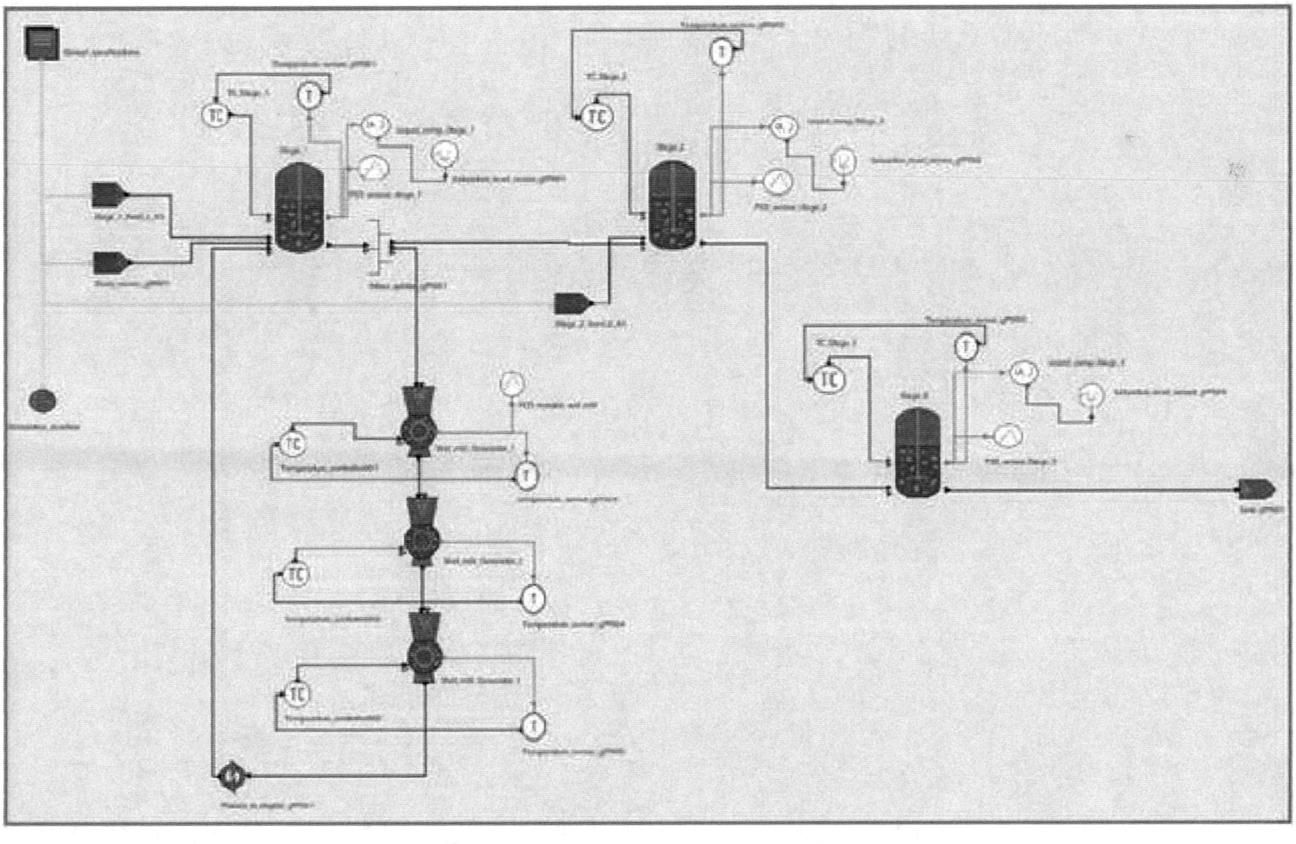

574658 A Methodology for Predicting the Performance of Batch Strippers in the Mass Transfer Limited Regime

Batch stripping of volatiles from liquids are employed widely in chemical industry in the form of simple stirred tanks with a stripping agent such as steam or nitrogen sparged below the agitator. The main advantages of such batch stripping units are: low capital cost and less complex operation, when compared to packed columns or other forms of continuous stripping units. Irrespective of the mode of operation, it is a challenging exercise to model and accurately predict the performance of stripping operations particularly in the mass transfer dominated regime, for example when stripping volatiles from highly viscous fluids. An equilibrium based approach in this case is not applicable and fails if used for predicting the stripping performance. It gets even more complicated to model a dynamic process such as a batch stripper where temperature, pressure (vacuum), stripping agent flow rate and volatiles composition can all change with time.

In this talk, we demonstrate a methodology for predicting the performance of dynamic plant scale batch stripping operations using simple lab experiments combined with the concepts of stripping factors and transfer units. Stripping volatiles from a commercially produced polyol in a 9000 gallon batch stripper is used as a case study. A concept of instantaneous stripping factor is used to gauge the stripping power at any given point of time. This is combined with the number of transfer units obtained from the volatiles concentration in the liquid phase before and after the stripping operation. A functional relationship between the two is established using lab scale experiments, which is then used to accurately predict the performance at the plant scale, as demonstrated in this study. Such an approach can be used to design new stripping operations or for rating existing ones for potential improvements.

588831 Learning from Field Inspection of Towers

Field inspection of towers is more than just making sure equipment is installed according to specification drawings. Equipment inspection also helps educate starting engineers. It also shows how equipment fails in the field and identifies possible problems not caught at the design stage. Examples are shown how equipment inspection has identified problems and what types of design, fabrication, and installation errors can make it all the way to a tower that's supposed to be ready to start up.

557870 Point Efficiency Data for Distillation Column Design at Elevated Liquid Viscosities

Liquid viscosity is considered to be one of the most important physical properties that influences distillation column efficiency. Although most of the distillation efficiency correlations that have been published thus far emphasize the importance of viscosity, they were developed or validated based on efficiency data at liquid viscosities less than 1 cP. For example, the O'Connell correlation was developed based on efficiency data at less than 0.5 cP. However, distillation with liquid viscosities greater than 1 cP is encountered in certain cases in the chemical process industry. Examples include separation of monomers from polymeric solutions, vacuum distillation of viscous crude oil, separations involving ionic liquids, and vacuum distillation of bio-oils to obtain biofuels. Therefore, distillation efficiency data at liquid viscosities greater than 1 cP are required to accurately quantify the effect of liquid viscosity on distillation efficiency and to develop improved efficiency correlations.

In scale-up of distillation column design, point efficiency is the foundation upon which tray and column efficiencies are calculated. Oldershaw columns have been proven to be effective in representing the point efficiency of commercial columns. In this paper, point efficiency data collected using an Oldershaw column for 4 different test systems, including a new viscous test system (cyclopentanol/cyclohexanol) that we recently established[1], at liquid viscosities ranging from 0.2 to 4.2 cP will be presented. The methodology used for quantifying the effect of liquid viscosity on point efficiency will also be presented. These new data will be useful in the design of distillation columns operating at elevated liquid viscosities.

1. Manivannan, R. G.; Mohammad, S.; McCarley, K.; Cai, T.; Aichele, C., A New Test System for Distillation Efficiency Experiments at Elevated Liquid Viscosities: Vapor–Liquid Equilibrium and Liquid Viscosity Data for Cyclopentanol + Cyclohexanol. *Journal of Chemical & Engineering Data* **2019,** 64, (2), 696-705.

557891 Process Development for High Efficiency Cryogenic Air Separation Unit

Due to a rise in electricity prices of past years, there are high expectations for improving efficiency of cryogenic air separation units (ASUs), which generally consume a large amount of energy to produce oxygen, nitrogen and argon. On the other hand, a basic process for ASUs has improved little for the past hundred years. ASUs have advanced in regards to safety, operability, load followability, etc. owing to development or improvement of comprised equipment (e.g., air compressor, air purifier, heat-exchanger, rectification column) and many process modifications. However, the specific power consumption of an ASU has remained nearly flat in the past two decades, especially for producing high purity oxygen (typically higher than 99.5 mole %). Hence, in this work we have developed a state-of-the-art ASU process for high purity oxygen product capable of reducing power consumption to break through this situation.

A conventional ASU has employed a classical double column process, which comprises a high-pressure column, a low-pressure column and an argon column as a side-cut column connected to the low-pressure column, and the bottom of the low-pressure column is thermally integrated with the top of the high-pressure column by means of an indirect heat exchanger, so called a condenser/reboiler.

In contrast, the developed ASU process comprises a high-pressure column, a low-pressure column and an argon column, and these three columns are thermally integrated in series. Liquid oxygen at the bottom of the low-pressure column is vaporized by condensing argon vapor at the top of the argon column, and liquid oxygen at the bottom of the argon column is vaporized by condensing nitrogen vapor at the top of the high-pressure column. This process cycle makes it possible to improve efficiency of the low-pressure column, and it can be reduced the required heat exchange duty to generate ascending oxygen vapor in the low-pressure column. Therefore, the power consumption of the ASU can also be reduced. In our case study under some practical conditions, it has been confirmed that this new ASU process is effective for saving energy to produce high purity oxygen, nitrogen and argon.

559983 Experimental and Numerical Study of the Mass Transfer Performances of Commercial Sulzer EX Structured Packing at High Driving Force

Introduction

Stripping and distillation columns are widely used at industrial scale to separate molecules with different boiling points. The trays traditionally used in distillation columns are being gradually substituted by structured packings offering good gas-liquid mass transfer performances with low pressure drop. Many structured packing solutions exist for the industrial scale. The choices are more limited at the laboratory scale. The Sulzer EX structured packing represents one solution for improving gas-liquid contact and mass transfer performances within a stripping column at laboratory scale. Although the EX structured packing is widely used for separation processes at the lab scale, available information about mass transfer correlations for this specific packing is reduced. One can mention the work from Roesler, J. et Muhammad, A., 2018[1], in which the authors present the mass transfer rate parameters for this particular packing within a column of 20 mm diameter. The authors determined correlations for both gas and liquid mass transfer coefficients and gas-liquid contact area as a function of the relevant dimensionless numbers (Reynolds and Schmidt). During the experiments, the superficial velocities of the fluids and the driving forces were extremely low. No studies were found in literature concerning the gas-liquid mass transfer phenomenon for EX structured packing at high driving force. In this particular case, both gas and liquid hydrodynamics might be affected by the mass transfer phenomenon. Thus, mass transfer rate as well as the dependence of mass transfer coefficients on superficial velocities may increase by means of the local instabilities generated within the fluids at high driving forces.

Objectives and methodology

The objective of this work was to establish new mass transfer parameters correlations for the Sulzer EX structured packing at high driving forces. Experiments carried out in a 20 mm ID column containing 1430 mm packing height for separating the 370 °C$^-$ boiling point fraction from the 370 °C$^+$ fraction present in a vacuum gas-oil were used for determining the different mass transfer rate parameters correlations and their dependence on gas and liquid velocities. The column was placed inside an oven constituted by six heating zones allowing the implementation of specific linear temperature profiles to improve separation. The laboratory tests were carried out at atmospheric pressure and temperatures in between 230 °C (top of the column) and 345 °C (bottom of the column). The bottom of the column was fed with hydrogen which was used for stripping purposes.

A dynamic mathematical model was developed to extract the mass transfer parameters from the experiments. The model considers both convective and diffusive mass transfer mechanisms as well as the gas-liquid equilibrium for all the pseudo-components (lumps) of the feed.

Results

The impact of the hydrogen flow on the 370 °C$^-$ cut mass fraction at the bottom of the laboratory column is illustrated in Figure 1. The gas flow improves the separation performances by increasing the driving force between the liquid and the gas phases. However, the model shown that the separation improvement is mostly related to the enhancement of mass transfer between phases. In fact, the mass transfer between phases is a strong function of the gas flow at these conditions. This is not the case at low driving forces, for which the gas phase mass transfer coefficient is substantially higher than the liquid phase mass transfer coefficient. Some flow instabilities are likely to appear at high driving forces. These instabilities increases the rate of surface renewal and consequently, the liquid side mass transfer coefficient. Thus, the mass transfer phenomenon can become dependent of the gas mass flow at these conditions.

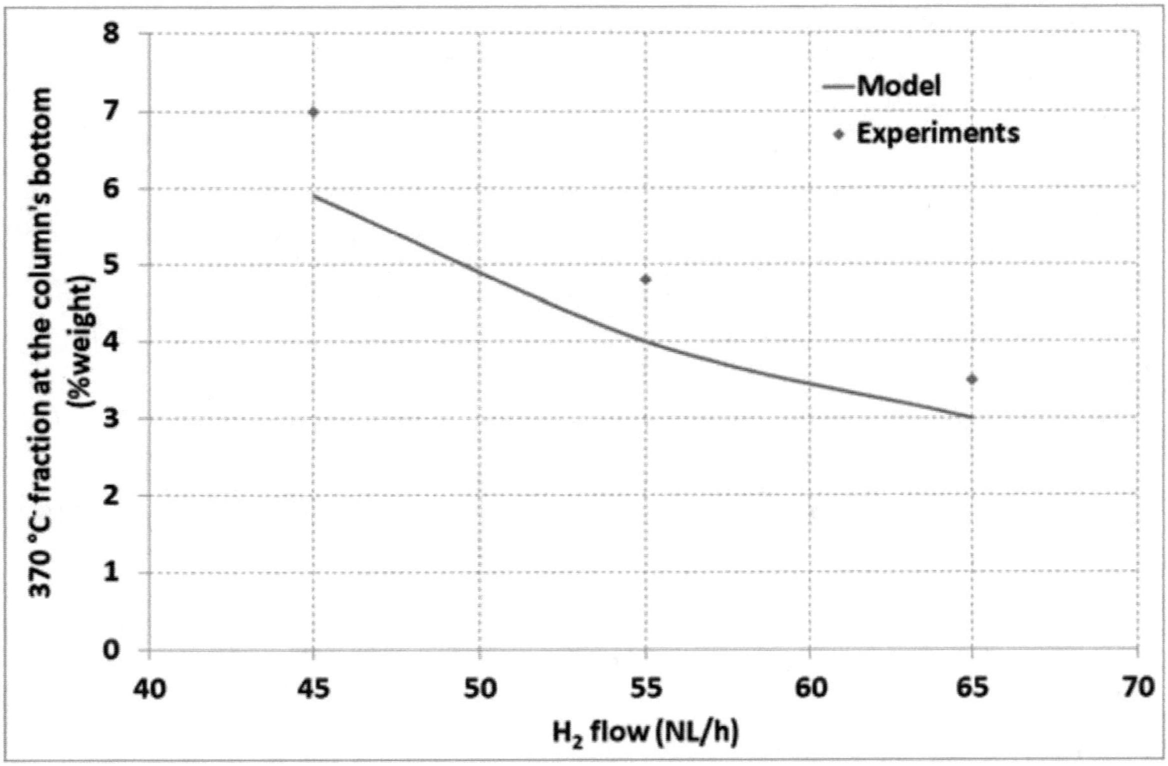

Figure 1 – Evolution of the 370 °C$^-$ cut mass fraction at the bottom of the laboratory column as a function of the H_2 flow. Model and experimental results are represented by curves and symbols, respectively.

Liquid flow – 340 g/h, T_{top} = 230 °C, T_{bottom} = 345 °C, 50 % of 370 °C$^+$ and 370 °C$^-$ fraction in a mass basis.

The evolution of the 370 °C$^-$ cut mass fraction at the bottom of the column as a function of the liquid flow to be fractionated is shown in Figure 2 along with the experimental point used for model validation purposes (gas flow of 100 NL/h and liquid flow of 534.6 g/h). The separation performance decrease is slight despite a considerable liquid flow increase, which is traduced by a lower liquid residence time in the column. The low residence time is compensated by the improvement on the gas-liquid mass transfer through the local instabilities generated in the liquid phase at high driving forces. Consequently, the liquid size mass transfer coefficient is substantially higher than the one obtained at the same liquid velocity and low driving force. This conclusion is in good agreement with the dependence of the mass transfer phenomenon on gas velocity.

The model was used afterwards in order to optimize the separation for a liquid flow of 534.6 g/h containing 50 % of 370 °C$^+$ and 370 °C$^-$ fraction in a mass basis. Thus, the needed gas flow to obtain less than 3 % (2.6 % precisely) of 370 °C$^-$ fraction in a mass basis at the bottom of the column was 100 NL/h, which perfectly agrees with the experimental result.

Finally, the model was also used to establish the influence of the temperature profile on separation performances. Both the model and the experiments show that the dependence of the separation on the temperature profile was low when the temperature at the top of the column was varied between 230 and 260 °C, and between 345 and 365 °C at the top of the column.

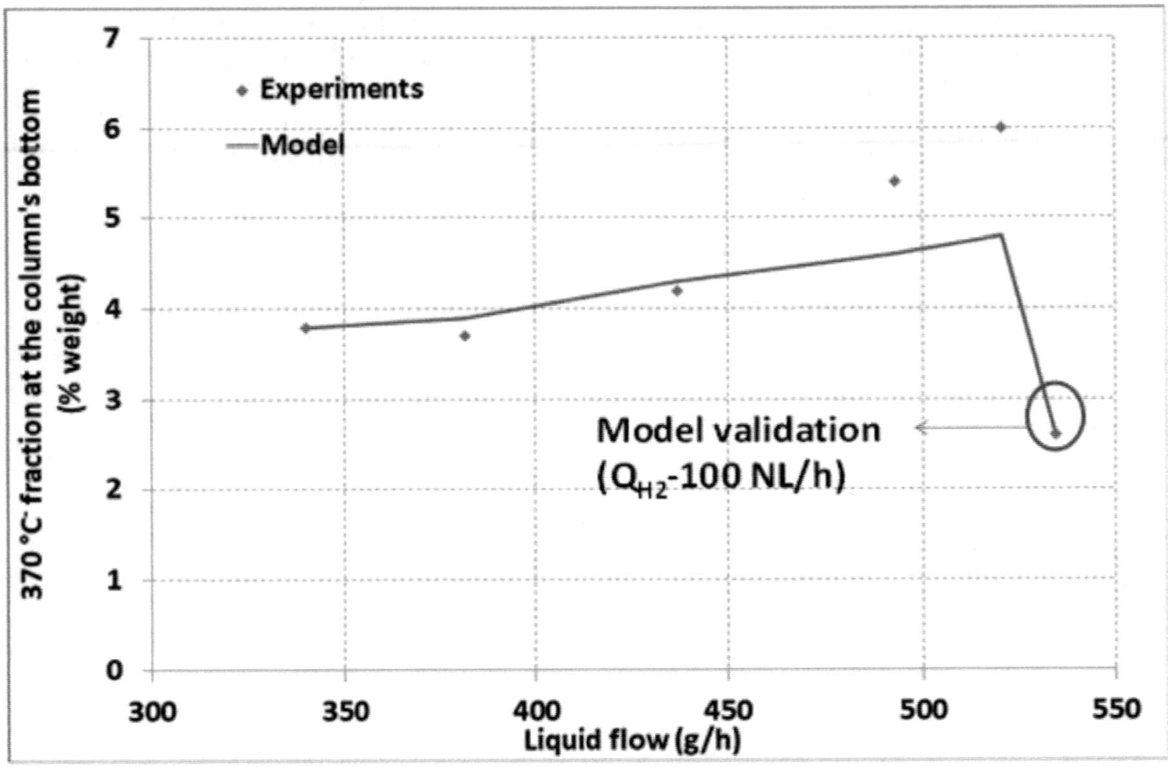

Figure 2 – Evolution of the 370 °C$^-$ cut mass fraction at the bottom of the laboratory column as a function of liquid flow. Model and experimental results are represented by curves and symbols, respectively. Gas flow – 75 NL/h, T_{top} = 230 °C, T_{bottom} = 345 °C, 50 % of 370 °C$^+$ and 370 °C$^-$ in a mass basis. Model validation at a gas flow of 100 NL/h and a liquid flow of 534.6 g/h.

Conclusion

The objective of this work was to understand the mass transfer phenomenon within a laboratory scale stripping column of 20 mm ID containing 1430 mm of Sulzer EX structured packing at high driving forces.

Both the gas and the liquid velocities influence gas-liquid mass transfer phenomenon. The increase of the gas phase velocity allows the driving force to be increased, as well as the global mass transfer coefficient. The liquid velocity increase is traduced in a slight decrease on separation performance, lower than expected. This is probably due to the occurrence of some liquid flow instabilities at high driving forces that increases the surface renewal rate and partially compensate the reduction on residence time. This conclusion is in agreement with the high dependence of the mass transfer on gas velocity at high driving forces, which is not the case at low driving forces.

A model taking into account mass transfer by convection and interphase mass transfer, as well as gas-liquid equilibrium was developed and validated within a wide range of operating conditions. The model was used to optimize the performance of the separation column after calibration, and the numerical results were in good agreement with the experimental data.

Finally, both the model and the experiments shown that the dependence of the separation on the temperature profile was negligible when the temperature at the top of the column was varied between 230 and 260 °C, and between 345 and 365 °C at the top of the column.

---

[1] Roesler, J., Muhammad, A., **Mass transfer rate parameters of Sulzer EX laboratory scale Gauze**

Structured Packing, *Chem. Eng. Trans.*, 69, **2018.**

566428 Separation Column Gamma Scans - Moving into a New Dimension

Gamma Scans are widely used to diagnose separation column hydraulics both in research settings and at commercial installations. Columns containing packing as well as trays are routinely studied with gamma scans. Typical gamma scan data is presented in graphical form in a plot of scan distance or column elevation versus radiation intensity. The radiation intensity measured is the amount of radiation passing through the column from the scanning radiation source on one side of the vessel to the radiation detector located on the opposite side of the vessel. The radiation intensity is an implicit measure of the process density between the radiation source and detector.

Using the results from gamma scans aid in diagnosing the hydraulic condition of operating separation columns. Differences in measured radiation intensity show how the liquid-vapor is behaving inside the column. Comparing these results to how the liquid-vapor should be acting, given the internals installed, allows for hydraulic issues to be diagnosed.

Recent developments have enhanced the usefulness of the basic gamma scan results by deriving calculated process densities from the gamma scan data. The methodology, the limitations and the realistic benefits of these calculations will be presented. Case studies from both research and commercial environments will be discussed.

569942 National Academies Report: A Research Agenda for Transforming Separation Science

This presentation will report on the recommendations and findings of a consensus study performed by The National Academies of Science, Engineering and Medicine (http://nas-sites.org/dels/studies/separations/) on developing a research agenda for the next era of separation science. It was sponsored by the Department of Energy, the National Science Foundation and National Institute of Standards and Technology. The study was conducted by an ad-hoc committee with expertise from academia, National Laboratories and the private sector. At the time of abstract submission the study was under review and will be published in May, 2019. The chair of the committee, Professor Joan F. Brennecke from the McKetta Department of Chemical Engineering, will be available to make the presentation. The statement of task for the committee is listed below.

The National Academies of Sciences, Engineering, and Medicine will convene an ad hoc committee of experts and scientific leaders to develop an agenda for fundamental research in separations science. The committee will write a report that will:

• Assess recent and ongoing research efforts to advance science that underpins separations of chemical mixtures and identify priority areas of research that will transform the field of separations science;

• Identify advances in chemical transformations, chemical and materials in-situ and operando characterization, computation and theory, and synthesis capabilities that can be employed to advance separations sciences and illustrate how they can be utilized to advance separations science;

• Address the intersections between chemistry, biochemistry, materials, physics, engineering, and information science that will be essential for scientific progress;

• Identify needs and opportunities for novel instrumentation and tools that will advance our understanding of separations processes, including laboratory scale to mid-scale to large scale, such as synchrotron light sources or neutron facilities;

• Identify the educational and human resource needs (including cross-sections of academia and industry) to enable advances in separations science; and

• Assess the potential impacts that fundamental research in separations can have on technologies and practices in industry.

The report will provide guidance to research sponsors, as well as to the research communities in academia and industry. The report's recommendations will focus on science needs and priorities rather than specific funding or organizational aspects.

572807 Challenges and Advancements in Extraction Processes for the Recovery and Purification of Cannabidiol from Hemp

Cannabidiol (CBD) consumer sales over the past 4 years has increased 500% and is expected to continue to increase to a $2 billion-dollar market in the next 5 years. CBD is 1 of over 100 different cannabinoids found in the Hemp plant (C. sativa). This class of compounds all have similar molecular structures and are heat labile, making separation difficult. The FDA is beginning the process of regulating CBD, which may invalidate most current processes for the extraction of CBD products. Given the projected market size the development of commercial scale, high recovery processes are required. Some of the challenges and breakthroughs in solvent extraction, crystallization, and liquid chromatography will be explored.

574617 The Biorefinery Separations Challenge - New Materials and Processes

Lignocellulosic biomass conversion is a key route for sustainable chemical production. As in the petrochemical industry, the overall process economics will be strongly influenced by the cost and efficiency of molecular separation processes in addition to the catalytic or enzymatic conversion processes. However, the development of systematic approaches for biorefining separations is in a state of infancy. Biorefinery streams contain complex molecules and highly multicomponent mixtures. Adsorptive and membrane separations based upon advanced materials are expected to be integral to successful biorefining, but a new generation of materials and processes is required to meet this 'high-resolution separation' challenge. This talk will highlight the above issues in three ways. First, a systematic analysis of biomass conversion routes reveals a number of opportunities for high-resolution separation materials and processes. In this context, we consider the current status, potential, and barriers faced by the main classes of nanoporous materials. Finally we will overview two ongoing areas of biorefinery separation research at Georgia Tech. The first involves the development of robust membranes to fractionate kraft black liquor for recovery and/or recycle of lignin and valuable organic and inorganic chemicals. The second involves the development and use of robust adsorbents for purification of mixtures containing furanic molecules.

560340 Shock Electrodialysis for Ion-Selective Water Purification

Shock electrodialysis is an emerging method of water desalination, based on separating fluid streams on either side of a deionization shock wave propagating through a charged porous medium under over-limiting current. This talk will introduce the basic physics of shock ED and demonstrate promising new applications in water treatment. For dilute feed streams, shock ED is capable of complete (>99.99%) salt removal with high (>85%) water recovery, which we apply to the removal of cobalt and cesium ions from nuclear waste. High (>90x) selectivity for multivalent ions is also illustrated by the continuous removal of lead and magnesium from dilute sodium chloride solutions. High selectivity for toxic metal ions at low salinity leads to ultra-low electrical energy cost (~\$0.01/m^3), which could lead to a practical solution to purify drinking water from corroded pipes or contaminated sources. This work also raises theoretical questions about extreme nonlinear electrokinetic phenomena in porous media, far above the diffusion-limited current.

571859 Thermodynamics and Chemo-Mechanics of Electrosorption Electrodes for Separation Applications

Integrating renewable energy with desalination technologies is an important challenge, in order to design resilient and decoupled water and energy infrastructures. One integrated technology capable of directly using renewable electrons for salt-based separations is capacitive deionization (CDI). CDI performs ion removal using an electrosorption-based process. Separation is achieved through storing ions within an electric double layer on a high surface area carbon-based electrode. Salt removal is largely related to the physical properties of the electrode material (i.e. surface area, pore size, etc.). Furthermore, the nature of the contaminant (anion, cation, or uncharged species) influences ion packing within the electrode. How species interact and order within the confined region of the electric double layer and the nano and micropores influences the thermodynamic efficiency of separation. This talk examines how chemo-mechanical electrode signatures can be used to describe ion dynamics and separation efficiencies in nonporous carbon and intercalation compounds. Model anionic (phosphate) and cationic systems (alkali ions) are used to systematically evaluate the role of valency and ionic size on separation properties in a wide range of faradaic and electric double layer materials. Separation efficiency is largely influenced by co-ion repulsion dynamics that occur during potential swings (i.e. charge/discharge steps).

565688 The Impacts of Fixed Chemical Charge Location and Abundance on the Performance and Cost of Capacitive Deionization Systems

Capacitive deionization (CDI) is an electrochemical separation technique that couples salt removal with electrochemical energy storage. CDI has recently received renewed interest as a potentially energy efficient method of desalting brackish groundwater. The charge efficiency (CE) of a CDI cell, defined as the ratio of adsorbed charge to applied charge, is a critical cell performance metric, as it directly relates to salt adsorption capacity (SAC) and thermodynamic energy efficiency. Several strategies have been proposed to improve CE by introducing complementary fixed charge groups to promote counter ion storage within capacitive electrode macro- and micropores. These strategies include placing ion exchange membranes (IEMs) between the electrodes and desalination channel, fixing charge directly onto the carbon surface, and incorporating charged polymers within the electrode macrostructure with binders. In this talk we will present simulated and experimental evidence that reveals how the location and density of fixed charge groups in conjunction with cell cycling behavior can dramatically impact CDI adsorption capacity and energy efficiency. We will also discuss these impacts in the context of capital and operating cost.

564151 Practical Applications of Electrokinetic Separations

Separations technologies applied in aqueous phase can be categorized into (i)electrokinetic-driven, (ii)pressure-driven, (iii)thermally driven and (iv)biological-related. In electrokinetic processes, concentrated ions are separated from the water solution in an electric field, and therefore, energy use is correlated with the quantity of ions removed. Electrokinetic processes can provide water desalination at a targeted salinity or preferential ions removal that is "fit-for-purpose". Electrokinetic processes only remove ionized or ionizable species and is not useful for removing organics or biological species. In pressure-driven processes, purified water is separated from the brine concentrate. Typically, pressure-driven processes cannot tune salinity for fit-for-purpose quality but are effective at organic and biological species removal. Electrokinetic processes encompass electrodialysis (ED), electrodeionization (EDI), capacitive deionization (CDI), cation intercalation desalination (CID), and ion concentration polarization (ICP).

Compared to the pressure-driven membrane separation technologies used most in industrial separations, applications of selective separations have increased in recent years and becomes important to address the challenges of technology adaptation to climate change. For examples, the production of biofuel and bio-products to reduce green-house gas emission from fossil fuel and the non-conventional water supply for water-energy nexus have required high energy efficient and cost effective separation technologies. Innovative electrochemical separations can provide transformational impacts in advancing selective separations for highly energy efficient, small capital footprints and low processing cost. It, thus, enables the paradigm-shift of using alternative fuels and water supplies for industrial applications. We will discuss the key process performance metrics, energy consumption and processing rate, of EDI and CDI applied in the fields of biorefinery, waste to energy and water energy nexus to separate charges species from "dilute" aqueous phase. The ions separation performance demonstrated from various aqueous streams include 1) inorganic and organic salts removal/capture from pretreated hydrolysate, and 2) organic acid capture/production from bioprocessing streams in biofuel production; 3) volatile fatty acid removal/capture from waste anaerobic digester for waste to energy; 4) selective desalination of hardness, alkalinity, silica and ammonia from impaired water for cooling water supply. Critical issues in process design and material property to achieve electrokinetic separation rate and energy efficiency for economic viability will be discussed.

569353 Electrochemically Modulated Mitigation of Acid Gas Emissions

Significant changes in global climate patterns and increasing ocean acidities with their negative impacts on the health of our planet have been ascribed to the ever-increasing rise in atmospheric CO_2levels, primarily attributed to anthropogenic sources such as the combustion of fossil fuels. The mitigation of these acid gas emissions is a daunting task, both because of the scale of the problem and because of the economic ramifications associated with the capture of the greenhouse gases and their subsequent utilization or subsurface storage. Effective means for the direct treatment of emissions with CO_2concentrations of 5 to 15% from a wide range of sources, such as in the power industries, at industrial facilities and from on-board vehicle exhausts, are sorely needed. Of late, there has also been some interest in the capture of CO_2directly from the atmosphere, at concentrations close to 0.04%, which offers its own challenges for implementation.

The traditional means for CO_2capture and release generally rely on either chemical or physical absorption in solvents with subsequent heating of the solution to release the captured $CO_2$and regenerate the solvent. The captured CO_2can then be compressed for injection and sequestration in subsurface geological formations. These processes require significant energy integration which adds complexity and cost to the overall capture operation. An example is the use of amine solutions where the capture is done at about 40-60°C, while regeneration at temperatures in the 120 to 150°C range is generally required. Isothermal operations that obviate or significantly reduce the heat integration requirements in these processes could have significant advantages over the traditional methods. One such approach is to exploit electrochemical technologies for the capture and release of CO_2, which can be operated isothermally, and can rely only on renewable energy resources, if desired.

We will describe two different approaches for CO_2capture from a wide range of sources. The first relies on an indirect approach, based on the electrochemical release of a metal ion to an anode chamber to displace the bound CO_2on an amine via a competitive complexation process; the CO_2can be recovered in a flash tank. The amines are regenerated in the cathode compartment of the cell, where the metal ion is reduced and plates out on the electrode, leaving the amine free for cycling back to the absorber. In the second approach, an electroactive moiety complexes *directly*with the CO_2upon activation by electrochemical reduction and releases the $CO_2$when it is re-oxidized on reversal of the applied cell voltage.

The general principles underlying these two electrochemically modulated acid gas separation processes will be outlined, with an emphasis on the thermodynamic and transport considerations required for their effective implementation in carbon capture, together with an initial economic evaluation of their performance.

579961 Mechanisms of Improvements in Virus Control By Electrocoagulation: Removal and Inactivation

Prof. Baltus and I have collaborated for over a decade on microorganism control from contaminated water supplies. We have coauthored papers on (hindered) convection and diffusion of bacteria and virus across microporous filters, but my group has been separately pursuing the idea of virus aggregation. The principal hypothesis from an environmental engineering standpoint is that virus removal is facilitated by increasing effective size using large-pored microfilters, which otherwise do not remove them. An additional advantage of coagulation is that it simultaneously removes other contaminants including natural organic matter and colloids thereby improving overall quality of our drinking water. My presentation will cover aspects of virus aggregation by externally added hydrolyzing metal-ion coagulants.

The principal objectives of this research are to elucidate underlying virus destabilization/removal mechanisms during electrocoagulation and identify possible intermediates leading to inactivation. We explore the electrochemical oxidation of sacrificial elemental aluminum and iron anodes to directly dissolve metal ion coagulant precursors that neutralize virus surface charge and upon hydrolysis sweep coagulate viruses. In all cases, the F-specific ssRNA coliphage MS2 was employed as the model virus. Atomic force microscopy was performed to directly measure adhesion forces between viruses and aluminum precipitates. Experiments were performed on synthetic buffered water at low and high pH as well as using saline background solutions to electrolytically produce free chlorine. Low chlorine concentrations combined with virus shielding and aggregation within flocs resulted in very slow disinfection rates in solutions with high salinity.

Inactivation mechanisms were pursued via attenuated total reflectance – Fourier transform infrared spectroscopy and quantitative reverse transcription polymerase chain reaction. Evidence for oxidative transformations of capsid proteins including formation of oxyacids, aldehydes, and ketones and RNA damage kinetics will be discussed. Finally, recent results on electrochemical iron inactivation and coagulation will be presented.

565782 Fabrication of Catalytic Water Treatment Membranes with Phosphorene-Based Pore Fillers

Photocatalytic membrane processes generate oxygen-reactive radicals under irradiation by light that could degrade pollutants in feed solutions and reduce fouling. Phosphorene is a two-dimensional material made by exfoliating black phosphorus, a monotypic van der Waals crystal. Unlike other common two-dimensional materials, under light irradiation, phosphorene can generate reactive oxygen species (singlet oxygen, hydroxyl radicals and superoxide radicals). Therefore, it shows promise as a photocatalyst. The goal of this study was to incorporate phosphorene into membranes made from a polymer blend of sulfonated poly ether ether ketone and polysulfone, then test these membranes in different media to examine their stability in various environment and study the method of adhesion of the nanomaterial in the membrane. In this research, phosphorene was first fabricated by the liquid exfoliation of black phosphorus. Experiments were done in an acidic, basic, neutral and in a common solvent media. Various membranes that composed of different wt.% of phosphorene were casted and the rate of leaching of the phosphorene nanoparticle was studied. Results obtained showed enhanced flux recovery for the phosphorene modified membranes. Fluorescent studies also revealed lower coverage of methylene blue on these membranes as compared to the unmodified membranes. Mechanism of photocatalysis will be investigated and reaction equations proposed.

574425 Tangential Flow for Analyte Capture (TFAC): A Method for the Purification of Nanoparticle Biomarkers Enabled By Ultrathin Silicon Membranes

Extracellular vesicles (EVs) are lipid vesicles ranging from 30 – 1000 nm in diameter secreted by cells to serve a variety of functions. Certain members of the EV family, such as exosomes, are thought to be important for intercellular communication. Exosomes contain cell specific biomarkers (i.e. miRNA, protein) that are useful for point-of-care (POC) diagnostics. A variety of techniques are used to isolate EVs, including ultracentrifugation, affinity-based capture, precipitation and filtration, but these often require large sample volumes (50 – 300 mL) which are not feasible for a POCsystem. Additionally, these isolations are often contaminated with protein, leading to difficulty in the identification of biomarkers. Therefore, an alternative method of isolating EVs from small volumes is requiredfor POC. Silicon nanomembranes are ultrathin (50-100nm) nanoporous membranes discovered during the rapid crystallization of pure silicon films. These membranes are manufactured with a variety of pore diameters (~ 5 nm – 100 nm) and have high porosities (20% – 40%) and exceptional permeabilities. These membranes are manufactured in a 'chip' format that is 5.4 mm x 5.4 mm, making them perfect for small volume applications such as microfluidic filtration. In this work, we present a novel exosome capture methodforsmall volume (~ 1 mL)samples.We callthis methodtangential flow analyte capture (TFAC). After demonstrating the capture ofexosomes from undiluted serum, wepresenta computational model for TFACthat is validated with 60 nm gold nanoparticleswith and without the addition of serum albumin (2mg/ml). We usethe model to define optimalcaptureconditions, and thendemonstrate the capture of > 75% of exosomes from defined protein/exosome sampleand compare these results to capture from raw urine.

579962 Advanced Membrane Materials for Environmental Liquid Separations

Our group focuses on fundamental understanding of the relationships between formation, structure, material properties, and material performance in advanced membranes for applications in environmentally relevant processes.

Surface and groundwaters have increasing concentrations of salts from agriculture and water softening activities—increasing the need for inland brackish water desalination and concentrate management. During inland brackish groundwater desalination, disposal of the concentrate can often constrain the design and overall feasibility. Reverse osmosis (RO) is typically used to achieve 75 to 85% recovery of the feed water and limited by sparingly soluble salts (e.g., calcium sulfate) or minerals (e.g., silica) and the overall concentration of the feed. Pervaporation is a membrane process that separates mixable liquids by a combination of liquid permeation through and vapor evaporation from a dense semi-permeable membrane. It uses hydrophilic dense nonporous (polymer membranes) or microporous (inorganic membranes), in which the driving forces are vapor pressure of the main permeating species and, indirectly, temperature. In pervaporation there is no limit on the salinity of the feed water. We will discuss our results to incorporate direct solar light to enhance the pervaporation process for desalination as well as our development of new membrane for pervaporation.

Liquid fossil fuels are a significant portion of the annual energy consumption of the United States. Butanol is a second generation biofuel whose production does not compete with food production and butanol is a suitable substitute for liquid fossil fuels. Using a condensation-cure method we have synthesized free-standing polydimethylsiloxane/zeolite-imidazolate framework-71 (PDMS/ZIF-71) mixed-matrix membranes with ZIF-71 loadings of up to 40%. Our membranes have the highest reported selectivity for pervaporation recovery of ethanol and butanol from dilute aqueous solutions (selectivity of 5.64 ± 0.15 for 2 wt% 1-butanol in water and a selectivity of 0.81 ± 0.04 for 2 wt% ethanol in water). Tensile testing indicates that the Young's modulus of the membranes increased with higher ZIF-71 loadings. We also demonstrate that the organic linkers found in ZIF-71 are incompatible with addition-cured PDMS films.

576709 Highly Selective Ultrafiltration – Exploiting Non-Steric Interactions for Bioprocessing

Ultrafiltration is a well established method for protein concentration and buffer exchange. Nearly all current biotherapeutics are formulated using ultrafiltration / diafiltration processes. However, recent advances in our understanding of membrane transport phenomena has led to exciting opportunities for the development of highly selective ultrafiltration processes for bioseparations. This talk will review these developments and how they have been used to develop membrane systems for the purification of conjugated vaccines and nucleic acid-based gene therapy agents. These results clearly demonstrate the significant potential of using ultrafiltration membranes for the commercial production of next generation biotherapeutics.

577326 Additive Manufacturing of Thin Film Composite Membranes

Our recent work on printing polyamide membranes demonstrated the ability to make desalination membranes of exceptional smoothness and controllable thickness. The approach required the electrospray deposition of an amine monomer and an acid chloride crosslinker onto a substrates where they react to form an aromatic polyamide. The technique allows for exceptional control of the reaction and makes it advantageous over conventional interfacial polymerization. Conventional interfacial polymerization is a relatively uncontrollable process and has few options for independently tailoring selectivity and permeance. We can adapt this approach to making membranes of novel polymers as well, many of which cannot be formed easily into thin film or composite structures with high permeance *and* high selectivity. We use this approach with several emergent polymer systems materials to demonstrate the value of this approach.

573196 Adsorption of Fluoroquinolones at Gas-Liquid Interface By Foam Fractionation

Adsorption of fluoroquinolones at gas-liquid interface by foam fractionation

Rajesh Ghosh[1*], Haritha Hareendran[2], Subramanium Pushpavanam[1*]

[1]Department of Chemical Engineering, Indian Institute of Technology, Madras, Chennai, 600036, India

[2]Department of Chemical Engineering, Alagappa College of Technology, Anna University, Chennai, 600025, India

*Contact: ghoshrajesh1226@gmail.com, spush@iitm.ac.in

In this work we investigate the adsorption of fluoroquinolone antibiotics at gas-liquid interface using ionic surfactants such as sodium dodecyl sulfate (SDS) and cetrimonium bromide (CTAB) using a foam fractionation technique. Ciprofloxacin (CF) is chosen as a model fluoroquinolone. We investigate the effect of surfactant head groups on the adsorption performance of the drug. CF being a zwitterionic compound showed affinity towards both cationic and anionic surfactants. Further, we probe the binding mechanism of the drug and surfactant using infrared spectroscopy and nuclear magnetic resonance. The effect of the drug on the foamability and production of bubbles (foam) is investigated using surface tension analysis. It was found that the presence of CF in the solution mixture increased the surface tension of the solution thus showing a trivial decline in the foamability. This was also corroborated in the adsorption experiments. The separation of the drug from the bulk mixture was most influenced by the pH of the feed solution. A separation efficiency of more than 96% was obtained along with an enrichment factor of 9.3. Overall, SDS showed a better separation efficiency compared to the cationic surfactant indicating that the binding of the anionic surfactant to the various protonated sites of the drug was a more dominant factor for separation and adsorption to the gas-liquid interface.

Keywords: foam fractionation, fluoroquinolones, ionic surfactants

553499 Sorbent Enhanced Reverse Water Gas Shift Reaction: Maximizing CO Yield

Solar fuels can replace fossil fuels in transportation vehicles. Alcohols or liquid hydrocarbons are preferred solar fuels in view of their combination of energy density and the worldwide existing markets and infrastructure. Hydrogen, which can be produced efficiently from solar energy through direct splitting of water or through combination of solar electricity production with electrolysis of water [1], is a very attractive intermediate as liquid hydrocarbons can be produced through hydrogenation of CO_2. The CO_2 is captured at fossil fuel based power plants or, in the near future, directly from the air [2]. The conversion of CO_2 to transport fuels with hydrogen has the potential of the integration of CO_2 capture and energy storage/conversion. The reaction of hydrogen with CO_2 is initiated by the reverse water-gas shift reaction (RWGS): $H_2 + CO_2 \Rightarrow CO + H_2O$ ΔH_0 41kJ/mol. The reaction can be combined with Fischer-Tropsch technology to produce liquid hydrocarbon fuels. The use of the RWGS reaction for the production of methanol and dimethyl ether (DME) fuels is another very interesting one. DME is highly volatile and has no C-C bonds. As a consequence, it combusts soot-free in a diesel-type engine, and thus resolves the inherent soot/NOx trade-off that is common for conventional diesel-type fuels. However, the synthesis of methanol and DME directly from CO_2 suffers from very low equilibrium conversion. Synthesis of methanol with CO enriched feed greatly improves the conversion. The RWGS is an endothermic equilibrium reaction, and the equilibrium lies on the product side only at high temperatures. The catalytic conversion at lower temperatures, 473-573K can be enhanced greatly if reaction product water is selectively removed. Carvill et al. [3] showed that pure CO could be produced at 523K in a sorbent enhanced RWGS (COMAX) process. For methanol and DME production it suffices if the COMAX step produces a mixture of CO, CO_2 and hydrogen (synthesis gas). In the present contribution, parameters governing the conversion of CO_2 into CO by in-situ removal of water, the optimization of catalytic materials, the functionalities that selectively form CO and prevent consecutive reaction at higher pressure, as well as the process configuration will be presented.

[1] Blankenship, R. E., Tiede, D. M.,Barber, J.et al. Science 2011, 332 (6031), 805.

[2] Haije,W.G. ,Geerlings, J.J.C. Environ. Sci. Technol.,2011, 45 (20), 8609.

[3] Carvill, B. T., Hufton, J. R., Anand, M., Sircar, S. , AIChE J. 1996, 42, 2765.

576059 Modeling and Optimization of a Moving Bed Process for Post-Combustion CO_2 Capture Using a Diamine-Appended Metal–Organic Framework

Current commercial technologies for post-combustion CO_2 capture are mainly based on amine-based solvents, but these technologies suffer from high energy penalty. Therefore there are considerable recent interests in the area of solid sorbent-based processes as they have been reported to have certain advantages over amine-based processes[1]. For these sorbent-based processes to be a viable alternative to the amine-based processes, efficient gas-solid contactors are absolutely necessary. While there has been much work on optimizing the design of the towers, their internals, and operating conditions for the solvent-based capture processes, there are very few publications in the open literature on optimal design of the contactor technology for solid-sorbent systems. In particular, a moving bed (MB) contactor can offer several advantages over other solid-gas contactors, such as a fixed bed or fluidized bed. In a MB, the solid particles enter at the top of the bed and travel downwards as the flue gas enters at the bottom of the bed and flows upward. This counter-current flow pattern results in improved driving forces for mass transfer when compared to a fixed bed or fluidized bed system. Achieving a high overall driving force is critical for 90% CO_2 capture, especially at the top end where the flue gas leaves the system with a very low partial pressure of CO_2. Because of the moderate hydrodynamics in MB contactors, particle attrition in these beds is much lower compared to the fluidized beds.

This work focuses on the optimal design and operation of the MB contactors for a novel diamine-appended metal organic framework (MOF). The diamine-appended MOF used in this work, diamine 2,2-dimethyl-1,3-diaminopropane (dmpn) Mg_2(dobpdc) (dobpdc^{4-} = 4,4′-dioxidobiphenyl-3,3′-dicarboxylate) [dmpn-Mg_2(dobpdc)], exhibits a step-shaped isotherm under low partial pressures of CO_2 which results in high working capacities when compared to traditional sorbents. This functionalized MOF also shows excellent long-term stability and maintains performance under humid conditions, both desirable attributes for flue gas CO_2 capture[2,3,4]. However for realizing the high potential of this MOF, rapid heat removal/addition must be achieved so that the step-shaped isotherm can be exploited for improving its economics.

In this work, a dynamic, one-dimensional, non-isothermal MB reactor model is developed and used to model a full scale temperature swing adsorption process using the functionalized MOF. The model simultaneously solves a set of partial differential equations for the mass and energy balances of the system, and accounts for internal and external mass transfer resistances, reaction kinetics, and heat transfer between the gas and solid phase. The model also accounts for heat transfer between the system and an embedded cooler in the adsorber and a heater in the regenerator. The embedded heat exchangers can provide/remove large amount of heat thus facilitating an operation that is close to the desired temperature profile. However, the added cost and lost volume due to the embedded heat exchangers should be accounted for. Therefore a mathematical programming problem is solved for simultaneous optimization of the design and operating conditions of the MB process. Traditional and various extended area designs were investigated for the embedded heat exchangers. In addition, various working fluids are investigated for the cooler/heater not only for optimal heat addition/removal but also for potential heat recovery from the hot sorbents thus improving the overall economics of the capture process.

References:

[1] Kim, H., Miller, D., Modekurti, S., Omell, B., Bhattacharyya, D., Zitney, S., Mathematical Modeling of a Moving Bed Reactor for Post-Combustion CO_2 Capture. *AIChE Journal.* 2016; 62 (11), 3899-3914

[2] Milner, P.J., Siegelman, R.L., Forse, A.C., Gonzalez, M.I., Runcevski, T., Martell, J.D., Reimer, J.A., Long, J.R. A Diaminopropane-Appended Metal-Organic Framework Enabling Efficient CO_2 Capture from

Coal Flue Gas via a Mixed Adsorption Mechanism. *Journal of the American Chemical Society.* 2017; 139 (38), 13541-13553

[3] Mcdonald, T.M., et al., Cooperative insertion of CO_2 in diamine-appended metal-organic frameworks. *Nature.* 2015; 519, 303-308

[4] Forse, A.C., Milner, P.J., Lee, J., Redfearn, H.N., Oktawiec, J., Siegelman, R.L., Martell, J.D., Dinakar, B., Porter-Zasada, L.B., Gonzalez, M.I., Neaton, J.B., Long, J.R., Reimer, J.A., Elucidating CO_2 Chemisorption in Diamine-Appended Metal-Organic Frameworks. *J. Am. Chem. Soc.* 2018; 140(51), 18016-18031

569849 Separation and Purification of Bioderived Furanic Molecules with Metal-Organic Frameworks

Furanics are important "platform chemical" precursors to polymers, pharmaceuticals, and fine chemicals, and can also be used as biofuels. Conversion of sugars or other feedstocks to furanics often leads to complex product mixtures in which the components are typically heat sensitive, have similar physicochemical properties, and form azeotropes, thus making conventional separation processes such as distillation unfavorable. Such "high-resolution" separations (wherein the molecules possess very similar characteristics) require approaches based upon precisely tailored nanoporous materials as separating agents. In particular, this work introduces the use of metal-organic framework (MOFs) materials to purify furanic products from liquid mixtures generated during processing of lignocellulosic feedstocks. MOFs are a group of newly explored nanoporous materials consisting of metal ions/clusters coordinated to organic linker molecules. Based upon their structural properties as well as preliminary experiments, two classes of MOFs (i.e., UiO and ZIF) are identified as potential adsorbents with high selectivity, capacity, and reusability. Our present focus is on developing two new separation processes for the purification of furfural and 2,5-dimethylfuran (DMF) respectively. A suite of characterization and modeling techniques including single-component adsorption, ideal adsorbed solution theory (IAST) calculations, multicomponent vapor/liquid mixture breakthrough experiments, and pulsed-field-gradient nuclear magnetic resonance (PFG-NMR) are used to build the key structure-property (i.e., adsorption/diffusion) relations. A recent advance (Chiang et al, *ACS Sus Chem Eng*, 6, 7931-7939, 2018) is the purification of DMF from n-butanol (BuOH) wherein the separation is limited by an azeotrope formation at 90% DMF at 1 bar and elevated temperature. Our current investigation shows that ZIF-8 (with high DMF/BuOH selectivity at low DMF content) and defect-engineered UiO-66 (with high BuOH/DMF at low BuOH content) adsorbents, when used in series, can upgrade dilute DMF feeds from 10 wt% to purified (99%+) DMF and also generate a high-purity (>95%) n-BuOH recycle. Finally, we will discuss the design and optimization of a simulated moving-bed (SMB) adsorptive separation process using ZIF-8 and UiO-66 adsorbents for DMF separation and BuOH recycle from a multicomponent feed mixture feed.

575271 Investigation of Rare Earths Exchanged on Zeolite Y for Adsorptive Desulfurization of Model Fuels in the Presence of Aromatics

Investigation of Rare Earths Exchanged on Zeolite Y for Adsorptive Desulfurization of Model Fuels In the Presence of Aromatics

Tyler B. Crowl, Kevin X. Lee, and Julia A. Valla

Department of Chemical & Biomolecular Engineering, University of Connecticut, 191 Auditorium Road, Unit 3222, Storrs, CT 06269-3222, USA,

Phone: +1-860-486 2508, e-mail: ioulia.valla@.uconn.edu

Sulfur compounds are found in hydrocarbon transportation fuels and when combusted pose a threat to the environment. The traditional Hydrodesulphurization (HDS) process requires high operating costs and excess hydrogen to deeply desulfurize the fuel and the severity of the operating conditions may degrade the quality of the fuel. Adsorptive desulfurization (ADS) using bimetallic ion exchanged zeolite Y is an attractive alternative to HDS which can be effective at even ambient conditions.[1] Cerium, a rare earth, has been shown to exhibit excellent capacity and selectivity for the removal of benzothiophene through the formation of p-complexation and direct s-bond, when ion-exchanged into zeolite Y.[2] Copper has been shown to have a synergistic relationship with Ce on zeolite Y that enhances the adsorption capacity and the overall desulfurization performance.[3] This study investigates other Rare Earths (RE) such as La, Sm, Nd, Pr, and bimetallic Cu-RE combinations to understand the effect of RE on ADS of benzothiophene.

Y Zeolite was ion-exchanged with different RE to 5wt% and materials were characterized by several techniques to verify successful loading of the metals. Mesoporous materials were synthesized using a top-down approach prior to ion-exchange with Cu-RE.[4] Surface acidity was determined by using pyridine diffuse reflectance infrared spectroscopy (DRIFTS-FTIR). Metal loadings on the zeolite were quantified using inductively couple plasma mass spectrometry (ICP-MS). The crystalline structure was determined by X-ray diffractometer (XRD) measurements.

Adsorption experiments were conducted in a fixed-bed column to study the continuous adsorption of sulfur compounds in model fuels. The mass of sorbent was varied between 0.2 and 0.3

grams to achieve a constant bed height of 3cm. The model fuels were prepared by combining 100 ppm sulfur of benzothiophene (BT) in octane and 20% benzene and 100ppm sulfur of dibenzothiophene (DBT) in octane with 1% naphthalene. The effluent was collected periodically and analyzed using a gas chromatograph sulfur chemiluminescence detector (GC-SCD). Figure 1a shows that PrY performed the best out of the new RE tested, but still did not outperform CeY. Thus, the order from highest capacity to lowest was found to be Ce>Pr>La>Nd>Sm. Figure 1b shows the BT breakthrough curves on bimetallic Cu-RE materials. The best performing bimetallic RE was CuCeY, with a capacity of 1.756 mg S g^{-1}, and the order from highest capacity to lowest was Ce>Nd>Sm>Pr>La. The addition of Cu enhances the BT adsorptive capacity for all of the RE tested, but did not show a significant increase for La and Pr. BT adsorption results show that Ce is the best performing RE and has the most synergistic performance.

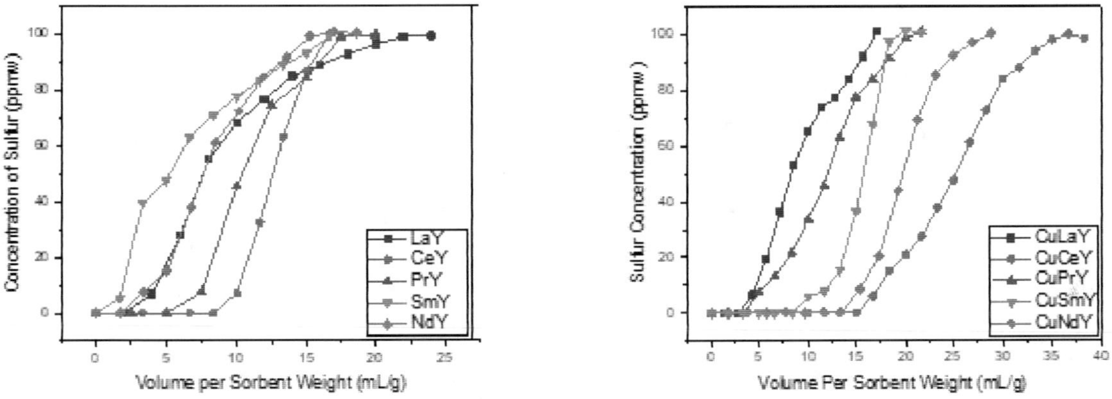

Figure 1: Breakthrough curves of BT on (a) REY and (b) Bimetallic RE Y with 20% benzene in octane.

In the presentation, corresponding RE exchanged with mesoporous zeolite Y will be discussed showing the results of the DBT adsorption. Overall trends between BT and DBT adsorption on RE will be also discussed.

References

1. Lee KX, Valla JA. Applied Catalysis B : Environmental Investigation of metal-exchanged

mesoporous Y zeolites for the adsorptive desulfurization of liquid fuels. "*Applied Catal B, Environ.* 2017;201:359-369. doi:10.1016/j.apcatb.2016.08.018

2. Velu S, Ma X, Song C. Selective Adsorption for Removing Sulfur from Jet Fuel over Zeolite-Based Adsorbents. 2003:5293-5304. doi:10.1021/ie020995p

3. Lee KX, Tsilomelekis G, Valla JA. Applied Catalysis B : Environmental Removal of benzothiophene and dibenzothiophene from hydrocarbon fuels using CuCe mesoporous Y zeolites in the presence of aromatics. Appl Catal B Environ. 2018;234(April):130-142. doi:10.1016/j.apcatb.2018.04.022

4. García-Martínez J, Johnson M, Valla J, Li K, Ying JY. Mesostructured zeolite y - High hydrothermal stability and superior FCC catalytic performance. Catal Sci Technol. 2012;2(5):987-994. doi:10.1039/c2cy00309k

567730 Catalytic Adsorptive Desulfurization of Model Fuel over Ti-Si Adsorbents at Low Ti Loading Under Ambient Conditions

Ultra-clean fuel production has attracted widespread attention globally due to intensified environmental and health concerns. Previously, we proposed catalytic adsorptive desulfurization (CADS) using bifunctional catalyst-adsorbent under ambient conditions for selective desulfurization of low-sulfur diesel for ultra-clean fuel production (Fuel, 2016, 174, 118-125; AIChE J., 2018, doi:10.1002/aic.16055). In this work, we explored CADS using a newly developed Ti-Si-rd adsorbents featured with fairly low Ti loading (0.01~2.5 wt.%). A series of Ti-Si-rd was synthesized by a facile impregnation method. Effect of Ti loading at low concentration range on CADS capacity in batch was elaboratively studied. The optimized Ti-Si-rd at low Ti loading showed higher CADS capacity compared to that upon Ti-Si adsorbents reported in the previous work, which reached up to 150 mg/g from a model fuel containing high DBT concentration up to 1500 ppm, at the sorbent-to-fuel ratio of 100. Additionally, Ti-Si-rd also maintained >95% of CADS capacity in five consecutive regeneration cycles, suggesting excellent regenerability and stability of the Ti-Si-rd adsorbent as a plus plausible feature. Furthermore, to elucidate the correlation between CADS capacity and physical-chemical properties of Ti-Si-rd, the adsorbents were further characterized by N2 adsorption, X-ray diffraction, FTIR spectroscopy, and inductively coupled plasma optical emission spectrometer. The chemical environments of Ti in various Ti-Si adsorbents was further studied to clarify the effect of Ti coordination and chemical environment on CADS.

572084 Desulfurization of Fuels Using Activated Carbon Prepared from Asphaltenes

Sulfur compounds which exist in liquid hydrocarbons have harmful effects to the environment and to the human health. Especially for the heaviest of these compounds (thiophenes, benzothiophenes, naphthothiophenes), it is difficult to be removed using the most common current industrial desulfurization process (hydrodesulfurization). Adsorptive desulfurization has, therefore, emerged as a promising alternative, as it is less energy intensive and it can more efficiently achieve the deep desulfurization of commercial fuels.

Asphaltenes are the most complex and the most polar fraction of crude oil, consisting of fused aromatic rings with short aliphatic chains. According to ASTM D3279-97, they are defined as the fraction of crude oil and bitumen which is soluble in toluene and insoluble in n-heptane. These compounds are associated with problems in the transportation and in the processing of crude oil. However, they contain a large proportion of carbon, which makes them attractive for the production of activated carbon.

In this work, asphaltenes were extracted from bitumen and they were subjected to stabilization and carbonization. Moreover, their porosity was enhanced through chemical and physical activation. The asphaltene-based activated carbon was used for the adsorption of sulfur compounds (thiophene, benzothiophene) from liquid fuels. These results will be presented here, together with a characterization of the properties of the precursor asphaltenes, as well as of the activated carbon.

567218 Screening of Polymorphs and Measurement of Growth Rates of L-Histidine at Controlled Supersaturation Using Continuous-Flow, Microfluidic Device

A continuous-flow, a well-mixedmicrofluidic device is presented here to measure solubility, as well as growth rates and polymorphs of L-Histidine (L-his) from ethanol-water mixtures at controlled supersaturations. While conventional tools such as microtiter plates, and droplet-based microfluidics measure polymorphs, nucleation and growth kinetics in depleting supersaturation, the continuous-flow microfluidic device can trap crystals to measure nucleation and growth kinetics while constantly feeding them with supersaturated solution. Here we use multi-inlet cyclone mixer design, similar to a mixed-suspension-mixed-product-removal (MSMPR) crystallizer, for anti-solvent crystallization using ethanol and an aqueoussolution of L-his. The supersaturation in such device is controlled by manipulating flow rates of ethanol and water solutions. Solubility data of L-his in the ethanol-watermixture is obtained by increasing the flow rate of anti-solvent until the crystals are formed and then gradually decreasing the flow rate so that crystals stop growing. At a fixed supersaturation, the evolution of crystal morphology can be processed to estimate face-specific growth rates. Since the stable and metastable forms of L-Histidine have distinct morphology, the polymorphs are directly identified using an opticalmicroscope. A multi-channel, multi-compartment microfluidic device is also shown for the continuousscreening of polymorphs of L-his.

558999 Homogeneous Nucleation and Size Distributions of Protein Crystals Nucleated and Grown in Weightlessness

The crystallization of bioproducts (especially proteins) for purification and, more recently and more importantly, for delivery demands smaller and more uniform crystal populations. This can only be achieved by maximizing the number of nucleation sites, which in turn requires that nucleation be homogeneous -- independent of pre-existing particles or surfaces. In traditional reactors, nucleation is achieved by rapid mixing under conditions of homogeneous isotropic turbulence (HIT). Earlier studies found, at first by accident and then deliberately, that protein crystals that nucleated, presumably homogeneously, in orbital space-borne reactors at reduced gravity occurred at higher numbers and greater uniformity than expected. Experiments were performed on the U. S. space shuttle and the Russian Mir space station using several crystallization reactors based on a "Fluids Processing Apparatus" (FPA) developed and provided by BioServe Space Technologies (University of Colorado, Boulder). Hen egg lysozyme, sodium chloride and 0.01 M acetate pH 4.0 were used as protein, precipitant and buffer, respectively. In each sample 1.5 mL of protein was injected into 2.0 mL of precipitant over a period of 1 minute. Identical experiments were performed on orbit and in the laboratory. Crystal size distributions were determined with the aid of ImageJ image analysis software. In all cases coefficients of variation of populations that were crystallized on orbiting spacecraft were less than those of populations crystallized on the ground. On the spacecraft, essentially all crystals nucleated in suspension and not on reactor vessel walls. Crystals formed in low gravity had especially smooth, shiny facets implying very low mosaicity. They were also very stable over time, seeming to remain unchanged for up to 22 years in mother liquor at ambient temperature. Apparently the lack of turbulence and lack of sedimentation in low gravity resulted in minimal crystal motion in any direction until the cessation of growth due to depletion of solute in the diffusion-governed environment. These orbital low-gravity studies suggest that, instead of turbulent mixing, which transports crystals through variable and unpredictable concentrations of additional protein molecules, a quiescent environment for combining protein and precipitant may lead to the enhanced size limitation and uniformity desired to achieve certain purification and/or formulation goals. Research supported by NASA Grants NAGW-1197 and NAG 8-1165.

563912 A New Molecular Templating Approach for Controlling the Nucleation of Peptides

A new molecular templating approach for controlling the nucleation of peptides

Frederik J. Link*[a], Jerry Y.Y. Heng[a]

[a]Department of Chemical Engineering, Imperial College London, South Kensington Campus, London SW7 2AZ, UK

*f.link18@imperial.ac.uk

Keywords: template, polypeptide crystallisation, nucleation

Abstract

The crystallisation of bio-active polypeptides has an important significance in a wide range of medical and industrial applications since they can be used as highly effective pharmaceutical compounds [1]. The bottleneck of crystallisation is the lack of controllability of the nucleation mechanism. Templates introduce a novel way of triggering nucleation, even at low supersaturation [2]. Therefore, templates are remarkable in controlling crystallisation. Recent studies showed that by using hard templates, such as silica based porous surfaces, the induction time of nucleation of polypeptides can be significantly accelerated even if the supersaturation is super low [3]. However, the impact of the surface properties on nucleation is barely understood. It is believed that porous surfaces trigger nucleation due to local increase in supersaturation within the pores attributable to confined space rather than non-covalent interaction between the polypeptide and the surface. The approach of using small organic molecules acting as "sticky-bridges", by linking polypeptides via non-covalent bonds, can provide new routes for crystallising polypeptides. This mechanism could reduce the surface entropy of the polypeptides rather than just increasing the supersaturation locally [4].

In this study, a simple model peptide which is known to crystallise and consists of a variety of different amino acids with different surface patches is used to investigate the influence of small organic molecules on the crystallisability. The optimal buffer-solution, pH, and precipitant have been adapted from literature and verified as optimal conditions before they were used in this study. To evaluate the impact of small organic molecules acting as seeds on the crystallisation, their chemical properties, the amount and the size of the seeds are varied. The crystallisability is investigated by means of measuring induction time, crystal yield and shape. For this purpose, hanging-drop vapour diffusion experiments will be carried out to evaluate the impact of the seeds on the nucleation.

The induction time of the seeded approach is significantly lower than the induction time for the unseeded approach, even at low supersaturation. This can be due to the specific interactions between the surface patches of the small organic molecules and polypeptides. As a result, it is believed that the chemical properties of the surface patches have a significant impact on controlling nucleation.

Further work will be done with regards to characterise the impact of the seeds with regards to surface chemistry, amount, and molecular size. Therefore characterisation of the seed surface patches and evaluation of the interaction between them and peptide will be done. Additionally, the impact of the seeds on the solubility of the peptide has to be investigated. All these findings should result ultimately in the ability of engineering seeds with application-tailored properties to specifically control the nucleation of polypeptides.

References

1. Kastin, A.J., *Handbook of Biologically Active Peptides*. 2 ed. 2013: Academic Press. 2032.
2. Shah, U.V., et al., *Heterogeneous nucleants for crystallogenesis and bioseparation.* Current Opinion

in Chemical Engineering, 2015. **8**: p. 69-75.
3. Shah, U.V., et al., *Crystallization of Proteins at Ultralow Supersaturations Using Novel Three-Dimensional Nanotemplates.* Crystal Growth & Design, 2012. **12**(4): p. 1772-1777.
4. Wlodawer, A., Z. Dauter, and M. Jaskolski, *ProteinCrystallography Methods and Protocols*. Methods in Molecular Biology, ed. J.M. Walker. 2017, New York: Springer New York : Imprint: Humana Press.

573938 Measuring the Rate of Co-Crystal Nucleation Utilizing Continuous Microfluidic Droplet Generation

Co-crystals provide a unique method for tuning the physical properties of crystalline products such as APIs. Many drugs with poor bioavailability can be made more effective with the addition of a co-crystal former. However large-scale production of co-crystal products requires an understanding of nucleation rate associated with the system. Previously, it has been difficult to directly measure the nucleation rate using macroscopic techniques. However, with the advent of droplet based microfluidic techniques, direct observation of nucleation statistics is possible by treating each droplet as an individual crystallization event. It is desired to observe as many droplets as possible simultaneously to improve the confidence of the gathered statistics. Continuous droplet generation permits the observation of very large droplet counts, without the need for a proportionally larger device or complex microscopy techniques. Our work uses a T-junction droplet generation nozzle connected to a serpentine residence track to demonstrate proof of concept of a continuous co-crystallizing device. We examine nucleation rates for the co-crystal system of caffeine and malic acid, which allows us to test the influence of various parameters including temperature step, ratio of co-formers, and presence of additives on the nucleation rate. The ability to efficiently study co-crystallizing systems allows for the exploration of previously untested compositions, and contributes to the development of new APIs.

564899 Impacts of Solvent and Polymer Additive Polyvinylpyrrolidone on the Ethyl Vanillin Crystal Morphology

Ethyl vanillin (EVA) is an important synthetic chemical which is widely used in the food additives, spices, cosmetics and pharmaceuticals. However, EVA crystal crystallized from various solvents by cooling crystallization shows thin flake shape which could be crushed during the downstream process. This study investigated the effects of solvents and additives on the morphology of EVA crystal from experiments and molecular dynamic simulation. By adding polymer additive polyvinylpyrrolidone (PVP) into the EVA solution, the morphology of EVA was modified from the thin flake to the thick flake with distinct side facets. In contrast, the morphology of EVA in the presence of the PVP monomer molecule and other similar additives didn't change. Molecular dynamic simulation revealed that the solvent molecules were trapped in the holes of {010} surface and hindered its growth process significantly resulting the {010} surface dominated the morphology. The strong interaction between PVP and {100} surface accounted for the growth inhibition along the <100> direction giving rise to the appearance of side facets. Furthermore, mean square displacement analysis showed that polymer additive PVP possessed the obvious lower displacement ability compared with other similar additives, besides, distinct mean square displacement difference was found between different crystal surfaces of EVA, these may affect solute approaching process. These simulation results were consistent with the experimental results.

568798 Impact of Polymers and Surfactants on Solid-State Crystallization at the Amorphous Pharmaceutical-Water Interface

Amorphous pharmaceutical formulations are an increasingly common strategy for dissolution and bioavailability enhancement of poorly soluble drug compounds. Stabilization against crystallization during storage as well as during dosing is usually achieved by dispersing amorphous drug in a polymeric matrix and has been well studied. However, during dissolution it is possible to generate drug-rich amorphous interfaces that are not intimately mixed with polymer. Solid-state crystallization at this amorphous drug-water interface is not well understood, nor is the role played by excipients in influencing this crystallization phenomenon. The work presented herein utilizes experimental and modeling tools to characterize and model solid-state crystallization at the amorphous-water interface as well as the impact on crystallization from polymer and surfactant species in solution. Atomic force microscopy (AFM) and scanning electron microscopy (SEM) measurements on the evolving interface reveal non-classical and highly anisotropic crystallization. Interpretive 2D lattice Monte Carlo simulations exhibit semi-quantitative agreement with experiments and elucidate the impact of the surface energy of the underlying amorphous material on the morphology of the crystalline phase. AFM and SEM data also demonstrate crystallization inhibition with polymer additives, with concentration-dependent crystal morphology. In contrast, surfactants promoted crystallization. Preliminary experiments explore these findings by connecting qualitative crystallization observations to the underpinning transport phenomena facilitating surface rearrangement. These results provide fundamental insight for ultimately mitigating this undesirable phase change.

574369 Continuous Protein Crystallization in a 3D Printed Airlift Crystallizer

Continuous crystallization has been receiving growing interest for processes with a small throughput such as those found in pharmaceutical and biotechnology industries. Continuous processing offers several potential advantages compared to conventional batch-wise processing such as easier scale-up, reduction in equipment footprint, and the ability to implement simpler control strategies.[1,2] Mixed-suspension, mixed-product removal (MSMPR) is one of the traditional modes of operation for continuous crystallization and is generally preferred for slow-growing crystals that require a long residence time or for crystals that have a high tendency for fouling.[3] Protein crystallization typically requires a long residence time and hence calls for the use of an MSMPR mode of operation. An additional advantage of using the MSMPR mode is that an existing batch crystallizer can easily be adopted for continuous crystallization,[4] which is important for multi-purpose production facilities. Airlift crystallizers are well suitable for MSMPR crystallization because of the ease of operating for long residence times and rigorous mixing.

An airlift crystallizer[5,6] is a pneumatically agitated column generally consisting of a riser and a downcommer. Gas is introduced into the riser through a sparger, which reduces the density of the materials in the riser. The resulting difference in the density between the contents of the riser and the downcommer leads to the liquid circulation in the airlift crystallizer, which keeps the crystals in suspension. Recently, we have developed a 3D-printed airlift crystallizer (ALC) for batch protein crystallization. This 3D printed airlift crystallizer showed a reduction in the induction time for nucleation and generally larger and less agglomerated crystals could be produced in the ALC compared to a conventional stirred tank crystallizer (STC) for the case of lysozyme crystallization.[7] The reduction in the induction time was expected and likely caused by heterogeneous nucleation supported by the gas-liquid surface. Generally, it is known that secondary nucleation due to attrition is dominant in an STC,[8] as protein crystals are known to be more fragile than crystals of small organic molecules.[9] Furthermore, different dominating mechanisms for secondary nucleation exists in both ALC and STC.[10] The technical feasibility of the airlift crystallizer for reactive[11] and cooling crystallization[12] in continuous flow for salts and small organic molecules has been demonstrated. However, airlift crystallizers in continuous flow have not been studied systematically yet. Since different nucleation mechanisms are present in either an ALC or a STC, it is likely that the steady state and transient behaviour of an ALC operated in continuous flow mode will be different compared to an STC. Therefore, in the present work, we have extended protein crystallization in an airlift crystallizer from batch-mode to continuous flow mode.

The objective of this work is to develop and characterise a 3D printed ALC for continuous crystallization of proteins in MSMPR mode and to compare its performance in terms of throughput and product quality with that of an STC operated in MSMPR mode. Lysozyme is used as the model compound with sodium chloride as the precipitant. The ALC is designed and fabricated using 3D printing (stereolithography). Continuous crystallization experiments will be carried out in the ALC and the performance will be compared to that of the STC operated in MSMPR mode. MSMPR operation is achieved by continuously feeding the crystallizer at a relatively low flow rate and removing the slurry from the crystallizer at a higher flow rate in intervals. The crystallizer is operated for a duration of 9 hours (about three residence time). The protein concentration reaches steady-state quickly in an STC, which justifies the relative short operation time and will be compared to the protein concentration as function of time in an MSMPR ALC. The possible classification of the crystals during product removal will be investigated. Finally, seeding will be explored to enhance the start-up of the ALC.

Acknowledgement: The work described in this abstract was supported by a grant from the Research Grants Council of the Hong Kong Special Administrative Region, China, Project No. 16242916.

References

(1) Lee, S. L.; O'Connor, T. F.; Yang, X.; Cruz, C. N.; Chatterjee, S.; Madurawe, R. D.; Moore, C. M. V; Yu, L. X.; Woodcock, J. Modernizing Pharmaceutical Manufacturing: From Batch to Continuous Production. *J. Pharm. Innov.* **2015**, *10* (3), 191–199.

(2) Lakerveld, R.; Benyahia, B.; Heider, P. L.; Zhang, H.; Wolfe, A.; Testa, C. J.; Ogden, S.; Hersey, D. R.; Mascia, S.; Evans, J. M. B.; et al. The Application of an Automated Control Strategy for an Integrated Continuous Pharmaceutical Pilot Plant. *Org. Process Res. Dev.* **2015**, *19* (9), 1088–1100.

(3) Quon, J. L.; Zhang, H.; Alvarez, A.; Evans, J.; Myerson, A. S.; Trout, B. L. Continuous Crystallization of Aliskiren Hemifumarate. *Cryst. Growth Des.* **2012**, *12* (6), 3036–3044.

(4) Wong, S. Y.; Tatusko, A. P.; Trout, B. L.; Myerson, A. S. Development of Continuous Crystallization Processes Using a Single-Stage Mixed-Suspension, Mixed-Product Removal Crystallizer with Recycle. *Cryst. Growth Des.* **2012**, *12* (11), 5701–5707.

(5) Soare, A.; Lakerveld, R.; Van Royen, J.; Zocchi, G.; Stankiewicz, A. I.; Kramer, H. J. M. Minimization of Attrition and Breakage in an Airlift Crystallizer. *Ind. Eng. Chem. Res.* **2012**, *51* (33), 10895–10909.

(6) Lakerveld, R.; Van Krochten, J. J. H.; Kramer, H. J. M. An Air-Lift Crystallizer Can Suppress Secondary Nucleation at a Higher Supersaturation Compared to a Stirred Crystallizer. *Cryst. Growth Des.* **2014**, *14* (7), 3264–3275.

(7) Mathew Thomas, K.; Lakerveld, R. A 3D-Printed Airlift Crystallizer for Protein Crystallization. *Submited* **2019**.

(8) Tait, S.; White, E. T.; Litster, J. D. A Study on Nucleation for Protein Crystallization in Mixed Vessels. *Cryst. Growth Des.* **2009**, *9* (5), 2198–2206.

(9) Tait, S.; White, E. T.; Litster, J. D. Mechanical Characterization of Protein Crystals. *Part. Part. Syst. Charact.* **2008**, *25* (3), 266–276.

(10) Anisi, F.; Kramer, H. J. M. Crystallization Kinetics in an Airlift and a Stirred Draft Tube Crystallizer; Secondary Nucleation Models Revisited. *Chem. Eng. Res. Des.* **2018**, *138*, 200–211.

(11) Gonzalez-Contreras, P.; Weijma, J.; Buisman, C. J. N. Bioscorodite Crystallization in an Airlift Reactor for Arsenic Removal. *Cryst. Growth Des.* **2012**, *12* (5), 2699–2706.

(12) Anisi, F. Control of the Key Phenomena in Continuous and Batch Crystallization Processes: Novel Process and Equipment Design, TU Delft, **2019**.

568977 Optimal Design and Control of Reactive Distillation Systems Under Uncertainty

Over the last few decades, there has been an increasing focus on process intensification in an effort to reduce environmental impact and to improve the economic performance of chemical processing plants. Reactive distillation combines reaction and separation into one single unit, offering significant energy and potentially capital savings as well as improvements in reaction selectivity and yield (Luyben and Yu, 2008 and Kiss, 2013). The combination of the two different phenomena in a single unit, however, makes the design and operation of the process more demanding due to differences in the operational window for reaction and separation (Harmsen, 2007). Although recent research has shed light on many aspects of reactive distillation, the extension of conventional distillation design techniques to reactive systems is still a challenge, particularly for its control, and rigorous methods for the design of reactive distillation processes under uncertainty, including their controllability evaluation, has not yet been fully established (Wang et al., 2010).

The concept of distillation column superstructures has been successfully applied in the past for the optimal design of simple distillation columns (Dünnebier, G. and Pantelides, 1999). This network representation method determines all the desired column design and operation specifications by solving a mathematically complex optimization problem. The objective function is typically a cost function, and there are constraints imposed either by the system itself (mass balances) and/or the user (product specifications). Panjwani and co-workers (Panjwani et al., 2005) applied the superstructure concept for the optimal design (structural decisions) of a single reactive distillation column. In this work, the concept of distillation column superstructures will be extended to general reactive distillation systems in order to determine both structural and operational decisions; including but not limited to, total number of stages, feed stages location, existence of side draw streams, reactive zones etc. In addition to a single reactive distillation column, the synthesis of more complex reactive distillation processes, e.g. including a pre-reactor, a sequence of reactive/non-reactive distillation columns etc., will also be considered.

The optimization problem considered is a Mixed-Integer Non-Linear Problem (MINLP), as the design variables are both continuous (e.g. reflux ratio) and discrete (e.g. number of stages). The solution of the MINLPs firstly determines the existence of the units and secondly, their design and operating parameters in case they exist. In this way, the optimal design of various process alternatives can be determined, thus allowing the selection of the most suitable process alternative based on the given objective function. The software used in this work is gPROMS ProcessBuilder (Process Systems Enterprise, 2019), selected due to its rigorous mathematical solvers and excellent optimization capabilities.

Different types of reactive distillation processes, with varying difficulty in terms of the separation task (easy/difficult separation) and different kinetics (fast/slow reaction), are considered. The MINLP considered for each case aims to minimize the total annualised (capital and operational) cost of the process, subject to constraints such as required product purity. Using the superstructure approach, the optimal design parameters for each case are determined and the impact of the process characteristics on the final design is investigated.

Once the optimal steady-state configuration has been determined, the optimal design is then processed in dynamic mode in order to investigate the controllability and stability of the process by considering various disturbances which could potentially be introduced (e.g. feed flow rate disturbance). This investigation will show how the optimal steady state design will perform during operation, and will establish whether any adjustments are needed to the design in order to render it more stable and/or controllable.

The optimal design is obtained by the rigorous optimization of a mathematical model formulated as an MINLP. This model, however, depends on a number or model parameters, for which the reaction parameters are key for reactive distillation. A robust design should tolerate a certain level of parameter uncertainty, and for this reason, the impact of uncertainty in the reaction kinetic parameters on the optimal design will also be investigated in order to determine exactly how sensitive the optimal designs are to these parameters.

It will be demonstrated how the separation parameters (given by the relative volatilities range) impact on the process synthesis and the reaction. Furthermore, it will be demonstrated that the existence of a slow or fast reaction within the column also has an impact on the structural and operational decisions for the design of a reactive distillation column. Of particular interest is the interaction between ease of separation and the characteristics of the reaction system, and it will be shown that the process with a slow reaction and more difficult separation has the highest total cost due to its more demanding process design & operation. Moreover, the dynamic analysis will demonstrate how reaction kinetics and separation parameters impact on the stability and controllability of a process, and how controllability can determine whether a process alternative is actually viable or not. The uncertainty analysis will also demonstrate how sensitive the optimal design is to input uncertainty, and will provide insight into how uncertainty may render an optimal design practically infeasible.

In conclusion, this work will provide a methodology for how to select the most suitable reactive distillation process design that is robust under dynamic operation as well as under input uncertainty, based on various (simple and more complex) process configuration alternatives as expressed by the superstructure.

References

Dünnebier, G. and Pantelides C.C. (1999). Optimal Design of Thermally Coupled Distillation Columns, *Ind. Eng. Chem. Res.*, 38, 162-176.

Harmsen, G. J. (2007). Reactive distillation: The front-runner of industrial process intensification: A full review of commercial applications, research, scale-up, design and operation, *Chem. Eng. Process.*, 46, 774–780.

Kiss, A. A. (2013). *Advanced Distillation Technologies: Design, Control and Applications*, Wiley, Chichester.

Luyben, W. L. and Yu, C. C. (2008). *Reactive Distillation Design and Control*, Wiley, New Jersey.

Panjwani, P. et al. (2005). Optimal design and control of a reactive distillation system, *Eng. Optimiz.*, 37, 733-753.

Process Systems Enterprise (2019), gPROMS, www.psenterprise.com/gproms.

Wang, S. J. et al. (2010). Plantwide Design of Ideal Reactive Distillation Processes with Thermal Coupling, *Ind. Eng. Chem. Res.*, 49, 3262-3274.

570465 Phase Equilibria of a Reactive Distillation Process for Isobutyl Acetate Production

Environmental concerns on the use of petrochemical solvents have boost the production and use of greener substitutes. Among these, biobased esters of short and medium chain alcohols stand out as direct or functional substitutes for a large variety of petrochemical solvents for different applications. However, the main challenge for a most biobased esters are the higher production costs compared with the traditional fossil chemicals. To overcome this drawback, there is need to developed more efficient and integrated processes for synthesis and separation.

In the last few years, reactive distillation has been applied to the synthesis of a large variety of bioderived esters. The synergic effect of combining reaction and separation helps overcoming chemical equilibrium limitations, improving process productivity and reducing energy consumption. Further improvement can be achieved by developing even highly integrated processing schemes such as the dividing wall reactive distillation, reactive HIDIC columns, reactive centrifuge distillation, etc. In any case, the conceptual development of such processes requires a reliable representation of the phase equilibria in the multicomponent system.

Amongst the large variety of biobased esters of industrial interest, isobutyl acetate has found a variety of applications in high value-added niche markets such as in fragrance, flavors and cosmetics enamels. Currently, isobutyl acetate is produced by Fischer esterification of isobutanol with acetic acid under batch operation. The process is limited by the chemical equilibrium, so continuous water removal during reaction is required. However, there are several azeotropes between all the components even between alcohol and acetic acid, so excess alcohol or a make-up alcohol stream are required to achieve high conversions. Thus, the implementation of a reactive distillation process seems suitable to overcome major limitations during isobutyl acetate synthesis. While the idea has been mentioned in the past, a suitable feasibility study has yet to be done.

Modeling of reactive distillation operations require a reliable thermodynamic model of the reactive mixture. In particular, the isobutyl acetate system is characterized for being highly non-ideal; the system exhibits different azeotropes (binary, ternary, and possibly quaternary), immiscibility of reactants, and dimerization in the vapor phase. Also, incomplete phase equilibria data are available in the literature, so process modeling can be affected by the unknown behavior of the mixtures. Then, the aim of this work was to do a complete review of available phase equilibria data of the mixtures involved in the synthesis of isobutyl acetate. Taking into account the missing data, LLE measurements were done for mixtures of acetic acid - isobutanol - water at 10, 20 and 30 °C. Then, by using results from the whole set of experiments, together with reported VLE and LLE data of binary and ternary mixture, a complete set of binary interaction parameters for an activity-based model were regressed. The model agrees reasonably well with fitted data, and I was used to create a residue curve map for the conceptual design of an isobutyl acetate reactive distillation process.

570636 Classification and Comparison of Dividing Wall Columns

Dividing wall columns (DWCs) are known to reduce capital cost of a distillation process by putting multiple columns into a single shell. However, in literature[1,2], DWCs are also said to reduce energy requirement, which is an incorrect statement since each DWC configuration requires same heat duty as its thermodynamically equivalent simple column configuration. For example, the well-known Wright's dividing wall column[3] is thermodynamically equivalent to the 3-component fully thermally coupled (FTC) configuration[4]. FTC configuration always has the lowest heat duty when all the columns are operated at similar pressures. Thus comparing Wright's dividing wall column with other three column configurations can often result in incorrect conclusion regarding heat duties.

Medenoor Ramapriya et al[4,5] presented a simple rule that, for the first time, enabled exhaustive enumeration of DWCs corresponding to any given thermally coupled distillation column-configuration in 2017. This work has enabled the synthesis of all possible DWCs for a given separation. However, use of "dividing wall column" continues to refer to only some common DWC configurations while ignoring several potentially useful DWC options.

In this research, all possible dividing walls are classified into 5 types, each type of dividing wall has its unique structural characteristics, such as the position of the dividing wall in the column, the number of breaking points on the column, the number of transfer streams, the number of condensers and reboilers, etc. These characteristics indicate the capital cost to build such a DWC. This classification is expected to simplify choices of dividing wall columns and provide a common language regarding DWCs.

A ternary separation is used as an example to illustrate how the classification works. The heat duty requirement of each ternary separation configuration has been calculated from the Global Minimization Algorithm[6] developed in the group. Combining energy consumption and capital cost, this research provides a guideline to choose the optimal DWC configuration for a ternary separation.

Literature Cited

1. Dividing-Wall Column for Fractionationof Natural Gas Liquids in Floating LiquefiedNatural Gas Plants. *Chemical Engineering Technology.* 2016;39(12):2348–2354.
2. Synthesis of dividing-wall columns (DWC) for multicomponent distillations—A systematic approach. *Chemical Engineering Research and Design.* 2011;89(8): 1281–1294.
3. Wright RO. Fractionation Apparatus. 1949. US Patent 2,471,134.
4. A Systematic Method to Synthesize All Dividing WallColumns forn-Component Separation—Part I. *AIChE J.* 2018;64(2):649-659.
5. Thermal Coupling Links to Liquid-Only TransferStreams: A Path for New Dividing Wall Columns. *AIChE J.* 2014;60(8):2949-2961.
6. Nallasivam, V.H. Shah, A.A. Shenvi, J. Huff, M. Tawarmalani, R. Agrawal, Global optimization of multicomponent distillation configurations: 2. Enumeration based global minimization algorithm, *AIChE J.* 62 (2016) 2071.

572568 Transesterification in the Reactive Dividing Wall Column – Energy Minimum Design and Operation

Process intensification has become an important concept to meet the challenges of an increasing global competition in the chemical industry. A promising possibility of process intensification is the integration of several primarily separated process operations into one single shell. Industrially applied examples are the reactive distillation or the dividing wall column. A further integration step leads to the highly integrated apparatus of the reactive dividing wall column (RDWC). The RDWC shows significant savings in investment as well as operational costs. Due to the high degree of integration and thus the resulting complexity, no industrial scale RDWC has been built and operated so far. Comprehensive studies on safe and reliable design methods as well as a steady operation are required to develop a profound process understanding for the complex apparatus of RDWC.

Our research aims to investigate the process of the RDWC holistically. As a promising application, a reaction system class of two reactants, which react reversibly to two products, is investigated. The products are the extreme boiling components of the reaction system, whereas the reactants are the middle boiling components. Additionally, a fifth inert component (intermediate boiling component) is present, which represents an impurity of the feed. This reaction system class is represented by the transesterification of hexyl acetate and butanol with the inert component heptanal. By the application of a systematic design method Pareto-optimal designs of the RDWC are developed and the robustness of the designs is investigated. From the results dependencies of the process parameters can be deduced and comprehensive process analysis can be carried out. Furthermore, the energy saving potential of RDWC compared to the process alternative of a reactive distillation sequence is determined. The operation of RDWC is investigated by dynamic simulations as well as experiments with a continuous RDWC pilot plant of 12 m height. Start-up strategies for a large number of operation points, steady-state operation and control strategies are investigated. From the results concepts for a safe operation of specified operating points and constant product purities during operation can be developed.

This oral presentation will show how to design and operate the reactive dividing wall column for a transesterification of butanol and hexyl acetate. First, the design method will be presented. Optimal process designs will be discussed in detail and dependencies of the process parameters will be pointed out. Additionally, pilot plant experimental results will be shown. Here, decisive factors for a steady and energy minimum operation will be discussed.

567105 The Separation of Multi-Glycols with Divided-Wall Column

Ethanediol (EG), propanediol (PG) and butanediol (BD) are all important platform compounds. They can be produced by catalytic hydrocracking of sorbitol obtained from cereal enzymolysis and often obtained as multi-glycol mixtures. It is difficult to separate these diols due to the similarity in molecular structures and properties, which limited their industrial production. Many methods have been studied to separate mixtures of glycols in literatures, such as azeotropic distillation, adsorption, extraction, etc. Still they have some specific weakness and wait for further study.

In this paper, distillation has been considered for the separation and purification of mixtures comprised of (1,2-PG, 1,3-BD, and 1,4-BD) due to its technical maturity and ease to scale up. A Divided-wall column (DWC), which is well known as low energy consumption and equipment cost due to its process intensification, has been studied systemically and compared with the normal distillation sequence process. First, the residue curve map (RCM) of the ternary system was calculated and there are no distillation boundaries in the ternary system, so the separation can be accomplished by distillations.

Second, two-column distillation sequence process was simulated with the UNIQUAC-HOC equation. The Hayden-O'Connell equation was taken in considering the non-ideality of the vapor phase, and the UNIQUAC model for the liquid phase. The optimized conditions and product specifications were determined for each column in the two-column distillation sequence process.

Third, the DWC was constructed in Aspen Plus, with which the detailed design variables and operation conditions were obtained. The initial design of the DWC was completed using the three-column method based on the Fenske-Underwood-Gilliland-Kirkbride equation. The DWC process was initially designed via the single-factor testing by sensitivity analysis taking the purity of products, capital cost, and heat duty in both the condenser and the reboiler into consideration. The appropriate conditions of the DWC process which includes number of main tower trays, number of prefractionator trays, reflux ration (r mole basis), feed stage, distillate rate (mole/s), side rate (mole/s) and location of side stream were obtained by sensitivity analysis at 1atm.

Then, the DWC configuration was compared with the conventional two-column distillation sequence in terms of the purity of products, energy consumption and total annual cost (TAC). Here, the column diameters were estimated with the function of Tray Sizing in Aspen Plus, and the tray space was taken as 0.61 m. A payback period of 3 years was assumed. The TAC was calculated with the method proposed by Luyben. The following results were obtained: The purity of 1,3-BD in mass could be increased from 90% to 96% with DWC configuration, that of 1,4-BD from 98% to 99%. Approximate 30.0% energy savings and 28.4% TAC savings could be achieved. Some parameters, such as the reflux ratio r, the internal liquid flow to the prefractionator l and the internal vapor flow to the prefractionator v, had great effects on the purity of products. They were optimized with the response surface methodology (RSM) on the basis of the Box-Behnken design (BBD) in terms of the purity of products and decrease energy requirements after preliminary DWC design. The results showed that heat duty in reboiler was mainly determined by the reflux ratio r. The energy consumption increased rapidly with the increase in the reflux ratio. The influences of l and v on the heat duty were small compared to that of r. The three parameters r, l and v had significant effects on the concentration of product of 1,3-BD. They were set as 6.0, 0.133 mole/s, 0.506 mole/s, respectively, by taking the purity of products and the energy requirements into consideration. The energy requirements and TAC of DWC configuration could be furtherly reduced after optimization with the RSM and showed much more benefits in terms of energy saving and hence cost reduction. The purity of products would be improved with energy saving up to 33.5% and TAC reduction by 32.0% with DWC configuration in contrast to the normal two-column distillation sequence process with the same feed. According to the above, it is feasible to employ DWC for the separation of multi-component glycols. DWC can be a promising technology for the production of bio-based glycols on large scale due to the big energy saving and thus TAC reduction.

579964 Opportunities and Challenges for Hydrogen Production Using Membranes

The generation of clean electric power and production of low-carbon fuels are urgent needs that must be addressed as the global concentration of anthropogenic carbon dioxide (CO_2) emissions is rapidly increasing. As shown in **Figure 1**, electricity & heat generation, transportation, and industrial sectors together produce more than 80% of the energy-related CO_2 emissions. The solutions proposed for mitigating the effects of climate change are deployment of low carbon content fuels, as well as improving the efficiency of the current technologies for energy conversion. Utilizing hydrogen as an alternative fuel in the industrial, residential, and transportation sectors can help achieve these goals.

Hydrogen fuel from low-carbon sources is viewed as a promising long-term option for tackling CO_2 emissions from distributed sources. A major benefit of utilizing hydrogen can be seen when it is fed to a proton exchange membrane fuel cell (PEMFC), where only water is produced as a byproduct of electricity production.

The global growth rate of hydrogen demand is expected to remain constant up to 2025 at the current rate of 3.5% per year. Current annual hydrogen production capacity in U.S. refineries is approximately 15.8 million tonnes. Of this volume, 3.3 million tonnes is the captive by-product hydrogen and 2 million tonnes is combusted for heat generation. Hence, the total hydrogen capacity from the conventional captive and merchant markets is approximately 11 million tonnes per year. Around 50 million tonnes of hydrogen is produced globally each year. Approximately less than 50% of this produced volume comes from steam methane reforming (SMR), 30% from oil/naphtha reforming, 18% from coal gasification, 3.9% from water electrolysis, and the remainder from other sources such as plasma processes. Since hydrogen is primarily produced from fossil fuels, such as natural gas, CO_2 is generated and subsequently released into the atmosphere. However, sustainable hydrogen production combined with carbon capture and storage (CCS) technology could be a possible strategy for addressing CO_2 generated from the transportation sector. For instance, using an alternative technology to SMR with pressure swing adsorption (PSA) for H_2 production, such as a membrane reactor (MR), has the potential to lead to H_2 production on demand with subsequent CO_2 storage. This technology can be suitable for distributed and/or mobile hydrogen.

In recent decades, MR technology for H_2 production has advanced significantly. As a special field of interest, metallic membranes, specifically Pd and its alloys, have been heavily investigated in the hydrogen separation field because hydrogen can selectively permeate through the dense metal layer reaching 99.999+% purity. These characteristic advantages could make metallic membranes and MRs suitable for onsite hydrogen production and separation to satisfy distributed hydrogen demands for mobile and small-scale systems.

The use of Pd-based membranes is also appropriate for smaller and emerging hydrogen markets, such as the semiconductor and fuel cell industries, due to the scales and purity requirements. First, these applications are small- to medium-scale with capacities ranging from 5 to 1,000 m^3/h, more than 100 times smaller than large-scale applications such as refining and ammonia production. The inherent compactness of membrane units would enable on-site production of hydrogen for applications on these scales. Alternatively, distributed small-scale users can source hydrogen from large-scale plants at a reasonable production cost. However, the high cost of distribution would make this option less attractive.

The distribution cost of hydrogen is 15 times higher than that of liquid hydrocarbon fuels on a mass basis primarily due to higher pumping costs. If CCS is considered, CO_2 distribution to sequestration sites would contribute to the cost of small-scale production units. In this case, Sjardin et al. concluded from a techno-economic study that at a hydrogen production scale below 40 MW (ca. 12,300 m^3/h), MRs with CCS ($19/GJ) would become competitive with centralized H_2 production ($18/GJ).

On-site hydrogen production would also provide self-sufficient supply and circumvent delivery delays as well as issues with storage safety. Nevertheless, higher production costs due to smaller product sizes and safety concerns over local production and handling of hydrogen as an explosive gas should be taken into consideration. Moreover, fuel cell and semiconductor applications require hydrogen of very high purities. In semiconductor manufacturing, hydrogen is used as a carrier gas for trace doping elements and hydrogen of near absolute purity is needed. For PEMFCs, CO impurities in the hydrogen feed could strongly adsorb onto the catalyst, leading to catalyst deactivation and cell performance deterioration.

Therefore, despite several decades of research, most metallic membranes have remained at the pilot scale due to uncertain economic feasibility and competitiveness at a large scale. Although the scientific literature on this topic is exceptionally rich, only few scientific studies have focused on the analysis of the costs and challenges associated with industrial feasibility.

559407 Functionalized and Responsive Membranes: An Overview

Membrane processes provide a highly flexible separation technique for water treatment and materials recovery. The special features for membrane processes that make them attractive are their compactness, ease of fabrication, operation, and modular design. The development of responsive, multifunctional materials and membranes for water applications requires a high level of control of both the characteristics of the base polymeric support layer, as well as, its corresponding pore configurations. Microfiltration membranes are primarily used for the removal of bacteria and particles, but on the other hand by appropriate pore functionalization with polymeric charge groups or integration of nanoparticles of these MF membranes (termed Functionalized Membranes) one can use these as high capacity metal capture membranes to catalysis and pH/temperature flux modulations with separation selectivity. The dependence of conformation properties of polyelectrolytes provides tunable separation and membrane flux control by pH and temperature-based stimuli responsive properties. Many current treatments for the reclamation of contaminated water sources are chemical-intensive, energy-intensive, and/or require post-treatment due to unwanted by-product formation. We demonstrate that through the integration of nanostructured materials within pore-functionalized synthetic membrane platforms (such as, PVDF membrane pores functionalized with poly-acrylic acid, and/or PNipam), we are able to conduct environmentally important reactions for toxic organic degradation and selective toxic material sorption from water without the addition of expensive or harmful chemicals. The presentation will include pH and temperature responsive properties of bench-scale to full-scale membranes, sorption/desorption, and catalytic nanoparticle creation in membrane pore domain for water and catalysis related applications. This research has been supported by NIH-NIEHS-SRP, NSF EPSCoR, and Chevron Corporation.

580186 Modeling of Gas Transport and Adsorption Considering Liquid Generation during the Removal of Mercury from Flue Gas

Mercury has been identified as a toxic substance that is regulated worldwide to control its emission from industrial flue gas stacks. Several technologies are being evaluated for coal fired power plants to address the decreasing emission regulations currently implemented in the USA and Europe. One removal option is the use of Sorbent Polymer Catalyst (SPC) composite membranes to simultaneously capture and convert flue gas components. W.L. Gore and Associates have developed a unique porous composite membrane based on the combination of adsorbent and catalyst particles with PTFE. This material is being utilized in the GORE™ Mercury Control System (GMCS) that captures Mercury and SO_2 from industrial flue gas by a selective sorption and catalytic process.

This paper describes the structure of this material and introduces a simple model to describe the mechanisms to capture mercury by chemisorption and convert SO_2 to liquid sulfuric acid. Experimental observations have confirmed that all adsorption and reaction processes can be approximated as first order reaction steps. This simplification leads to an explicit solution to the differential equations at the various length scales including all steps for transport and reaction/adsorption. Two different Thiele moduli result on the two length scales within the active particles and within the SPC membrane that are used to assess the utilization of the particles and the limiting factors for the removal process. The overall removal efficiency is estimated by a mass balance along the flow channels of the modules.

Model results are compared to experimental observations from the lab environment to simulated flue gases to observations of pilot plants using actual flue gases. A comparison between the model results illustrates the shift in limiting process steps between these conditions. This insight can be utilized to optimize the structure and performance of the material under these conditions. Finally, the paper will conclude with illustrating the limitations of this simple model and the steps to improve the model description are described.

574512 Understanding Fouling Mechanisms in Virus Filtration of Biotherapeutic Proteins

Virus filtration is an important step in the purification of monoclonal antibodies to ensure the safety of the product. Filters are designed to remove viruses down to approximately 20 nm in diameter from protein solutions, in which the product of interest is approximately 8 nm. Because of the small size difference between the product and the virus, virus filters can be prone to fouling by impurities in the process stream, such as protein aggregates. Understanding the mechanisms of fouling enables better design of both filters and filtration processes.

To better understand the mechanisms driving filtration performance, two feeds were identified: one dominated by size-based fouling and one with a significant pressure or velocity dependent fouling component. Fits of the filtration data to a combined blocking-adsorption fouling model confirmed a difference in the balance between pore blocking and inferred adsorptive fouling between the two feeds. The effects of both membrane characteristics (including structure and chemistry) and filtration process conditions were investigated to determine the impact on throughput performance and on the location and extent of fouling for the two feeds.

The fouling behaviors exhibited by each of the feeds are relevant for commercial therapeutic proteins, highlighting the importance of having a good understanding of process fluids in order to determine the best strategies to mitigate fouling.

579963 Interfacial Junctions Control Ion Transport through Charge Patterned Membranes

Unique transport mechanisms emerge when nanostructured membranes are patterned with multiple chemistries. For example, charge-patterned mosaic membranes possess surfaces functionalized with discrete domains of both positive and negative charge. These oppositely-charged domains provide pathways for both the cation *and* anion from a dissolved salt to permeate through the membrane without violating the macroscopic constraint of electroneutrality. As such, charge mosaic membranes are capable of transporting whole salts more rapidly than neutral molecules of comparable or smaller sizes. Here, by using ink jet printing to create a family of membranes with systematically varied charge patterns, we elucidate the molecular interactions that promote the transport of salts under the action of pressure-driven flow. For patterns that consist of equivalent areal coverages of positively-charged and negatively-charged domains, the effects of the geometric parameters were encapsulated in a single variable, the interfacial packing density, that quantified the fraction of the membrane surface covered by junctions between oppositely-charged domains. Experimentally, the transport of symmetric electrolytes (*i.e.*, KCl and $MgSO_4$) increased with the value of the interfacial packing density, while the interfacial packing density did not significantly affect the transport of asymmetric electrolytes (*i.e.*, K_2SO_4 and $MgCl_2$). Simulations of the electrical potential near the membrane surface demonstrate that for symmetric electrolytes, the structural charge heterogeneity reduces the barrier to ion partitioning thereby promoting salt transport through the membranes. For asymmetric electrolytes, the charge heterogeneity skews the local availability of ions from the stoichiometric ratio of the salt thus hindering salt transport. These findings demonstrate the promise of accessing transport mechanisms that could revolutionize chemical separations and sensing applications through chemical-patterning of membranes.

579965 My Career: Discs, Rods, Flexible and Deformable Particles, and Some Side Trips Along the Way

This presentation will be a 'tour' of my career, from my early PhD research investigating the hindered diffusion of petroleum asphaltenes to recent work involving diffusion of tailed viruses and filtration of single stranded DNA and 'squishy' bacteria. The common thread through much of this work has been a focus on the impact that particle and molecular shape can have on hindered diffusive and convective transport through porous membranes.

Theoretical efforts to understand hindered diffusion and convection date back several decades (Anderson and Quinn (1974), Brenner and Gaydos (1977), Anderson (1981), Ennis (1996) and Deen (1987)) and were primarily focused on spherical particles in cylindrical pores, often with only short range (steric) interactions governing particle transport. In my presentation, I will discuss some of the research my students and I have pursued with less 'ideal' systems. These efforts have involved both experimental as well as theoretical investigations.

In addition to discussing some of our work focused on particle and macromolecular transport in microfiltration and ultrafiltration membranes, I will also present results from my work involving gas separations with room temperature ionic liquids and supported ionic liquid membranes.

576419 Water Vapour Capture on MIL-125(Ti)_NH_2 By (P)TSA Process

Introduction

Water (H_2O) scarcity is a global problem and it might worsen in the future, especially in arid and semi-arid areas. To cover this lack of potable water, several techniques are being developed. One of them is the extraction of H_2O from atmospheric air by adsorption based technologies using Metal Organic Frameworks (MOFs). Some MOFs present great potential and stability under H_2O adsorption/desorption cycles. In the present work, the potential of the MIL-125(Ti)_NH_2 for water capture from air is assessed.

Experimental

MIL-125(Ti)_NH_2 was synthesized at the Korea Research Institute of Chemical Technology (KRICT). The preparation of granulated materials was performed by wet granulation method. The adsorption equilibrium isotherms were measured in a Rubotherm magnetic suspension microbalance (precision of 0.01 mg), in batch mode. Prior to adsorption equilibrium isotherms measurements, the material was activated at 423 K during 12h under vacuum. CO_2, N_2, O_2, and H_2O vapour adsorption equilibrium isotherms were performed at 283/293 K, 303 K, and 323 K. Adsorption breakthrough experiments were conducted on a bench-scale fixed-bed unit at 298 K and 1 bar. The stainless-steel column was packed with the MOF and both ends of the column were capped with glass wood. MOF was activated by heating up to 423 K under helium for 12 h. Diffuse reflectance infrared Fourier transform (DRIFT) spectra measurements were performed using Nicolet 510-P FTIR spectrometer were performed to assess adsorption kinetics. The samples were analysed without any dilution. In a typical experiment, the sample was placed inside the chamber, which was then sealed and continuously purged with helium. MIL-125(Ti)_NH_2 was activated under He flow at 423 K for 3h. Then, the sample was cooled to 298 K and a He flow saturated with water was passed over it. Infrared spectra were recorded periodically. The spectrum of pure He at ambient temperature (298 K) was used as background. Finally, a phenomenological model was developed to describe the adsorption dynamic behavior in the adsorber, packed with the shaped MIL-125(Ti)_NH_2 and was validated against the bench-scale breakthrough experimental results.

Results and discussion

CO_2, N_2, O_2, and H_2O vapour adsorption equilibrium isotherms on MIL-125(Ti)_NH2 were attained. H_2O vapour isotherms show type V behaviour, according IUPAC classification and is not visible any hysteresis loop, though it is difficult to confirm this without higher resolution at low partial pressures. The O_2 and N_2 adsorption equilibrium isotherms present an almost linear shape while the CO_2 adsorption equilibrium isotherms are of Type I according to the IUPAC classification.

In the experimental H_2O breakthrough curve on MIL-125(Ti)_NH_2, H_2O vapour was carried with pure He, at about 50% RH. Adsorption breakthrough curve of water vapor presented a dispersive front up to P/P_o=0.21 and then a compressive front up to P/P_o=0.5. A similar pattern was also observed during the desorption with a dry flow of helium through the bed: dispersive from P/P_o=0.5 to 0.23 and then compressive, as expected from the shape of the water adsorption isotherms. This behaviour was well predicted by the developed fixed bed model,

The water breakthrough curve in co-adsorption with CO_2 on MIL-125(Ti)_NH_2 was also assessed. The water vapor is transported by a stream of CO_2, at 40% RH, and fed into column, mixed with a dry helium (50%) stream. It starts with a dispersive front up to P/P_o=0.18 followed by front a compressive up to P/P_o=0.40. During desorption step, a dispersive front from P/P_o=0.4 to 0.24 is presented and then compressive front, similar to the experiment without CO_2. The fixed bed model can predict well the dynamic behaviour of water adsorption in the presence of CO_2, without taking in consideration any competition between both. This can be explained by the strong water adsorption.

The DRIFT spectrum of dry MIL-125(Ti)_NH2 exhibited three typical vibrational bands in the 3800–3200 cm^{-1} region, peaking at 3500 cm^{-1}, 3380 cm^{-1}, and 3680 cm^{-1}. After several cycles of adsorption/regeneration, the DRIFT spectra of the neat MIL-125(Ti)_NH_2 remained mostly unchanged, which indicates that the material didn't suffer any major structural modification upon cyclic water adsorption. This conclusion is corroborated by the textural characterization performed to the sample used in the microbalance to access the adsorption equilibrium.

To evaluate the potential of the shaped MIL-125(Ti)_NH_2 as adsorbent in a type 2 Atmospheric Water Vapor Processing (AWVP) technology, a column with about 1 m^3 of volume (L – 1.25 m, Ø – 1 m) was considered to operate under adiabatic conditions, packed with 345 kg of MOF, and a bed porosity of 0.4. With the aim of designing a technology of easy implementation, only two steps were considered, feed and co-current purge. The regeneration temperature was considered to be 333 K, 353 K, and 373 K, while the condensation of the purge outlet stream was proposed to be 283 K, 288 K, and 298 K, for the TSA processes. The different designs were compared in terms of their productivity (l/day). The feed was considered to be at 298 K, 1.1 bar, and a RH of 50 %. As expected the highest productivity (320 $l \cdot day^{-1} \cdot ton^{-1}$) was obtained for the regeneration at the highest considered value (373 K), and condensation at the lowest considered value (283 K).

Conclusions

H_2O isotherms present a Type V isotherm shape, not being visible any hysteresis. The amount adsorbed measured was 20.2 $mol \cdot kg^{-1}$ at P/P_0=0.83. The most adsorbed component in this adsorbent was CO_2, followed by O_2, and N_2. H_2O breakthrough history has the expected behaviour according with H_2O adsorption equilibrium isotherms. First, presents a dispersive front at low P/P_0, and then a compressive front. The dynamic adsorption behaviour of water adsorption in the fixed bed was well predicted by the developed model. DRIFT analysis revealed that MIL-125(Ti)_NH_2 structure remained stable after several adsorption/regeneration cycles. Additionally, water adsorption kinetics seems to be fast enough, and equilibrium is achieved in about 100 min. The (P)TSA proposed process shows that MIL-125(Ti)_NH_2 is a promising material for water harvesting from atmosphere, with productivity between 40 $l \cdot day^{-1} \cdot ton^{-1}$ and 300 $l \cdot day^{-1} \cdot ton^{-1}$, depending on the regeneration conditions.

556960 Dynamic Behavior of Pressure Swing Adsorption Process for Nitrous Oxide Recovery from the Off-Gas of Adipic Acid Production

Most of the adipic acid is used to manufacture synthetic polymers such as nylon 6/6 and polyurethanes. The off-gas of adipic acid production mostly consists of N_2O, O_2, N_2, CO_2, and H_2O. A significant amount of N_2O (at a typical world scale plant: 13-26 $MtCO_2e$ per year) is emitted from the process. Since the global warming potential of N_2O is about 330 times that of CO_2, N_2O is one of the representative non-CO_2 greenhouse gases. Therefore, it should be recovered from the off-gases for the mitigation of global climate change. The recovered N_2O can be widely applied to many industrial areas such as medicine, automotive, and food and beverage. In particular, high purity N_2O is a high value-added chemical used in chemical vapor deposition process in semiconductor production. Proper adsorbent selection is required for the design of adsorption separation process. In this study, we evaluate the performance of adsorbents and examine the adsorbents are applicable to the adsorption equilibrium separation and kinetic separation process. Adsorption isotherms of N_2O, O_2, and N_2 were measured on various carbon-based (activated carbon and carbon molecular sieve) and silica-based (zeolite and silica) adsorbents by a volumetric method. Adsorption amount, adsorption affinity and working capacity are compared among the adsorbents for equilibrium separation. Also, adsorption kinetics for each adsorbent is also evaluated by applying either the non-isothermal model or isothermal dual resistance model to experimental uptake curves. The parameters in each model considers the physical properties and adsorption characteristics of adsorbent-adsorbate systems. The validity of the equilibrium and kinetic results are confirmed by comparing the experimental breakthrough curves with the dynamic simulation results using the obtained values. Based on the breakthrough results, the simulations of the pressure swing adsorption process are conducted.

576730 Equilibrium Theory Analysis of Equalization Tanks in a PSA Process

Pressure equalization between beds and also between beds and tanks in pressure swing adsorption (PSA) processes are critical steps for achieving higher enrichment of heavier gas in the heavy product and higher recovery of light gas in the light product. However, only minimal attention has been given to the utilization of an empty tank in the pressure equalization step. This empty tank does not contain any adsorbent and is simply used as a medium for internal gas storage. This kind of equalization step is referred to as a bed-to-tank-to-bed equalization step. Equalization tanks can be useful to decrease the number of the beds and to decrease the number of idle steps in PSA cycle schedules.

To begin to understand the role of tanks in a PSA process, an isothermal equilibrium theory analysis is carried. Analytic expressions are derived based on linear isotherm and an arbitrary binary feed stream to describe the performance of a simple PSA process at the periodic state that produces a pure heavy component at high recovery. To observe the effect of the number and size of the tanks, the performances are analyzed in varying number of equalization steps and volume of tanks in terms of the different concentrations and velocity profiles exhibited during various cycle steps. The latest result from this study will be presented.

571589 An Equilibrium-Based Shortcut Temperature Swing Adsorption (TSA) Model for Screening Adsorbents for Heavy Component Recovery

1. Introduction

The chemical industries are seeking efficient separation technologies that can improve resource- and energy-efficiency, reduce greenhouse gas emissions, maximise process yield and increase profit. Metal organic frameworks (MOFs) are attracting attention as adsorbents that could outperform conventional technologies such as distillation. However, with over 80,000 possible MOFs to choose from, the development of novel adsorption processes applying MOFs is challenging. The performance of adsorption processes depends significantly on the chosen adsorbent, and in turn on the process configuration and operating conditions. To unlock the full potential of adsorptive processes for separation, an integrated approach for identifying appropriate adsorbents, separation configurations and their operating conditions is required.

The literature on adsorbent screening focuses on pressure (or vacuum) swing adsorption (P/VSA) systems, rather than temperature swing adsorption (TSA) which present opportunities for heat integration but generally have lower productivity than PSA processes because of the time required for heating and cooling the bed. The lack of screening approaches for TSA adsorbents means that novel TSA processes cannot easily be identified. This work addresses this gap in capabilities by developing a shortcut TSA model for quick screening of adsorbents for recovering the more strongly-adsorbing component (A) of a binary (A/B) mixture. The model may also be used for parametric analysis to identify suitable operating conditions. The model is applied to screen adsorbents for the recovery of CO – a major feedstock in the production of acetic acid – from syngas mixtures primarily containing CO and N_2.

2. Methods

The new shortcut model considers a four-step TSA cycle with heating, cooling, pressurisation and adsorption steps. The model mirrors the model of Maring and Webley (2013) for screening adsorbents for application in pressure swing adsorption processes.

The initial condition is a bed saturated with gas at the feed composition (after breakthrough). This initial condition is selected because it is the point at which most information is known about the bed (i.e. temperature and gas-phase composition). The heating step follows, where one end of the column is open and the bed is heated (indirectly) to the desorption temperature. This step is modelled as a 'frozen' front: the step is divided into a discrete number of temperature steps without considering the rate of change of temperature. Material and energy balances are solved simultaneously for each temperature step to determine the amount of A and B recovered in each step, and the gas-phase composition. These values provide the initial guess for the next temperature step, until the final desorption temperature is reached. The cooling step which follows is modelled similarly to the heating step. Both ends of the bed are closed while the bed is cooled to the adsorption temperature. As no material flows into or out of the bed, the bed pressure reduces as cooling progresses. A pressurisation step follows, where the feed mixture enters the bed at one end, restoring it to the adsorption pressure. This step is also solved as a frozen front; the material and energy balances are solved simultaneously to determine the bed temperature, gas-phase composition and amount fed at each intermediate step. Finally, in the feed step, both ends of the bed are opened and the feed is introduced to the bed; the weakly adsorbed component (B) is recovered until breakthrough occurs. As the bed conditions at the start and end of the feed step are known, the amount of fresh feed and of component B recovered during the feed step can be calculated from a material balance. The solution procedure for the shortcut model is implemented in MATLAB.

3. Results and discussion

To validate the shortcut model, post-combustion capture of CO_2 from a binary CO_2/ N_2 mixture, earlier investigated by Joss et al., (2017), is modelled using a detailed column model. The CO_2 purity and recovery predicted by the shortcut and detailed models are within 5%, suggesting that the shortcut model provides a meaningful representation of the actual process performance.

The shortcut model is also used to screen several MOFs for CO/N_2 (A/B) separation. Several combinations of adsorption and desorption temperatures were simulated in the range 300–550 K to recover CO from a feed at atmospheric pressure comprising 60 mol% CO. The performance of each candidate material was compared to that of a commercial MOF, HKUST-1, which achieved a maximum CO purity and recovery of 78% and 68% respectively in the investigated range. Ni-MOF-74, a 1-D hexagonal-structured MOF, performed best, recovering up to 95% of CO at purities of up to 99 mol%. The better performance of Ni-MOF-74 correlates with the higher density of open-metal sites to which CO binds preferentially.

4. Conclusions

The new shortcut TSA model enables candidate adsorbents to be screened and ranked for recovering a strongly adsorbing component from a binary mixture, considering process conditions. The shortcut model is shown to predict process performance reasonably well, with predictions within 5% of those of a detailed model. This work, for the first time, provides a tool for quick, systematic screening of adsorbents for application in TSA processes, which can offer benefits over PSA in some cases, but also still present significant engineering challenges. The model could be coupled with high-throughput molecular simulation and materials engineering for large-scale screening and to aid the discovery of new materials with optimal properties.

References

- J. Maring and P.A. Webley, 2013, A new simplified pressure/vacuum swing adsorption model for rapid adsorbent screening for CO2 capture applications, Int. J. Greenh. Gas Control, 15: 16–31.
- Joss, L., Gazzani, M., Mazzotti, M., 2017. Rational design of temperature swing adsorption cycles for post-combustion CO_2 Chemical Engineering Science 158 (2017) 381–394

572059 Sub-Ambient Pressure Swing Adsorption Enabling High Capacity CO_2 Capture

Adsorption of CO_2 from post-combustion flue gas is one of the leading candidates for low-cost carbon capture systems. While a number of challenges face the widespread adoption of this technology, one key issue will be to reduce the massive demand for sorbent materials[1]. This talk focuses on the use of structured sorbent materials exhibiting high capacity for sub-ambient pressure swing adsorption. We have found flue gas compression and cooling is counterintuitively energy efficient when extensive heat integration and energy recovery is utilized, and this concept enables high operating capacity for a variety of MOF sorbents[2]. Current rigorous technoeconomic estimates reveal parasitic loads as low as 22% and total costs of CO_2 as low as \$41/tonne.

In this talk, the highly heat integrated sub-ambient pressure swing adsorption process that enables high operating capacities using robust metal-organic frameworks will be discussed. Synthesis and formation of advanced composite materials capable of exceeding 10 $mol_{CO2}/kg_{sorbent}$ will be considered, as will methods of improving sorbent performance via management of sorption enthalpy[3]. The effect of cold operation on breakthrough performance, particularly front spreading due to cold operation will be considered. Single and dual bed PSA results will be presented for the sub-ambient PSA process for both fiber and pellet beds to demonstrate its economic feasibility feeding the PSA results back into the economic model.

References

1. Yu, C.-H.; Huang, C.-H.; Tan, C.-S., A Review of CO2 Capture by Absorption and Adsorption. *Aerosol and Air Quality Research* **2012,** 12, (5), 745-769.
2. Park, J.; Lively, R. P.; Sholl, D. S., Establishing upper bounds on CO2 swing capacity in sub-ambient pressure swing adsorption via molecular simulation of metal-organic frameworks. *Journal of Materials Chemistry A* **2017**.
3. DeWitt, S. J. A.; Rubiera Landa, H. O.; Kawajiri, Y.; Realff, M.; Lively, R. P., Development of Phase-Change-Based Thermally Modulated Fiber Sorbents. *Industrial & Engineering Chemistry Research* **2018**.

574856 Experimental Validation of PSA Optimization Techniques

Advancements in numerical and optimization techniques, and the availability of cheap computational power have had a significant impact on PSA design. It is now very common to see rigorous optimization techniques being used in conjunction with detailed process models. These techniques are being employed to compare different process configurations, screen materials etc. However, there is a dearth of experimental studies that validate these optimization results. This presentation seeks to bridge this important gap.

In this work, the vacuum swing adsorption (VSA) separation of CO_2 from a mixture of 15% CO_2 and N_2 on Zeolite 13X is considered. Volumetric and gravimetric methods were used to measure the pure component isotherms, while breakthrough experiments were used to measure the competitive isotherms. Based on these input parameters that were obtained from a few milligrams of material, a detailed model was developed to simulate two cycles: A simple 4 step cycle consisting of feed/co-current blowdown/counter-current blowdown/ feed pressurization; and feed/co-current blowdown/counter-current blowdown/ light product pressurization. A genetic algorithm based optimization algorithm was used to calculate the Purity-Recovery Pareto curve. The Pareto curves clearly demonstrated the superiority of the cycle with the light-product pressurization over feed-pressurization. Thee points from each Pareto curve were chosen and the operating parameters, as predicted by the model, were translated to an experimental two-column PSA rig that contained 324g of adsorbent. The results showed that the performance indicators (Purity and recovery) and the transients of various process variables (flows, compositions, temperatures) predicted by the model were well reproduced by the experiments. The study demonstrates the reliability of optimization techniques for the design of PSA systems.

572521 Evaluation of Structured Adsorbents for Carbon Capture Applications

Carbon capture and storage has been advocated as a near term technology for capturing CO_2 from large stationary sources like power plants[1]. Different process like adsorption, absorption and membrane separation are being evaluated as potential candidates for capturing and concentration CO_2 from these sources. Regarding adsorption, the research focus can be classified into two broad categories: 1: Synthesizing novel materials such as metal organic frameworks (MOFs), amine supported adsorbents, zeolite imidazolate frameworks (ZIFs) with high CO_2 capacity and selectivity, and 2: Cycle design and process optimization of adsorption processes. A traditional adsorption process employs packed columns containing adsorbent pellets or extrudates which are few millimeters in size. The main challenges associated with packed beds are large pressure drops and mass and heat transfer limitations at high flowrates. This would result in significant increase in energy consumption and a decrease in productivity of the capture process. In recent years, structured adsorbents have gained significant interest owing to the potential advantages of lower pressure drop and faster mass transfer over conventionally shaped adsorbents, allowing significant improvement in energy consumption and productivity.

Most of the studies on these structured adsorbents are confined to adsorption equilibrium, kinetic studies and pressure drop tests and to the best of our knowledge, we have not seen in the open literature, how much of an improvement in productivity and energy consumption can be achieved over conventional pelletized sorbents. This information can be obtained from rigorous process optimization studies. The goal of the present work is therefore to evaluate the performance of structured sorbents against conventionally shaped pellets. The structured adsorbents used in this study were obtained by 3D printing. The advantage of this method is that, it is possible to obtain different shapes of adsorbents, with controlled channel sizes and geometries. We have considered two cases with different CO_2 concentrations 1. Post-combustion CO_2 capture from a coal fired power plant containing 15% CO_2, 5% H_2O and 80% N_2 and 2. The PSA tail gas of the hydrogen purification unit which contains 51% CO_2, 10-15% CO,10-15% CH_4 and 20-25% H_2 and a small amount of H_2O.

In this study, we have used a six-step vacuum swing adsorption (VSA) cycle which has been studied in literature[2], to evaluate the performance of the aforementioned two types of adsorbents. We have chosen supported amine sorbents in this study, owing to their tolerance to moisture in comparison with zeolites and MOFs. The amines were first grafted on to the silica beads and 3D-printed silica structure. In the next step, single component isotherms of the different gases were obtained using a commercial volumetric apparatus and dynamic column breakthrough experiments were carried out to measure the kinetic constants. Detailed optimization of the VSA cycle was then carried out to arrive at operating conditions satisfying 95% CO_2 purity and 90% recovery constraints with maximum productivity and minimum energy consumption.

References:

1. *IPCC, 2014: Climate Change 2014: Synthesis Report. Contribution of Working Groups I, II and III to the Fifth Assessment Report of the Intergovernmental Panel on Climate Change*; IPCC, Geneva, Switzerland.
2. Khurana, M.; Farooq, S., Simulation and optimization of a 6-step dual-reflux VSA cycle for post-combustion CO2 capture. *Chemical Engineering Science* **2016,** 152, 507-515.

570533 Lab-on-a-Chip Device for Shear Enhanced Purification of Macro Biomolecules

Protein manufacturing required a multitude of sequential steps to achieve a desirable product. Briefly, the process spans two essential processes, namely, the formation of the target biomolecule (upstream) and the purification of the target protein (downstream). For most therapeutic applications, FDA recommends a high level of purity (99.99%) and hence, the purification processes consist of many non-trivial steps. Thus, protein purification is cumbersome, expensive and requires the meticulous application of each step to reduce potential loss of product. Furthermore, post purification, the process equipment often requires extensive, costly maintenance that needs harsh chemicals. Here we present a new methodology of using shear forces that offers an alternative approach to protein purification that reduces the number of steps, making it easier, greener, and less expensive. Furthermore, our method can be scaled according to the needs of the final product. Analogous to affinity chromatography, a stationary phase – Carbon Nanotubes (CNTs) - will be used in conjuncture with a mobile phase in lab-on-a-chip device setup. We have combined the lab-on-a-chip device with microelectrodes to perform Electrical Impedance Spectroscopy (EIS). EIS will be used to monitor any changes within the system (attachment/no attachment). Protein A is attached to CNTs using EDC-NHS chemistry. Post attachment, CNT's are tightly packed in a microfluidic channel (akin to a packed bed reactor) and are held in place (against the fluidic drag force) by a transverse electric field from non-planar interdigitated electrodes. Preliminary data clearly shows that the CNTs are capable of generating shear forces close to hydrogen bond strength depending on several factors such as the microfluidic architecture and the flow velocity. At a low flow rate, the target proteins (monoclonal antibodies in our experiments) attach to Protein A while the rest of the solution flows by (filtration). The flow rate is then suitably modified to increase the shear force enough to release the biomolecules from the CNT complex into an elution buffer while maintaining their bio-functionality (purification). Hence, the shear force approach drastically reduces the number of steps required for the desired purification. Further post elution buffer, prolonged washing with a solution allows us to regenerate the CNT's for the next round of purification. We have developed optimal conjugation protocols for the process, and have studied in details the effect of fluid flow, pH, and temperature on the purification efficacy of our platform. We believe that this technology not only benefits the protein manufacturing scene in general (i.e., the production of mainstream proteins such as insulin), but also other contemporary, specific applications such as using monoclonal antibodies from cancer treatment to sensors.

575922 Heme Extraction and Separation for the Preparation of Apohemoglobin

Background: In this study we outline the development of a procedure to extract and separate the iron containing heme group from the blood protein hemoglobin (Hb) to produce apohemoglobin (apoHb) using a tangential flow filtration (TFF) extraction and purification procedure. ApoHb is a protein derived biomolecule currently being investigated as a delivery vehicle for hydrophobic molecules targeting the CD163 clearance pathway in monocytes and macrophages. Previous strategies to prepare this biomolecule use either an acidified-acetone precipitation or a methyl ethyl ketone liquid-liquid extraction. Unfortunately, these procedures require large volumes of highly flammable, chemically incompatible solvents that necessitate the use of expensive separations equipment.

Methodology: To facilitate scale-up of apoHb production, we developed a procedure for apoHb extraction that combines extraction with purification within a single liquid phase. In addition, during the development of this process, we also developed methods to analyze these materials with an assay to target the functionality of the hydrophobic heme-binding pocket of Hb. Here, an acidified ethanol-based solution was used to protonate the proximal histidine residue, partially unfold the Hb protein, and extract the heme group. The resulting material then underwent diafiltration to continue extraction by separating the apoHb from the heme solution. After separation, total protein and active protein yields were quantified via the molar extinction coefficient of apoHb and a heme-binding pocket activity assay, respectively. Analysis of the Soret peak was used to determine residual heme content. Mass spectrometry and SDS-PAGE were used to confirm the chemical identity and purity of the apoHb. The storage stability of the apoHb product was analyzed at varying temperature. Size exclusion chromatography was used to verify haptoglobin binding was similar to native Hb. Finally, heme was reincorporated into apoHb to produce reconstituted Hb (rHb) for biophysical characterization of the O_2 equilibrium curve, O_2 dissociation and CO association kinetics of the rHb.

Results and Conclusions: The new apoHb-TFF process resulted in total protein yields of more than 95% and active protein yields of approximately 75%. The apoHb had less than 1% residual heme and SDS-PAGE indicated more than 99% purity of apoHb with an identical globin chemical profile to Hb, except for the absence of heme. Storage studies demonstrated no loss of heme-binding activity in solutions at 4 °C, -80 °C, and in lyophilized form over extended periods of time. Size exclusion chromatography confirmed that apoHb bound to haptoglobin. Furthermore, the mass of the a and ß subunits match the theoretical mass of the a and ß subunits of human Hb. Biophysical properties of rHb were approximately the same as native Hb. Overall, this study describes a novel and improved method to produce apoHb. This scalable and effective method for purification can prepare sufficient materials for much preclinical and clinical studies that aim to use apoHb as a drug encapsulation and delivery biomolecule.

577512 Separating Early Stage Amyloid Aggregates Using Microchannel Electrophoresis

Amyloid proteins are involved in a variety of diseases including Alzheimer's Disease (amyloid beta protein) and Type 2 Diabetes (amylin protein). Several technical challenges exist for the quantitative analysis of different sizes of amyloid proteins present during the early stages of aggregation and thus new technologies are required for enhanced protein separation. Electrophoretic and spectroscopic techniques including SDS-PAGE, western blotting, matrix-assisted laser-desorption ionization mass spectrometry (MALDI-MS), and electrospray ionization (ESI-MS) have been tested for their ability to detect and separate oligomeric Aβ, but have shown limited effectiveness due to oligomeric transience, low concentrations, and difficulty of isolation. In addition, these and other analysis techniques often require the use of excessively high concentrations, denaturing agents, or other components that can alter the native aggregation species. Microchannel electrophoretic offers a fast and efficient separation technique for measuring early Aβ aggregates with little interference to the native aggregation process. A protein's characteristics, including its tendency to aggregate, are determined largely by its primary sequence, and thus variations in an amyloid's primary structure can be correlated to changes in its aggregation process. Here we show the results of microchannel electrophoretic analysis of native amyloid beta compared to several primary sequence variants. During native amyloid beta aggregation, monomers and multiple oligomeric species were observed. Comparing variants to native amyloid beta showed that changes in protein flexibility affect protein aggregation rates and propensity to form specific oligomeric species.

577908 Characterization of Protein-Polymer Conjugates Using Aqueous Two-Phase Systems

Covalent attachment of polymers to proteins to produce bioconjugates with improved properties such as increased stability to external stimuli and longer circulation times in vivo has been accomplished using various methods over the last few decades. Traditionally, a linear or branched polyethylene glycol chain is attached to a protein to give the conjugate these improved properties. Many different polymers, polymer structures, and combinations of polymers may also be used to give the conjugate not only improved stability but also specific functionality. As more complex protein-polymer bioconjugates are developed and utilized therapeutically and industrially, there is an increasing need to characterize and purify the conjugated products of the reaction. This characterization may help spur future developments in the areas of bio-therapeutics, biosensors, biotechnology, and enzyme ruggedization among others. Extraction using aqueous two-phase systems presents an opportunity to characterize conjugates in a mild environment, retaining conjugate function and allowing future use of the conjugate.

In this work, aqueous two-phase systems were evaluated for the separation of protein-polymer bioconjugates on both benchtop and millifluidic scales. Protein-polymer conjugates were formed by atom transfer radical polymerization from initiation sites attached to the protein. This polymerization method allows for the creation of polymers with different structures in aqueous conditions and leads to simple purification of the conjugate from the reaction mixture. A millifluidic device was used to form two-phase droplets of controlled composition surrounded by an immiscible carrier fluid for quicker analysis and less material usage than on the benchtop scale. This high-throughput method allowed many conditions to be tested while minimizing the number of conjugate reactions required. Selected benchtop experiments were performed to validate the millifluidic approach. By varying the phase-forming compositions and components, the protein to which polymers were attached, and the attached polymer identity and length, a series of partition coefficient data was obtained. The results suggest that protein-polymer conjugates partition much differently than the native protein at the same conditions. Partitioning of high polymer grafting density conjugates is compared to the partitioning of an equivalent amount of polymer in an aqueous two-phase system and the results are presented here. We also consider multi-step extraction to purify conjugates with similar physical structure.

556557 Novel Enzymes and Separations in Biotechnology-Identifying a Process Tailored Extraction System-

The development of industrial scale bioprocesses has come into high scientific and industrial interest over the last decades. Especially in the red and white biotechnology, process costs are often dominated by 50 - 80 % by the downstream processing. Thus, novel separation technologies such as aqueous two-phase extraction using the TAPPIR Technology have come into focus. Extractions based on aqueous two-phase system (ATPS) were shown to have an enormous potential for the extraction of biomolecules due to a high biocompatibility and low operating costs. The most common ATPS consist of a hydrophilic polymer and a salt serving as phase-forming components. The distribution of the target protein within these systems can either be influenced by the choice of the phase forming components, and additionally by a choice of a displacement salt. It is essential to identify a suitable ATPS using profound knowledge on the molecular interactions in solution.

One promising and fast measure accounting for complex interactions of biological macromolecules is the second osmotic virial coefficient (B_{22}) characterizing the interactions between the biological macromolecules in solution and the cross virial coefficient (B_{23}) characterizing interactions of the biological macromolecules with the salts / impurities. Assisted by a thermodynamic modeling approach, identifying the interactions of the low molecular weight components in the extraction system, allows for a fast and reliable first estimate on an appropriate extraction system.

B_{22} as well as B_{23} data was measured using composition gradient multi angle light scattering (GC-MALS) for different proteins and salts in aqueous solution.

One additional advantage using this approach is the fact, that the partition coefficient can be estimated using a hybrid ShortCut Modell based on a minimal set of experimental data. The results will help to improve the development and optimization of ATPS as well as the TAPPIR Technology used for the design of an efficient downstream process.

563491 Optimization Opportunities of Virus Purification in Different Modes of Aqueous Two-Phase System

Lack of platform technologies for downstream processing of viral based vaccines and gene therapy vectors have led to a boom in developing new technologies. Aqueous two-phase systems (ATPS) have been in the forefront of the unit operation candidates to provide a high throughput and high yield unit operation. ATPS, formed by two partially miscible aqueous solutions, are known for their inexpensive, mild, and environmentally friendly advantages. However, the discrepancy in understanding the partitioning mechanism of biomolecules is hindering industrial implementation. Using two non-enveloped viruses – porcine parvovirus and human rhinovirus, the partitioning mechanism in a polyethylene glycol (PEG)-citrate system was explored. The preferential partitioning of viruses to the PEG-rich phase was heavily influenced by the virus surface characteristics (surface charge and hydrophobicity), and relative phase characteristics (electrostatics and hydrophobicity). Furthermore, the opportunities in enhancing the driving forces by manipulating the system characteristics and compositions will be discussed. We have developed two separate methods to improve the driving forces: 1) osmolyte addition and 2) affinity peptide ligand addition. Osmolytes, naturally occurring compounds, changed the water structure in the system which resulted in 100% recovery of the viral particles in the PEG-rich phase at lower system compositions compared to systems without osmolytes. The other method involved a virus specific affinity peptide ligand conjugated to PEG. The affinity study performed with a peptide ligand specific to PPV showed a relative increase in the recovery as compared to the original systems. Overall, this talk will provide a scope to enhance the partitioning and guidelines to generalize the virus recovery process within a confined experimental space.

573414 Using Droplet-Based Millifluidic Platform to Separate Biomolecules in Aqueous Two-Phase Systems

Aqueous two-phase systems (ATPS), formed by mixing two polymers or a polymer and a salt, has been used as a separation and purification tool for more than 50 years. With nearly 80% water content, ATPS can achieve separations with minimum damage to particles with delicate structures such as proteins. Because of its potential for continuous operation and process integration, ATPS is particularly promising to meet the downstream processing needs created by the fast-growing production rate of biomolecules. The partitioning behavior involved, however, is complex and difficult to predict due to the number of factors that can alter the process. The application of ATPS is therefore limited because of the inability to easily optimize the desired partitioning behavior in very large parameter space. Here, we propose using a droplet-based millifluidic platform to systematically investigate the impact of multiple factors on the partitioning behavior in an ATPS. The precise manipulation of micro-scale droplets allows for higher compositional resolution, shorter settling times, and lower sample consumption, making this technique an efficient and flexible tool for rapidly exploring a large parameter space. In millifluidic droplets, the partitioning behavior of hemoglobin in PEG-potassium phosphate salt two-phase system is characterized as a function of tie-line length, the PEG/salt concentrations, and the concentration of added salt. The amount of protein in each phase is inferred from absorbance monitored by an in situ spectrophotometer. The degree of partitioning can be manipulated so that more proteins go to the preferred phase. The condition for maximum partitioning is further explored to facilitate separation of two proteins in the same ATPS.

565444 Highly Selective MOF Thin-Film Nanocomposite (TFN) Membranes for Water Purification

The rapid increase in global demand for high purity water for municipal, agricultural, and industrial applications has stimulated substantial research in highly selective membranes for water purification and desalination. However, state-of-the-art polyamide thin-film composite (TFC) membranes show relatively low rejections particularly to small neutral contaminants such as boron. Aiming to improve the selectivity of the current TFC desalination membranes, amine-functionalized metal-organic frameworks (MOFs), such as UiO-66 and MIL-101(Cr), with good water stability, small particle size, high surface area, and good crystallinity were successfully synthesized and incorporated into the polyamide selective layer of a traditional TFC membrane to fabricate novel thin-film nanocomposite (TFN) membranes. Compared to the unmodified benchmark TFC membrane, the incorporation of amine-functionalized MOF nanoparticles significantly changed the properties of the polyamide selective layer in terms of morphology, hydrophilicity and chemistry, which lead to a significant improvement of membrane separation properties. The best performing TFN membrane shows a relatively high pure-water permeability of 0.56 LMH/bar and a high NaCl rejection of 96.7% at 150 psi. This performance is much higher than that of the benchmark membrane and even better than the state-of-the-art seawater desalination membranes. This study has demonstrated the promising potential of MOF-polymer composite materials for applications related to water purification.

582487 Design of Ultrafiltration (UF), Nanofiltration (NF) and Low-Pressure Reverse Osmosis (LP-RO) Hollow Fiber Membranes for Pressure Retarded Osmosis (PRO) Pretreatment

Pressure retarded osmosis (PRO) has been widely investigated to harness the osmotic energy from salinity gradient. Seawater desalination brine (SWBr) and wastewater retentate (WWRe) are employed as the draw and feed solutions respectively due to a large salinity difference between them and reuse of wastewater, but the fouling caused by WWRe on PRO membranes is too severe. In order to mitigate fouling on PRO membranes, one-step spun ultrafiltration (UF), thin-film composite (TFC) nanofiltration (NF) and TFC low-pressure reverse osmosis (LP-RO) hollow fiber membranes were designed and fabricated into 1-inch modules to pretreat the WWRe. The filtrate from each pre-treatment method was used as the feed solution in the subsequent PRO tests. Results showed that the newly developed NF and LP-RO were effective to remove foulants and scalants. Using NF-filtrates, the PRO process could produce a stable power density of 7.3 W/m^2. It was a 10.7% drop from the initial power density. Using LP-RO-filtrates, the PRO process could produce a stable power density of 8.4 W/m^2, which was only a 7% drop from the initial power density. With the help of Separation Technologies Applied Research and Translation (START), the LP-RO hollow fiber membrane module was successfully scaled up to 4 inch in diameter. The LP-RO membrane fabrication was also optimized and the RO fluxes of both 1 inch and 2 inch LP-RO modules were more than 3 LMH/bar with NaCl rejection more than 99%.

557872 Creation of Well-Defined "Mid-Sized" Micropores in Carbon Molecular Sieve Membranes

Carbon molecular sieve (CMS) membranes are promising candidates for the separation of organic molecules due to their stability, ability to be scaled at practical form factors, and the avoidance of expensive supports or complex multi-step fabrication processes. A critical challenge is the creation of "mid-range" (e.g., 5-9 Å) microstructures that allow for both facile permeation of organic solvents and selection between similarly-sized guest molecules. Here, we create these microstructures via the pyrolysis of a microporous polymer (PIM-1) under low concentrations of hydrogen gas. The introduction of H_2 inhibits aromatization of the decomposing polymer network and ultimately results in the creation of a well-defined bimodal pore network with the smallest micropores existing at 5.1 Å. The H_2 assisted CMS dense membranes show a dramatic increase in *p*-xylene ideal permeability (~15 times), with little loss in *p*-xylene/*o*-xylene selectivity (18.8 vs. 25.0) when compared to dense PIM-1 membranes pyrolyzed under a pure argon atmosphere. This approach is extended to hollow fiber membranes, which exhibit *p*-xylene permeances up to $5.2\times10^{-10}$with permselectivities of 8.9 for equimolar *p*-xylene/*o*-xylene mixtures, highlighting the potential of this approach to be translated from the laboratory to the field.

575325 Interfacial Hard Templating of Hierarchically Structured, Ultrathin Carbon Molecular Sieve Membranes

Inorganic membrane materials such as carbon molecular sieves (CMS) show promise in transforming gas and liquid separations, however, they struggle from the well-known trade-off between molecular selectivity and permeability[1,2]. This motivates efforts to reduce defect-free membrane thickness while simultaneously tailoring membrane texture and function. In this talk, we present a versatile silica colloidal crystal based nanocasting approach[3,4] that yields ultra-thin (15-300 nm) microporous polyimide-derived CMS self-supported on three-dimensionally ordered mesoporous (3DOm) carbon layers. We show how the large area associated with the template-replica interface in our 3DOm carbon films offers a previously unexploited handle for simultaneously tailoring microstructure of the replica phase. Specifically, we elucidate how temperature-tunable silanol surface chemistry of the template can be used to tailor the graphitic character (e.g., graphitic content relative to turbostratic carbon) of CMS replicas under mild carbonization conditions (600-900 ^{o}C) and in the absence of specific metal catalysts. The interfacial origin of this effect is confirmed through studies showing a sensitivity of graphitic character to film thickness. Studies also show enhancement of graphitic character in thin carbon films as compared with untemplated carbon molecular sieves, with maximum sensitivity near the beginning of the carbonization process (around 600 ^{o}C). We hypothesize that the interface-mediated tunability of the replica graphitic character derives from adsorption-induced orientation of pyromellitic dianhydride-co-4,4'-oxydianiline (PAA) precursors once the surface silanol density is reduced to the point of approximate registry (e.g., hydrogen-bonding) with PAA molecular dimensions. Taken together, this templating approach offers a strategy for multi-scale control over film structure. Using permeation of small gases (e.g., CO_2, N_2, O_2, H_2), we will demonstrate how the template interface-mediated structural control and reduction of film thickness can be exploited for high-flux, high selectivity separations.

References

[1] H. Foley, *Micropor.Mater.,* **4** (1995), 95.

[2] L. Robeson, *J. Memb. Sci.,* **62** (1991), 2.

[3] Z. Tian, M. A. Snyder, *Langmuir,* **30** (2014), 12411.

[4] M. Sharma, M. A. Snyder, *in preparation.*

[5] M. Inagaki, New Carbons, Elsevier, (2014).

561986 Thin-Film Composite Hollow Fiber Membrane with Inorganic Salt Additives for High Mechanical Strength and High Power Density for Pressure-Retarded Osmosis

Using high salinity draw solutions in pressure retarded osmosis (PRO) demands novel membranes with a higher withstanding pressure and a higher power density to take full advantage of the increased osmotic gradient for greater osmotic energy production. This study highlights novel strategies to design an inner-selective thin-film composite (TFC) PRO hollow fiber membrane with a high operating pressure and a high power density by adding an inorganic salt, $CaCl_2$, to the dope solution and fine-tuning the spinning conditions. An optimal $CaCl_2$ dosage can not only effectively increase the dope viscosity but also reduce the mean pore size and narrow the pore size distribution of the polyethersulfone (PES) hollow fiber substrates. A high flowrate ratio of dope solution to bore fluid greatly enhances the mechanical strength of the PES substrates. The combination of these two strategies results in a more permeable, better-supported and less defective polyamide selective layer after interfacial polymerization. The best newly developed TFC hollow fiber has a power density of 38 W/m^2 at 30 bar by using 1.2 M NaCl and deionized water as the feed pair. To our best knowledge, these are the highest power density and operation pressure ever achieved by PRO hollow fiber membranes in the literature. The improved performance will advance the PRO technology closer to commercialization and make osmotic energy more competitive with other renewable energies. This membrane will be implemented on a 250 m^3/day PRO pilot with reverse osmosis seawater brine and wastewater concentrate as the draw and feed pair, to demonstrate osmotic power generation in Singapore.

565202 High Performance Thin Film Composite Membranes with Well-Aligned Nanofiber Substrates for Forward Osmosis Applications

Forward osmosis (FO) membranes with ultrahigh water flux and low reverse salt flux is of urgently demand in FO process. However, the severe internal concentration polarization (ICP) effect in FO is one of the important problems to be overcome in order to maintain a high FO performance of thin film composite (TFC) membranes. In this work, a highly aligned nanofiber-based thin film composite membrane was reported for the first time to develop the high-performance TFC-FO membrane. By adjusting the rotating speed of the collector via the electrospinning process, the prepared aligned nanofiber substrates possess higher pore size and better pore interconnectivity than the random ones. Polyamide selective layer were successfully formed via interfacial polymerization on aligned and random nanofiber substrates. The surface structure, water permeability, salt rejection and forward osmosis performance of resultant thin film composite TFC membranes were systemically studied. The ICP effect is significantly eliminated, which may be due to the higher pore size and better pore interconnectivity of the aligned nanofiber substrate. Consequently, the prepared aligned nanofiber based TFC membrane exhibits a high water flux (J_w) of 50.7 LMH and a low reverse salt flux (J_s) of 6.8 gMH in FO mode, as well as J_w of 62.9 LMH with J_S of 8.7 gMH in PRO mode, using deionized (DI) water and 1 M NaCl aqueous solution as the feed and draw solutions respectively. This work may provide a new orientation in developing high-performance TFC membranes with high selectivity.

577314 Effect of ZIF-8 and Lignin Nanoparticles in Pervaporation Separation of *N*-Butanol from Aqueous Solutions Using PDMS Based Mixed Matrix Membranes

Global energy demands are primarily met through non-renewable sources such as coal, natural gas, and oil. However, the scarcity and rising prices of fossil fuels, coupled with environmental challenges lead researchers to find out a microbial alternative to address these issues. Biofuel has emerged as an attractive solution which can be effectively used in the current scenario. Among the probable list of alternatives, biobutanol is an important renewable biofuel, which can be easily integrated into existing fuels due to its better performance and other enviable advantages over other biofuels. It can be directly used in existing car engines without requiring any modifications.

Butanol is produced via acetone-butanol-ethanol (ABE) fermentation from renewable resources such as agricultural by-products. Therefore, it avoids the food vs. fuel debate. However, there are significant challenges to produce this alcohol as an economically viable biofuel. The most important challenge is improving fermentation performance to achieve higher butanol concentration and higher productivity. To improve the productivity in ABE fermentation process, the in-situ removal of the fermentation products, especially butanol as the most toxic one, from fermentation broth is essential.

In this study, pervaporation, a membrane-based process was studied for in situ separation of butanol. This technique has the greatest potential due to its high selectivity, low energy requirement and high efficiency. The primary objective of this study was to improve the performance of the Polydimethylsiloxane (PDMS) membrane for the pervaporation separation and the recovery of butanol by adding fillers into its matrix to make mixed matrix membranes (MMM). Lignin, the second most abundant natural polymer resin, and in situ synthesized Zinc-based Metal Organic Frameworks (ZIF-8) were used as fillers. Different lignin percentages and different membrane thicknesses along with the different sizes of ZIF-8 nanoparticles were used and the separation performances of these MMMs were compared to the neat PDMS. Different characteristics of lignin and ZIF-8 with their impact on the performance of the host membrane were discussed. Results showed that increasing lignin particles up to 10 wt% will improve MMM performance by 280% and 60% for flux and selectivity, respectively. This enhancement was reasonable due to the presence of hydrophobic porous lignin, which enlarged the free volume within the PDMS matrix and provided more permeation pathways for penetrants. On the other hand, by increasing the filler particle amount to 15 wt%, flux was improved by 40% but selectivity was reduced dramatically by 70%. The lignin agglomeration and low filler-polymer bonds at excessive loadings would be responsible for the water molecules (with smaller kinetic diameter than butanol) to more easily bypass the hydrophobic lignin particles, then diffuse to the next layer and make the membrane less selective. Also, by decreasing the membrane thickness, an improvement on flux was observed. In this scenario, in spite of the decreased selectivity, pervaporation separation index of the membrane improved.

Furthermore, separation performance of different ZIF-8/PDMS MMMs emphasized the effect of particle sizes and interfacial bond between particles and polymer so that decreasing the particle size up to 50% resulted in 92% and 218% improvement in the flux and selectivity, respectively.

564030 Crystallization Kinetics of Cephalexin in the Presence of Cephalexin Precursors

Crystallization kinetics of cephalexin in the presence of cephalexin precursors

Matthew A. McDonald*, Andreas S. Bommarius*, Martha A. Grover*, and Ronald W. Rousseau*

School of Chemical & Biomolecular Engineering, Georgia Institute of Technology, Atlanta, USA

Keywords: impurity incorporation, ß-lactam antibiotics, process analytical technology

Cephalexin is an important antibiotic that has seen growth across all markets over the past decade as use of other ß-lactam antibiotics, such as amoxicillin, has decreased or plateaued. The nucleation and growth kinetics are determined with online process analytical technologies (PAT). Effects of two precursors to cephalexin, 7-aminodesacetoxycephalosporanic acid (7-ADCA) and phenylglycine methyl ester (PGME), on nucleation and growth are examined. Primary nucleation of cephalexin was unaffected by both precursors. Growth of cephalexin, however, was hindered by 7-ADCA, even in small quantities (<10 mM), and to a lesser extent by PGME. Possible mechanisms of growth inhibition are discussed; a model wherein the precursors slow growth by adsorbing to the crystal surface and pinning advancing growth steps best explains the data. Impact on crystal purity is also discussed. Secondary nucleation also appeared to be affected by the precursors but not to the same extent as growth. Offline image analysis indicated that secondary nucleation is dominant over primary nucleation and that breakage of high-aspect ratio crystals may play a role in crystal breeding. The needle-like shape of cephalexin monohydrate also resulted in two interesting phenomena: (1) at low stir rates dead zones formed when heterogeneous nuclei on baffles grew into a sherbet-like unmixed slurry and (2) at high stir rates the incorporation of air into the slurry resulted in the accumulation of a stable foam that did not interact with the liquid slurry. These kinetic data, PAT tools, and process insights are discussed in the context of a continuous manufacturing scenario [1].

[1] McDonald, M. A., Marshall, G. D., Bommarius, A. S., Grover, M. A., and Rousseau, R. W. (2019) Crystallization kinetics of cephalexin monohydrate in the presence of cephalexin precursors, *Crystal Growth & Design*, SUBMITTED

573695 Towards Enhanced Understanding of Sono-Crystallization for Pharmaceuticals

Sono-crystallization involves application of ultrasound waves to enable nucleation and/or breakage of crystals. It poses an alternate technology for in-situ seed generation and particle engineering in conjunction with the crystallization. Usage of this technology has been limited due to the lack of process understanding and the absence of appropriate scale-up rules along with the appropriate process equipment. In this study, breakage kinetics is measured as a function of sonication parameters on crystal slurries in saturated solutions. Particle size distribution (PSD) and chord length distribution (CLD) are measured as a function of time in a recirculation loop and single pass configuration of the ultrasonic horn. The relationship between sonication parameters and breakage kinetics is established by regressing the breakage model parameters by fitting the experimental data using population balance equations (PBE). The presentation will elaborate on the enhanced understanding of sono-crystallization process and its scale-up based on the combination of experiments and modeling using model compounds, Carbamazepine and Ibuprofen.

All authors are employees of AbbVie and may own AbbVie stock. AbbVie sponsored and funded the study; contributed to the design; participated in the collection, analysis, and interpretation of data, and in writing, reviewing, and approval of the final publication.

572999 Technoeconomic MINLP Optimisation of M.S.M.P.R. Cascades for Continuous Crystallisation of Melitracen Hydrochloride, an Antidepressant API

ABSTRACT

Continuous crystallisation is an emerging field of research due to its potential to enhance process flexibility, efficiency and product quality consistency compared to current batch techniques.[1] Development of robust continuous crystalliser designs is imperative for continuous pharmaceutical manufacturing (CPM) campaigns, which have received significant attention from academic, industry and regulatory bodies in the past decade.[2] Crystallisation is a key unit operation in pharmaceutical manufacturing given the significant quantity of solid dosage forms in pharmaceutical products.[3] Mixed suspension, mixed product removal (MSMPR) crystallisers are a widely studied continuous design for their simple operation, low maintenance requirements, avoidance of rapid fouling typical of continuous tubular designs and ease of adaptation from existing stirred tank designs. A variety of experimental and theoretical studies have already investigated the design, simulation and optimisation of steady-state and dynamic operations of MSMPR crystallisers for different active pharmaceutical ingredients (APIs) and other pharmaceutically relevant compounds.[4] Rapid screening of candidate flowsheet configurations of MSMPR cascades for different APIs can facilitate process development towards end-to-end CPM.[5]

Mixed Integer Nonlinear Programming (MINLP) has been widely implemented in various applications to rapidly screen for optimal design configurations.[6] Mixed integer problems contain both continuous decision variables, e.g., unit operating and design parameters) as well as integer (binary or otherwise) decision variables, e.g., denoting existence of a unit operation or deciding recycle stream feed location. Optimisation of MSMPR cascades has yet to be implemented for pharmaceutical process design.

Melitracen HCl is a tricyclic antidepressant API available as single drug preparations and also in combinative therapies. The continuous crystallisation of the API in MSMPR crystallisers was recently demonstrated,[7] facilitating its process modelling and optimisation. Existing work investigates the effects of different MSMPR design and operating parameters on crystal size for a maximum of two MSMPRs, showing significant variation in process performance and crystal product attributes when varying the number of implemented crystallisers.[7] Investigating the effect of longer cascades and implementing recycle variations can further improve crystallisation yields in MSMPR design.[8] Economic considerations are equally important as meeting specific product attributes when designing crystallisation processes, especially when the effect of production scale is considered.[9]

This work implements MINLP optimisation of MSMPR cascades for continuous melitracen crystallisation to screen for cost optimal flowsheet configurations subject to constraints on crystal product attributes. The MINLP superstructure for MSMPR cascades considers varying numbers of implemented crystallisers and recycle options, including considerations of constrained recycle crystal size distributions. The steady-state process model describes the crystallisation process by simultaneous solution of population balances, crystallisation kinetics and mass balance equations. The MINLP optimisation problem is solved at different plant capacities. Minimum total cost components, optimal flowsheet configurations and crystallisation yields and crystal product average sizes and distributions are compared for different problem instances.

REFERENCES

(1) Wang, T.; Lu, H.; Wang, J.; Xiao, Y.; Zhou, Y.; Bao, Y.; Hao, H. Recent progress of continuous crystallization. *J. Ind. Eng. Chem.* **2017**, *54*, 14–29.

(2) Teoh, S.K.; Rathi, C.; Sharratt, P. Practical assessment methodology for converting fine chemicals processes from batch to continuous. *Org. Process Res. Dev.* **2015**, *20*(2), 414–431.

(3) Chen, J.; Sarma, B.; Evans, J.M.B.; Myerson, A.S. Pharmaceutical crystallization. *Cryst. Growth Des.* **2011**, *11*(4), 887–895.

(4) Zhang, D.; Xu, S.; Du, S.; Wang, J.; Gong, J. Progress of pharmaceutical continuous crystallization. *Engineering* **2017**, *3*(3), 354–364.

(5) Baxendale, I.R.; Braatz, R.D.; Hodnett, B.K.; Jensen, K.F.; Johnson, M.D.; Sharratt, P.; Sherlock, J.P.; Florence, A. Achieving continuous manufacturing: technologies and approaches for synthesis, workup, and isolation of drug substance. *J. Pharm. Sci.* **2015**, *104*, 781–791.

(6) Caballero, J.A.; Grossmann, I.E. Rigorous design of complex liquid-liquid multi-staged extractors combining mathematical programming and process simulators. *Comput. Aided Chem. Eng.* **2009**, *27*, 981–986.

(7) Capellades, G.; Joshi, P.U.; Dam-Johansen, K.; Mealy, M.J.; Christensen, T.V.; Kiil, S. Characterization of a multistage continuous MSMPR crystallization process assisted by image analysis of elongated crystals. *Cryst. Growth Des.* **2018**, *18*(11), 6455–6469.

(8) Diab, S.; Gerogiorgis, D.I. Technoeconomic optimization of continuous crystallization for three active pharmaceutical ingredients: cyclosporine, paracetamol and aliskiren. *Ind. Eng. Chem. Res.* **2018**, *57*(29), 9489–9499.

(9) Li, J.; Trout, B.; Myerson, A.S. Multistage continuous Mixed-Suspension, Mixed-Product Removal (MSMPR) crystallization with solids recycle. *Org. Process Res. Dev.* **2016**, *20*(2), 510–516.

573994 Controlled Manipulation of the Size and Shape of Needle-like Crystals in a Cyclic Process

Crystallization from solution leads to a wide range of particle sizes and shapes, which has an impact on various downstream processes. This emphasizes the fact that there is a need to develop processes and operating strategies to obtain equant shaped crystals with a narrow distribution. Experimental studies in the literature have often dealt with the manipulation of size, but not shape, due to the lack of commercially available devices capable of characterizing both the size and the shape of crystals. Recent advances in online size and shape monitoring[1], have enabled understanding fundamental phenomena and providing a means to explore complex size and shape modification processes. For instance, the cyclic 3-stage process, involving a growth, a breakage, and a dissolution stage, has been modeled and experimentally validated. Its purpose is to transform a needle-like seed population to equant shaped particles.[2,3] This process was previously operated at predefined operating conditions, and depending on the choice of these conditions and of the number of cycles, final product populations with different average aspect ratios (defined as the ratio of the average particle length to the average particle width) were obtained. However, operating a process in this fashion would not guarantee consistent product quality in terms of the final particle size and shape distribution (PSSD) due to the presence of unmodeled phenomena and process disturbances. This issue can be circumvented by integrating feedback controllers into the overall process.

The availability of a quantitative, online size and shape monitoring tool, the μ-DISCO,[1,4] and three control strategies that have been developed recently for the three stages of the process, enable operating the entire 3-stage process in a fully automated and controlled fashion. The controller for the growth stage drives seed populations to predefined target orthants under growth-dominated conditions.[5,6] The controller for the breakage stage reduces the average length of the particles – and thereby their aspect ratio – by subjecting them to breakage in a rotor-stator wet mill.[7] Finally, the controller for the dissolution stage dissolves a certain fraction of the total solid volume by altering the temperature of the process, with the goal of dissolving fines.[8] The robust performance of these controllers are successfully demonstrated in thorough experimental campaign.

These controllers are then implemented within the context of the 3-stage process to drive needle-like seed populations to different targets in the size and shape space. Experimental outcomes reveal that the process control strategy manages to repeatedly reach the same final PSSD. The experimental campaign clearly underlines the robustness of the controllers, which helped eliminating the need to specify operating conditions for individual stages. It is worth noting all the control laws operate by solely relying on the online monitoring capabilities of the μ-DISCO and on solubility data. To conclude, this study highlights the effectiveness of simple model-free feedback control strategies exploiting online monitoring to address issues related to manipulation of particle shape.

References

1. Rajagopalan, A. K.; Schneeberger, J.; Salvatori, F.; Bötschi, S.; Ochsenbein, D. R.; Oswald, M. R.; Pollefeys, M.; Mazzotti, M. A Comprehensive Shape Analysis Pipeline for Stereoscopic Measurements of Particulate Populations in Suspension. *Powder Technol.* **2017**, *321*, 479–493.
2. Salvatori, F.; Mazzotti, M. Manipulation of Particle Morphology by Crystallization, Milling, and Heating Cycles - A Mathematical Modeling Approach. *Eng. Chem. Res.* **2017**, *56* (32), 9188–9201.
3. Salvatori, F.; Mazzotti, M. Manipulation of Particle Morphology by Crystallization, Milling, and Heating Cycles: Experimental Characterization. *Eng. Chem. Res.* **2018**, *57* (45), 15522–15533.
4. Bötschi, S.; Rajagopalan, A. K.; Morari, M.; Mazzotti, M. An Alternative Approach to Estimate Solute Concentration: Exploiting the Information Embedded in the Solid Phase. *Phys. Chem. Lett.* **2018**, *9* (15), 4210–4214.
5. Bötschi, S.; Rajagopalan, A. K.; Morari, M.; Mazzotti, M. Feedback Control for the Size and Shape Evolution of Needle-like Crystals in Suspension. I. Concepts and Simulation Studies. *Growth Des.*

2018, *18* (8), 4470–4483.

6. Rajagopalan, A. K.; Bötschi, S.; Morari, M.; Mazzotti, M. Feedback Control for the Size and Shape Evolution of Needle-like Crystals in Suspension. II. Cooling Crystallization Experiments. *Growth Des.* **2018**, *18* (10), 6185–6196.
7. Rajagopalan, A. K.; Bötschi, S.; Morari, M.; Mazzotti, M. Feedback Control for the Size and Shape Evolution of Needle-like Crystals in Suspension. III. Wet Milling. *Growth Des.* **2019**.
8. Bötschi, S.; Rajagopalan, A. K.; Morari, M.; Mazzotti, M. Feedback Control for the Size and Shape Evolution of Needle-like Crystals in Suspension. IV. Modeling and Control of Dissolution. *submitted for publication.*

573909 Digital Design and Operation of Continuous Crystallization Processes Via Mechanistic Modelling

Digital design and operation of continuous crystallization processes via mechanistic modelling

Niall Mitchell[1], John Mack[2], Furqan Tahir[2], Cameron Brown[3], Tariq Islam[3], John Robertson[3], Eduardo Lopez-Montero[2], Alastair Florence[3]

[1] *Process Systems Enterprise (PSE) Ltd. London, UK*

[2] *Perceptive Engineering Ltd, Daresbury, UK*

[3] *CMAC, University of Strathclyde, Glasgow*

Mechanistic models are becoming more commonly applied for Research and Development in the pharmaceutical sector to gain process understanding, enable process design and operation. Traditionally, the output from this activity is a validated mechanistic model, which is capable of quantitatively predicting the behaviour of the various Critical Quality Attributes (CQA) for typical batch or continuous pharmaceutical processes for a wide range of Critical Process Parameters (CPP). However, these tools are almost exclusively employed in an offline manner currently to enable digital design efforts, primarily aimed at assessing process robustness and variability, with very little subsequent online application of the mechanistic model to enable control or soft sensing.

Model Predictive Control (MPC) is an established technology in the process industries. It uses a statistical model of the process to capture the dynamic relationships between the inputs (CPPs) and outputs (CQAs) of the process. Using this statistical model, it predicts impact of known disturbances on operation and controls the process through co-ordinated moves on multiple inputs. The MPC exploits all opportunities to reduce variability in the CQAs whilst compensating for measured and unmeasured disturbances.

The statisitical process model is built from process response test data at production scale using techniques such as Pseudo-Random Binary Sequence (PRBS) or step-tests. The PRBS test is a relatively non-invasive technique compared to traditional step-tests as it allows the product to remain within specification whilst generating statistically rich information for modelling. Although PRBS testing is suitable for many industries it cannot be used in the Pharmaceutical sector as product generated during testing cannot be utilised for clinical or commercial supply. Consequently, the cost to generate the statistical control model would be significant, presenting a barrier to uptake of the technology. This drawback can be overcome by the integration of mechanistic models, developed using laboratory scale data with MPC system, such as Perceptive Engineering's PharmaMV platform via a digital design approach.

In this work we outline, the application of an advanced process modelling tool, namely gPROMS FormulatedProducts, to describe a number of pharmaceutical crystallization processes. The mechanistic process model and the mechanistic model kinetic parameters were validated using process data gathered from the literature and from lab-based experiments. The lab-based mechanistic model was subsequently used to predict the behaviour of the full scale production scale.

The validated mechanistic model was subsequently integrated with PharmaMV to develop and tune the MPC against the mechanistic simulation of the process, by using the mechanistic model as a Digital Twin or Virtual Plant as follows:

1. gPROMS: Build mechanistic model
2. gPROMS: Small scale parameterisation experiments & mechanistic model validation
3. gPROMS + PharmaMV: Validate/check mech model against full scale data
4. gPROMS + PharmaMV: Build MPC using mechanistic model as a digital twin
5. PharmaMV: Transfer MPC to live process and test

With this approach, the MPC derived from the mechanistic model was utilized to accurately control the defined CQAs, such as final particle attributes (PSD, yield) for continuous crystallization processes, with reduced material wastage at the production scale

573335 Quality By Control (QbC) Guided Quality By Design (QbD) for the Optimal Operation of Integrated Crystallizer-Wet Mill Systems

The simultaneous control of crystal shape and size is particularly important in fine chemical and, especially, pharmaceutical crystallization. These two quantities not only influence significantly the dissolution rate and bioavailability of final drug products, but also contribute to the manufacturability and efficiency of downstream operations. The manipulation of crystal shape, however, is difficult since it requires the decoupled growth rate control of individual crystal faces. The supersaturation and solvent system dependency of these rates are often not strong enough to enable impactful crystal shape control through temperature and/or solvent composition variation. Promising results have been achieved by using tailored growth rate modifier additives in the case of model systems, however, this does not provide a general framework for controlling the crystal shape.[1]

If the crystal size or shape cannot be achieved in the crystallizer - because the target is out of the achievable domain of the crystallization process – downstream correcting steps are involved. The most often used processes are granulation and/or milling. Recently, the wet-mill process was integrated with the crystallizer in different ways to increase the achievable crystal size domain. It was observed that the integrated operation leads to significant improvements in the produced crystal size.[2] However, similar studies have not been carried out yet for crystal shape control problems.

The aim of this work is to analyze and evaluate the possibilities of controlling the bivariate size distribution (which by definition involves crystal size and shape information) of high-aspect ratio crystals in an integrated crystallizer-external wet mill system. It is considered that in the crystallizer primary and secondary nucleation, crystal growth and dissolution along the length and width axes occur, whereas in the wet mill – in addition to the aforementioned mechanisms - fragmentation and attrition is also considered. The advanced hydrodynamic description enables the direct implication of stirrer revolution speed in the model. The generated system of hyperbolic partial differential and ordinary differential equations is solved by the fully discretized high resolution finite volume method, involving GPU acceleration, which was showed to bring significant computational time reduction for bivariate population balance equations.[3] The dynamic optimization of wet-mill rotation speed, crystallizer temperature and recirculation flowrate revealed that the target bivariate crystal size distribution can be much better achieved with the integrated system than manipulating only the crystallization conditions.

The optimum operating profiles obtained with respect to various constraints, objectives and initial conditions show clear similarities, which suggests that the general operating pattern is system independent. After this discovery enabled by the modelling and process optimization, or, in broader concept, by the QbC approach, the methods of experimental design, or, on broader concept, the QbD, can be applied to determine the exact values of heating/cooling rates as well as the stirring rate of the internal/external wet mill for a particular crystallization system. The experimental validation of the method is in progress.

References

1. Klapwijk, A. R., Simone, E., Nagy, Z. K. & Wilson, C. C., Cryst. Growth Des. 16, 4349–4359 (2016).
2. Yang, Y., Song, L., Gao, T. & Nagy, Z. K., Cryst. Growth Des. 15, 5879–5885 (2015).
3. Szilágyi, B. & Nagy, Z. K., Comput. Chem. Eng. 91, 167–181 (2016).

572330 Model Based Optimization of Cooling Crystallization of Active Pharmaceutical Ingredients Undergoing Thermal Degradation

Crystallization is one of the most important separation processes in the pharmaceutical, food and chemical industries. In the pharmaceutical industry, it serves as both a separation unit operation to produce high-purity crystalline products and as a particulate design process. Cooling crystallization is the most widely used crystallization strategy to purify a new drug molecule under development. However, many drugs degrade at high process temperatures which leads to reduced product yield. Reduction of the initial process temperature decreases the rate of degradation, but simultaneously decreases the process yield. In this work a model-based optimization problem is formulated for unseeded cooling crystallization of Aspirin, an industrially relevant API. Two different optimization objectives are considered in this study viz. maximization of product crystal size and minimization of degradation product concentration, which give widely different optimal cooling profiles. Investigation of the fundamental mechanisms viz. particulate nucleation and growth reveals major differences between the two optimal scenarios. Experiments are also carried out in the laboratory using online Process Analytical Technology tools viz. Focused Beam Reflectance Measurement, PVM and online Ultra-High Performance Liquid Chromatography to validate the model simulations. It is found that there is substantial plant-model mismatch for one of the optimal scenarios, which necessitated model reformulation to include the adsorption of degradation impurity onto the product crystals.

572227 Resolution By Crystallization: The Quest for Chiral Purity

Pharmaceutically active compounds and their related intermediates are often produced as a mixture of stereoisomers, both as enantiomers and as diastereomers from their achiral raw materials. While stereoisomers of pharmaceutically active compounds bear the same chemical structures, many compounds of altered chirality possess differentiated biological activities. In many cases the desired bioactive stereoisomer (Eutomer) is produced as a mixture of stereoisomers of varied weight %. The concept of stereoisomer resolution is therefore of great interest in the process development and manufacturing of chiral pharmaceutical products.

The present work demonstrates the application of ternary phase diagrams for the early phase process definition and for advanced process understanding of stereoselective crystallization operations. Two case studies will be presented focusing on the initial crystallization design work flow, methodology, theory and engineering details associated with journey towards the resolution of stereoisomers.

572129 Reactive Agglomeration: Improving Processability of Amorphous Solids in Precipitation Process

Production of amorphous solids using fast solvent evaporation methods or anti-solvent precipitation pose significant challenges in terms of processability, thermal stability and yield. Fast solvent evaporation using a spray dryer or a rota-vapour are limited by scale and require volatile solvents with adequate solubility. Alternatively, amorphous solids are generated through anti-solvent precipitation, wherein high supersaturation leads to formation of fine particles and create operational difficulties for filtration as large scales. Researchers have explored spherical agglomeration process to enlarge the size of precipitating solids; however it has limited choice of bridging solvents and critical process susceptibility to multiple variables such as wettability and residual solvent content of the solids etc. These challenges necessitate development of a large-scale feasible process for generation of amorphous solids.

We present here an innovative approach of coupling "anti-solvent precipitation with reactive crystallization" for generation of amorphous solids. In the present study, water-insoluble drug was dissolved in a weak acid (water-miscible) followed by gradual dropwise addition to a weak aqueous base solution. Each droplet, act as a template, wherein (I) diffusion/mixing of acid to aqueous base solution, (ii) reaction of acid and base and (iii) precipitation of the drug occurs instantaneously within the limiting boundary of receding droplet. These simultaneous events enable high-localized supersaturation and uneven precipitation throughout the droplet template. Diffusing solvent from the droplet template acts as solution channels linking the precipitated solids in the close vicinity. Drug precipitation in those channels forms bridges among the solids resulting in larger agglomerates of irregular shapes. These larger agglomerates formed at high-localized supersaturation are amorphous in nature and easy to filter. Use of weak acid and base decrease the probability of forming salts.

As a case study, a poorly water-soluble drug was selected for which generation of an amorphous form is desirable as it improves the bioavailability. Preliminary solubility screening was performed to identify suitable acid and its corresponding base. Drug was dissolved in the acid and thereafter added dropwise in an aqueous solution of weak base. Base quantity was stiochiometrically determined and a higher molar equivalent was used to ensure complete acid-base reaction. PATs (Process Analytical Technologies) such as FBRM (Focused Beam Reflectance Measurement) and PVM (particle vision microscopy) enabled real-time tracking of particle evolution and their size distribution. In comparison with reverse anti-solvent addition and spherical agglomeration, larger aggregates and lesser fines were observed by reactive agglomeration. The obtained solids were also characterized using several off-line analytical tools such as SEM (scanning electronic microscope), PXRD (powder x-ray diffraction) and MDSC (modulated-differential scanning calorimetry) which indicate the formation of large, irregular shaped aggregates with amorphous nature. The filtration rate of the agglomerates were measured using a standard filter apparatus and resulting filtration profiles showed less cake resistance and faster filtration rate. The feasible process thus developed was further optimized and scaled up to manufacturing scale.

559074 Surprisingly Complex Phase Behavior of 3,5,5-Trimethylpyrrolidin-2-One Affecting Its Chiral Puirty Enrichment through Crystallization

The separation of enantiomers is of great interest to the pharmaceutical industry. One of the tools utilized for the production of a pure chiral active phamaceutical ingredient (API) is enrichment of enantiomers by crystallization. Resolution of a chiral compound strongly depends on the kind of solid state crystal lattcie that it forms. Three common types of racemates include racemic compound, conglomerate and solid solution.

In this work, thorough solid state characterizations of (S)-3,5,5-trimethylpyrrolidin-2-one is presented. While macro-scale composition of the solution phase and the solid phase suggested a simple racemic solid solution, the detailed crystal structure of the solid phase proved to be much more complex. Multiple solid phases exist; however, the system still seems to be on a solid solution continuum.

Based on the phase diagram, the yield and chiral purity trade-off was established to guide the optimization of crystallization process. It is also found the yield and chiral purity relationship at equilibrium remains the same in different solvent systems.

572416 Mechanistic Investigation of Spherical Crystallization of Active Pharmaceutical Ingredients Using in-Situ PAT and Mathematical Models

With increasing R&D costs for drug companies, the focus on Pharmaceutical Manufacturing (PM) cost has gained centerfold. Process Intensification (PI) is one of the most widely used strategies to reduce the cost of PM. Spherical Crystallization is an innovative PI strategy which leads to the formation of spherical crystalline agglomerates, thereby enabling control over manufacturability and bioavailability in a single unit operation. Spherical Crystallization also enables formulation of drug-excipient co-agglomerates, which can potentially lead to direct compounding of drug products leading to complete elimination of Drug Product Manufacturing. However, controlling the size of the primary particles that form the agglomerates, along with the agglomerate size remains a challenging task. In this work, mechanistic mathematical models are first constructed for the SC process which gives the time course evolution of the droplet population and crystal population within the droplets. It is found that there are major differences in primary crystal size distribution for two different agglomerate sizes, which ultimately would lead to differences in bioavailability of the product agglomerates. Experiments are also carried out in the laboratory using Process Analytical Technology tools (FBRM) to find the evolution of the droplet population within the crystallizer. It is found that from the dynamical evolution of the FBRM data, the fundamental rate processes of crystal nucleation and growth within individual droplets could be deduced. This work would ultimately provide a framework for implementation of Quality By Design for PI in PM involving multi-scale rate processes.

575868 Continuous Crystallization of Carbamazepine in Oscillatory Baffled Crystallizer

Continuous crystallization is a hot topic in continuous pharmaceutical manufacturing. In general, there are two major approaches of the continuous crystallization processes: mixed suspension mixed product removal (MMSPR) and plug-flow (PF) crystallizations. Continuous oscillatory baffled crystallizer (COBC) is a relatively new type of PF crystallizer. Recently, kinetic[1,2] and dynamic[3] studies of COBC, as well as in-situ monitoring and characterization of PF crystallizers[4] have been well demonstrated. Here, from the perspective of process robustness and product quality, we will share our experience on developing a continuous crystallization process of carbamazepine in ethanol. Process temperature, oscillation amplitude, frequency, and flow rates were studied. Offline characterization and in-line process analytical technology (PAT) tools were applied to analyze the CBZ particle shape, size, polymorph, and yield.

*This presentation only reflects the views of the authors and should not be construed to represent FDA's views or policies

References

1. Brown CJ, Ni X-W. Evaluation of Growth Kinetics of Antisolvent Crystallization of Paracetamol in an Oscillatory Baffled Crystallizer Utilizing Video Imaging. Cryst. Growth Des. American Chemical Society; 2011; 11:3994–4000. Available from: https://pubs.acs.org/doi/10.1021/cg200560b
2. Onyemelukwe II, Benyahia B, Reis NM, Nagy ZK, Rielly CD. The heat transfer characteristics of a mesoscale continuous oscillatory flow crystalliser with smooth periodic constrictions. Int. J. Heat Mass Transf. 2018; 123:1109–19. Available from: https://linkinghub.elsevier.com/retrieve/pii/S0017931017351177
3. Ni X, Jian H, Fitch A. Computational fluid dynamic modelling of flow patterns in an oscillatory baffled column. Chem. Eng. Sci. Pergamon; 2002; 57:2849–62. Available from: https://www.sciencedirect.com/science/article/pii/S0009250902000817
4. Ferguson S, Morris G, Hao H, Barrett M, Glennon B. In-situ monitoring and characterization of plug flow crystallizers. Chem. Eng. Sci. 2012; 77:105–11. Available from: https://linkinghub.elsevier.com/retrieve/pii/S0009250912000991

571648 Kinetic Assessment for Continuous Crystallization Process Development of an Early Development Compound

Acquisition of reliable kinetic correlations that describe how rates of nucleation and growth vary over the crystallization design space is necessary for enabling robust prediction of crystallizer performance. The aim of this study was to generate a standardised workflow approach for the generation of crystallization kinetics data using population balance modelling (PBM), and to assess the batch crystallization process for feasibility of continuous processing for an early development compound. Initially, the crystallization growth kinetics were estimated by means of seeded isothermal desupersaturation experiments in a batch reactor. PAT tools such as Attenuated Total Reflectance-Fourier Infrared (ATR-FTIR) spectroscopy and Focused Beam Reflectance Measurement (FBRM) were utilised for *in situ* measurement of solute concentration and for detection of occurrence of nucleation events, respectively. The initial crystal size distributions (CSDs) and desupersaturation profiles were used to estimate the growth kinetics as a function of temperature and supersaturation. Two solvent systems were tested to extract growth kinetics and were used for comparison. Activity coefficients as a function of temperature and composition were considered in the kinetic model, which enabled to directly evaluate the solvent dependency effect. Next, a continuously operated single-stage mixed suspension mixed product removal (MSMPR) crystallizer was used to extract the secondary nucleation kinetics. During the continuous crystallization study, a recycling system was employed to significantly reduce the waste of raw materials during the MSMPR study prior to the onset of steady state, which can be very useful during the early stages of process development where limited amount of product is available. Furthermore, problems such as blocking of transfer line and feed inlet were eliminated, by applying a rapid intermittent withdrawal of slurry via dipped pipe[1], and by using a tube-in-tube configuration to insulate the feed inlet line. Finally, a novel 3D printed jacketed nozzle was attached at the end of the feed inlet line to avoid the saturated solution from crystallizing which can lead to blockage. Based on PBM and utilisation of the kinetic data[2], CSD was simulated and residence time distribution, operating temperature and productivity of the MSMPR were optimised and finally experimentally verified. Overall, a standardised workflow was developed for the kinetic assessment of an early development compound for continuous crystallization.

1. Hou, G., Power, G., Barrett, M., Glennon, B., Morris, G. & Zhao, Y. (2014) 'Development and Characterization of a Single Stage Mixed-Suspension, Mixed-Product-Removal Crystallization Process with a Novel Transfer Unit', Crystal Growth & Design, 14(4), pp. 1782-1793.
2. Power, G., Hou, G., Kamaraju, V. K., Morris, G., Zhao, Y. and Glennon, B. (2015) 'Design and optimization of a multistage continuous cooling mixed suspension, mixed product removal crystallizer', Chemical Engineering Science, 133, pp. 125-139.

564084 Professor Linda Wang's Research Group – the Early Years

As Prof. Wang's third Ph.D. graduate, I was involved with early establishment of lab and modeling capabilities. Most of my experimental work was determining isotherms and mass transfer by chromatographic breakthrough for amino acids, peptides, and proteins on ion exchange resins. As computing power was quite limited in the late 1980s, modeling of these complicated chromatographic processes required many simplifying assumptions. However, Prof. Wang had a vision of being able to simulate these systems to help design more efficient purifications for the biochemical and pharmaceutical industries. I took leadership for development of the new process simulator. I will review the development history of the simulator, which we named VERSE. The program, which is still in use a full 30 years later, gained both fundamental features (isotherms, kinetic models, surface diffusion, reactions) as well as process capabilities (breakthrough, gradient elution, Simulated Moving Bed). Once I was at Air Products, we collaborated on non-biochemical process development. VERSE supported our design of counter-current moving bed processes for efficient lithium ion exchange on zeolites. Along with the technical story, I'll share some photos and memories about the group.

559947 Simulated Moving Bed- Learning from the Past and Shaping the Future

Simulated Moving Bed (SMB) technology developed by UOP inventors Broughton, Gerhold (1) and Carson (2) under the name of Sorbex processes is used in many sectors from the petrochemical industry (PAREX process for separating p-xylene from C8 isomers), sugar industry (SAREX process for production of HFCS) to more recent applications in the pharmaceutical industry for chiral separations (3).

I first learned about SMB during my PhD studies in Nancy back in early seventies and I found the concept of simulating the solid movement by shifting the inlet/outlet ports periodically very bright. When going from "old" applications to "new" applications one can see differences on number of columns used and particle size but also similarities in terms of aspect ratio L/D around 10 (4). These new applications include the production of drugs as Zyrtec, Prozac, insulin, etc. New operating modes have emerged and we should mention VARICOL (5) where asynchronous shits are introduced which allows reducing the number of columns in many pharma applications from 6 to 5. These non-conventional methods are easily implemented in the FlexSMB unit built in our lab and shown in Figure 1b)

Figure 1 SMB Licosep 12x26 from Novasep (a) and FlexSMB (b)

Design methods for SMB started with triangle theory (6), expanding to separation volume (7) and standing wave theory (8). In the classical application of p-xylene separation the nature of desorbent changed along the years from heavy desorbent as para diethylbenzene (PDEB) to low affinity desorbent as toluene (LD Parex process). Detailed modeling of industrial Parex unit has been presented elsewhere (9).

An interesting problem in chiral separations is the separation of nadolol isomers since there are two pairs of racemates and various strategies can be envisaged to address that problem (10). All these applications are related to liquid-solid systems. Applications for vapor phase and gas phase systems are scarce. We have been working at LSRE-LCM on the separations of light olefins/paraffins (one of the "seven separations to change the world") using different adsorbents (binderless 13X zeolie, ZIF-8, MOFs) (11).

In our lab we are involved in Process Intensification coupling reaction and adsorptive separation by using Simulated Moving Bed Reactors (SMBR) and even coupling SMBR with membrane permeation in PermSMBR (12). These technologies have been tested in the production of acetals (DEE, DME, DBE, DEB, GEA), esters (ethyl lactate, butylacrylate (13)).Here we address the use of SMBR for p-xylene production.

The current production of *p*-xylene is based on a separation-isomerization loop with limited yield due to the thermodynamic equilibrium, which results in large cycle loops to achieve the desired amount of p-xylene. A large loop along with gas phase conditions within the isomerization unit increases significantly the energy consumption within the process. The energy, together with raw materials, required in the process can be significantly reduced by the use of multifunctional reactors. By combining reaction and separation in the same unit, the thermodynamic equilibrium can be shifted towards the target product; in this case, in-situ removal of *p*-xylene shifts the isomerization of xylene towards the *para*-isomer. The most employed process for the separation of *p*-xylene is based on the SMB technology. The proposed SMBR uses the same principle as the SMB and incorporates the reaction section, the catalyst, within the adsorption columns. In order to combine both processes, isomerization of xylenes must be conducted in liquid phase. Although the conversion may be lower, it brings other advantages such as better thermal control and longer catalyst life, which allows for off-site catalyst regeneration and therefore easier control of pollution. The presence of the catalyst throughout the columns contaminates the *p*-xylene in the extract port due to the reverse reaction. Therefore, the SMBR must be combined with a crystallization unit to obtain pure *p*-xylene. Moreover, since *p*-diethylbenzene (the most employed desorbent in the SMB) isomerizes in the presence of acid catalysts, other aromatic compounds (e.g., toluene) must be considered as desorbent for the SMBR.

The successful implementation of any reactive-adsorptive process depends on the determination of the fundamental reaction and adsorption data governing those processes. Even though adsorption of xylenes is well-known at industrial conditions (about 180 °C), studies at higher temperatures are not found in the literature. Since the optimum temperature for the SMBR is expected to be above 200 °C, batch experiments above 180 °C were carried out. The adsorbent was faujasite-type zeolite exchanged with Ba, similar to what is used in industry. Since catalysts and adsorbents are mixed throughout the unit, the isotherms were determined in the absence of water to prevent the catalyst from going through dealumination, which irreversibly affects its acid sites. *p*-Xylene was still the more adsorbed compound although the selectivity was lower and decreased at higher temperatures.

The industrial isomerization of xylenes is carried out in gas phase over zeolite ZSM-5 with a noble metal such as Pt. In liquid phase, diffusion constraints may reduce significantly the conversion of xylene isomers over ZSM-5; therefore, zeolites with larger pores must be studied. Fixed-bed catalytic experiments were conducted over zeolites ZSM-5, Beta, and Mordenite with different acidity also at above 180 °C. Acid catalysts in pellet form are limited to few zeolites with specific silica-to-alumina ratio, which defines the acidity of the solid. To overcome that, several zeolites in powder form were shaped to obtain appropriate pellets for fixed-bed studies through an extrusion-spheronization process. Large-pore zeolites Beta and Mordenite exhibited higher conversion and longer catalyst life than ZSM-5. Furthermore, side reactions were analyzed under the conditions studied in the presence of toluene. Contrary to what is found in the literature, transalkylation of toluene with ethylbenzene were observed in addition to disproportionation of ethylbenzene. Disproportionation of toluene was not detected confirming that that reaction goes through a different mechanism.

The literature reports several SMBR optimization studies. However, the majority of these investigations focus on the SMBR exclusively, without considering its integration with the subsequent desorbent recovery units. Thus, a short economical evaluation of the SMBR with the associated distillation columns was conducted. Since all units are connected, they shall be optimized all together. The objective function to be maximized was the price of the product minus the operating costs. The distillation columns were modeled as simple as possible (i.e., shortcut methods) but validated against commercial software to determine the heat duty. Since the SMBR consists of 24 columns (the same as the existing SMB), the unit was modeled with the true moving bed reactor approach. The variables to be manipulated during the optimization involved the arrangement of columns and flow rates in each zone of the SMBR, switching time, temperature, desorbent consumption, catalyst-to-adsorbent ratio, among others. The constraints included the minimum purity in the extract and maximum pressure drop per bed. The SMBR offering the best performance consisted of 4, 3, 15, and 2 columns in zones 1, 2, 3, and 4 respectively, operating at about 280 °C, and resulting in a reduction of the cycle loop in about 30 %.

(1) DB Broughton and C Gerhold, Continuous sorption process employing fixed bed of sorbent and moving inlets and outlets, US Patent 2985589 (1961)

(2) DB Carson and FV Purse, Rotary valve US Patent 3040777 (1962)

(3) RM Nicoud, The separation of optical isomers by simulated moving bed chromatography, Pharm.

Technol. Eur 11, 36-44 (1999)

(4) P Sá Gomes, M Minceva and AE RodriguesSimulated moving bed technology:old and new, Adsorption, 12, 375-392(2006)

(5) P Adam et al, Process and device for separation with variable length, US Patent 6136198 (2000)

(6) Storti et al, Optimal design of multicomponent countercurrent adsorption separation processesinvolving nonlinear equilibria,, Chem Eng Sci, 44, 1329-1345 (1989)

(7) D Azevedo and AE Rodrigues, Design of simulated moving bed in presence of mass transfer resistances, AIChEJ, 45, 956-966 (1999)

(8) Z Ma and NHL Wang, Standing wave analysis of SMB chromatography: linear systems, AIChEJ, 43, 2488-2508 (1997)

(9) MSP Silva et al, Modeling and simulation of an industrial scale Parex process, AIChEJ, 61, 1345-1363 (2015)

(10) R. Arafahet al, Separation of Nadolol Stereoisomers Using Chiralpak IA Chiral Stationary Phase, Chirality 28 (2016) 399-408.

(11) Vanessa F. D. Martins, et al, " Development of gas-phase SMB technology for light olefin/paraffin separations**",** *AIChEJ* **62**(7) 2490-2500 (2016).

(12) Carla Pereira, Viviana Silva and A.E. Rodrigues, "Coupled PermSMBR- Process design and development for 1,1 dibutoxyethane production, *ChERD*, **92**, 2017-2026 (2014)

(13) Dânia S Constantino, et al "Simulated Moving Bed Reactor for Butyl Acrylate Synthesis: from Pilot to Industrial Scale", *Chem Eng and Processing* **97**, 153-168 (2015)

555582 Potential Roles of a Simulate-Moving-Bed Chromatography As a Final-Step Purification Process for the Production of High-Valued Products from Marine Algae Biomasses

There have been a continued interest in utilizing land or marine biomass for the production of biofuel ingredients or high-valued chemical feedstock components. Between the two biomasses with different sources, the use of marine biomass has recently been gaining more attention because land biomass could compete with agricultural crops for land or fresh water, and also have an undesirable effect on environmental protection due to the excessive exploitation of wood resources. To maximize the industrial applicability and usefulness of marine biomass, it is essential to establish highly efficient downstream processes for recovering target components with high purities from the marine-biomass hydrolysate. The most important step in such downstream processes is a final-stage purification step, which is more true particularly if the usage of marine biomass is oriented toward the production of high-valued pharmaceuticals, cosmetics, human milks, and prebiotics. In this presentation, it will be shown that a simulated-moving-bed (SMB) chromatography will be highly suitable as a final-stage purification process for facilitating the industrial utilization of marine biomass. This will be exemplified by the recent research results about the production of fucose and other high-valued products from microalgae and macroalgaes, whose industrial applications have been in progress or under consideration.

574624 Design of Bypass-Simulated Moving Bed Process for Reduced Purity Requirements

Simulated moving bed separations are typically studied for cases where high target product purities are required. However, there are situations where reduced purities could be desired: e.g., hybrid separations and products where lower purities are tolerated. In such situations, the bypass SMB- a process configuration where the SMB over-purifies a portion of the feed which is then mixed with the feed to obtain the target purity can be considered. In fact, the manufacture of high-fructose corn syrup uses this approach.

In this presentation the design of BP-SMB is considered. For systems that follow linear isotherms, based on the equilibrium theory of chromatography, explicit expressions to compare the productivity of the BP-SMB with respect to the classical SMB and the desorbent requirements are derived. Numerical simulations are used to confirm these results. For systems following non-linear isotherms, numerical simulations are used to study the potential advantage of the technique. It is shown that the BP-SMB is preferred for systems where the selectivity is high, column efficiencies are high. The manufacture of HFCS is considered and it is shown that in this separation, while there are no productivity gains to be obtained using the BP-SMB, there are operational advantages that improve the robustness of the process.

573781 The Rigid Adsorbent Lattice Fluid Model Applied to the Correlation of Liquid Phase Partition Coefficients of Different Solvents in MFI Zeolites

In this contribution we extend the use of the Rigid Adsorbent Lattice Fluid (RALF) model to the correlation of liquid phase partition coefficients. This is achieved through the formulation of the problem as a multicomponent equilibrium calculation where both the solid and the solvent are explicitly taken into account. The liquid phase fugacity can be calculated either from the Sanchez-Lacombe equation of state or directly from the knowledge of the infinite dilution activity coefficients of the solutes and the vapour pressure of the solvent.

The methodology is applied to the very interesting behaviour of liquid phase adsorption in MFI zeolites. The homogeneous RALF model is shown to reproduce the observed trends for smaller solvent molecules (methanol and n-hexane). For longer chain hydrocarbons this breaks down due to the difference in behaviour between adsorption in the straight and sinusoidal channels and channel intersections of MFI zeolites. A modified RALF model that takes into account the different adsorption sites is parametrised using molecular simulations.

577544 Improving Indoor Air Quality: Determining Adsorption Properties of VOC's Using Concentration Pulse Chromatography

Sick Building Syndrome (SBS) is caused by indoor air pollution and is a major issue for people who spend most of their time indoors. This condition is frequently attributed to chemical contaminants such as formaldehyde, toluene, xylene, benzene, ammonia, and trichloroethylene which accumulate indoors and are of particular concern in places with inadequate ventilation. In general, these components are not present in acute exposure concentrations but over prolonged exposure, these components can have chronic health effects such as allergic sensitization, increased cancer risks, and respiratory disease. Adsorption has been a strategy to reduce the amount of volatile organic compounds using commercially available materials, but there is yet to be a cyclical process in place to continuously remove these contaminants from indoor settings.

This presentation will summarize the investigation into finding a promising adsorbent among silicas, activated aluminas, zeolites, activated carbons, and polymers for removal of formaldehyde, toluene, xylene, benzene, ammonia, and trichloroethylene from air. Concentration pulse chromatographic technique has been used to determine Henry's law constants and heat of adsorption values of these compounds with above-mentioned adsorbents. This study is important for future work into gas masks and indoor filters to purify air from ammonia and volatile organic compounds to alleviate people who suffer from SBS.

572918 Upgrading of Low-Value Carbon Precursors to Porous Carbon Materials for Greenhouse Gas Capture

Global warming and the related climate change issues, attributed to the increase of greenhouse gas emissions, have attracted significant attention worldwide. To reduce the greenhouse gas emissions from the energy sector, many studies have been conducted on the development of CO_2 capture and storage (CCS) technology. Adsorption using solid materials is promising for CO_2 capture as this approach exhibits advantages of low energy requirement for regeneration, scale-up feasibility, and mild operating conditions. Among various adsorbents, activated carbon is attractive because of advantages such as cost-effectiveness, chemical and mechanical stabilities, and tunable pore structures. Activated carbon can be obtained from several carbon-containing precursors. In this study, low-value carbon precursors were used to produce high value-added porous carbon for capturing greenhouse gases including CO_2. From the experimental results, the relation between porosities and greenhouse gas adsorption uptakes were investigated and the pore size that is crucial for greenhouse gas capture was determined. The prepared porous carbons not only exhibited a high greenhouse gas uptake, but also good selectivity, simple regeneration, excellent cyclic stability, and rapid adsorption-desorption kinetics.

581347 Chemical Engineering Innovations for a Renewable Economy

Chemical engineering evolved in the twentieth century with rise in the use of fossil resources and myriad of products derived from then. An array of these products required great innovations and made huge impact on the human race, and such examples include urea, transportation fuels, polymers, water purification chemicals, etc. Going forward as we make transition from fossil resources to renewable resources such as solar energy, wind and sustainably available biomass, we again have great opportunities for innovation that would not only make great impact but redefine chemical engineering. My talk will focus on my own experiences with challenges and opportunities presented to us as chemical engineers. I will share an overall perspective on the use of biomass as sustainable source of carbon, harvesting solar energy in conjunction with growing food without any land constraint, and creating solution processed solar cells that utilize chemical engineering principles along with knowledge from other engineering disciplines.

581350 Engineering Materials for Clinical Application

Dr. Schmidt's research has resulted in numerous patents that are licensed to companies, including AxoGen, Alafair Biosciences, Smith and Nephew, and Siluria Technologies. Comments for this panel will focus on peripheral nerve applications and one key success that is the foundation for the Avance Nerve Graft from AxoGen. In addition, along the way to uncovering strategies for nerve regeneration, Dr. Schmidt and her team also discovered a biomaterials processing approach that led to a separate application of protecting tendon's after surgery, resulting in the VersaWrap Tendon Protector from Alafair Biosciences, a start-up company in Austin, Texas.

Nerve Repair: Dr. Schmidt's technology for processing decellularized nerve tissue was licensed to AxoGen in Alachua, Florida (http://www.axogeninc.com) for the AVANCE™ nerve graft. Many tens of thousands (~100,000) of AxoGen's AVANCE™ grafts have been implanted in patients at hundreds of hospitals globally.

Post-surgical Wound Management: Using a novel in situ crystallization processing technique, Dr. Schmidt's team created mechanically robust hyaluronic acid-based hydrogel films that can be easily handled, sutured, and delivered laparoscopically. This research is the basis of Alafair Biosciences, a start-up company in Austin, TX (http://www.alafairbiosciences.com), which is run by Dr. Schmidt's PhD graduate and a neurosurgeon collaborator. Alafair's first product, VersaWrap™ Tendon Protector, has been implanted in more than 500 patients since December 2017.

563695 Bioinspired Pervaporation Membranes

Inspired by the exceptional composition, structure, formation process and functions of natural materials, a variety of bioinspired pervaporation membranes have been designed and fabricated using cell membrane, mussel, nacre as representative prototypes and biomineralization, bioadhesion, self-assembly as major methods. These membranes have exhibited fascinating properties and outstanding performances in alcohol dehydratin, gasoline desulfurization. Bioinspiration, as the versatile platforms for sustainable development of membrane technology, have great implications in designing membrane materials and intensifying membrane processes. This presentation will give a brief overview about the preparations, applications as well as perspectives of bioinspired pervaporation membranes.

569080 Membrane Protein Nanosheet Based Membranes

Biological membranes containing membrane protein (MP) channels and MP-mimic functionalized materials have emerged as significant platforms to develop separation membranes with specific solute selectivity. Biomimetic membranes incorporating these pore structures are expected to exhibit high permeability and selectivity since they possess channels with well-defined pore geometry and functionality to exclude or pass specific solution components. Nevertheless, MP-based biomimetic membranes studied so far suffer from low protein insertion into biomimetic matrix and the use of vesicular morphologies of channel-incorporated liposomes. These challenges have resulted in much lower than anticipated improvements.

Passive β-barrel channel proteins can be regarded as ideal pore geometries since they have transport properties targeted towards specific small molecule separations necessary to maintain cellular functions. MPs in various organisms are used for exclusion of large molecules while allowing or inducing ion and small molecule passage. Additionally, the stability and mutation tolerance of these MPs make them ideal candidates for engineering channel based scalable membranes. The β-barrel structure is a robust scaffold with high thermodynamic stability and can be immobilized in an oriented and functionally-active form within amphiphilic polymers and lipid bilayers. These precisely-sized and stable MP channels were used in this work for constructing high performance biomimetic membranes.

In this presentation, we will present a comprehensive approach to construct scalable biomimetic membrane with rapid fabrication of highly porous β-barrel MP two-dimensional (2D) crystals and nanosheets. These highly packed crystalline structures and well-ordered nanosheets were fabricated by self-assembling MPs in a poly(butadiene)-b-poly (ethylene oxide) (PB-PEO) di-block copolymer matrix using an innovative solvent evaporation method that can be completed within 2-hours. The prepared 2D unique hybrid materials were deposited onto commercial porous substrates as high-performance and defect-free selective layers by a modified layer-by-layer approach. Three different pore-forming β-barrel MPs, outer membrane protein F (OmpF), a mutated version of a bacterial ferrichrome outer membrane transporter (designated FhuA ΔC/Δ4L), and channels formed from the *Staphylococcus aureus* toxin, alpha-hemolysin (αHL) were selected to demonstrate this approach. These proteins were chosen since they possess unique elliptical/cylindrical pore cross-section of 0.8×1.08 nm, 1.31×1.62 nm, and 1.50×1.50 nm for OmpF, FhuA ΔC/Δ4L, and αHL, respectively. Achieving high porosity, tuning such small pore sizes, as well as maintaining uniformity in pore sizes are quite difficult in current membrane models.

The three MP biomimetic membranes synthesized showed significant improvements in water permeability and maintained designed selectivity, as a result of high packing densities and unitary pore shapes of MP channels. These biomimetic membranes demonstrated 20-1000 times greater water permeability than commercial membranes with comparable molecular exclusion ratings. These water permeabilities were 293 $\pm$ 51 L m^{-2} h^{-1} bar^{-1} (denoted as LMH bar^{-1}), 793 $\pm$ 226 LMH bar^{-1}, and 2,107 $\pm$ 235 LMH bar^{-1} (all values mean $\pm$ s.d., $n \geq 3$) for OmpF, FhuA ΔC/Δ4L, and αHL-based biomimetic membranes, respectively. In addition, these membranes demonstrated the molecular exclusion performance inferred from their barrier pore sizes with molecular weight cut-offs (MWCOs) of ~480, ~1,130, and ~930 Da for OmpF, FhuA ΔC/Δ4L, and αHL-based biomimetic membranes, respectively. These results exhibit the potential of utilizing β-barrel MPs to tailor membrane selectivity within a critical separation range, while enhancing water permeability.

In summary, we have fabricated densely packed MP-block copolymer based biomimetic membranes. The three MP-based membranes demonstrate greater water permeability while maintaining specific small solute selectivity than commercial membranes with similar separation ranges. This work presents a new class of materials combining high selectivity and permeability of MPs with block copolymers to synthesize stable membranes that could contribute to the energy reduction in chemical and biological separations.

564924 A pH-Responsive TFC NF Membrane with Rechargeable Anti-Bacterial Property and High Efficient Dye/Monovalent Salt Separation Via Catechol/Boronate Complexation

Membrane decorated with biocides was an effective way to surpass biofilm growth. However, their immediate biocidal effect usually suffered from significant decline due to the irreversible consumption of the biocides. Here we reported an innovative loose nanofiltration membrane with rechargeable antibacterial capability that fabricated by performing a simple interfacial polymerization via 3-aminophenylboronic acid (3-APBA) and trimesoyl chloride (TMC) on a polysulfone substrate. A natural biocadal – tannic acid (TA) – can be grafted onto the membrane surface under natural/alkaline condition and then released from the surface under acidic environment, due to the pH-responsive property of catechol/boronate complexation. The resultant membrane exhibited integrated properties of fast bacterial inactivating efficiency against *Staphylococcus aureus*, rechargeable antibacterial capability and significantly expanded duration. In addition, the achieved membrane showed remarkable separation efficiency to dye/monovalent salt system. The successful synthesis of this novel NF membrane with rechargeable anti-bacterial property could provide new insights into the development of pH-responsive and sustainable antibacterial membranes.

557619 Expected Performance and Potential Design Strategies for Highly Selective Biomimetic Desalination Membranes

Expected Performance and Potential Design Strategies for Highly Selective Biomimetic Desalination Membranes

Jay R. Werber,[1,2] Cassandra J. Porter,[2] and Menachem Elimelech[2]

[1] Department of Chemistry, University of Minnesota, Minneapolis, MN

[2] Department of Chemical and Environmental Engineering, Yale University, New Haven, CT

Reverse osmosis is a central technology in desalination and the potable reuse of municipal wastewater. The need for increased selectivity, compared to conventional thin-film composite (TFC) membranes, has led to interest in biomimetic membranes. In these membranes, the selective layers mimic the structure and performance of cell membranes by placing biological or artificial water channels (e.g., aquaporin and carbon nanotube porins) within a bilayer of lipids or amphiphilic block copolymers. Aquaporin in particular is nearly perfectly selective, yet highly permeable to water. Despite much work on designing water channels and fabricating biomimetic membranes, a fundamental understanding of the expected performance of this class of membrane materials is lacking. In this work, we characterize expected biomimetic membrane performance using the resistance model, which considers individual components in a membrane as analogs to resistors in a circuit. We first consider the selective layer, which comprises channels and the surrounding bilayer. While channels have been well-characterized, permeability data for block copolymer bilayers was scarce. As such, permeabilities of water, neutral solutes, and ions were measured using lipid and block copolymer vesicles. Using these permeabilities, we show that a defect-free biomimetic selective layer could have water/salt selectivities ~100 billion times greater than current membranes, resulting in near-perfect salt rejection. However, hydrophobic organic micropollutants relevant to wastewater treatment would be highly permeable, leading to poor rejections. As a possible solution, we next consider composite membrane structures, particularly the use of TFC membranes as a support for biomimetic selective layers. We find that the size-based permeability of TFC membranes would allow for high rejection of micropollutants. Additionally, TFC membranes as support layers would minimize the negative impact of defects in the biomimetic selective layer, allowing for highly selective performance even with substantial defect areas (>1%). This work provides fundamental insight into molecular transport through biomimetic membranes, as well as a strategic path towards the fabrication of ultra-selective desalination membranes.

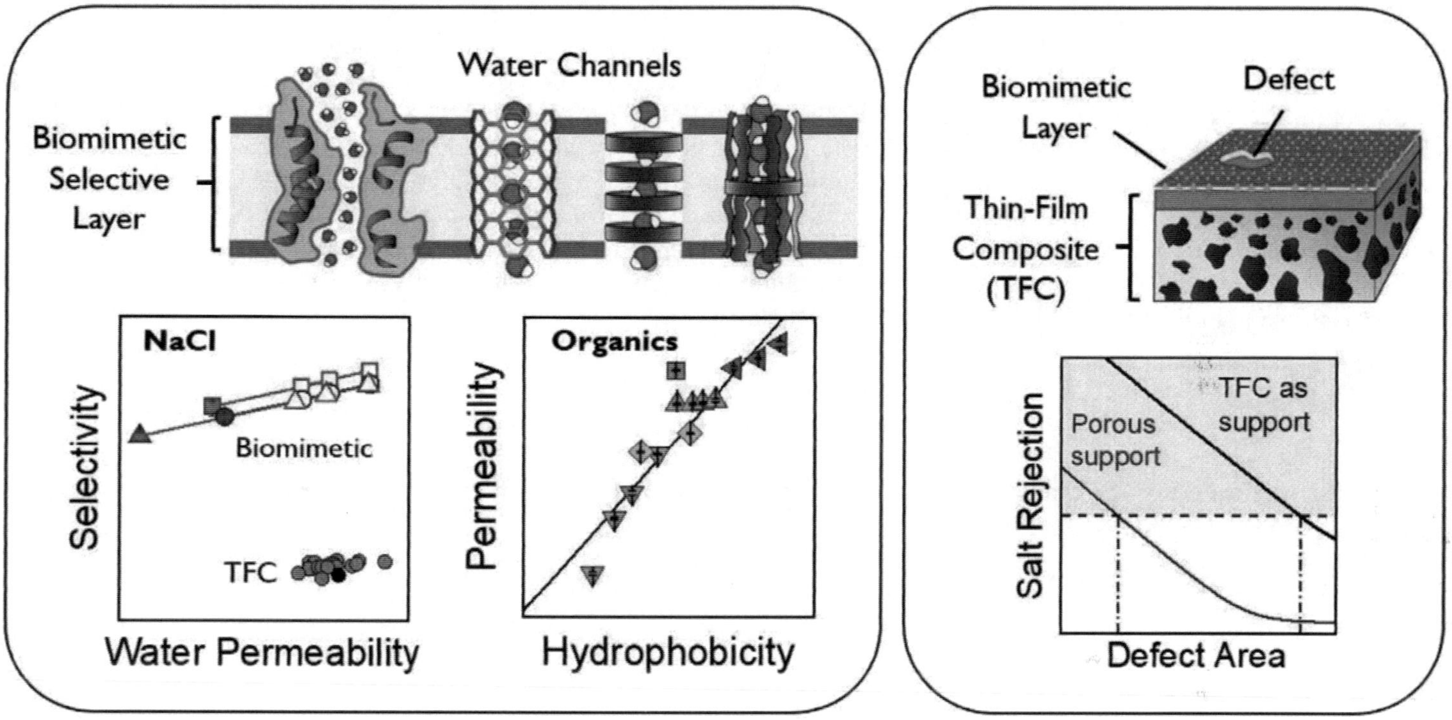

References:

1. Werber, J.R.; Deshmukh, A.; Elimelech, M. The Critical Need for Increased Selectivity, Not Increased Water Permeability, for Desalination Membranes. *Environ. Sci. Technol. Lett.* **2016**, 3, 112-120. DOI: 10.1021/acs.estlett.6b00050
2. Werber, J.R; Elimelech, M. Permeability and Selectivity Limits of Biomimetic Desalination Membranes. *Science Advances* **2018**, 4, eaar8266. DOI: 10.1126/sciadv.aar8266
3. Werber, J.R.; Porter, C.J.; Elimelech, M. A Path to Ultra-Selectivity: Support Layer Properties to Maximize Performance of Biomimetic Desalination Membranes. *Environ. Sci. Technol.* **2018**, 52, 10737-10747. DOI: 10.1021/acs.est.8b03426

568089 Bioinspired Membrane Pores Containing Phospholipid Bilayer for Immobilized Enzymatic Catalysis

Enzymes, immobilized within membrane pores, offer improved kinetics due to pressure-driven convective flow. On the other hand, phospholipid bilayer (PLB), owing to its fluidity and native "cell-membrane" environment, facilitates the activity and stability of enzymes. We reported integration of the biomimetic characteristics of PLB and convective flow through porous membrane within the same configuration (1). Novel bioinspired membranes were developed by incorporating PLB within membrane pores using two different approaches – (i) direct deposition of lipid bilayer within bare membrane pores, and (ii) immobilization of a polymer cushion prior to lipid bilayer within membrane pores. PLB-functionalized membrane pores were examined for enzymatic catalysis by electrostatically immobilizing glucose oxidase (GOx) enzyme. In-depth activity study revealed that the bioinspired membrane with polymer cushion supported PLB and with convective mode of flow exhibited benefits in terms of activity, stability and operating range of pH. Besides, we reported the reusability of the membrane matrix as well as the PLB by exploiting the reversible attachment-detachment sequence of the functional moieties (2). Characterization of the detached phospholipid confirmed retention of the original structural and functional properties as exhibited before attachment. The study was further extended for the development of immobilized bi-enzymatic (GOX-HRP) and multi-enzymatic (Glucoamylase-GOX-HRP) systems. Development of the bioinspired membrane, characterization and in-depth performance analysis (activity, stability and reusability) studies will be discussed in the presentation.

References: 1. A. Kumari, **S. Datta***, Phospholipid bilayer functionalized membrane pores for enhanced efficiency of immobilized glucose oxidase enzyme, *J. Mem. Sci.*, **539** (2017) 43-51

2. A. Kumari, L. Rekhie, **S. Datta***, Reversibly attached phospholipid bilayer functionalized membrane pores, *Langmuir*, **34** (2018) 14395-14401

578287 Modelling of Fluid Transfer in Nanoporous Carbons with Molecular Dynamics Simulation

The separation of CO_2/CH_4 is important in both biogas upgrading and natural gas sweetening. In the past two decades, gas membrane separation technology has become much more advanced due to interest in the technology's reliability, low-energy consumption, and minimal capital investment. However, most membranes used in industry today are still limited due to an inherent upper bound limitation on their performance, showing a conventional trade-off behaviour between permeability and selectivity[1].

It is believed that membrane process is a compromise in competition between two dominant mechanisms: the permeability mechanism and the selectivity mechanism[2]. In essence, the system behaviour is dominated by the complex structure of the membranes and the complex interaction between the gas and the membranes. However, no quantitative results of the permeability and selectivity with the structure structural properties and the interaction were obtained, makes it different to revealed the inherent mechanism to break the Robeson upper bound.

In this work, a novel method based on linear nonequilibrium thermodynamics[3] and equilibrium molecular simulation was developed to quantitatively predict selectivity and permeability for CO_2/CH_4 mixtures based on the structural properties of materials such as effective contact surface area (S_{eff}). By adjusting the potential parameters (ε, σ) artificially, different effective contact surface area was obtained and the optimal region to break the Robeson upper bound was obtained.

References:

[1] H. B. Park, J. Kamcev, L. M. Robeson, M. Elimelech, B. D. Freeman, "Maximizing the right stuff: The trade-off between membrane permeability and selectivity", Science, 2017, 356, 1137.

[2] J. H. Li, "Exploring the Logic and Landscape of the Knowledge System: Multilevel Structures, Each Multiscaled with Complexity at the Mesoscale", Engineering, 2016, 2, 276-285.

[3] W. L. Xie, X. Y. Ji, X. Feng, X. H. Lu, "Mass-Transfer Rate Enhancement for CO2 Separation by Ionic Liquids: Theoretical Study on the Mechanism", AIChE Journal, 2015, 61, 4437-4444.

576338 Forming Water-Wire Networks and Achieving Aquaporin-like Water Permeability and Water/Salt Selectivity Using Artificial Water Channels

Achieving both high water permeability and water/salt selectivity is of great interest in desalination membrane research field and, inspired by outstanding water transport property of biological water channel aquaporins, artificial water channel system has emerged as promising platform for desalination membrane developments. However, the water transport performance of current artificial water channels is still unsatisfactory compared to aquaporins. Here, we present unprecedented artificial water channel structure that enables to achieve billions of water molecule conduction per second per channel architecture that is comparable to aquaporins. This outstanding single channel permeability could be achieved through spontaneous channel cluster formation inside lipid bilayer membranes and water-wire network formation. Also, quantitative molecular transport study demonstrated $\sim 10^9$ water/ion selectivity property of this channel structure, of which value exceeds the selectivity requirements for desalination membrane ($\sim$16,000) developments by several orders of magnitude. Consequently, this cluster-forming channel structure is expected as highly promising pore architectures that would bring significant stride for desalination membrane developments.

557926 Separation of Monoclonal Antibody Charge Variants with Continuous Chromatography and Mixed-Mode Resins

For monoclonal antibodies (mAbs) expressed by mammalian cells, the generation of charge variants is inevitable, which would lead to lower potency and stability. It is necessary to separate charge variants. However, it is quite difficult to obtain target mAb with high purity and high yield due to the similarity of charge variants with mAb.
To improve the resolution and trade-off between purity and recovery, two new bioseparation technologies, mixed-mode chromatography and twin-column continuous chromatography, were combined. The mode of multi-column counter-current solvent gradient purification (MCSGP) was investigated with the recycle of the peak overlap between target mAb and charge variants. With cation exchange resins as the control, various factors including mixed-mode ligands, resin particle sizes, elution modes, gradient slopes and flow rates were tested. The results indicated that the resolution of charge variant with mixed-mode resins was higher than that with normal cation exchange resins due to the combination of different molecular interactions. In addition, MCSGP with mixed-mode chromatography could improve the recovery significantly with high purity. Near 100% purity of target mAb could be obtained after the process optimization. The chromatographic models were used to aid the process development. The results demonstrated that twin-column continuous chromatography with mixed-mode resins is a promising technology for mAbs charge variants separation with high purity and high recovery.

Keywords: Monoclonal antibody, Charge variants, Continuous chromatography, Mixed-mode chromatography, Cation exchange chromatography

571657 Quantifying Host Cell Protein Interactions with Antibodies Using Interaction Chromatography

Protein aggregation and impurities such as host cell proteins, HCPs, are critical problems for the safety of biopharmaceuticals as they are linked to adverse immunologically related responses in patients. Much effort has been made to gain a better understanding of aggregation, however, the mechanisms leading to protein aggregation are still not fully understood. Similarly, the removal of host cell proteins and aggregates from the product is associated with very high costs leading to lengthy and costly manufacturing processes for biopharmaceuticals.

Protein-protein molecular interactions in solution are known to be involved in protein solution aggregation behaviour and are a common issue for the manufacturing of biopharmaceuticals such as monoclonal antibodies (mAbs). The osmotic second virial coefficient (B_{22}) is a fundamental physicochemical property that describes the protein-protein interactions in solution that has been linked to predict aggregation, protein phase behavior and solubility in previous research. Self-Interaction Chromatography (SIC) and Cross-Interaction Chromatography (CIC) have recently become popular tools to determine, B_{22}, and protein-protein interactions. These tools are based on quantifying the interactions between one protein that is immobilized on resins and packed into a column and another which is injected through the column (same protein for SIC and different protein for CIC).

SIC has shown successes in previous research predicting protein aggregation in downstream processing and bioprocess formulation. However, limited research using these tools have been done to study the interactions that exist between host cell proteins and purified proteins for us to better understand the interface between upstream and downstream operations and get a better prediction of aggregation early on in the process.

In this study we utilized CIC and characterized the interactions using cell culture samples containing different amounts of host cell proteins and purified mAbs. These interactions were then compared with specific interactions between two purified proteins. Early results displayed that samples with HCPs and other impurities displayed much stronger interactions and more negative virial coefficients than samples with less HCPs and impurities. Next phase will be to evaluate specific HCPs and HCP loads and the interactions associated with these. This will then be able to provide direct guidelines on difficult-to-remove HCPs and added mechanical insights that could aid future downstream process development.

576079 Bovine Serum Albumin and Myoglobin Separation By Size Exclusion Simulated Moving Bed Technology

Introduction

Proteins are found in bacteria and viruses as well as in complex organisms, vertebrates and higher mammals such as humans [1]. Protein functions depend on its structure, which is influenced by physical and chemical parameters. Additionally, proteins are already included in our daily life, being present in detergents, food, cosmetics, perfumes and others such as pharmaceutical products. They can also be used in organic chemistry, for enzymatic catalysis [2]. Therefore, proteins must be produced and purified industrially, at large scale, to be implemented into their applications. To obtain a protein, in its pure form, from a mixture of others can be a laborious and expensive task. Indeed, to develop a process and determine the optimal conditions to purify proteins is the major challenge in the area. Separation and purification of proteins by SMB started with the size-exclusion simulated moving bed (SE-SMB) [3]. This kind of separation is simple; the working pH is not relevant; neither the ionic strength nor the composition of the buffer. This is mainly because these aspects do not influence the size, which is the parameter that is responsible for the separation [4]. In this kind of separation, a mixture of proteins with different molecular masses, the path that the compounds take will be different. The smallest will be able to enter in the pores, and as a result, their concentration wave velocity in the column is slowed. The larger ones, rarely enter in the pores or if they are very large, they are not able to enter at all, being its elution in the column easier and faster [1].

Materials and Methods

The main objective of this work is to study the separation of proteins, bovine serum albumin (BSA) and myoglobin (Mb) by size-exclusion simulated moving bed.

Bovine Serum Albumin belongs to albumins family, which includes other proteins such as Human Serum Albumin and ovalbumin. Albumin is the most predominant plasma protein, and it is produced in the liver [5]. Its molecular weight is 65.4 kDa. Myoglobin is a cytoplasmic hemoprotein and belongs to the globin family, as well as hemoglobin (Hb), neuroglobin (Ngb) and cytoglobin (Cygb) [6, 7]. Its molecular weight is 16.89 kDa. Therefore, the fractionation range of the selected resin, Sephadex G-50, is from 1 to 30 kDa, which excludes the BSA totally, since its molecular weight is higher than the upper exclusion limit.

To achieve the objective of separating the two proteins, fixed bed experiments were performed to obtain the adsorption breakthrough curves in Sephadex G-50, as well as the binary breakthrough curves. A phenomenological model was developed and used to predict the obtained experimental results. SMB experiments were carried out.

Results

The results of the fixed bed experiments, in terms of the retention times of the two proteins, showed that retention time of Mb is higher than BSA, which was expected since its molecular weight is within the fractionation range of the adsorbent, so it is possible for Mb to enter in the pores. The developed mathematical model was validated with fixed bed experiments, and it was possible to conclude that these two proteins can be separated by size exclusion chromatography.

An SMB model was used to determine the separation region, for raffinate and extract purity of 99.9 %. An operating point within the mentioned region was selected to perform the separation experimentally. Thus, the appropriated flowrates and switching time were selected. Afterward, SMB experiments were performed in the FlexSMB® unit, and the model was validated. It was concluded that the SMB model could predict the experiments and that the system can separate the BSA and Mb efficiently.

Acknowledgments

This work is a result of: Project "AIProcMat@N2020 - Advanced Industrial Processes and Materials for a Sustainable Northern Region of Portugal 2020", with the reference NORTE-01-0145-FEDER-000006, supported by Norte Portugal Regional Operational Programme (NORTE 2020), under the Portugal 2020 Partnership Agreement, through the European Regional Development Fund (ERDF); Associate Laboratory LSRE-LCM - UID/EQU/50020/2019 - funded by national funds through FCT/MCTES (PIDDAC), and by FCT/PD/BD/137891/2018.

References

1. Whitford, D., *Proteins: Structure and function*, ed. Wiley. 2005.
2. Kumar, P. and S. Sharma, *An overview of purification methods for proteins.* International Journal of Applied Research, 2015. **1**: p. 450-459.
3. Houwing, J., H.A.H. Billiet, and L.A.M.v.d. Wielen, *Mass-transfer effects during separation of proteins in SMB by size exclusion.* AIChE Journal, 2003. **49**.
4. biotech, A.p., *Gel filtration - Principles and Methods* 8th ed.
5. Raoufinia, R., et al., *Overview of Albumin and Its Purification Methods.* Adv Pharm Bull, 2016. **6**(4): p. 495-507.
6. Hendgen-Cotta, U.B., M. Kelm, and T. Rassaf, *Myoglobin functions in the heart.* Free Radic Biol Med, 2014. **73**: p. 252-9.
7. Ordway, G.A. and D.J. Garry, *Myoglobin: an essential hemoprotein in striated muscle.* J Exp Biol, 2004. **207**(Pt 20): p. 3441-6.

577463 Polymeric Adsorbent for Endotoxin Removal from Biopharmaceutical Solutions

The presence of endotoxin, also known as lipopolysaccharides (LPS), as a side product appears to be a major drawback for the production of therapeutic biomolecules that are essential for research, pharmaceutical, and industrial applications. In the biotechnology industry, gram-negative bacteria (*e.g.*, *Escherichia coli*) are widely used to produce recombinant products such as proteins, plasmid DNAs and vaccines. These products are contaminated with LPS, which may cause side effects when administered to animals or humans. Purification of LPS often suffers from product loss. For this reason, special attention must be paid when purifying proteins aiming a product as free as possible of LPS with high product recovery. Although there are a number of methods for removing LPS, the question about how LPS removal can be carried out in an efficient and economical way is still one of the most intriguing issues and has no satisfactory solution yet. In this work, polymeric poly-e-caprolactone (PCL) nanoparticles (NPs) (~ 780 nm) were synthesized using solvent evaporation technique at a relatively low cost and demonstrated to possess sufficient binding sites for LPS adsorption and removal with ~100% protein recovery. The PCL NPs removed greater than 90% LPS from protein solutions suspended in water using only one milligram (mg) of NPs, which was equivalent to ~ 1.5 million endotoxin units (EU) per mg of particle. The LPS removal efficacy increased to a higher level (~96%) when phosphate buffered saline (PBS containing 137 mM NaCl) was used as a protein suspending medium in place of water, reflecting positive effects of increasing ionic strength on LPS binding interactions and adsorption. The results further showed that the PCL NPs not only achieved 96 % LPS removal but also ~100% protein recovery for a wide concentration range from 10 – 1000 mg/ml of protein solutions. The NPs were highly effective in different buffers and pHs. This is the first demonstration of endotoxin separation with high protein recovery using polymer NPs which provide strong adsorptive interactions for LPS removal from protein solutions. Additional features of these NPs are biocompatible (environmental friendly), recyclable after repeated elution and adsorption with no significant changes in LPS removal efficiencies. The results indicate that PCL NPs are an effective LPS adsorbent in powder form, which have a great potential to be employed in large-scale applications.

578863 Flow-through Chromatography of Proteins As a Continuous and Integrated Separation Method

Continuous manufacturing including process integration is expected to increase the productivity of protein drug production. However, it is not easy to operate the continuous process especially for downstream processing as many unit operations (chromatography and membrane filtration) are involved.

An operation method known as flow-through chromatography (FTC) is considered to be an efficient method for separating a target protein as the flow is continuous. In FTC, a target protein is eluted from the chromatography column without adsorption whereas contaminants are strongly bound. Since at least two different modes of chromatography are needed in order to remove contaminants, two FTC columns have to be connected in order to build the continuous process. This is not an easy task since the mobile phase properties (pH, salt, buffer ions) are different for the two columns.

In this paper, we investigated how FTC can be designed based on linear gradient elution (LGE) data based on our LGE model. As a model separation system removal of dimer from BSA monomer was chosen. Model calculation showed that FTC is very sensitive to a small change in salt concentration and/or pH of the mobile phase. We also examined the efficiency of FTC processes in terms of impurity removal efficiency from the cell culture broth containing monoclonal antibody. It was found that two FTC (anion exchange chromatography - cation exchange chromatography) can remove impurities efficiently when the mobile phase pH and conductivity were properly chosen. It was also shown that the two columns can be connected for a pseudo continuous FTC operation.

561920 Preparation of a Novel Tetrapeptide Biomimetic Chromatographic Resin and Its High Antibody-Binding Capacity

The separation and purification of protein drug, especial antibody is a big challenge in biopharmaceutical process. Biomimetic affinity chromatography with short peptide ligands, as a promising bioseparation technique, has great potential to protein separation and purification.

In this work, we focused on the exploring the new biomimetic chromatographic resin preparation and properties with new tetrapeptide as ligand. A tetrapeptide library with critical residues of natural ligands (ProA, ProG) to hIgG was constructed and a ligand (Ac-FYWR) with high LibDock score was selected by molecular docking. Then, the selected ligand was evaluated by static/dynamic adsorption. It was found that ligand Ac-FYWR has the high binding capacity and selectivity for hIgG. The results showed the Q_m of Ac-FYWR-4FF resin to hIgG was 110.7 mg/g resin and the K_d of the resin was 0.10 mg/mL at pH 8.0, while the resin almost could not adsorb hIgG at pH 5.0. Moreover, at pH 8.0, $Q_{10\%}$ of Ac-FYWR-4FF resin was 26.7 mg/mL for hIgG but just 1.0 mg/mL at pH 5.0, which presented high selectivity of the screened resin by absorbing at pH 8.0 and eluting at pH 5.0. Finally, the Ac-FYWR-4FF resin was applied to separate mAb from CHO cell culture supernatant with the purity of 97.34% by only one-step separation.

The results show that the screened tetrapeptide resin (Ac-FYWR-4FF resin) could be well used in the antibody separation and purification, be help to develop the biomimetic chromatographic and give the new direction for the design of the peptide ligand with high adsorption for target proteins.

Keywords: biomimetic affinity chromatography; adsorption behavior; Ac-FYWR-4FF resin; CHO cell culture supernatant;

576895 Production and Isolation of Phospholipids from Microalgae for Formulation of Green Cosmetics

Phospholipids, essential ingredients of various cosmetics formulations, are currently obtained from food sources, such as soybean and egg yolk. However, sustainability concerns have been raised due to competition with food uses. Microalgae, a promising source of natural ingredients, can produce phospholipids on potentially a more sustainable basis. The marine microalga *Nannochloropsis oculata* was cultivated in a vertical flat panel photobioreactor in batch and continuous mode. The macronutrient ratio of nitrates-to-phosphates was varied to enhance phospholipid synthesis. Cell harvesting was followed by solvent extraction to solubilize and recover the microalgal phospholipids. For quantification of the individual phospholipids (phosphatidylcholine, phosphatidylethanolamine, phosphatidylinositol and lyso-phosphatidylcholine) recovered from the microalgal biomass, we employed P-31 NMR spectroscopy. Finally, using a semi-preparative HPLC column, we fractionated and isolated the individual phospholipid compounds.

580451 Addressing the Dual Challenge through Advanced Separations: From Materials to Process Scale-up

The dual challenge of supplying energy to a growing middle class while reducing emissions is shaping the future of energy globally. Large-scale hydrocarbon separation technologies account for roughly 45% of energy consumption in gas processing, refining and petrochemicals manufacturing. In response to these challenges, ExxonMobil has been developing advanced separation concepts focused on some of the highest impact streams such as olefin/paraffin and xylene isomer separations natural gas processing and carbon capture. Development of novel materials, fundamental and data driven process models and lab-bench-pilot scale evaluation are often simultaneously required to take these applications from a concept to technical demonstration. The talk discusses three diverse examples that have the potential to dramatically reduce the energy/carbon and physical footprint of critical operations not feasible via conventional processing.

Flexible pure silica zeolite such as ITQ-55 can kinetically separate ethylene/ethane with an unprecedented selectivity of ~100, owing to their distinctive pore topology with large heart-shaped cages and framework flexibility. Carbon Molecular Sieve (CMS) materials in a hollow fiber membrane format can be used in an organic solvent reverse osmosis (OSRO) process to separate xylene isomer molecules that differ in size by one-tenth of a nanometer at room temperature and without requiring a phase change. Finally pilot testing and scale-up approach for an adsorption-based natural gas dehydration and CO_2 removal technology is discussed that has the potential to reduce the overall gas treating physical footprint by ~50%.

579782 Engineering Low-Energy-Intensive Large-Scale Gas Separations

Large-scale separation and purification processes transform low value resources into valuable commodities; however, they also tend to consume a great deal of energy. Growing demand and the desire for higher worldwide standards of living are intensifying needs for improved separation approaches to minimize process CO_2 footprints and energy use. Advanced membrane and sorbent approaches can meet these needs through combined advanced materials, device formation and process engineering. While already used for water purification, more efficient approaches cannot yet be used widely, due to the lack of a full spectrum of advanced materials suitable for use with non-aqueous feeds. Recent developments in membrane and sorbent materials now appear likely to extend the low energy intensity separation revolution beyond water to include the full spectrum of large-scale feeds. Crosslinked glassy polymers, polymer-selective nanoparticle hybrids and carbon molecular sieve (CMS) materials cover different areas in this new separation landscape. This presentation will consider examples of these new materials, devices based on them, as well as their manufacturing and the savings they enable in advanced gas separations.

577218 Ionic Liquids for Chemical Separations

Ionic liquids (ILs) present intriguing possibilities for numerous chemical separations because their functionality can be tuned to enhance specific interactions. We will describe four examples, two involving gas separations and two on liquid separations. ILs are particularly attractive for gas separations because their exceedingly low vapor pressures mean that they will not contaminate the purified gas stream. First, we will focus on designing IL systems for post-combustion CO_2 capture. Even by physical absorption, many ILs provide sufficient CO_2 selectivity over N_2, O_2, CH_4 and other gases. However, when CO_2 partial pressures are low, the incorporation of functional groups to chemically react with the CO_2 can dramatically increase capacity, while maintaining or even enhancing selectivity. Guided by quantum calculations, molecular modeling and process modeling, we have been able to design, synthesize and test ILs that reversibly bind up to one mole of CO_2 per mole of IL, with the desired enthalpy of reaction. Currently we are developing supported IL membranes and composite IL/polymer membranes for olefin/paraffin separations, which are necessary in the dehydrogenation of light hydrocarbons to make polymers, fuels and chemicals. While most ILs do not have sufficient inherent olefin/paraffin selectivity, the addition of carriers to achieve facilitated transport can enhance the selectivity to very attractive levels. We will show that the advantages of using ILs for olefin/paraffin separations goes well beyond the stability and non-volatility of the IL membrane. For liquid separations, we will show how ILs can be designed to extract aromatics from aliphatics with very little loss of the solvent to the raffinate phase, along with a novel method using CO_2 as an antisolvent to separate the aromatic from the IL. The final example will demonstrate the use of ILs as an entrainer to break azeotropes. Overall, we hope to show the versatility of this unique materials platform for a wide variety of industrially relevant chemical separations.

555880 Sorption Kinetics in MFI- the First Hydrophobic Zeolite

Although preceded by several silica clathrasils, MFI, first synthesized and named ZSM-5 by Mobil Corp. in 1972 (U.S. Patent 3,702,886) is probably the first hydrophobic zeolite. The Al free analog (silicalite) was synthesized by Union Carbide at about the same time. In independent publications in 1978[1,2] the structures of these materials were shown to be the same.

During the next decade numerous applications of these materials emerged both as catalysts and selective adsorbents. Since then the molecular sieve properties and the sorption kinetics have been studied in considerable detail. Some of these results are reviewed and discussed in the light of more recent data. The pore size is about 5.5Å which is large enough to admit molecules as large as cyclohexane or o-xylene, albeit with significant hindrance. Diffusion in a silicalite crystal is non – isotropic, with almost equal diffusivities in both transverse directions (straight and sinusoidal channels) and a significantly lower diffusivity in the direction of the longest axis (z). Microscale measurements (PFG NMR and QENS) show generally good agreement but measurements at greater length scales often yield lower diffusivity values, apparently due to the presence of structural defects leading to dislocation of the pores. However, CO_2 (a small fast diffusing molecule) shows the opposite effect with higher apparent diffusivities measured over the length scale of the crystal, possibly due to preferential transport along the twin planes.

Recent studies by IFM and frequency response have revealed a transition from intra-crystalline diffusion control in large silicalite crystals to surface resistance control in small (micron sized) crystals. It has also been shown that while the intra-crystalline diffusivity is reasonably constant the surface resistance can vary by orders of magnitude between different crystals within the same sample.

It has recently been suggested that surface resistance may lead to a difference in the average diffusional path length between adsorption and desorption. However, a recent MD study shows that this hypothesis is incorrect; the mean path length and the path length distribution are the same for adsorption and desorption with or without a surface barrier, in conformity with the Principle of Microscopic Reversibility.

1. G.T. Kokotailo, S.L.Lawton, D.H.Olson, W.M. Meier *Nature* **272,** 437 (1978)
2. E.M.Flanigen, J.M.Bennet, E.W.Grose, J.P.Cohen, R.L.Patton, R.M.Kirchner, J.V.Smith *Nature* **271**, 512 (1978)

557372 Process Intensification in Multicomponent Distillation

Distillation is, and continues to be, the predominant choice for separation of mixtures in chemical and petrochemical industries. Given its ubiquity and energy intensive nature, it is surprising to note that the basic distillation flowsheet for a majority of separations involving multicomponent mixtures remain almost the same for decades. However, the recent push for *process intensification* is driving the research towards designing novel distillation flowsheets that reduce the overall energy requirement and equipment size. In this talk, we consolidate recent advancements and present an easy-to-follow recipe for industrial practitioners to design novel *intensified distillation flowsheets*.

Though the term 'process intensification' is relatively new, a handful of intensified distillation arrangements have been reported in the literature. Well known examples include Wright's [1] dividing wall column (DWC), Agrawal's [2] DWC arrangements for side-rectifier and side-stripper, Brugma's [3] configuration, Kaibel's [4] column, etc. Nevertheless, a systematic approach to draw such arrangements is missing in the literature. Here, we discuss the following aspects of process intensification for a general zeotropic n–component distillation: (1) Improving operability of thermally coupled columns by means of eliminating vapor streams in thermal couplings with only liquid transfers or column section rearrangement; (2) double and multi-effect arrangements for thermally coupled configurations; (3) simultaneous heat and mass integration to reduce column sections and overall heat duty; (4) systematic consolidation of column shells to draw novel operable DWC configurations with shells ranging between 1 to $n - 2$. We demonstrate these aspects through examples and illustrate how they lead to compact, easy-to-operate, energy-efficient and cost-effective multicomponent distillation system designs.

1. Wright, R. O. Fractionation Apparatus. U.S. Patent 2,481,134, 1949.
2. Agrawal, R., 2001. Multicomponent distillation columns with partitions and multiple reboilers and condensers. *Industrial & engineering chemistry research*, *40*(20), pp.4258-4266.
3. Johan, B.A. Process and device for fractional distillation of liquid mixtures, more particularlypetroleum. U.S. Patent 2,295,256, 1942.
4. Kaibel, G., Miller, C., Klass, K., Kindler, A., Aquila, W., Dobler, W., Breuer, K., Pommer, A. and Dirnsteiner, T., BASF SE. Dividing-wall column with feed in whole or in part in the vaporous state and/or side stream take off in whole or in part in the vaporous state. U.S. Patent 7,090,748, 2006.

579826 Developing Adsorption Fundamentals for Gas Separation Processes

Rapid cycle swing adsorption processes are an attractive method to selectively separate large volumes of gases. It continuously attracts broad interests to develop separation applications meeting desires for low energy, low capex, and low carbon emission processes. The design and performance of these adsorption processes are highly dependent upon the interplay between adsorbent material compositions, synthesis, and transport fundamentals which include multicomponent adsorption equilibria and mass transfer kinetics. Therefore, developing fundamental understanding provides basis for adsorption processes optimization and intensification. Illustrative examples of both experimental and theoretical capabilities we have developed to obtain this type of transport information will be presented.

572310 Rapid Screening of Solvents for Crystallization Using Group Contribution Method

Regardless of the technique of crystallization – cooling or anti-solvent, the choice of solvent can change the outcome of crystallization often at the same operating conditions. The changes in the intermolecular interactions between solute and solvent can either prevent the attachment of solute molecule at certain configurations or can encourage the interactions between solute-solute molecules. The specific molecular interactions can be studied with molecular simulations, but it requires compatible topology file of the solute and solvent molecules coupled with sufficient computational power. However, the complexity of the problem can be reduced by understanding contribution of each functional group on the solute and solvent molecule towards energy of solvation. Both of the crystallization techniques require the use of solvents which maintain the crystallization in a regime where desired crystal structure is obtained. The regime can be influenced by the energy of solvation. The dynamics of these solvation shell can be used to optimize the choice of solvent along with the operating conditions to optimize the process of crystallization. Technique of group contribution allow classifying specific molecular interactions based on the functional groups. For crystallization in pharmaceutical industries, the solvents must be chosen from a list of approved solvents and hence the technique of group contribution can help rapidly screen the solvents to maintain the crystallization in a desired regime based on the necessary functional groups.

573677 Dynamic Modelling, Simulation and Optimisation of Batch Enzymatic Amoxicillin Production

ABSTRACT

Antibiotics are societally essential pharmaceutical products, making previously untreatable illnesses such as pneumonia and tuberculosis curable, thus revolutionising modern medicine.[1] Access to essential medicines in low- to middle-income countries as well as antibiotic shortages in developed countries remain an important issue.[2,3] The complex molecular structures of antibiotics imply that they are generally expensive due to the multistep, materially-intensive syntheses required to make such molecules. Designing cost-effective antibiotic manufacturing routes is imperative.

The family of β-lactam antibiotics are some of the most important pharmaceutical products, with cephalosporins and semi-synthetic penicillins corresponding to approximately 65% of global production of antibiotics.[4] Amoxicillin is one β-lactam antibiotic used to treat respiratory and urinary tract infections and meningitis and listed as a World Health Organisation (WHO) essential medicine. The batch enzymatic synthesis of amoxicillin involves a kinetically controlled network of series-parallel reactions of desired amoxicillin synthesis and undesired hydrolysis of both amoxicillin and the substrate.[5] The demonstrated enzymatic synthesis of amoxicillin and kinetic model in the literature paves allows elucidation of optimal design and operating parameters via modelling and optimisation.[6]

While studies into modelling and simulation of synthesis and crystallisation of β-lactam antibiotics have been recently demonstrated,[7] dynamic modelling and optimisation of the described batch enzymatic synthesis of amoxicillin has yet to be implemented. Specifically, establishing optimal temperature control profiles for the batch reactor to meet a specific production objective for amoxicillin synthesis will allow for elucidation of optimal heating policies and possible reduction of total batch run times.[8]

This work implements dynamic modelling and optimisation of the batch enzymatic synthesis of amoxicillin.[5] First, the enzymatic synthesis pathway is presented with the kinetic model. Temperature dependence of kinetic parameters is introduced by regression of Arrhenius constants from experimental data, followed by isothermal simulation for design space investigation. A constrained dynamic optimisation problem is then described to minimise the consumption of key substrate subject to different constraints on product amoxicillin concentrations, comparing resulting optimal temperature profiles.

REFERENCES

(1) Zaffiri, L.; Gardner, J.; Toledo-Pereyra, L.H. History of antibiotics. From salvarsan to cephalosporins. *J. Investig. Surg.* **2012**, *25*(2), 67–77.

(2) Balkhi, B.; Araujo-Lama, L.; Seoane-Vazquez, E.; Rodriguez-Monguio, R.; Szeinbach, S.L.; Fox, E.R. Shortages of systemic antibiotics in the USA: how long can we wait? *J. Pharm. Heal. Serv. Res.* **2013**, *4*(1), 13–17.

(3) Pulcini, C.; Beovic, B.; Béraud, G.; Carlet, J.; Cars, O.; Howard, P.; Levy-Hara, G.; Li, G.; Nathwani, D.; Roblot, F.; Sharland, M. Ensuring universal access to old antibiotics: a critical but neglected priority. *Clin. Microbiol. Infec.* **2017**, *23*(9), 590–592.

(4) Giordano, R.C.; Ribeiro, M.P.A.; Giordano, R.L.C. Kinetics of β-lactam antibiotics synthesis by penicillin g acylase (PGA) from the viewpoint of the industrial enzymatic reactor optimization. *Biotechnol. Adv.* **2006**, *24*(1), 27–41.

(5) Gonçalves, L. R. B.; Fernández-Lafuente, R.; Guisán, J. M.; Giordano, R. L. C. The role of 6-aminopenicillanic acid on the kinetics of amoxicillin enzymatic synthesis catalyzed by penicillin g acylase

immobilized onto glyoxyl-agarose. *Enzyme Microb. Technol.* **2002**, *31*(4), 464–471.

(6) Diab, S.; McQuade, D.T.; Gupton, B.F.; Gerogiorgis, D.I. Process design and optimisation for the continuous manufacturing of nevirapine, an active pharmaceutical ingredient for H.I.V. treatment. *Org. Process Res. Dev.* **2019**, *23*(3), 320–333.

(7) McDonald, M.A.; Bommarius, A.S.; Rousseau, R.W.; Grover, M.A. Continuous reactive crystallization of β-lactam antibiotics catalyzed by penicillin G acylase. Part I: Model development. *Comput. Chem. Eng.* **2019**, *123*, 331–343.

(8) Rodman, A.D.; Gerogiorgis, D.I. An investigation of initialisation strategies for dynamic temperature optimisation in beer fermentation. *Comput. Chem. Eng.* **2019**, *124*, 43–61.

573793 Dynamic Simulation and Optimisation of Ampicillin Crystallisation

ABSTRACT

Crystallisation is an integral process in the pharmaceutical industry. It is commonly used for the separation of intermediates or as the final process in the production of an active pharmaceutical ingredient (API). The final size, shape and form of the crystals influences not only downstream operations but also the physical and chemical properties of the solid product.[1] The Food and Drug Administration (FDA) and other regulatory agencies constantly set new regulations to ensure the safety and stability of pharmaceutical products. The precise control of crystallisation is of the outmost importance to ensure that the final product has the desired attributes while satisfying the international regulations.

Ampicillin is a semi-synthetic penicillin derived β-lactam antibiotic commonly used to treat various bacterial infections such as urinary and respiratory tract infections, it is one of the ten most consumed antibiotics worldwide.[2] In a recent report the World Health Organization (WHO) characterized penicillins as critically important and high priority antimicrobials essential for the world healthcare system.[3] For this reason, it is paramount to ensure that the global production of ampicillin is sufficient, within international regulations and economically viable for the pharmaceutical companies. Optimal design and operating parameters are necessary to ensure optimal manufacturing of this societally important antibiotic.

Recent works published by our groups and others show the utility and versatility of conducting conceptual studies for pharmaceutical processes in order to ascertain the benefits and drawbacks of different design scenarios as well as to elucidate the optimal operating conditions of the processes.[4] Furthermore, work performed for the dynamic optimisation of bioprocesses illustrates the potential of the use of dynamic optimisation for the analysis of batch bioprocesses, which can establish optimal control variable trajectories to meet some design or operational objective.[5] Dynamic optimisation for the batch crystallisation of ampicillin has not yet been implemented. A recently published article[6] provides a crystallisation kinetic model for ampicillin suited for dynamic crystallisation optimisation. Size independent kinetics for primary, secondary nucleation and crystal growth were fitted from experiments which took place under different scenarios, providing parameters that are able to predict system behavior under a wide range of different conditions. The population balance is solved with the method of moments. The behaviour of ampicillin as a function of pH allows crystallisation by pH-variation.[7] The article utilizes extended Pitzer parameters to model ampicillin solubility.

In this work dynamic simulation under a variety of conditions for design space investigation is performed. The crystallisation of ampicillin is simulated for different linearly decreasing pH profiles varying the pH change rate and the initial and final pH values. For each scenario, the mother solute concentration as a function of time and the final mean crystal size are computed. The objective is to minimise the crystallization batch time subject to various constraints, such as the target mean crystal size and final mother liquor solute concentration. The optimisation studies are performed using the orthogonal collocation on finite elements method.[8] The IPOPT (Interior Point Optimizer) algorithm is employed to solve the Non-Linear Programming (NLP) problem. Results from each optimisation strategy are obtained and discussed. Finally, the economic potential and the environmental impact of the different designs are investigated.

REFERENCES

(1) Chen, J.; Sarma, B.; Evans, J.M.B.; Myerson, A.S. Pharmaceutical crystallization. *Cryst. Growth Des.* **2011**, *11* (4), 887–895.

(2) WHO Regional Office for Europe. Antimicrobial medicines consumption (AMC) network; 2014.

(3) WHO list of critically important antimicrobials for human medicine (WHO CIA list); 2017.

(4) Diab, S.; Gerogiorgis, D.I. Technoeconomic optimization of continuous crystallization for three active pharmaceutical ingredients: cyclosporine, paracetamol, and aliskiren. *Ind. Eng. Chem. Res.* **2018**, *57* (29), 9489–9499.

(5) Rodman, A.D.; Gerogiorgis, D.I. An investigation of initialisation strategies for dynamic temperature optimisation in beer fermentation. *Comput. Chem. Eng.* **2019**, *124*, 43–61.

(6) Encarnación-Gómez, L.G.; Bommarius, A.S.; Rousseau, R.W. Crystallization kinetics of ampicillin using online monitoring tools and robust parameter estimation. *Ind. Eng. Chem. Res.* **2016**, *55* (7), 2153–2162.

(7) Franco, L.F.M.; Mattedi, S.; Filho, P.D.A.P.A new approach for the thermodynamic modeling of the solubility of amino acids and β-lactam compounds as a function of pH. *Fluid Phase Equilib.* **2013**, *354*, 38–46.

(8) Čižniar, M.; Salhi, D.; Fikar, M.; Latifi, M.A. Matlab dynamic optimization code DynOpt, user's guide. **2006**, 1–7.

575181 Kinetic Monte Carlo Simulation of Anisotropy in $CsPbBr_3$ Nanocrystals

$CsPbBr_3$ nanocrystals have recently garnered a lot of attention owing to its applications in solar cells, next-generation displays and other optoelectronic devices[1] that can surpass the performance of currently existing semiconductor nanocrystals. The photoluminescence peaks are readily tunable by both size-tuned quantum confinement and chemical tuning of the bulk bandgap via facile anion exchange. However, to reap the benefits of the fascinating properties of $CsPbBr_3$ it is imperative to have a narrow particle size distribution. Furthermore, previous studies have shown that nanocrystals of different anisotropic shapes exhibiting varying anisotropic photophysical properties are possible for $CsPbBr_3$ nanocrystals viz. nanoribbons (NRs), nano-platelets (NPs) and nanowires (NWs)[2] which means that the crystal growth rate is not uniform in all directions despite near-cubic lattice symmetry. This anisotropic crystal behavior depends on the temperature of operation, the precursor concentration and the choice of different ligands[3]. The observation that one of the dimensions of this crystal follows kinetic control while the size of the other two dimensions are controlled by thermodynamics has made it very difficult to predict the morphology of these NCs[4]. The aforementioned reasons make it of extreme importance to understand the crystallization mechanism which can be employed to predict the size and morphological anisotropy of these NCs and tailor them for use in a specific optoelectronic device or other application. As of now, there are rarely any studies in this direction. Consequently, there is a paucity of modelling parameters for simulating this crystallization process. The goal of this research is to address these lacunae by proposing a probable mechanism of the crystal growth which will explain the anisotropic crystal growth and estimate the model parameters of the $CsPbBr_3$ crystallization process.

In this study, a single crystal with simultaneous growth on (001), (010) and (100) faces has been modeled by implementation of a kinetic Monte Carlo (KMC) algorithm. This KMC is based upon a solid-on-solid[5] model for crystal growth which considers three types of event viz, adsorption, migration, and desorption with the fundamental particle being a single unit-cell of $CsPbBr_3$. The growth on each face has been modeled using a 50 by 50 2-Dimensional lattice. On each of these three lattices, adsorption, migration and desorption events are being considered. For the adsorption event, it is assumed that each lattice site is energetically equal to the other and is available for the attachment, and thus the attachment rate is uniform over the lattice. Conversely, the rates of the migration of a particle and the dissolution of the particle in the solution are dependent on the number of the nearest neighbors of the particle. The local environment of each particle can have up to 6 neighbors, and this allows us to classify each lattice site into a separate class depending on the number of nearest neighbors that each lattice site has. This book-keeping technique helps us increase the computational efficiency. The rates for these events can be calculated based on the average binding energy of the lattice (φ), average bonding energy per bond between the neighbors (E_{pb}) and the attachment coefficient (K_0^+). To explain the anisotropic crystal growth, an anisotropic factor (γ) has been introduced in the direction of (001) face. Extensive simulations were performed to determine these four parameters and γ by corroborating with the experimental results. The experimental results are being generated by our experimental collaborator to facilitate precise determination of parameter[4]. Extensive simulation studies are carried out to evaluate the influence of temperature and supersaturation on the shape and size $CsPbBr_3$ NC at the end of a batch.

References

1. Wei, Y., Cheng, Z. & Lin, J. An overview on enhancing the stability of lead halide perovskite quantum dots and their applications in phosphor-converted LEDs. *Chemical Society Reviews* **48**, 310–350 (2019).
2. Peng, L., Dutta, A., Xie, R., Yang, W. & Pradhan, N. Dot–Wire–Platelet–Cube: Step Growth and Structural Transformations in $CsPbBr_3$ Perovskite Nanocrystals. *ACS Energy Letters* **3**, 2014–2020

(2018).
3. Zhang, J. *et al.* Growth mechanism of CsPbBr3 perovskite nanocrystals by a co-precipitation method in a CSTR system. *Nano Research* **12**, 121–127 (2019).
4. Dong, Y. *et al.* Precise Control of Quantum Confinement in Cesium Lead Halide Perovskite Quantum Dots via Thermodynamic Equilibrium. *Nano Letters* **18**, 3716–3722 (2018).
5. Nayhouse, M., Sang-Il Kwon, J., Christofides, P. D. & Orkoulas, G. Crystal shape modeling and control in protein crystal growth. *Chemical Engineering Science* **87**, 216–223 (2013).

571106 Population Balance Modeling of Cooling-Antisolvent Crystallization to Derisk Tech-Transfer

We have developed a population balance model (PBM) for BMS Compound A crystallization process, using gPROMS FormulatedProducts (gFP), and apply the model to predict the PSD for four ~35 kg scale plant batches, with good agreement to actual data. The PBM is initially constructed from lab scale (~10 g) crystallizations on the basis of particle size distribution (PSD) measurements and solution concentration measurements of the active pharmaceutical ingredient (API) throughout the crystallization process. The kinetic parameters associated with secondary nucleation are then refined using PSD data from four initial plant runs during process-scaleup, keeping other parameters identical to those from the laboratory experiments. The updated PBM is used to predict the PSD for another set of four plant batches, run with a different agitator type, and accurately predicts an increase in particle size. Analysis of the crystallization kinetics, simulated using gFP, indicates that secondary nucleation due to attrition is the predominant mechanism influencing the PSD in the crystallization process for compound A, thus explaining why agitator type has a strong influence on PSD. The PBM is used to predict PSD from crystallizations run under different conditions, different scales, and in different reactors using the global sensitivity analysis (GSA) module in the software. The model helps to provide a more robust particle size control strategy by employing fundamental crystallization kinetics to predict PSD, which is useful to anticipate changes upon tech transfer or future scale-up.

557932 A Mathematical Model of the Ultrasound-Assisted Crystallization of Aspirin in a Continuous Tubular Reactor

The crystallization of organic nano- and microparticles is significant in all kinds of industries, products, and applications, such as organic micron-sized crystals of active pharmaceutical ingredients (APIs) and nanodispersions of coating resins and microcapsules with active ingredients [1]. Currently, only batch processes are implemented at a sufficiently large scale to meet the market needs, as batch crystallization has been widely studied. However, the production of particles in batch processes is associated with problems. The most common limitations in batch reactors are the difficulties in the control of heat and mass transfer, factors that affect not only the average particle size and particle size distribution (PSD), but also batch to batch variations. The drawbacks of batch crystallization led researchers and industry initiatives to explore continuous processing technology, maintaining higher product standards [2].

As for continuous processes, the most important advantages are their ability to produce particles with specific particle size and smaller PSD. Additionally, the polymorphism can be controlled [3]. They produce quantities on demand, minimizing the use of raw materials and equipment. Furthermore, mixing energy and quantity of solvent could be decreased, and, last but not least, they are capable of being scaled up [4]. Ultrasound can be used for improving continuous crystallization processes because the generated bubbles act as a cluster attachment sites, forming finer crystals with low average size and narrower PSD. However, the nucleation mechanism that takes place during ultrasound is still unclear [5].

Concerning mathematical modeling, population balance models (PBMs) are used extensively to describe the crystallization processes because they are available to capture the changes in the number and size of crystals over time. In case of a tubular reactor, also a spatial PBM that describes the PSD at different axial length position can be formulated. Impressive advances have been obtained during the previous years in capturing the crystal production in reactors used for continuous crystallizers (plug flow crystallizers [6] and mixed suspension mixed product removal (MSMPR) reactors [7]), and batch systems with [8] or without ultrasound [9]. However, there is an unmet need to investigate novel and high-fidelity mathematical models for scaling-up the continuous crystallization of APIs under ultrasound. This work proposes the development of the first mathematical model of the continuous antisolvent crystallization of aspirin in water and ethanol in an experimental plug flow reactor with ultrasound. Heat generation due to mixing and ultrasound, which increases crystal solubility and affects the yield of the system subsequently, is also included in the model. Prior to this work, the population balance model's growth and nucleation rates in the studied antisolvent system at different temperatures and antisolvent content were estimated under silent conditions [10]. Global sensitivity analysis of our ultrasound crystallization system reveals which are the most critical parameters for capturing the crystal properties, such as the volume based distribution parameters Dv10, 50, and 90, and the system yield. Dv50 is the median for a volume distribution. Dv10, and Dv90 show the sizes at which the cummulative volume distribution reaches 10%, and 90%, respectively. Estimation of the most important model parameters and validation of the model's predictive capability were successfully performed for continuous crystallization experiments with different antisolvent volume flowrates and initial supersaturations.

This work shows that mathematical modeling is undoubtedly useful in order to capture the crystal growth in continuous crystallizers under the effect of ultrasound and subsequently it will provide insight into the design of new crystallizers.

References

[1] Z. Gao, S. Rohani, J. Gong, J. Wang, Recent developments in the crystallization process: Toward the pharmaceutical industry, Engineering 3 (3) (2017) 343–353 (2017).

[2] X. Yang, D. Acevedo, A. Mohammad, N. Pavurala, H. Wu, A. L. Brayton, R. A. Shaw, M. J. Goldman, F. He, S. Li, et al., Risk considerations on developing a continuous crystallization system for carbamazepine, Organic Process Research & Development 21 (7) (2017) 1021–1033 (2017).

[3] J. S.-I. Kwon, M. Nayhouse, G. Orkoulas, P. D. Christofides, Crystal shape and size control using a plug flow crystallization configuration, Chemical Engineering Science 119 (2014) 30–39 (2014).

[4] I. R. Baxendale, R. D. Braatz, B. K. Hodnett, K. F. Jensen, M. D. Johnson, P. Sharratt, J.-P. Sherlock, A. J. Florence, Achieving continuous manufacturing: Technologies and approaches for synthesis, workup, and isolation of drug substance. may 20–21, 2014 continuous manufacturing symposium, Journal of pharmaceutical sciences 104 (3) (2015) 781–791 (2015).

[5] J. Jordens, E. Canini, B. Gielen, T. Van Gerven, L. Braeken, Ultrasound assisted particle size control by continuous seed generation and batch growth, Crystals 7 (7) (2017) 195 (2017).

[6] Q. Su, B. Benyahia, Z. K. Nagy, C. D. Rielly, Mathematical modeling, design, and optimization of a multisegment multiaddition plug-flow crystallizer for antisolvent crystallizations, Organic Process Research & Development 19 (12) (2015) 1859–1870 (2015).

[7] Q. Su, Z. K. Nagy, C. D. Rielly, Pharmaceutical crystallisation processes from batch to continuous operation using msmpr stages: Modelling, design, and control, Chemical Engineering and Processing: Process Intensification 89 (2015) 41–53 (2015).

[8] A. Kordylla, T. Krawczyk, F. Tumakaka, G. Schembecker, Modeling ultrasound-induced nucleation during cooling crystallization, Chemical Engineering Science 64 (8) (2009) 1635–1642 (2009).

[9] J. D. Ward, D. A. Mellichamp, M. F. Doherty, Choosing an operating policy for seeded batch crystallization, AIChE Journal 52 (6) (2006) 2046–2054 (2006).

[10] C. Lindenberg, M. Krättli, J. Cornel, M. Mazzotti, J. Brozio, Design and optimization of a combined cooling/antisolvent crystallization process, Crystal Growth and Design 9 (2) (2008) 1124–1136 (2008).

578008 Crystal Morphology and Process Control of Multiple High-Salinity Wastewater Via Membrane Distillation Crystallization

Crystal Morphology and Process Control of Multiple High-salinity Wastewater via Membrane Distillation Crystallization

Xiaobin Jiang, Gaohong He*, Xuehua Ruan

State Key Laboratory of Fine Chemicals, Research and Development Center of Membrane Science and Technology, School of Chemical Engineering, Dalian University of Technology, Dalian, P. R. China

*Corresponding author: Email: hgaohong@dlut.edu.cn

Abstract

Nowadays, the separation of water recovery and salinity is mainly carried out by evaporative crystallization (ECr) [1]. However, ECr has the disadvantages of high operation temperature and uncontrolled supersaturation. Therefore, it is necessary to introduce a new separation technology [2,3].

This study is based on existing research work, and propose treatment of multiple high-saline wastewater with evaporation membrane distillation crystallization (VMDC). Coal chemical industry wastewater is a typical multiple high-saline wastewater, which is mainly composed of NaCl and $MgCl_2$. The influence of different mass fraction of NaCl and $MgCl_2$ in the feed fluid on the permeation flux of membrane has been studied. Besides, the effect of impurity $MgCl_2$ on the NaCl crystal in the process of MDC has also been investigated. The results indicated that, as the mass fraction of $MgCl_2$ in the feed increases from 0% to 100%, the viscosity of the solution gradually increases, and permeation of the water of the membrane distillation gradually decreases(shown in Fig. 1),. In addition, as the mass fraction of $MgCl_2$ in the feed increases continuously, crystal surface appearance defects has been obviously (shown in Fig. 2),and average crystal size has been gradually increased(shown in Fig. 3).

The study has been demonstrated the possibility of VMDC in the treatment of multiple high-saline wastewater. Compared with traditional evaporative crystallization, it provides a theoretical basis for the practical application of VMDC.

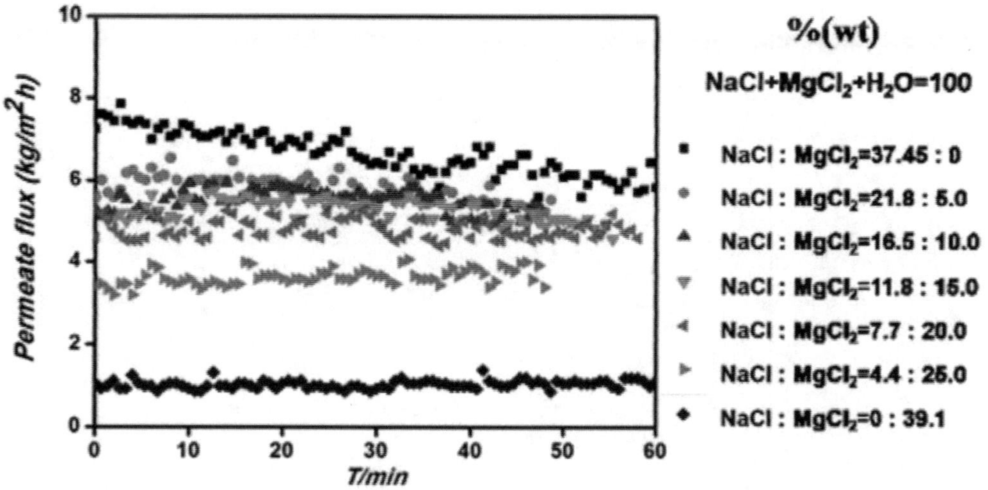

Fig. 1 The permeate flux changing with different NaCl and $MgCl_2$ mass fraction

Fig.2 The crystal surface morphology with different NaCl and $MgCl_2$ mass fraction

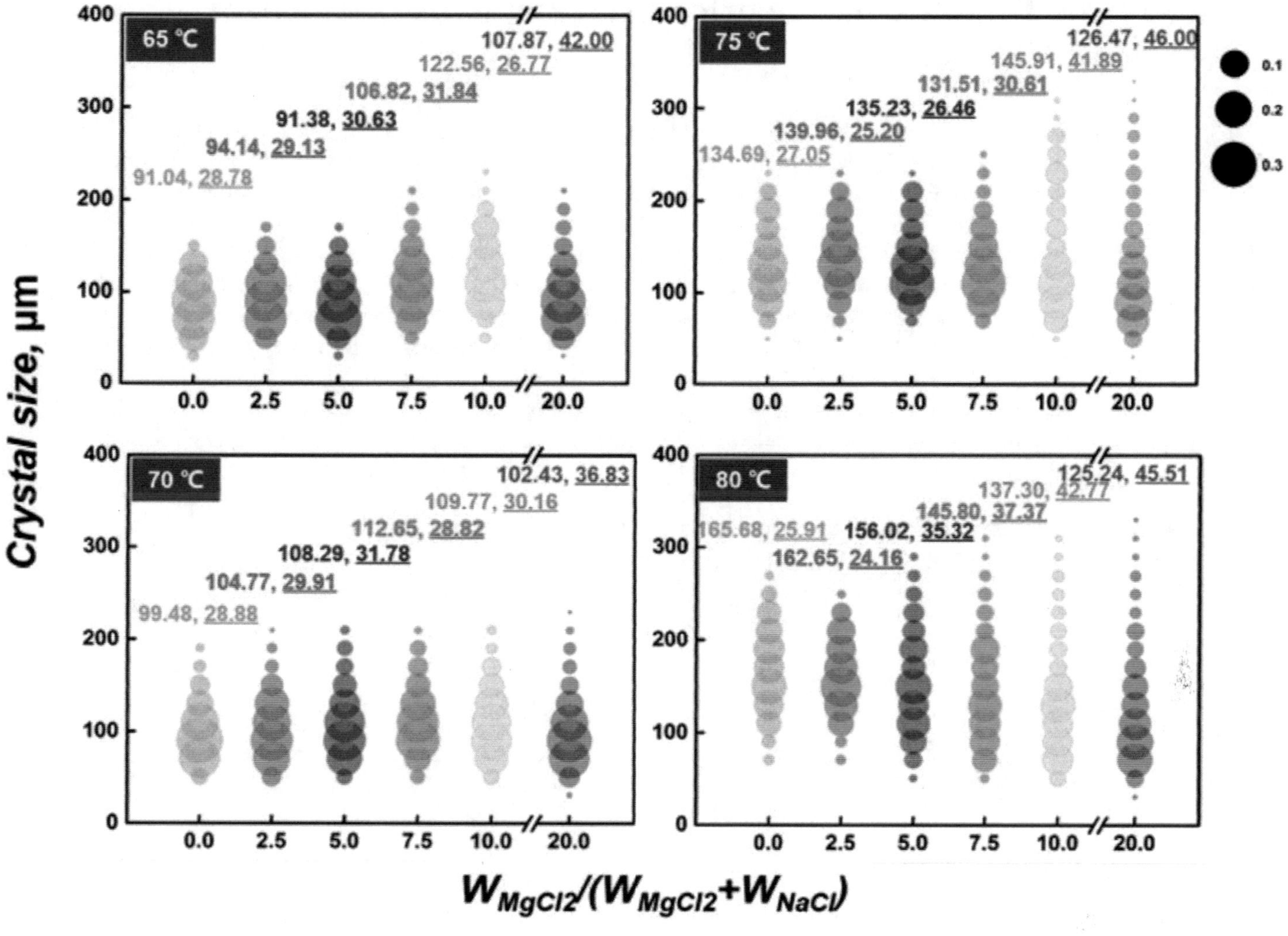

Fig. 3 The crystal size distribution with different NaCl and $MgCl_2$ mass fraction

References

[1] Lu H, Wang J, Wang T, et al. Crystallization techniques in wastewater treatment: An overview of applications[J]. *Chemosphere*, 173: 474-484 (2017).

[2] Zhang Y, Peng Y, Ji S, et al. Review of thermal efficiency and heat recycling in membrane distillation processes[J]. *Desalination*, 367: 223-239(2015).

[3] Chabanon E, Mangin D, Charcosset C. Membranes and crystallization processes: State of the art and prospects[J]. *J. Mem. Sci.*, 509: 57-67(2016).

572529 Applicability of Generic Force Fields for Adsorbent Screening

The availability of accurate, transferable force field parameters is critical in computational screening and discovery of next-generation adsorbent materials, such as metal-organic frameworks (MOFs), for energy and environmental applications. Currently, generic force fields, such as Universal Force Fields (UFF) and Dreiding Force Fields (DFF), are widely used to model the van der Waals interactions between framework atoms in MOFs. Although these force fields have been suggested in the literature to provide similar ranking of materials than more sophisticated DFT-derived force fields, the applicability of "off-the-shelf" force fields to predict experimental ranking of adsorbent materials has not been examined. In this work, we examined the applicability of "off-the-shelf" force fields in scoring the adsorbent materials for on-board methane storage and delivery applications. Although the adsorption properties computed based on UFF and DFF provide similar ranking of adsorbent materials, we find that these "off-the-shelf" force fields do a poor job in predicting the ranking of the adsorbent materials based on experimental isotherms. We briefly discuss our force field development efforts based on the adsorption isotherm from NIST-ISODB.

575848 Bottom-up Approach to the Coarse-Grained Surface Model: Effective Solid-Fluid Potentials for Adsorption on Heterogeneous Surfaces

Coarse-grained surface models with a low-dimension positional dependence have great advantages in simplifying the theoretical adsorption model and speeding up molecular simulations. In this work, we present a bottom-up strategy, developing a new two-dimensional (2D) coarse-grained surface model from the "bottom-level" atomistic model, for adsorption on highly heterogeneous surfaces with various types of defects. The corresponding effective solid-fluid potential consists of a 2D hard wall potential representing the structure of the surface, and a one-dimensional (1D) effective area-weighted free-energy-averaged (AW-FEA) potential representing the energetic strength of the substrate-adsorbate interaction. Within the conventional free-energy-average (FEA) framework, an accessible-area-related parameter is introduced into the equation of the 1D effective solid-fluid potential, which allows us not only to obtain the energy information from the fully atomistic system, but also to get the structural dependence of the potential on any geometric defect on the surface. Grand canonical Monte Carlo (GCMC) simulations are carried out for argon adsorption at 87.3 K to test the validity of the new 2D surface model against the fully atomistic system. We test four graphitic substrates with different levels of geometric roughness for the top layer, including the widely used reference solid substrate Cabot BP-280. The simulation results show that adding one more dimension to the traditional 1D surface model is essential for adsorption on the geometrically heterogeneous surfaces. In particular, the 2D surface model with the AW-FEA solid-fluid potential significantly improves the adsorption isotherm and density profile over the 1D surface model with the FEA solid-fluid potential over a wide range of pressure. The method to construct an effective solid-fluid potential for an energetically heterogeneous surface is also discussed.

570084 Adsorption-Induced Deformation of Faujasite Zeolites: Molecular Simulation Study

Adsorption-induced deformation is a phenomenon of contraction or expansion of solids caused by adsorption of gases [1]. Deformation in microporous materials is especially peculiar, and the adsorption strain isotherms may even qualitatively differ for the two close pore sizes. In these cases, when the adsorption isotherms for two materials are very close, but the strain isotherms measured during progressing adsorption exhibit very different behavior, measuring the adsorption-induced deformation can be an additional tool for the characterization of nanoporous materials [2].

Additionally, adsorption-induced deformation of some materials, such as zeolites, can influence significantly their performance in many technological processes: the adsorption-induced stress can change the permeability of zeolite membranes and thus influences the efficiency of molecular-sieve separation of adsorbed fluids [3]. Therefore, understanding of relationships between framework deformations and progressing adsorption in detail is important for the development of zeolite membranes.

In recent years there was significant progress in modeling of adsorption-induced deformation. In particular, a thermodynamic model of Ravikovitch and Neimark [4] provided the predictions of strains induced by adsorption of noble gases in CaNaX zeolite, which matched the experimental measurements quite well. This model represented the zeolite pores as uniform spheres, and predicted the strain based on the solvation pressure calculated for the rigid pore. Here we revisit the problem of adsorption-induced deformation of zeolites, modeling the zeolite structure using all-atom representation. We simulate adsorption-induced deformation using a combination of the grand canonical Monte Carlo (GCMC) and molecular dynamics simulations (MD), following the approach from [5] to study adsorption-induced deformation of these materials. We showed the overall consistency between the two models.

[1] G. Y. Gor et al., *Appl. Phys. Rev.,* **4** (2017) 011303.

[2] D. W. Siderius et al., *Adsorption,* **23** (2017) 593.

[3] T. Bowen et al., *J. Membrane Sci.,* **245** (2004) 1.

[4] P. I. Ravikovitch and A. V. Neimark, *Langmuir,* **22** (2006) 10864.

[5] S. M. Rogge et al., *Adv. Theory Simul.,* (2019) 1800177.

571516 Impact of Flexiblity on Gas Diffusion in Interpenetrated Metal Organic Frameworks

Impact of flexibillity on gas diffusion in interpenetrated metal organic frameworks

M Moshref - Javadi[a] **and** **R. Babarao**[a,b]

[a] School of Science, RMIT University, Melbourne, Victoria 3001, Australia

[b]Commonwealth Scientific and Industrial Research Organisation (CSIRO) Manufacturing, Clayton, Victoria 3169, Australia

Abstract:

Interpenetrated metal organic frameworks have evolved as a novel group of porous materials for gas adsorption and separation. They exhibit improved stability, selective adsorption, increased gas sorption, and tunability of physical and chemical properties. In most of the simulation studies, the frameworks were considered as rigid structures with fixed atomic positions. However, in reality, these structures show various types of flexibilities such as vibrations and linker rotations. In this work, by examining the dynamic behaviour of gas molecules in a series of interpenetrated frameworks with fully rigid versus fully flexible structures, novel insights into the characteristics of porous interpenetrated frameworks were presented. The results showed that gas - framework interactions, and gas diffusion highly depend on the linker type, gas loading, temperature, and most importantly dynamics of the lattice. The results obtained are of significance to the design and manipulation of porous interpenetrated metal organic frameworks with optimal properties for gas adsorption/separation applications.

References

1. Ramsahye, N. A., et al. "Adsorption and diffusion of light hydrocarbons in UiO-66 (Zr): a combination of experimental and modeling tools." The Journal of Physical Chemistry C118.47 (2014): 27470-27482.
2. Babarao, R, et al. "Interpenetrated zirconium–organic frameworks: small cavities versus functionalization for CO_2 " The Journal of Physical Chemistry C 120.24 (2016): 13013-13023.

574557 Molecular Models with Accurate Descriptions in Electrostatic Potential for Molecular Simulations of Adsorption

Nanoporous materials, such as zeolites or metal-organic frameworks, have shown great potential as adsorbents for the removal or separation of gas mixtures in energy or environment-related applications. However, while such a large database of nanoporous materials provide great opportunities, it has also made the discovery of optimum materials unprecedentedly challenging for a given application. To this end, computational materials discovery by employing classical molecular simulations can play a key role in facilitating materials discovery. To ensure the reliability of the predictions, accurate descriptions of the framework-molecule and molecule-molecule interactions are critically significant. For this reason, a molecular model that is able to accurately describe the molecule of interest is of utmost importance. Thus, we herein present a collection of molecular models of a variety of gas molecules, which their separation or purification is critically essential in energy- and environment-related processes. These gas molecules include carbon monoxide (CO), sulfur dioxide (SO_2), carbon dioxide (CO_2), nitrogen (N_2), hydrogen sulfide (H_2S), carbonyl sulfide (COS), nitrous oxide (N_2O), and ammonia (NH_3). For these models, the number of massless sites, the location of massless sites, and the point charges are first fully optimized to reproduce the electrostatic potential (ESP) surrounding the molecule determined by density functional theory (DFT) calculations, while their van der Waals (vdWs) interaction parameters are subsequently fitted by reproducing the experimental vapor liquid equilibrium (VLE). Thanks to the optimization of key model parameters as noted above in reproducing the ESP, a variety of properties in both the adsorbed and the bulk liquid phase can be well predicted. We also identified that different vdWs interaction parameters with a given number and location of massless sites and the corresponding point charges are able to resemble the experimental VLE curve equally well. However, these variations in the vdWs parameters can still yield consistent predictions in various properties, such as binding energy and geometry and bulk liquid density, emphasizing again the importance of reproducing ESP. On the other hand, we found that radial distribution function in the bulk liquid phase is sensitive to these variations in the vdWs parameters. This suggests that a relative orientation between the molecules can serve as a good secondary reference in addition to the VLE curve in the optimization of the vdWs interaction parameters. Overall, we have developed a set of molecular models with accurate description of the electrostatic potential for small molecules that are highly relevant to a variety of energy- and environment-related separation applications, which can be readily implemented to facilitate the computational discovery of optimum adsorbents.

574397 2D Phase Transitions of Fluid Adsorbed in Cylindrical Mesopores: Influence of Adsorbent-Adsorbate Interactions

Recently, molecular simulations of methane adsorption in mesoporous cylindrical silica identified a two-dimensional phase transition associated with non-trivial densification of the adsorbed liquid, a high-density local maximum in the isosteric heat of adsorption, and hexagonal rearrangement of annular layers of adsorbed fluid [1]. This observed rearrangement of adsorbed methane was proposed as a plausible explanation for the densification of methane adsorbed in MCM-41 and SBA-15 that was observed in small-angle neutron scattering (SANS) experiments [2]. The proposed phase transition was, however, unique to the methane-silica adsorption systems, not being observed in, for example, nitrogen-silica, argon-silica, or argon-carbon systems. To better understand this unique phase transition and why it occurs for the methane-silica systems in contrast to others, we perform additional Monte Carlo molecular simulations to identify key characteristics of the methane-silica system that give rise to its appearance. Specifically, we measure adsorption isotherms, heats of adsorption, and bond-orientational order parameters for the benchmark systems that do not exhibit such phase transitions, and then smoothly transform those benchmark systems into the methane-silica system to clearly identify the boundaries in adsorbent-adsorbate parameter space that trigger this type of phase transition. Lastly, we discuss engineering consequences of the observed phase transition and how it may impact other thermophysical properties of the adsorbed fluid.

[1] D.W. Siderius, W.P. Krekelberg, W-S. Chiang, V.K. Shen, and Y. Liu, *Langmuir*, 33:14252 (2017).
[2] W-S. Chiang et al., *Langmuir*, 32:8849 (2016).

575297 The Importance of Pore Connectivity in Gas Adsorption Atomistic Simulations for Porous Polymers

Porous membranes are essential to many industrial applications; for instance, separation of hazardous volatile organic compounds used in chemical processing or carbon sequestration from the burning of fossil fuels. The ability to design porous polymeric membranes with enhanced properties (selectivity, permeability, total loading) has the potential to help combat adverse environmental effects that many industrial processes contribute to. To aid in the design of these membrane materials Monte Carlo (MC) and molecular dynamics (MD) simulations have been proven to be efficient tools for prescreening and predicting membrane performance for a range of porous materials and gasses with reasonable fidelity. However, a few challenges still exist to increasing the predictive capabilities of these molecular simulations for amorphous polymeric membranes, one such challenge being accounting for pore accessibility during MC simulations. This work seeks to address the issue of pore accessibility through combining the analysis available in Zeo++, an open source software capable of characterizing porous material structure and connectivity, with pysimm, an open source software package that can facilitate a combined MC/MD approach to predict gas adsorption for porous materials.

We present a simulation study aimed at improved predictions of the adsorption behavior of a polymer of intrinsic microporosity (PIM-1). We first perform Voronoi decomposition of the PIM-1 structures with Zeo++ and the pore accessibility is characterized from the connectivity in the Voronoi network. The porous space within in the network that is characterized as inaccessible is then restricted from MC steps. Pysimm is then utilized to facilitate a MC simulation run followed by a MD simulation in order to account for sorption induced plasticization effects. The rearranged framework is then recharacterized with Zeo++ to assess if new pores have become accessible, and the process repeats iteratively. The effects of incorporating pore connectivity on the predictions from simulation is illuminated by comparing adsorption isotherms against MC/MD simulations without accounting for pore accessibility and standard GCMC simulations.

576208 Exploiting Structured Bed Thermal Conductivity for Enhancing the Performance of Temperature Swing Adsorption Processes

It is well established that the major limitation of temperature swing adsorption processes is its inherent slow cooling times as consequence of the low effective thermal conductivity of adsorbent beds. As a result, current large scale TSA processes require the use of internal heat exchangers that render the design of these units both complex and costly. With the advent of new thermally conductive metal microfibrous adsorbent structures, such as those being developed by IntraMicron, the design of new and simpler TSA beds requiring only external jackets for heat exchanging becomes a possibility.

In this presentation, the role of both radial and axial thermal conductivity on the performance of TSA processes will be evaluated via 2D and 3D CFD models using the Comsol Multiphysics® platform. This study will also show the impact of two different packing configurations, namely, having the adsorbent particles trapped and intermixed within a metallic microfibrous media, or, having interlayers of the same media with pelletized ones. The latest results from this study will be presented.

577044 On the Use of Structured Adsorbents in Ptsa Cycles for CO_2 Removal from Spacecraft Cabins

The overall goal of this project is to design a pressure temperature swing adsorption (PTSA) process that utilizes a structured 13X adsorbent. This TSA cycle design would be perhaps a drop-in replacement for the PTSA unit on board the ISS or it may be a grass roots design for future applications. This design is based on simulations being carried out with the USC dynamic adsorption process simulator (DAPS) and validated experimentally using the USC single bed apparatus in a PTSA mode.

Two different structured adsorbents are being considered. One consists of a steel foil support coated with a layer of 13X crystals that is being developed by Catacel and the other consists of a metal microfibrous media with entrapped ground and sieved 13X adsorbent particles. These tow structured adsorbents are being contrasted against traditional beaded 13X. This study is evaluating the performance in terms of CO_2/N_2 inside the bed and outside the bed breakthrough curves, and CO_2 removal during PTSA cyclic operation. The latest results of this project will be presented.

568573 Amine-Functionalized MIL-101 Monoliths for CO_2 Removal from Enclosed Environments

Amine-Functionalized MIL-101 Monoliths for CO_2 Removal from Enclosed Environments

Shane Lawson, Connor Griffin, Kambria Rapp, Ali A. Rownaghi, Fateme Rezaei*

Department of Chemical & Biochemical Engineering, Missouri University of Science and Technology, Rolla, MO 65409-1230, United States

Abstract

Increasing CO_2 concentrations from 400 to 2500 ppm decreases cognitive test scores from the 85th to the 25th percentile. In high-risk environments, such as space craft or submarines, this diminished cognitive functioning could have life-threatening ramifications. One way to reduce these levels is by using an amine incorporated solid adsorbent. Amino-solids are particularly effective sorbents for enclosed applications because of their high CO_2/N_2 selectivity value at dilute concentrations. Although these materials can reduce enclosed CO_2 amino-solid powders should be formed into structured contactors to eliminate particulate scattering and equipment contamination. In our earlier works, [1–4] we demonstrated that 3D-printing is a facile approach to formulate structured monolithic contactors for adsorption applications. Monoliths were produced with a variety of supports, including aminosilica, zeolites, and metal-organic frameworks (MOFs), and exhibited high CO_2 adsorption capacities under many conditions. Using the techniques culminated previously, we have recently formulated 3D-printed, amino-MIL-101 monoliths, for selective CO_2 adsorption from enclosed environments. In this work, [5] a tailored version of MIL-101 was impregnated with 85 wt. % Tetraethylenepentamine (TEPA) and polyethyleneimine (PEI-800) to form the powder precursors. The powders were then formed into pastes and printed into monoliths. Pristine MIL-101 monoliths were also printed and secondarily impregnated for comparison. From CHN measurements, the pre-impregnated TEPA and PEI monoliths contained 3.5 and 5.5 mmol N/g amine content, respectively. Over five adsorption/desorption cycles, this corresponded to average adsorption capacities of 1.6 and 1.4 mmol/g, respectively, at 3000 ppm CO_2 concentration. The post-impregnated monoliths contained ~50 wt.% less amine content; leading to 0.3 and 0.6 mmol/g CO_2 adsorption capacities, for TEPA and PEI, respectively. The adsorbent stabilities were also examined. PEI's low volatility led to exceptional stability throughout the five cycles for all materials while the TEPA-impregnated powder and post-impregnated TEPA monolith leached heavily and showed reduced CO_2 uptakes by cycle two. These effects were not observed in the pre-impregnated TEPA; which exhibited near zero capacity loss by cycle five. The high stability was attributed to the covalent tethering of TEPA to the MOFs chromium center during paste densification; where ultrasonic bath's heat forwarded the reaction mechanism. When examining the adsorption kinetics, the post-impregnated monoliths showed the steepest kinetics; however, this was partially attributed to their lower CO_2 uptakes. The next-fastest kinetics were observed in the impregnated powders; followed by the pre-impregnated monoliths. Typically, structured contactors improve upon powder kinetics; however, these effects were not observed here. Because the CO_2 was dilute, the poor adsorption kinetics were attributed to a slow adsorbate diffusion rate through the support walls. Whereas the powder was entirely accessible to the CO_2, the monolith required the gas to permeate the internal structure; which acted as a rate limiting step. Improving this rate limiting step will be a target of future works. Overall, though, this work provides a novel approach to formulate structured amino-MOF contactors for CO_2 removal from enclosed environments and could be highly promising with slight modification.

References

1. H. Thakkar et al. *ACS. Mater. Interfaces* **8** (41), 27753 (2016).
2. H. Thakkar et al. *ACS. Mater. Interfaces* **9** (8), 7489 (2017).
3. H. Thakkar et al. *ACS. Mater. Interfaces* **9** (41), 35908 (2017).
4. S. Lawson et al. *ACS. Mater. Interfaces* **22** (10), 19076 (2018).
5. S. Lawson et al. *ACS. Energy Fuels* **33** (3), 2399 (2019).

571803 3D-Printed ZIF-8 Structured Adsorbents for Biobutanol Recovery

Adsorptive separations using hydrophobic materials are an interesting alternative to traditional unit operations for the recovery of biobutanol. In most industrial applications, product recovery by adsorption is carried out using a fixed-bed of (spherical) adsorbent particles. To improve mass & heat transfer characteristics, the use of particles with a smaller diameter is beneficial. Yet, decreasing the particle diameter also leads to an increase in bed pressure drops, thus limiting the particle diameter which can be used in industrial systems.

The use of adsorbent monoliths allows to uncouple the relationship between mass & heat transfer and bed pressure drop. 3D-printing to synthesize these adsorbent monoliths gives a much larger design freedom, compared to traditional methods. In this contribution, the development of 3D-printed ZIF-8 adsorbent monoliths for the recovery of renewable alcohols will be discussed. Characterization of the monoliths was carried out via X-ray diffraction, infrared spectroscopy, scanning electron microscopy and the measurement of vapor phase alcohol isotherms. The separation efficiency of these monoliths was evaluated via the performance of breakthrough experiments using model fermentation mixtures. The effect of gas velocity on butanol adsorption, as well as the effect of the regeneration temperature on the adsorption cycle time was investigated.

573241 Significance of Flow Mal-Distribution and Channel Imperfections in Structured Monolith-Based Adsorption Processes

Structured adsorbents offer the advantage of extremely low-pressure drops, even at high gas superficial velocities. However, channel imperfections and inlet-flow mal-distribution often add to the intrinsic dispersion in structured adsorbent systems. Traditionally, these structured adsorbent systems are assumed to consist of identical channels and in order to match the breakthrough curve, an 'effective' diffusivity is often estimated. Since the root cause of the excess dispersion is not addressed, employing the effective diffusivity while modelling a complete cycle results in erroneous predictions. Here, these effects are modelled by considering a corrugated-structured, monolithic adsorbent system to be composed of a finite number of sectors and each sector to further consist of a finite 'types' of channels. The different sectors and the types of channels allow us to account for inlet flow mal-distribution and channel imperfections, respectively. The flow distribution and fraction of different type of channels are determined in order to match the observed breakthrough curve. Only introducing both flow distribution effects and different channel types it is possible to reproduce the experimental response. This is to be contrasted with using a simple column model and determining an effective mass transfer coefficient to capture the spread of the breakthrough curve.

Full cycle predictions are then performed for CO_2 capture from a dried flue gas stream. Comparisons between the 'real' monolith and an 'ideal' monolith with the actual mass transfer coefficient and with the apparent coefficient that matches the breakthrough curve are discussed. The full cycle simulations of the 'real' monolith are shown to reproduce correctly experimental results[1].

[1]Mohammadi, N., 2017. CO_2 capture from flue gas by a PSA process using a novel structured adsorbent. *Doctoral Thesis*. The University of South Carolina

572182 Development of a 3D-Printed Honeycomb Monolith Adsorbent for CO_2 Capture with Electric Swing Adsorption Process

Introduction and Objectives

3D printing technology was developed in 1980, but at the beginning of this century, its development increased significantly. Depending on the application and on the raw materials, the objects can be printed by different methods, such as Direct Ink Writing method. This method allows to print structures with solid free-form fabrication from an ink with high viscosity. This printing is done with a pressure delivery of an ink through one or multiple capillaries or syringes. The great flexibility in the raw materials and in the type of structure's and dimensions are advantages of this method. Recently, additive manufacturing gained worldwide attention in the development of adsorbents for gas separation processes, especially in monolithic shaped adsorbents production. There are only a few studies about the preparation of adsorbent materials by 3D printing processes, in particular, for application in CO_2 capture from flue gases by Electrical Swing Adsorption (ESA) process. As a type of Temperature Swing Adsorption process, known since 1970, in the ESA process the adsorbent is regenerated by increasing the temperature, but this heat is generated by applying an electric current directly in the adsorbent, and the temperature increases through the Joule effect. This type of process could presents main advantages, such as, higher efficiency, minimization of the lost heat, smaller systems needed, purge gas flow can be controlled independently of the heating rate, among others.

In this context, the development of a new material, with electrical conductivity and high CO_2 adsorption capacity properties, is the aim of this work.

This work reports the development of a honeycomb monolith with 70% of zeolite 13X and 30% of activated carbon with dimensions of 30×30×43 mm by 3D-printing process.

Results and Discussion

A mixture with 21.1 g of 13X zeolite (Chemiewerk, Kostritz), 9.1 g of activated carbon (Maxsorb MCS-30, Kansai Coke and Chemicals Co Ltd) and 1.5 g of CMC (VWR) of dried powders was prepared in a mixing chamber of a Multi Lab Extruder (Caleva Process Solutions Ltd, England). After that, water was added to form a homogeneous wet mixture. The mixture was loaded in a syringe, which was placed in a Discov3ry Extruder module (Structur3D Printing, Canada) to be printed, using the monolith design constructed in SolidWorks 2017® software. The Discov3ry is adjusted in order to be capable of exerting the required pressure to push the ink through the syringe orifice.

Before the 3D printing step, the prepared ink was analysed in terms of its rheological behaviour. Rotational parallel plate and oscillatory measurements were carried out using a rheometer (Anton Paar GmbH MCR 92, Austria). These rheological measurements demonstrated that the ink has a shear-thinning behaviour and high near-zero viscosity (15701 Pa.s), promoting the deposition process in the printing step and the extrusion with pressure, respectively. For a successful printing step, the important parameters defined were layer height (0.6 mm), wall thickness (0.84 mm), infill density (100%) and print speed (1.5 mm/s). In addition, two fans were used to dry the material during the printing step.

Textural characterization of the honeycomb monolith were done by several techniques, such as, N_2 adsorption at 77 K, CO_2 adsorption at 273 K, mercury porosimetry and SEM/EDS analyses. This characterization allowed the estimation of the surface area of the material (2028 m^2/g, with Langmuir method) and the volume of micropores (0.93 $cm^3{}_{STP}/g$, with Density Functional Theory). CO_2 isotherm showed type I behaviour, according to the IUPAC classification, characteristics of microporous material. Considering the weight proportion of zeolite 13X and AC in the 3D-printed material, it was possible to estimate the loss of CO_2 adsorption capacity. At 0.15 bar, the expected value for the CO_2 adsorption capacity was of 4.13 mol/kg (i.e., adding 66.7% and 28.7% of CO_2 adsorption capacity of zeolite 13X and activated carbon powders, respectively, being 4.6% of binder content). The real capacity measured in 3D-printed honeycomb monolith was 3.49 mol/kg, representing only a loss of about 15% (at 0.15 bar).

Adsorption equilibrium data of pure gases, CO_2 and N_2, at three different temperatures in the pressure range of 0 to 1.5 bar were measured by gravimetric method using a magnetic suspension balance (MSB, Rubotherm®, Bochum, Germany).

The experimental data was fitted considering the multicomponent Dual-Site Langmuir model. The parameters of Dual-Site Langmuir model were obtained for each adsorbate minimizing the sum of squared absolute errors between the calculated and experimental values. A good fit for modelling the carbon dioxide and nitrogen adsorption on honeycomb monolith was obtained, as can be seen from the comparison between isotherm model and experimental data.

Dynamic studies of the 3D-printed monolith were carried out by single (CO_2) and binary (CO_2/N_2) breakthrough experiments at 298 K and 1.3 bar. A mathematical model, including mass, energy and momentum balance was proposed and implemented in gProms software. Simulations results were validated with the experimental breakthrough results.

Acknowledgments: This work was financially supported by: Associate Laboratory LSRE-LCM - UID/EQU/50020/2019 - funded by national funds through FCT/MCTES (PIDDAC) and national funds through FCT (scholarship programme with reference PD/BD/105981/2014).

577899 Utilizing Gel Electrophoresis to Separate Aggreate after Photo-Crosslined Proteins

A key protein involved in Alzheimer's disease is amyloid beta protein (Aß). This disease effects more than 5 million US citizens to date. When amyloid beta starts to misfold and misbaheve, then the disease takes over. The misfolding of this protein leads to aggregation of proteins that cluster together to form different sized species. It has been researched that some of these have been shown to be neurotoxins. In order to properly treat the disease, understanding when these aggregations occur is crucial. By photo-crosslinking proteins, different aggregates can then be studied. Here we show that differences in size aggregates can be separated by gel electrophoresis after being photo-crosslinked to stabilize their forms.

578015 Pilot-Scale Design of Downstream Processing for Viral Therapeutics Using Aqueous Two-Phase Systems

There is an increasing demand for new viral-based vaccines and therapies to treat and prevent infectious diseases. Due to the increasing need and limited yields with current downstream unit operations for viral therapeutic production, alternative and high throughput technology developments are necessary. Also, the focus on cost reduction while increasing throughput is driving the desire for continuous processing in end-to-end manufacturing. However, the complexity, fragility, and lack of complete characterization of virus surfaces are hindering the development of platform processes for downstream operations. To address the yield and throughput challenges, we have explored a non-conventional method, aqueous two-phase system (ATPS), to improve current unit operations. ATPS is formed by mixing two partially miscible solutions that provide inexpensive, mild, and environmentally-friendly advantages over conventional methods. In our previous work with a polyethylene glycol (PEG) and citrate system, we have determined the key influential driving forces necessary for optimization to achieve near-complete separation of viruses and contaminants within a compact experimental space. A better understanding of the partition mechanism has made this method viable which was a key factor in restraining industrial implementation of ATPS. In this study, we will address the engineering design considerations (e.g. mixing rates, settling rates, settler sizing) of ATPS to operate at a large scale and in continuous mode. This work will supplement the drive to use ATPS as a continuous unit operation at industrial scale to enhance the throughput of viral therapeutic products at reduced production costs.

564681 Optimization of Hydrodistillation Extraction Process of Tea Tree Oil Using Response Surface Methodology

The optimization of hydrodistillation extraction of tea tree oil (TTO), i.e. the essential oil of *Melaleuca alternifolia*, was carried out with the response surface methodology. Nonionic surfactants, Tween 20 and Tween 80, were added in the still to help the solubilization of TTO to the liquor and, consequently, to increase the extraction yield of TTO by the hydrodistillation process. In principle, this study evaluated the relevance of several independent parameters, including the surfactant concentration, the extraction time, and liquid/solid ratio, with respect to TTO yield with a design of experiment (DOE) based on response surface methodology. Central composite design (CCD) was used to optimize the conditions of TTO extraction process. The chemical compositions of tea tree oil were analyzed and quantized by GC-FID and were refereed with the international standard ¡§ISO 4730¡¨. Additionally, TTO obtained from optimal condition was examined on the stability of its microemulsion formulations as well as the antibacterial property, compared with the commercial TTO.

Keywords: *Melaleuca alternifolia*; Tea Tree Oil; Extraction; Design of experiment (DoE); Response Surface Methodology (RSM)

Figure 1: Effect of surfactant (Tween 20) concentration on the hydrodistillation extraction kinetics of tee tree oil.

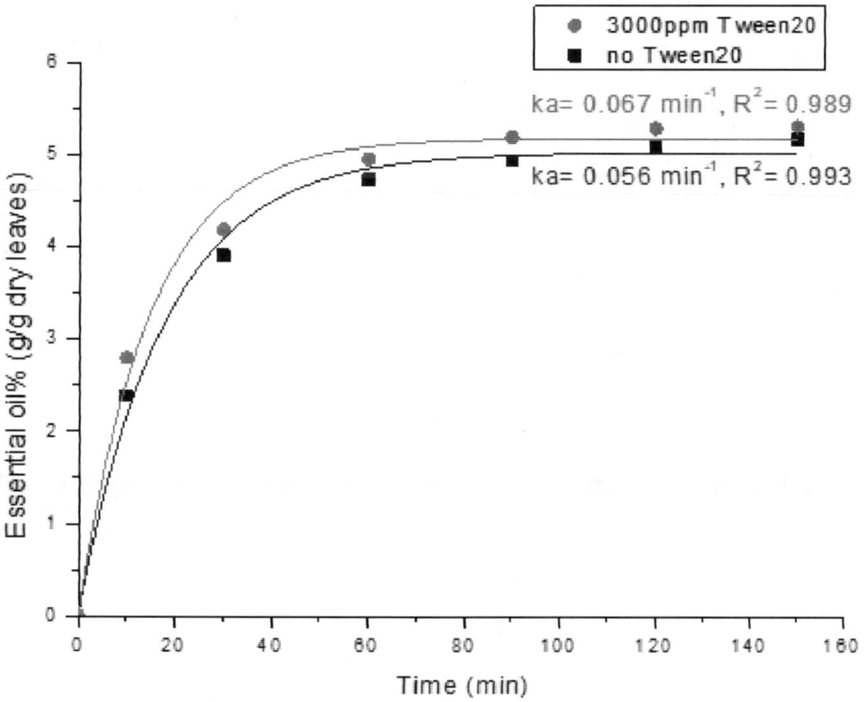

568070 Magnetic Separation of Apoptotic Red Blood Cells for Better Blood Transfusion Practices

In 1936, Pauling and Coryell observed that human RBCs in different chemical state express different magnetic moment due to the number of unpaired electrons between the hemoglobin and the stored iron. This difference in magnetic moment has led many researchers to explore the possibility of magnetically separating the RBCs. In 2017, Chalmers et al. has found that RBCs lose an average of 17% of their hemoglobin during ex vivo storage using magnetic characterization method. This finding has brought many attentions around the blood community for the need of better blood transfusion practices. In this study, we focus on attempting to find the relationship between the loss of hemoglobin in RBCs and the translocation of phosphatidylserine (PS, a phospholipid commonly used for detecting apoptosis of animal cells) and investigate the feasibility of magnetically separating the apoptotic RBCs from the fresh, healthy RBCs for blood transfusion.

Five expired human RBC units were obtained and were oxidized with sodium nitrate solution into paramagnetic RBCs. After labelling the RBCs with Annexin-V FLUOS (a fluorescent marker that would bind to translocated PS), the sample was sorted into four population based on the intensity of the marker expression using flow cytometer. The sorted samples were then magnetically characterized using the cell tracking velocimetry system (CTV), an analysis system that uses scientific camera and magnetic microfluidic channel to measure/quantify the magnetic susceptibility of cells. Then, the magnetic deposition microscopy (MDM), which is a microfluidic magnetic separation system, has been numerically modelled using a combination of commercial computational fluid dynamics software (ANSYS-Fluent) and custom algorithm that uses the magnetic characterization data from CTV. This numerical simulation method was used to calculate the theoretical separation site of the paramagnetic RBCs.

The magnetic data of the sorted RBCs were converted into pgHb/cell, a common unit measure used in the clinic to detect low iron deficiency anemia, and high PS expressing sub-population was shown to be low in hemoglobin concentration compared to the other sub-populations. The numerical model of the MDM system has also shown that different magnetic moment of entities can lead to different location of the deposition location, implying that the different magnetic moment of high PS-expressing (damaged) RBCs and PS-negative (healthy) RBCs could be used to magnetically separate healthy and damaged RBCs.

Annually, an average of 21 million units of human blood components are transfused in the US for various medical purposes. However, the blood transfusion efficiency (the rate of transfused RBCs' survival rate in a human recipient after 24 hours) is known to be about 75%, which can lead to serious clinical effects such as impaired oxygen delivery, iron overload, and many more. With the improvement in transfusion practices through these findings in research, more effective blood transfusion will be possible.

556053 Isolation and Purification of Biosurfactant Mannosylerythritol Lipids from Fermentation Broth with Methanol/Water/*N*-Hexane

High-performance separation and purification present essential obstacles for the application of biosurfactant mannosylerythritol lipids (MEL) in many fields. In this study, we developed an extraction process, which efficiently isolates MEL from fermentation broth with the solvent system methanol/water/*n*-hexane. At pH 2, extraction with methanol/water/*n*-hexane at a ratio of 2:1:1 recovers about 80% MEL, and the following extraction with methanol/water/*n*-hexane at ratio of 3:1:1 can further recover about 14% MEL. After that, MEL were purified using a methanol/*n*-hexane mixture with a ratio of 1:1. The whole presented extraction process recovers over 90% MEL with high purity, excellent surface activities, excellent surface activity (reduce the surface tension of water to below 28 mN/m at a concentration of 15 mg/L), and high emulsification activity (E_{24} of 85-96%). This efficient and easy-to-operate separation process has the potential to contribute to industrial applications of MEL in many fields requiring high purity.

582858 Optimizing the Separation of Oleanolic and Ursolic Acids by Simulated Moving Bed Chromatography using a C30 Adsorbent

Optimizing the separation of oleanolic and ursolic acids by simulated moving bed chromatography using a C30 adsorbent

Ivo S. Azenha[1], José P.S. Aniceto[1,*], Adélio Mendes[2], Carlos M. Silva[1]

[1] CICECO, Department of Chemistry, University of Aveiro, 3810-193 Aveiro, Portugal;

[2] LEPABE-Faculdade de Engenharia, Universidade do Porto, 4200-465 Porto, Portugal

*joseaniceto@ua.pt

1. Introduction

Eucalyptus globulus is a predominant species in the Portuguese forest [1] and a vital resource for the pulp and paper industry. The bark residues are usually burned for energy production without any further valorization. Recently, the bark has been identified as a source of triterpenic acids (TTAs) such as oleanolic and ursolic acids [2]. These compounds are known to possess a wide spectrum of bioactivities, including anti-oxidative, antitumoral, anti-inflammatory, anti-hyperlipidemic, and anti-microbial effects [3]. Thus, under the scope of the biorefinery concept, *E. globulus* bark is a potential candidate to extract high-value compounds such as TTAs. However, their separation after extraction is challenging as oleanolic and ursolic acids are two structurally related isomers and occur simultaneously in the same natural matrix.

Simulated moving bed (SMB) chromatography is a continuous adsorption technique, which appears as an efficient alternative to batch elution chromatography. The SMB continuous countercurrent mode of operation maximizes the mass transfer driving force, providing improved productivity and reduced solvent consumption [4]. Thus, an SMB process may be a potential candidate for TTAs separation.

2. Methods

A series of impulse experiments were conducted to select appropriate mobile phases for the separation of oleanolic and ursolic acids. Experimental breakthrough curves were measured in a custom laboratorial installation. Breakthrough of pure components were conducted to determine equilibrium and mass transport parameters, by fitting a chromatographic model to the experimental data. The model considers an axial dispersion plug flow pattern with the internal and external mass transfer resistances lumped into a global linear driving force coefficient. Parameters were validated through the successful prediction of breakthrough assays of binary mixtures.

3. Results and discussion

From a series of several impulse tests methanol/water 95/5 (%, v/v) emerged as the most favorable mobile phase to conduct their continuous separation by SMB providing a value of selectivity of 1.08. The C30 column demonstrated a remarkable separation capacity for the triterpenic acids, enabling simultaneously higher selectivities and faster analysis times, when compared with previous results using a C18 packing material and the same mobile phase [5]. Equilibrium and mass transport parameters obtained from breakthrough experiments of pure acids were then successfully validated for the prediction of breakthrough experiments of binary mixtures of oleanolic and ursolic acids with errors (AARD) of 5.2 % and 11.9%, respectively. These parameters were then used to design the SMB separation of a representative mixture of oleanolic and ursolic acids from a natural extract of *E. globulus*.

A classical SMB scheme was simulated and optimized with optimal operating conditions determined using a Design of Experiments – Response Surface Methodology (DoE-RSM) approach previously presented by the authors [5]. This methodology allows SMB optimization with a low number of simulation runs, resulting in a small computational load. Purity requirements were defined while the productivity was maximized. Rigorous phenomenological simulation results, of which the concentration profile in the SMB at cyclic steady state is presented in Figure 1, demonstrated that the SMB is possible to attain purities levels of 99.9 %, for both extract and raffinate outlets, and productivities of 1.705 kg/(m3adsorbent day) with 2-2-2-2 configuration. This is a significant achievement as previous results with a C18 stationary phase showed that purities of 99.4 % and 99.1 % for ursolic and oleanolic acids, respectively, were achievable at the expense of using three columns per section (3-3-3-3), and consequently at the expense of extremely low productivities [6]. The work presented here with the C30 column represents important improvements towards the successful chromatographic separation of these triterpenic acids.

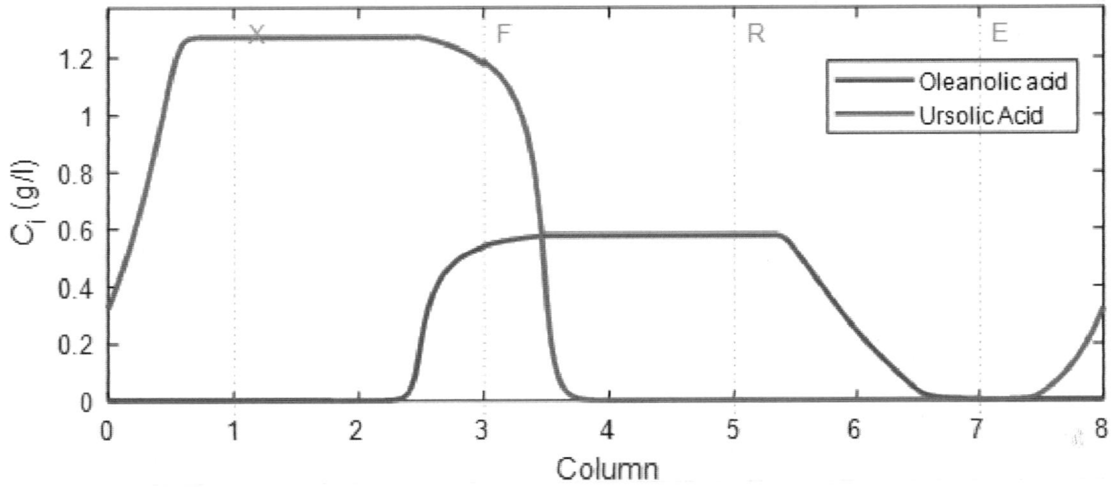

Figure 1. Simulation results of SMB operation at cyclic steady state for the isolation of oleanolic (blue line) from ursolic acid (orange line). E – eluent; F – feed; X – extract; R – raffinate.

4. Conclusions

The separation of oleanolic and ursolic acids by SMB was enhanced by applying C30 columns and a methanol/water 95/5 (%, v/v) mobile phase. The SMB unit with two columns per section, which was optimized combining the design of experiments and response surface methodologies with phenomenological computer simulations, attained purities of 99.9 %. This represents a significant improvement in terms of purity and productivity when compared with previous results obtained with C18 columns, which required a 3-3-3-3 configuration to attain purities of 99.4 %.

Keywords: chromatography, simulated moving bed, triterpenic acids

Acknowledgements

CICECO (FCT Ref. UID/CTM/50011/2019); Multibiorefinery project (POCI-01-0145-FEDER-016403);

LEPABE - POCI-01-0145-FEDER; I.S. Azenha thanks PhD grant SFRH/BD/126509/2016.

References

[1] J.S. Uva, Instituto da Conservação da Natureza e das Florestas (2013) p. 3.

[2] M.M.R. de Melo, E.L.G. Oliveira, A.J.D. Silvestre, C.M. Silva, J. Supercrit. Fluid., 70 (2012), pp. 137-145.

[3] J. Liu, J. Ethnopharmacol. 49 (1995) 57–68.

[4] J.P.S. Aniceto, C.M. Silva, Sep. Purif. Rev., 44 (2013), pp. 41-73.

[5] J.P.S. Aniceto, S.P. Cardoso, C.M. Silva, Comput. Chem. Eng. 90 (2016) 161–170.

[6] J.P.S. Aniceto, I.S. Azenha, F.M.J. Domingues, A. Mendes, C.M. Silva, Sep. Purif. Technol. 192 (2018) 401–411.

582273 Adsorption of Amoxicillin in Contaminated Water from Canebagasse Ash

Drugs have become a growing concern for the population, because many of them reach drinking water sources through wastewater or through their inappropriate disposal. This results in an increase in microbial resistance and possible effects on the endocrine system. The cane is processed mainly to obtain alcohol and sugar. Bagasse ash is obtained from incineration as a flying material in the form of ultrafine particles. One of its applications is its use as an adsorbent material. In the adsorption molecules that can be in liquid or gaseous state adhere on the surface of a solid. The objective of this study was to determine the adsorption capacity of cane bagasse ash to remove antibiotics from drinking water samples. 9 different amounts of ash were used (0.05, 0.1, 0.2, 0.5, 1, 2, 5, 10 y 20g) kore white. For each of the samples, 50 ml of the antibiotic solution (amoxicillin) was added. The percentage of antibiotic removal was measured after 1, 3, 7, 14, 21, 28 days of agitation (120 RPM). For the activation of the ash, 300 g of ash were added in a 2L beaker with 272g of zinc chloride, shaked for 2 hours, then left to stand for 48 hours and finally dried in a furnace for 72 hours. 105 ° C. The pH was brought to 7. El mejor porcentaje de remoción para la ceniza sin activar fue del 95.64% con 20 g para el día 7. However, on day 3 a very close removal of 95.3% was obtained. The ash of 10 g presented a percentage of removal very similar to that of 20 g during all measurements; and the 5 g ash achieved equally close removals on day 28. For the case of the activated ashes the best percentage of removal was 100% with the ash of 10 g for day 7. However, from day 1 a similar removal was obtained, of 98.9%. The ash of 10 g presented a better performance to the amount of 20 g. The ash of 5 g achieved removals close to those of the previous two, on day 7. After day 7 an effect of desorption of the antibiotic was found, for which it was found that for the case of the ashes without activation the best retention time is day 3 while for the activated ash it is day 1. Likewise, the activated ash achieved higher removal percentages than the non-activated ash, but its main advantage is evidenced in its ability to remove large amounts of antibiotic with much shorter retention times than with the nonactivated ash.

576546 Incorporating a Vacuum Pump Performance Curve in Cyclic Adsorption Process Simulation

Pressure varying steps, and depressurization steps in particular that utilize vacuum, are key to achieve optimum performance of pressure swing adsorption (PSA) processes. Recovery, enrichment, throughput and separation energy are intimately dependent on how and to what extent depressurization in the cycle is achieved. To gain a clearer understanding of its role in PSA processes, it is necessary that models provide adequate representation of the dynamics involved during vacuum depressurization.

One key area of interest is to properly represent how pressure and flow interact at the exit of a PSA bed. To date, most models utilize a valve equation to artificially represent the dynamics of a vacuum pump. This may lead to completely erroneous performance results because the dynamic response of the valve equation may not at all represent the dynamics of any existing vacuum pump that may be used in a PSA process. To this end, an actual pump pressure-flow relationship has been incorporated into the dynamic adsorption process simulator (DAPS) at the exit of the bed with and without mediation of a valve equation to capture the role of a vacuum pump on the performance of a vacuum depressurization step. The latest result from this work will be presented.

572019 Process Simulation of Removal Methyl Sulfide from Nitrogen Using Pressure Swing Adsorption Based on Aspen Adsorption

With the increasingly strict global environmental problems and the booming demand of chemical industry for raw material purification, desulfurization technology has gradually attracted researchers' attention. Selective adsorption desulfurization was widely used due to its low energy consumption and ultra-deep desulfurization. Based on Aspen Adsorption process simulation software, a fixed bed adsorption model of methyl sulfide-nitrogen two-component system was established for pressure swing adsorption (PSA) process with Cu^{2+} modified NaY zeolite as adsorbent. The adsorption isotherms were obtained by Monte Carlo simulation. The given operating pressure and temperature were 1 bar and 301 K, respectively. The concentration of methyl sulfide was 2000 ppmv and the average diameter of adsorbent particles was ca. 1.0 mm. The dynamic data of gas phase composition, outlet concentration and pressure in the single tower process were obtained by simulation and the corresponding breakthrough time was 1380 min. Fixed-bed adsorption experiments were carried out under the same conditions. The results show that the simulation is well consistent with the experimental data. It demonstrates that the fixed-bed adsorption model established in this study can satisfactorily describe the purification process of methyl sulfide-nitrogen adsorption.

A two-column PSA model was established, on the basis of single column analysis. The concentration distribution of CH_3SCH_3 in the two-column process and the change of adsorbent adsorption amount with time were studied. The variation of gas phase concentration and the movement of mass transfer zone along the axis of the adsorption bed in the step of boosting, adsorption, depressurization and flushing were revealed. The effects of mass transfer coefficient, temperature and pressure on the purification efficiency of PSA were systematically analyzed. A more reasonable and efficient time sequence table of the two-tower PSA process was designed via optimizing the process conditions based on the specific physical parameters, which was expected to provide reliable prediction of the adsorption breakthrough process for the selection and design of separation schemes for practical adsorption and desulfurization projects.

Key words: Pressure swing adsorption; Process simulation; Aspen Adsorption; Desulfurization; Gas purification

* Corresponding author. E-mail: yonghou.xiao@dlut.edu.cn, hgaohong@dlut.edu.cn.

559346 Evaluation and Optimization of Vpsa Processes with Nanostructured Zeolite NaX for Post-Combustion CO_2 Capture

Global warming resulted from greenhouse gases emission has been an issue of great concern. Concentration of greenhouse gases in the atmosphere has been continuously increasing over the last few decades due to the strong energy dependence of fossil fuels. Among the greenhouse gases, carbon dioxide is considered the main contributor of the global warming due to its huge emission amount. Thus, reducing anthropogenic CO_2 emission and controlling CO_2 concentration have become one of the most urgent global environmental issues. The carbon capture and storage technology (CCS) requires the reduction of carbon emissions primarily from large stationary point, such as coal-based plants. It is necessary to separate CO_2 from N_2 in post-combustion emissions before full utilization of CO_2.

Though CO_2 capture can be achieved by absorption with amine, the process is energy costly due to the high temperature requirement for regeneration. Adsorption stands out to be an attractive operation over amine capture due to its relatively low cost, flexible operating and more efficient regeneration. Vacuum pressure swing adsorption (VPSA) is an emerging technology that applies pressurized gas to the adsorption process as well as a vacuum during the desorption stage. VPSA is efficient when dealing with flue gas due to the low partial pressure of CO_2 and high partial pressure of N_2 in the flue gas stream.

The key challenge in CO_2 capture by adsorption technique is to find a suitable solid adsorbent material with good separation performance, as the adsorbent plays a critical role in the overall process performance. Many conventional adsorbents have been used for N_2/CO_2 separation, such as zeolites, activated carbons and modified mesoporous silica. However, the adsorption selectivity of CO_2 to N_2 and adsorption capacity of CO_2 are not high enough as current commercial adsorbents, which makes the adsorption process less competitive. An ideal adsorbent for CO_2 capture usually have a high CO_2 adsorption capacity, a large pore dimension that enables fast mass transfer, a decent CO_2 selectivity over $N_{2,}$ and good stability.

In this work, we present a comprehensive experimental and simulation study on a low-cost and efficient CO_2 capture technique using nanostructured zeolite NaX in VPSA process. The nanostructured zeolite NaX adsorbent has a high CO_2 adsorption capacity, a relatively high adsorption and desorption rate, and a large CO_2 selectivity over N_2. Therefore, it is suitable for post combustion CO_2 capture from dry flue gas. Current work first presents a material study for a group of nanostructured zeolite NaX samples prepared with different binder ratios and sintering temperatures. The porosity properties, mechanical properties, adsorption isotherm, kinetics, IAST selectivity, and column dynamics for the group of nanostructured zeolite NaX samples are compared. The results demonstrated that nanostructured zeolite material prepared with a low ratio of binder that sintered at 500°C has the best separation performance for a CO_2/N_2 mixture. In addition, the results indicate that the nanostructured zeolite NaX samples lead to a better separation performance compared with the commercial microsized zeolite NaX. To apply the nanostructured zeolites for CO_2 capture by VPSA, the operating conditions of the VPSA process is desired to be preliminarily designed and optimized before running the process in an actual pilot plant, which is both costly and time-consuming.

It is clear that product purity, recovery, productivity and energy consumption are four important performance indicators of the VPSA process, and they varied as the operating conditions changed. In addition, the performance of the VPSA process depends on the CO_2 concentration of the flue gas. To obtain over 90% CO_2 purity with a recovery of 90% for CO_2 capture from a flue gas containing 15% of CO_2, a deep vacuum (<0.05 bar) is often required. However, the deep vacuum involves multistage pump units which will dramatically increase the capital cost and energy consumption, which becomes the major obstacle of applying the CO_2 capture process by adsorption to the power plants. There is a great need to employ an advanced adsorbent for the VPSA process for CO_2 capture, and to optimize the process to lower the energy consumption while meeting CO_2 purity and recovery target.

Simulation and optimization studies of a two-bed six-step VPSA process was then performed in gPROMs environment for post-combustion CO_2 capture from dry flue gas using the optimal nanostructured zeolite NaX and a commercial microsized zeolite NaX. The simulation model has been validated with both breakthrough experimental data and reported data from literature. The objective is to minimize the energy consumption of the VPSA process for the specified CO_2 purity and recovery target by altering the operating conditions. The decision variables include the feed pressure, blowdown pressure, evacuation pressure, feed flow rate, and length to diameter ratio of the adsorption bed. The effect of cycle time was investigated independently due to the need for synchronization of the multi-bed configuration. The optimization results indicate that the energy consumption of the process with nanostructured zeolite is about 30% lower while achieving a higher CO_2 purity and productivity compared with a process employing a commercial microsized zeolite. This is due to the more moderate conditions (e.g. higher blowdown pressure, higher evacuation pressure and lower feeding pressure) the process required to meet specified CO_2 purity and recovery target. In short, the optimization results have shown great potential for employing nanostructured zeolite NaX by VPSA process for CO_2 capture from the flue gas with low CO_2 concentrations, as a CO_2 purity of over 95% and a recovery of 90% was obtained with a practical vacuum level (>0.1 bar). Lower vacuum levels or multi-stage operations may be possible to achieve products with higher purity.

565199 Flexible MOFs for Gas Separation – a Case Study Based on Static and Dynamic Sorption Experiments

Porous metal organic frameworks (MOF's) can be categorized into rigid and structurally flexible compounds. The latter change their structure influenced by external stimuli. Whether the corresponding phenomena such as gate opening or breathing could be utilized for gas separation applications, or not, is subject to ongoing research and controversial discussions.

In this work, two structurally flexible MOFs were selected for exploring guest-induced structural changes by adsorption of C4-hydrocarbons. Pure gas sorption isotherms of n-butane, isobutane, 1-butene and isobutene reveal a stepwise increase of the surface excess loading associated with gate opening. Based on these isotherms, dynamic sorption experiments with one- and two-component adsorptives have been carried out on a packed bed at different inlet concentrations suggested by the gate opening pressure determined in the isotherm measurements. This leads to stepwise breakthrough curves being affected by the entry concentrations. We discuss the chances and limitations originating from a flexible network transition on potential separation applications.

References

1. M. Lange, M. Kobalz, J. Bergmann, D. Lässig, J. Lincke, J. Möllmer, A. Möller, J. Hofmann, H. Krautscheid, R. Staudt and R. Gläser, Structural flexibility of a copper-based metal– organic framework: sorption of C4-hydrocarbons and in situ XRD, J. Mater. Chem. A, 2 (2014) 8075-8085 2. Lange, M. Sorption of C4-hydrocarbons on structurally flexible, porous coordination polymers; University of Leipzig, Doctoral Thesis, 2015.
2. M. Kobalz, J. Lincke, K. Kobalz, O. Erhart, J. Bergmann, D. Lässig, M. Lange, J. Möllmer, R. Gläser, R. Staudt, H. Krautscheid, Paddle Wheel Based Triazolyl Isophthalate MOFs: Impact of Linker Modification on Crystal Structure and Gas Sorption Properties, Inorg. Chem., 55 (2016) 3030-3039

Acknowledgements
We gratefully acknowledge the synthesis of the materials and the scientific collaboration with Prof. Dr. Harald Krautscheidt and his group at the University of Leipzig.

566145 Computational Modelling of CO_2 and CH_4 Diffusion in Bulk ZIF-8

Computational Modelling of CO_2 and CH_4 Diffusion in Bulk ZIF-8

Hari P. Paudel, Wei Shi, David Hopkinson, Yuhua Duan

Zeolite imidazolate frameworks (ZIFs) have selective adsorption properties and are useful materials for gas separation. As compared to traditional chemical sorbents (mostly amines), ZIFs have (1) high CO_2 loading capacity, (2) lower-energy consumption, (3) high thermal and chemical stability, (4) fewer mechanical requirements and lower corrosion rates (Banerjee et al, Science 2008, **319**, 939; Park et al, Proc. Natl. Acad. Sci. U.S.A. 2006, **103**, 10186, Nordin, et al **RSC Adv.**, 2014,**4**, 52530-52541). There have been several studies, both theoretical and experimental, that reveal that ZIF-8 material is highly responsive to changes of the external physical and chemical environments due to its hybrid metal/linker type structure. Interaction of sorbent materials with gas molecules plays a significant role in identifying materials' nature of selective adsorption and diffusion for some gas molecules over others. Therefore, understanding of ZIF-8 interaction with different gas molecules plays crucial role for further experimental and theoretical studies while considering the bulk, surface and interface composite model designs. With the understanding of gas adsorption and diffusion in bulk, surface and at interface of ZIF-8 helps in screening sorbent materials of similar or with no significant difference in their properties.

Several experimental and theoretical studies have been performed to understand the gas adsorption and diffusions in ZIF-8. While those reported theoretical studies are helpful for assessing the computational methodology and atomistic models at large, the basic theoretical understanding of adsorption processes and gas diffusion selectivity using quantum mechanical calculations is still in question. The preferential adsorption of gas molecule of one type in a gas mixture by sorbent material can hinder the diffusion of other type of gas molecule. On the other hand, due to a variation in size of gas molecules, diffusion process is affected for different type of gas molecules with different measure as they diffuse through the pores. Overall effect on transport and selectivity demands a deep understanding of interaction picture of framework with gas molecules.

Computational studies based on the density functional theory (DFT) can capture the fundamental properties of interactions and gas diffusion mechanisms that can be directly correlated with the experimental observations. We present our studies based on DFT calculations on CO_2 and CH_4 gas adsorption and diffusion in the bulk of ZIF-8. We evaluate the structural and electronic properties of bulk ZIF-8, determine the strong adsorption sites for gas molecules, and calculate the diffusion barriers for CO_2 and CH_4. Our calculations include long-range dispersion interactions. We analyze adsorption properties by fixing as well as relaxing both the framework atoms and gas molecules, examine their effects on gas diffusion properties, and compare with the experimental results where available. We find that the CO_2 and CH_4 adsorption energies at the strongest adsorption sites are 5.01 and 4.65 kcal/mol, respectively. The diffusion energy barriers for CO_2 and CH_4 are attractive, and are calculated to be -3.65 and -4.01 kcal/mol, respectively. With the above value of diffusion barrier, the diffusion coefficient for CO_2 is calculated to be 1×10^{-10} m^2/s.

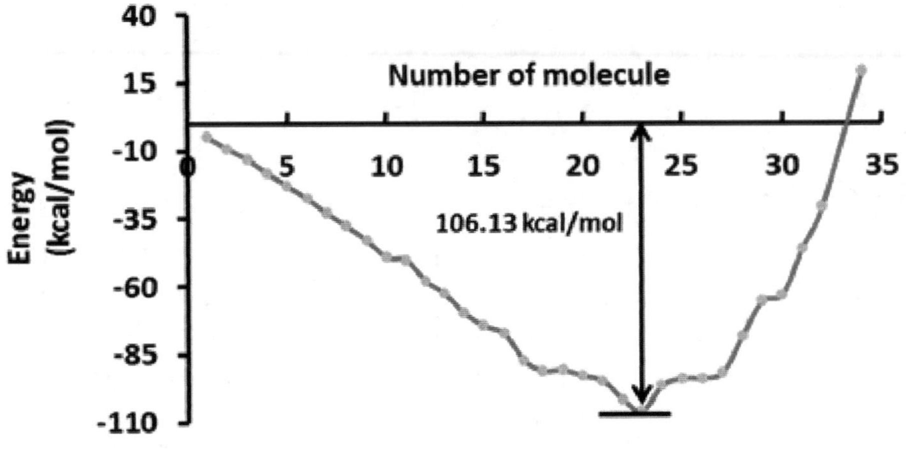

CO_2 adsorption energy

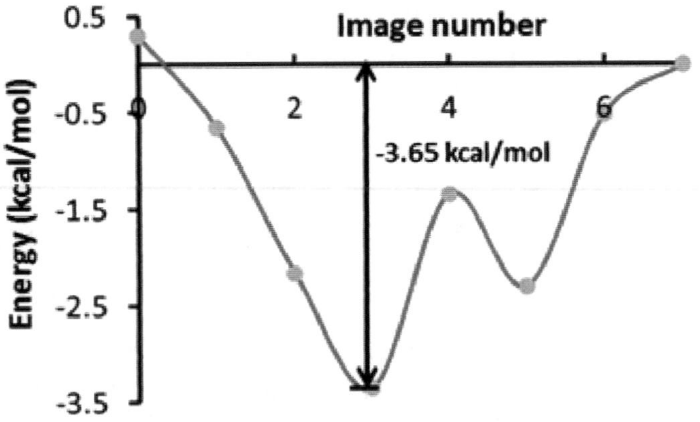

CO_2 diffusion barrier

567135 Adsorption Kinetics of CO_2 By Amine Groups

Amine functionalized adsorbents use for CO_2 capture is a promising technique due to its several advantages that include least cost, ease of operation and its ease of regeneration ability. This study aims to investigate the adsorption of CO_2 using adsorbents modified with amine group. In addition, the adsorption kinetic models that have been used in describing CO_2 adsorption by the amine groups is provided. It is found that the adsorption rate of CO_2 onto adsorbents material modified with amine groups is highly dependent on the interaction of CO_2 with the amine groups forming carbamate and bicarbonate (in the presence of water vapor). In addition, the mass transfer of CO_2 from the bulk gas phase into the active surface sites of the adsorbent also contribute significantly to the adsorption rate.

Single and binary component adsorption of 1-alcohols from an alkane using various activated alumina adsorbents

Catharine E. Bosman[1] and Cara E. Schwarz[1]*

[1]*Department of Process Engineering, Stellenbosch University, Banghoekweg, Stellenbosch, 7600 South Africa*

cschwarz@sun.ac.za +27 21 8084487

1. Introduction

Surfactants are compounds used to reduce surface tension between two or more immiscible mediums. Surfactants usually consist of a hydrophobic backbone and a hydrophilic tail. The first step in the production of surfactants is to graft a hydroxyl functional group onto an alkane resulting in an alcohol molecule. Complete conversion is not achieved which brings about the necessity of alcohol-alkane separation to enhance efficiency of surfactant production. Distillation and other techniques are used to separate these materials; however, some alcohol contaminants remain in the alkane stream after distillation impeding recycling of the alkane stream. The purification of an alkane stream through the process of adsorption with activated alumina has been shown to be a technically viable process [1]. A study by Groenewald investigated the single component adsorption of 1-alcohols from an alkane using activated alumina in a limited temperature range. There is however little knowledge on the binary component adsorption behaviour of an alcohol-alkane system using activated alumina as adsorbent.

The aim of this study is therefore to extend knowledge on single component adsorption to a wider temperature range and to gain insight into binary component adsorption of alcohol contaminants from an alkane stream through (i) the investigation of single component adsorption behaviour of alcohols (1-hexanol, 1-octanol and 1-decanol) from an alkane (n-decane) using various activated alumina adsorbents at 45°C to complement previously measured data at 25°C to 35°C; (ii) the investigation of binary adsorption behaviour of alcohols (1-hexanol, 1-octanol, 1-decanol) form an alkane (n-decane) using various activated alumina adsorbents; and, (iii) the comparison of single and binary component adsorption behaviour.

2. Methodology

Adsorption data was obtained using a bench-scale batch adsorption system. The set-up consists of beakers mounted inside a water bath at constant temperature. Beakers containing a previously determined quantity of alcohol-alkane solutions were immersed in the water bath and allowed to heat up to the specified temperature. A known amount of activated alumina adsorbent (Activated Alumina F220, Selexsorb CDx® or Selexsorb CD®) was added to the solutions. The alcohol-alkane solutions comprised n-decane with a known concentration of either 1-hexanol, 1-octanol or 1-decanol, or an equimass binary mixture of two of the mentioned alcohols. The initial alcohol concentration was in the range of 0.3 to 3 mass% and was varied to determine the effect of initial adsorbate concentration. The effect of temperature was investigated by varying the temperature from 25°C to 45°C and comparing the adsorption behaviour at the different temperatures. In addition, the effect of alcohol chain length in a single component system, and combination of different alcohols in a binary component system were also investigated. The effect of the different variables was investigated based on adsorption kinetic behaviour as well as adsorption equilibrium behaviour.

3. Results and Discussion

The three different adsorbents that were investigated exhibited relatively similar adsorption behaviour both with regards to kinetics as well as equilibrium, with Selexsorb CDx® and Selexsorb CD® achieving slightly greater removal efficiencies (Figure 1).

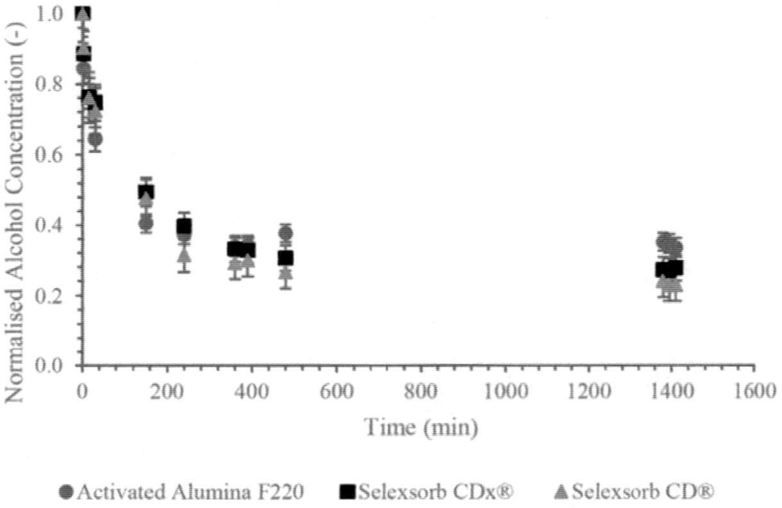

Figure 1: Concentration decay plots for the adsorption of 1-hexanol (initial concentration of 1 mass%) onto various activated alumina adsorbents, at 45°C

As expected, Selexsorb CDx® and Selexsorb CD® achieved greater removal efficiencies as these adsorbents had both surface areas as well as micropore volumes notably greater than Activated Alumina F220. Activated Alumina F220, Selexsorb CDx® and Selexsorb CD® had micropore volumes (Density Functional Theory analysis) of 0.049, 0.114 and 0.090 cm^3/g respectively. Activated Alumina F220, Selexsorb CDx® and Selexsorb CD® had BET surface areas (Brunauer-Emmett-Teller analysis) of approximately 334.34, 466.01 and 408.28 m^2/g, respectively.

The adsorption capacities of the various adsorbents are depicted in Figure 2. It can be observed that the adsorbents had very similar capacities with Selexsorb CD® having a slightly greater adsorption capacity than the other two adsorbents.

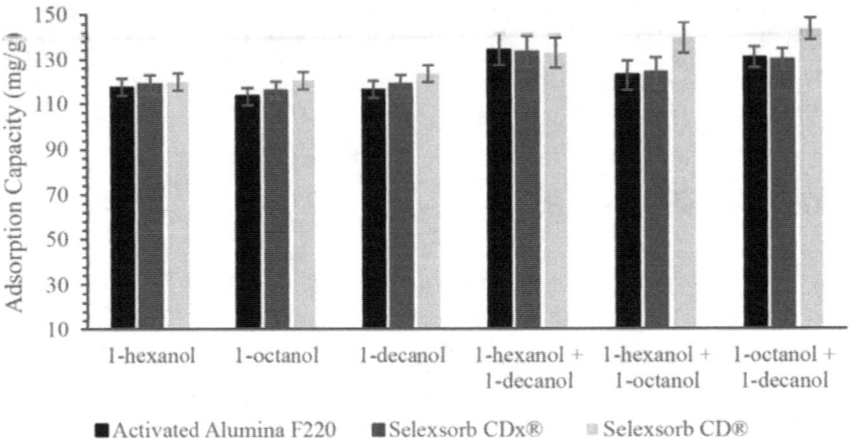

Figure 2: Equilibrium adsorption capacity (total alcohol) for different adsorbate systems, at 45°C

Temperature had a positive effect on the adsorption kinetics of all systems investigated, thereby increasing the adsorption rate with an increase in temperature. This is expected and is widely reported in various adsorption studies. Temperature also had a notable effect on the equilibrium adsorption capacities of the various single and binary component systems with equilibrium adsorption capacity exhibiting an increase with increasing temperature, in the range 25°C to 45°C. This suggested that adsorption was occurring either through chemical bonds alone, *i.e.* chemisorption, or through chemical bonds as well as physical forces, *i.e.* a combination of chemisorption and physisorption.

Initial alcohol concentration proved to be significant in the equilibrium adsorption behaviour. For an increase in initial alcohol concentration from approximately 0.3 to 1 mass%, equilibrium adsorption capacity increased notably. However, for initial concentrations greater than 1 mass% the capacity remained relatively constant and it is believed that the saturation capacity of the adsorbent was reached (Figure 3).

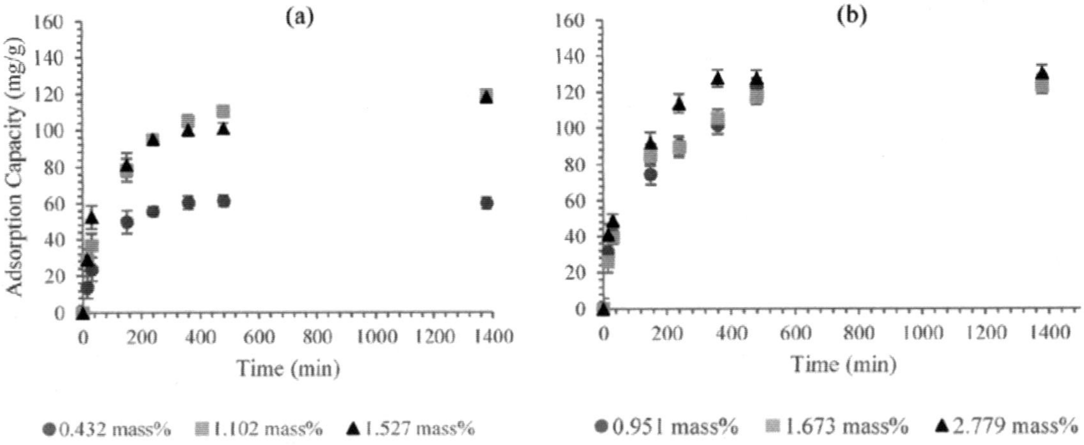

Figure 3: Time profiles for (a) the single component adsorption of 1-octanol onto Selexsorb® CD; and, (b) the binary component adsorption of a 50:50 mixture of 1-octanol and 1-decanol onto Selexsorb® CDx (T = 45°C)

The alcohol carbon chain length did not have such a pronounced effect on adsorption for the alcohols investigated. There was at times a slight decrease in equilibrium adsorption capacity for the longer chain length alcohols in the single component systems. In the binary component systems, the alcohols mostly adsorbed in the same ratio as initially in the solution, with all three adsorbents exhibiting a slight selectivity for the shorter chain alcohol.

The kinetics of alcohols in both a single component system as well as a binary component system were investigated and compared. It was found that the rate of adsorption of a single component remains relatively unchanged in a binary component system when compared to a single component system. The initial alcohol uptake in the single component system however was notably faster than in the binary component system and may be ascribed to possible adsorbate-adsorbate interaction.

Similar to the adsorption kinetics, the equilibrium behaviour of the single and binary component systems was also investigated and compared. The most significant difference between the single and binary component systems was found to be the equilibrium adsorption capacity. For equal initial concentrations of an alcohol, the uptake of that specific alcohol in a single component system proved to be significantly greater than that of the same alcohol in a mixture of two different alcohols (50:50 by mass) (Figure 4). This suggested antagonistic adsorbate-adsorbate interaction with the alcohol molecules competing for the same active adsorption sites. According to the interaction effect parameter, the adsorption of longer chain length alcohols was inhibited more by the presence of a shorter chain length alcohol in a binary component system, than the adsorption of a shorter chain length alcohol was inhibited by the presence of a longer chain length alcohol.

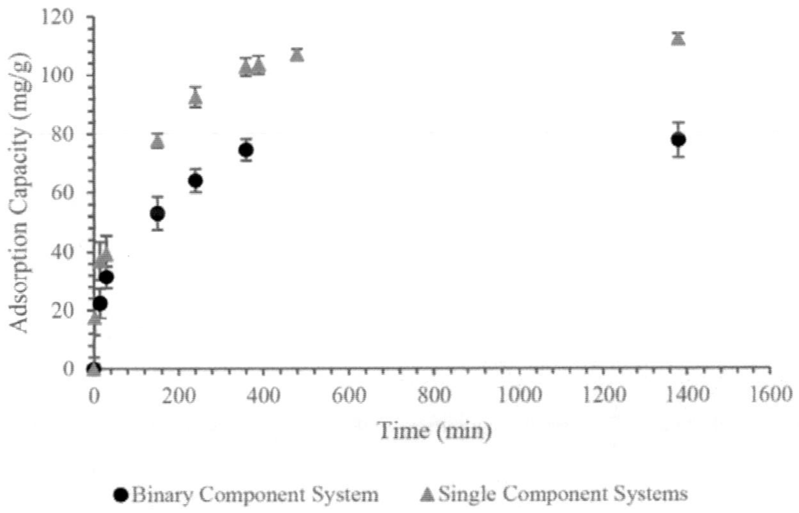

Figure 4: Comparison of the kinetic adsorption behaviour of 1-hexanol from n-decane in a single component system and in a binary component system consisting of 1-hexanol and 1-decanol (50:50 by mass), onto Selexsorb CDx® at 45°C (initial concentration for 1-hexanol alone in both the single and binary component systems was 1 mass%)

4. Conclusions

In conclusion, adsorption by activated alumina was shown to be a viable solution for the purification of alkane streams in the surfactant production process. The kinetic and equilibrium behaviour of both single and binary component adsorption systems were investigated, and it was found that initial alcohol concentration had the greatest effect on the adsorption behaviour within the alcohol concentration range investigated, with antagonistic adsorbate-adsorbate interaction present in the binary alcohol systems.

Equilibrium adsorption capacity increased with increasing temperature in the temperature range of 25°C to 45°C and the shorter chain 1-alcohols were adsorbed slightly better than the longer chain alcohols. It was also found that adsorption of the longer carbon chain alcohols was inhibited more by the presence of a shorter chain alcohol than vice versa.

The next step in the study would be to model the adsorption kinetics and equilibrium behaviour by use of kinetic models such as the pseudo-second-order model and the intra-particle diffusion model, and equilibrium isotherm models such as the single and multicomponent Langmuir and Freundlich isotherms. This will allow for further investigation of the adsorption behaviour of these systems.

References

[1] J. Groenewald, "Evaluation and comparison of the ability of three industrially relevant adsorbents to remove alcohol contaminants from an alkane solvent," University of Stellenbosch, Stellenbosch, 2019.

579203 Chitosan Functionalized with Histidine to Increase Metal Ions Adsorption

Chelating nanoparticles able to cross the blood-brain barrier (BBB) can be an effective pharmacological treatment to Alzheimer's disease (AD). Chitosan microparticles (CM) associated with histidine (composed by imidazole functionalities, to mimetize the Aß peptide) were used to chelate copper, zinc and iron. CM (prepared by spray-drying) and CM-histidine were characterized by SEM, XRD and potenciometric titration. The equilibrium data of metallic ions adsorbed onto chitosan-histidine systems were described by Langmuir and Sips models. The estimation of the mass transfer parameters was carried out using the Pore-Surface Diffusion Model. Histidine presented highest adsorption capacity for copper ions, which increased the adsoption capacity for chitosan-histidine compared to chitosan system. This is a very interesting finding for metal ions separation, with chelating potential to treat heavy metal poisoning.

571714 Successful Recovery and Concentration of 5-Hydroxymethylfurfural from Aqueous Solutions Onto Various Polymeric Adsorbent with Different Functional Groups

5-hydroxymethylfurfural (5-HMF) is a high value-added platform chemical that can be obtained from glucose, fructose, and directly biomass via chemical catalysis technology or biorefinery. Although a great deal of research about 5-HMF has been published in the past decade, the effective separation and purification of 5-HMF is still a drawback for its sustainable utilization for producing other high value-added chemicals on a commercial scale. In order to address the drawback, chromatographic adsorption separation technology has received more attention for the selective liquid adsorption of 5-HMF from aqueous solution or actual biomass hydrolysate in recent years.

Here, we investigate the separation behavior of glucose, 5-HMF and its follow-up products LA and FA from aqueous solutions onto various polymeric adsorbent with different functional groups (XAD-4, XAD7HP and XAD761 resins). The result showed that XAD761 resin exhibit highest selectivity and capacity for 5-HMF. The maximum adsorption capacity of 5-HMF onto XAD761 was 92.2 mg g^{-1} wet resin at 288 K. The adsorption equilibrium was better fitted by the Freundlich isotherm model at the studied range of 5-HMF concentrations. The adsorption process of XAD761 resin for 5-HMF was the spontaneous and exothermic, and the activation energy suggested physical nature of the adsorption process. The kinetic regression results showed that the kinetic data of 5-HMF was accurately followed by the pseudo-second-order kinetic model. In conclusion, the present adsorption studies of 5-HMF onto XAD761 resin revealed the potential of phenol-formaldehyde resin with hydroxy group to be applied as an adsorbent for aldehydes organic compounds.

565671 Development of Highly Selective Composite Adsorbents for Selective Recovery of Gold (Au^{3+}) and Silver (Ag^+) Ions from Electronic Wastes (E-wastes)

Production of electrical and electronic devices with short life cycles and lack of suitable recycling technologies has led to the surging electronic waste generation worldwide. E-wastes such as spent mobile phones contain higher amount of precious metals compared to their mineral ore sources hence are attractive secondary sources for recovery and recycling. Many adsorbents derived from biomass, modified magnetic nanoparticles, and MOFs have been developed for the recycling/recovery of these valuable metals like gold (Au^{3+}) and silver (Ag^+) from e-wastes. However, most of these adsorbents are not selective towards Au^{3+} or Ag^+. Meanwhile, crown ethers (CEs) are attractive macrocyclic complexing agents as their cavity structures can be tailored to accommodate the ionic size of a precious metal and their heteroatoms (i.e. O, N, or S) can be selected to afford their affinity towards Au^{3+} and Ag^+.

In this study, 13-, 15-, 17- and 19- membered CEs with sulfur donor atoms (thia-CEs) were synthesized and demonstrated to form stable complexes with Au^{3+} and Ag^+. The thia-CEs were prepared by intermolecular cyclization of bulky epoxide intermediates with 1,2-benzenedithiol. These thia-CEs were subsequently immobilized on porous polypropylene membrane (PP) and mesoporous silica (SBA-15) to fabricate thia-CEs @PPs and thia-CEs @SBA-15 composite adsorbents, respectively. The developed composite adsorbents (thia-CEs @PPs and thia-CEs @SBA-15) were thoroughly characterized by both computational (DFT) and experimental (high-resolution spectroscopy) methods such as FTIR, ^1H and ^{13}C NMR, HRMS, TGA, BET, XRD, SEM and ICP-MS. The full characterization results showed that the composite adsorbents were successfully prepared. The immobilization of the thia-CEs onto solid support materials (PP and SBA-15) enhanced the reusability of the active ionophores (thia-CEs). Moreover, DFT result showed that the thia-CEs have different cavity sizes (13-thia-CE = 1.31−1.94 Å; 15-thia-CE = 2.19−2.52 Å, 17-thia-CE = 2.12−2.21 Å and 19-thia-CE = 2.27−2.43 Å) which are responsible for the metal-ion specificity. The 19-thia-CE@PP is highly selective towards Ag^+ ions (2.30 Å), whereas the 15-thia-CE@PP is highly selective towards Au^{3+} ions (1.36 Å). Adsorption experimental result reveled that 19-thia-CE@PP has higher adsorption capacity (q = 206 mg Ag^+/g) and selectivity towards Ag^+($\alpha_{Ag^+/M^{n+}}$= 25 – 287) as compared to other cations (M^{n+}= Cu^{2+}, Ni^{2+}, Zn^{2+}, and Pb^{2+}) present in leached PCB e-waste. Similarly, Au^{3+} adsorption experiments using thia-CEs@SBA-15 were conducted and preliminary results showed that the 15-thia-CE@SBA-15 has high selectivity towards Au^{3+} in the presence of Pd^{2+} and Pt^{4+} ions. Overall results demonstrate the suitability of the composite adsorbents (thia-CEs@PPs and thia-CEs@SBA-15) for valuable metal recovery applications. Furthermore, the synthesis strategy described in this work has great potential for a wide range of applications such as metal sensing, metal recovery and catalysis. This research was supported by the National Research Foundation of Korea (NRF) under the Ministry of Science and ICT (No. 2018R1D1A1B07047503, No. 2017R1D1A1B03028102, and 2017R1A2B2002109) and the Ministry of Education (No. 22A20130012051(BK21Plus)).

573465 High Vying $Ti_3C_2T_x$ Mxene Nanosheets with Mehpa to Release Mercury from a Loaded Oil Phase

Solvent extraction is a method of selectively removing metal ions present in very low concentration in an aqueous medium. In recent years there has been an increased interest in using the technique in wastewater treatment[1-4]. Because metals are usually not soluble in organic solvents, the process requires the introduction of an extractant that will combine with the metal ion to form an organic soluble complex. The loaded solvent is then stripped with a suitable acidic aqueous phase, and the metal is transferred from the oil phase to the desired aqueous solution or stripping phase. The advent of nanotechnology has provided a great opportunity for the invention of desired nanomaterials with large surface-to-volume ratios and unique functionalities to treat pollutants. Nanomaterials now play a key role in environmental remediation and are used for the treatment of natural waters, sediments, soils, and industrial and domestic waste water[5-9]. Two-dimensional (2D) nanomaterials have been found to be efficient adsorbents for heavy metals[10-12]. Their abundant active sites, negative zeta potential, excellent dispersity in water, and large surface areas have made them ideal adsorbents for environment remediation[11-23]. The appealing features of 2D nanomaterials have motivated us to use them as oil-phase strippers in solvent extraction. Sorption of mercury removal from aqueous solutions has been studied by many researchers[24-27]. However, heavy metals stripping from an oil phase has not been reported yet.

In this paper, we present results from our study of mercury stripping from a loaded oil phase. First, we conducted solvent extraction studies to transfer mercury ions from drinking water to an organic phase (kerosene) by using di(2-ethylhexyl) phosphoric ethyl hexyl) phosphoric acid (MEHPA) as the extractant. We determined the mechanism for the formation of MEHPA-Hg complex by assuming a thermodynamically ideal extraction system and using the Ritcey and Ashbrook model[28-29]. We then used different 2D nanomaterial suspensions like GO, MoS_2, MnO_2 and $Ti_3C_2T_x$ MXene as stripping reagents. The 2D material suspensions were added to a mixture of the loaded organic phase (kerosene containing MEHPA–Hg complex) obtained from solvent extraction. The effects of the parameters such as pH, the contact time, concentrations of extractant and 2D nanomaterial, and temperature were investigated. The equilibrium stripping data were evaluated using the Langmuir isotherm model.

Keywords: Mercury, 2D Nanomaterials, Wastewater Treatment, MXene, Stripping Reagent

References

1. Laki, S.; Shamsabadi, A. A.; Kargari, A., Comparative solvent extraction study of silver (I) by MEHPA and Cyanex 302 as acidic extractants in a new industrial diluent (MIPS). *Hydrometallurgy* **2016,** *160*, 38-46.
2. Ikeda, S.; Mori, T.; Ikeda, Y.; Takao, K., Microwave-Assisted Solvent Extraction of Inert Platinum Group Metals from HNO3(aq) to Betainium-Based Thermomorphic Ionic Liquid. *ACS Sustainable Chemistry & Engineering* **2016,** *4* (5), 2459-2463.
3. Vander Hoogerstraete, T.; Onghena, B.; Binnemans, K., Homogeneous Liquid–Liquid Extraction of Metal Ions with a Functionalized Ionic Liquid. *The Journal of Physical Chemistry Letters* **2013,** *4* (10), 1659-1663.
4. Wang, Y.; Liu, H.; Fan, J.; Liu, X.; Hu, Y.; Hu, Y.; Zhou, Z.; Ren, Z., Recovery of Lithium Ions from Salt Lake Brine with a High Magnesium/Lithium Ratio Using Heteropolyacid Ionic Liquid. *ACS Sustainable Chemistry & Engineering* **2019,** *7* (3), 3062-3072.
5. Ray, P. Z.; Shipley, H. J., Inorganic nano-adsorbents for the removal of heavy metals and arsenic: a review. *RSC Advances* **2015,** *5* (38), 29885-29907.
6. Alipour, V.; Nasseri, S.; Nodehi, R. N.; Mahvi, A. H.; Rashidi, A., Preparation and application of oyster shell supported zero valent nano scale iron for removal of natural organic matter from aqueous solutions. *Journal of Environmental Health Science and Engineering* **2014,** *12* (1), 146.
7. Hu, Q.; Sun, D.; Wu, Q.; Wang, H.; Wang, L.; Liu, B.; Zhou, A.; He, J., MXene: a new family of

promising hydrogen storage medium. *The Journal of Physical Chemistry A* **2013,** *117* (51), 14253-14260.

8. Lukatskaya, M. R.; Mashtalir, O.; Ren, C. E.; Dall'Agnese, Y.; Rozier, P.; Taberna, P. L.; Naguib, M.; Simon, P.; Barsoum, M. W.; Gogotsi, Y., Cation intercalation and high volumetric capacitance of two-dimensional titanium carbide. *Science* **2013,** *341* (6153), 1502-1505.
9. Naguib, M.; Halim, J.; Lu, J.; Cook, K. M.; Hultman, L.; Gogotsi, Y.; Barsoum, M. W., New two-dimensional niobium and vanadium carbides as promising materials for Li-ion batteries. *Journal of the American Chemical Society* **2013,** *135* (43), 15966-15969.
10. Naguib, M.; Gogotsi, Y., Synthesis of two-dimensional materials by selective extraction. *Accounts of chemical research* **2014,** *48* (1), 128-135.
11. Shahzad, A.; Rasool, K.; Miran, W.; Nawaz, M.; Jang, J.; Mahmoud, K. A.; Lee, D. S., Two-Dimensional Ti3C2T x MXene Nanosheets for Efficient Copper Removal from Water. *ACS Sustainable Chemistry & Engineering* **2017,** *5* (12), 11481-11488.
12. Peng, Q.; Guo, J.; Zhang, Q.; Xiang, J.; Liu, B.; Zhou, A.; Liu, R.; Tian, Y., Unique lead adsorption behavior of activated hydroxyl group in two-dimensional titanium carbide. *Journal of the American Chemical Society* **2014,** *136* (11), 4113-4116.
13. Guo, J.; Peng, Q.; Fu, H.; Zou, G.; Zhang, Q., Heavy-metal adsorption behavior of two-dimensional alkalization-intercalated MXene by first-principles calculations. *The Journal of Physical Chemistry C* **2015,** *119* (36), 20923-20930.
14. Mi, X.; Huang, G.; Xie, W.; Wang, W.; Liu, Y.; Gao, J., Preparation of graphene oxide aerogel and its adsorption for Cu2+ ions. *Carbon* **2012,** *50* (13), 4856-4864.
15. Chandra, V.; Park, J.; Chun, Y.; Lee, J. W.; Hwang, I.-C.; Kim, K. S., Water-dispersible magnetite-reduced graphene oxide composites for arsenic removal. *ACS nano* **2010,** *4* (7), 3979-3986.
16. Tiwari, J. N.; Mahesh, K.; Le, N. H.; Kemp, K. C.; Timilsina, R.; Tiwari, R. N.; Kim, K. S., Reduced graphene oxide-based hydrogels for the efficient capture of dye pollutants from aqueous solutions. *Carbon* **2013,** *56*, 173-182.
17. Ying, Y.; Liu, Y.; Wang, X.; Mao, Y.; Cao, W.; Hu, P.; Peng, X., Two-dimensional titanium carbide for efficiently reductive removal of highly toxic chromium (VI) from water. *ACS applied materials & interfaces* **2015,** *7* (3), 1795-1803.
18. Naguib, M.; Mochalin, V. N.; Barsoum, M. W.; Gogotsi, Y., 25th anniversary article: MXenes: a new family of two-dimensional materials. *Advanced Materials* **2014,** *26* (7), 992-1005.
19. Anasori, B.; Xie, Y.; Beidaghi, M.; Lu, J.; Hosler, B. C.; Hultman, L.; Kent, P. R.; Gogotsi, Y.; Barsoum, M. W., Two-dimensional, ordered, double transition metals carbides (MXenes). *Acs Nano* **2015,** *9* (10), 9507-9516.
20. Mashtalir, O.; Naguib, M.; Mochalin, V. N.; Dall'Agnese, Y.; Heon, M.; Barsoum, M. W.; Gogotsi, Y., Intercalation and delamination of layered carbides and carbonitrides. *Nature communications* **2013,** *4*, 1716.
21. Mashtalir, O.; Lukatskaya, M. R.; Zhao, M. Q.; Barsoum, M. W.; Gogotsi, Y., Amine-assisted delamination of Nb2C MXene for Li-Ion energy storage devices. *Advanced Materials* **2015,** *27* (23), 3501-3506.
22. Rasool, K.; Mahmoud, K. A.; Johnson, D. J.; Helal, M.; Berdiyorov, G. R.; Gogotsi, Y., Efficient Antibacterial Membrane based on Two-Dimensional Ti 3 C 2 T x (MXene) Nanosheets. *Scientific reports* **2017,** *7* (1), 1598.
23. Li, G.; Tan, L.; Zhang, Y.; Wu, B.; Li, L., Highly Efficiently Delaminated Single-Layered MXene Nanosheets with Large Lateral Size. *Langmuir* **2017,** *33* (36), 9000-9006.
24. Li, B.; Zhang, Y.; Ma, D.; Shi, Z.; Ma, S., Mercury nano-trap for effective and efficient removal of mercury (II) from aqueous solution. *Nature communications* **2014,** *5*, 5537.
25. Sun, Q.; Aguila, B.; Perman, J.; Earl, L. D.; Abney, C. W.; Cheng, Y.; Wei, H.; Nguyen, N.; Wojtas, L.; Ma, S., Postsynthetically Modified Covalent Organic Frameworks for Efficient and Effective Mercury Removal. *Journal of the American Chemical Society* **2017,** *139* (7), 2786-2793.
26. Huang, N.; Zhai, L.; Xu, H.; Jiang, D., Stable Covalent Organic Frameworks for Exceptional Mercury Removal from Aqueous Solutions. *Journal of the American Chemical Society* **2017,** *139* (6), 2428-2434.
27. Qu, Z.; Yan, L.; Li, L.; Xu, J.; Liu, M.; Li, Z.; Yan, N., Ultraeffective ZnS Nanocrystals Sorbent for Mercury(II) Removal Based on Size-Dependent Cation Exchange. *ACS Applied Materials &*

Interfaces **2014,** *6* (20), 18026-18032.
28. Gamino Arroyo, Z.; Stambouli, M.; Pareau, D.; Buch, A.; Durand, G.; Avila Rodriguez, M., Thiosubstituted organophosphorus acids as selective extractants for Ag (I) from acidic thiourea solutions. *Solvent Extraction and Ion Exchange* **2008,** *26* (2), 128-144.
29. Baba, A. A.; Adekola, F. A., Beneficiation of a Nigerian sphalerite mineral: Solvent extraction of zinc by Cyanex® 272 in hydrochloric acid. *Hydrometallurgy* **2011,** *109* (3-4), 187-193.

570549 Capture of Alkynes from Hydrocarbon Mixtures with Multi-Functional Metal-Organic Framework By Combining Molecular Recognition and Size-Sieving

Capture of alkynes from hydrocarbon mixtures with multi-functional metal-organic framework by combining molecular recognition and size-sieving

Qingju Wang; Xili Cui; Zhaoqiang Zhang; Lifeng Yang; Huabin Xing*

College of Chemical and Biological Engineering, Zhejiang University, Hangzhou 310027

E-mail: xinghb@zju.edu.cn

Selective one-step removal of all the alkynes from pyrolysis gases is a challenging but important technology in the production of polymer-grade olefins. The currently existed selective removal technology, front-end hydrogenation, is unable to achieve the deep removal of several kinds of alkynes simultaneously due to the complex compositions of pyrolysis gases (including C1-C5 fractions). Therefore, the revolution to the current separation technology to further simplify the process is necessary and urgent. Adsorptive separation technology, with the characterizations of energy-efficient and deep removal, would be a promising technology. However, the conventional adsorbents can hardly specifically capture multi-type alkynes and propadiene from mixed cracking gas. In addition, the existing adsorbent can also adsorb higher hydrocarbons easily. Hence, it remains a daunting challenge to design an ideal material that can not only selectively adsorb multi-type alkynes and propadiene, but also exclude higher hydrocarbons. Here we prepare a multi-functional metal-organic framework with high-density of anions and suitable pore size for the capture of multi-type alkynes and propadiene from mixed cracking gas. The unique hydrogen-bonding properties of inorganic anions can achieve the specific recognition of the acetylene, propyne and propadiene, meanwhile, the exclusion of higher hydrocarbons by precisely tuning of pore cavity size. The adsorption isotherms and dynamic breakthrough experiment all confirm the unprecedented separation performance. The underlying interactions between gas and porous material are explored by modeling studies.

Acknowledgement: This research was supported by the National Natural Science Foundation of China (No. 21725603), Zhejiang Provincial Natural Science Foundation of China (No. LZ18B060001), and the National Program for support of Top-notch Yong Professionals (H. X.)

References:

1. Z. Q. Zhang, Q. W. Yang, X. L. Cui, L. F. Yang, Z. B. Bao, Q. L. Ren, H. B. Xing. Angew. Chem. Int. Ed. 2017, 56, 16094.
2. L. F. Yang, X. L. Cui, Z. Q. Zhang, Q. W. Yang, Z. B. Bao, Q. L. Ren, H. B. Xing. Angew. Chem. Int. Ed. 2018, 57, 13145.

566725 An Alternative Synthetic Method for Porous Organic Cage CC3 in Dimethylformamide

The prototypical Porous Organic Cage, CC3, was synthesized in the presence of dimethylformamide as solvent. A combination of stirring effects and solvothermal treatment conditions led to distinctive morphological, structural and textural properties of the resulting CC3 crystals.

561616 Study of Lanthanum Carbonate Modified Microfibrous Composite for Phosphate Adsorption in Wastewater

In this study, a novel lanthanum carbonate modified microfirous composite (LC-MC) adsorbent is synthesized for phosphate adsorption in wastewater. Box-Behnken design method is used to optimize the parameters of preparation process and a linear model is demonstrated to describe the relationship between adsorption capacity and preparation parameters. It was predicted that under conditions of 0.19 mg/L initial La concentration, 93.13 molar ratio of urea and La and 20.92 h synthesis time, the best adsorption capacity is 54.63 mg/g, similar to the actual value (55.2 mg/g). The phosphate adsorption experiments on LC-MC in batch reactor show that 6-9 is the best pH range for phosphate adsorption with low La leakage, and ionic strength has no obvious effect on the performance of LC-MC. The adsorption isotherm data are fitted well by Freundlich isotherm and the adsorption kinetic data are described well with the pseudo-second order model, and the adsorption achieves equilibrium at about 300 min. The adsorption can be divided into three steps according to Intraparticle diffusion model, including intra-particle diffusion, surface film diffusion and equilibrium. The presence of co-existing anions has little effect on the removal. The LC-MC exhibited an ideal regeneration performance in phosphate adsorption with 80% adsorption capacity remained after 5 cycles. Finally, the adsorption mechanism indicates that electrostatic attraction and ligand exchange are both involved in the phosphate adsorption process.

573126 Room-Temperature Synthesis of Carboxylated-Mxene Nanosheets for Mercury Scavenging

Toxic metals in drinking water affect millions of people worldwide. They are ubiquitous in the environment and pose a harmful effect on human health, especially for children. For example, mercury ions are converted to methylmercury (CH_3Hg) in the aqueous phase, which is one of the most toxic forms of mercury[1]. The U.S. Environmental Protection Agency has regulated the level of mercury in wastewater effluents[2]. Water treatment can be used to remove mercury from wastewater effluents. There are a few methods that have successfully been used for this purpose, but their energy and chemical costs are high. Adsorption has been regarded as one of the easy, effective, and economical techniques for the removal of heavy metals from contaminated streams[3-5].

Recently, $Ti_3C_2T_x$ MXene has been explored as an efficient sorbent for heavy metals[6-8]. Its abundant active sites, negative zeta potential, high conductivity, excellent dispersity in water, and a large surface area have made it an ideal sorbent for environment remediation[7-19]. However, pristine MXenes usually tends to oxidize in aqueous media, which makes them unstable. One of the surface functional groups of the MXene is the hydroxyl group, which allows for MXene surface modification and thus for changing the MXene properties.

In this work, to overcome the $Ti_3C_2T_x$ MXene oxidization in aqueous media and improve its potential for water treatment, we modified the MXene with a chloroacetic acid solution at room temperature and prepared carboxylated-MXene (MXene–COOH). The MXene–COOH was characterized by means of SEM, EDX, TEM, FTIR, XRD, and zeta potential measurements. Not only has MXene–COOH high stability and dispersity in aqueous media but also negative zeta potential, attracting positively charged pollutants. We conducted a batch-adsorption process to understand the influence of different parameters on the adsorption of Hg^{2+}onto the MXene and MXene–COOH nanosheets. The removal efficiency of Hg^{2+} for the MXene and MXene–COOH were found to be 72 and 99.4% at pH 5.5 over 60 min, respectively. In addition, we calculated the thermodynamic parameters such as ΔG°, ΔH°, and ΔS°.

Keywords: Adsorption, Mercury Ions, MXenes, Carboxylated-MXene, Wastewater Treatment.

References

1. Hu, X.; Mu, L.; Lu, K.; Kang, J.; Zhou, Q., Green Synthesis of Low-Toxicity Graphene-Fulvic Acid with an Open Band Gap Enhances Demethylation of Methylmercury. *ACS Applied Materials & Interfaces* **2014,** *6* (12), 9220-9227.
2. Bolger, P. T.; Szlag, D. C., An electrochemical system for removing and recovering elemental mercury from a gas stream. *Environmental science & technology* **2002,** *36* (20), 4430-4435.
3. Efome, J. E.; Rana, D.; Matsuura, T.; Lan, C. Q., Insight studies on metal-organic framework nanofibrous membrane adsorption and activation for heavy metal ions removal from aqueous solution. *ACS applied materials & interfaces* **2018,** *10* (22), 18619-18629.
4. Zhang, Q.; Dan, S.; Du, K., Fabrication and Characterization of Magnetic Hydroxyapatite Entrapped Agarose Composite Beads with High Adsorption Capacity for Heavy Metal Removal. *Industrial & Engineering Chemistry Research* **2017,** *56* (30), 8705-8712.
5. Cao, C.-Y.; Qu, J.; Yan, W.-S.; Zhu, J.-F.; Wu, Z.-Y.; Song, W.-G., Low-Cost Synthesis of Flowerlike α-Fe2O3 Nanostructures for Heavy Metal Ion Removal: Adsorption Property and Mechanism. *Langmuir* **2012,** *28* (9), 4573-4579.
6. Naguib, M.; Gogotsi, Y., Synthesis of two-dimensional materials by selective extraction. *Accounts of chemical research* **2014,** *48* (1), 128-135.
7. Shahzad, A.; Rasool, K.; Miran, W.; Nawaz, M.; Jang, J.; Mahmoud, K. A.; Lee, D. S., Two-Dimensional Ti3C2T x MXene Nanosheets for Efficient Copper Removal from Water. *ACS Sustainable Chemistry & Engineering* **2017,** *5* (12), 11481-11488.

8. Peng, Q.; Guo, J.; Zhang, Q.; Xiang, J.; Liu, B.; Zhou, A.; Liu, R.; Tian, Y., Unique lead adsorption behavior of activated hydroxyl group in two-dimensional titanium carbide. *Journal of the American Chemical Society* **2014,** *136* (11), 4113-4116.
9. Guo, J.; Peng, Q.; Fu, H.; Zou, G.; Zhang, Q., Heavy-metal adsorption behavior of two-dimensional alkalization-intercalated MXene by first-principles calculations. *The Journal of Physical Chemistry C* **2015,** *119* (36), 20923-20930.
10. Mi, X.; Huang, G.; Xie, W.; Wang, W.; Liu, Y.; Gao, J., Preparation of graphene oxide aerogel and its adsorption for Cu2+ ions. *Carbon* **2012,** *50* (13), 4856-4864.
11. Chandra, V.; Park, J.; Chun, Y.; Lee, J. W.; Hwang, I.-C.; Kim, K. S., Water-dispersible magnetite-reduced graphene oxide composites for arsenic removal. *ACS nano* **2010,** *4* (7), 3979-3986.
12. Tiwari, J. N.; Mahesh, K.; Le, N. H.; Kemp, K. C.; Timilsina, R.; Tiwari, R. N.; Kim, K. S., Reduced graphene oxide-based hydrogels for the efficient capture of dye pollutants from aqueous solutions. *Carbon* **2013,** *56*, 173-182.
13. Ying, Y.; Liu, Y.; Wang, X.; Mao, Y.; Cao, W.; Hu, P.; Peng, X., Two-dimensional titanium carbide for efficiently reductive removal of highly toxic chromium (VI) from water. *ACS applied materials & interfaces* **2015,** *7* (3), 1795-1803.
14. Naguib, M.; Mochalin, V. N.; Barsoum, M. W.; Gogotsi, Y., 25th anniversary article: MXenes: a new family of two-dimensional materials. *Advanced Materials* **2014,** *26* (7), 992-1005.
15. Anasori, B.; Xie, Y.; Beidaghi, M.; Lu, J.; Hosler, B. C.; Hultman, L.; Kent, P. R.; Gogotsi, Y.; Barsoum, M. W., Two-dimensional, ordered, double transition metals carbides (MXenes). *Acs Nano* **2015,** *9* (10), 9507-9516.
16. Mashtalir, O.; Naguib, M.; Mochalin, V. N.; Dall'Agnese, Y.; Heon, M.; Barsoum, M. W.; Gogotsi, Y., Intercalation and delamination of layered carbides and carbonitrides. *Nature communications* **2013,** *4*, 1716.
17. Mashtalir, O.; Lukatskaya, M. R.; Zhao, M. Q.; Barsoum, M. W.; Gogotsi, Y., Amine-assisted delamination of Nb2C MXene for Li-Ion energy storage devices. *Advanced Materials* **2015,** *27* (23), 3501-3506.
18. Rasool, K.; Mahmoud, K. A.; Johnson, D. J.; Helal, M.; Berdiyorov, G. R.; Gogotsi, Y., Efficient Antibacterial Membrane based on Two-Dimensional Ti 3 C 2 T x (MXene) Nanosheets. *Scientific reports* **2017,** *7* (1), 1598.
19. Li, G.; Tan, L.; Zhang, Y.; Wu, B.; Li, L., Highly Efficiently Delaminated Single-Layered MXene Nanosheets with Large Lateral Size. *Langmuir* **2017,** *33* (36), 9000-9006.

574039 A Comparison of Gas Adsorption Capacities of Isostructural Metal Organic Frameworks: [{Cu_2(abtc)}$_3$] Versus [{Cu_2(reduced abtc)}$_3$]

A Comparison of Gas Adsorption Capacities of Isostructural Metal Organic Frameworks: [{Cu_2(abtc)}$_3$] versus [{Cu_2(reduced abtc)}$_3$]

Prudhviraj Medikonda[1], Sastri Chivukula[2] and Sasidhar Gumma[1*]

[1] Department of Chemical Engineering, Indian Institute of Technology Guwahati, India

[2] Department of Chemistry, Indian Institute of Technology Guwahati, India

Corresponding author's e-mail: *s.gumma@iitg.ac.in*

Metal-organic frameworks (MOFs) can be defined as crystalline micro porous materials made by coordinatively organic linkers using metal clusters. Different types of aromatic polycarboxylate organic linkers such as 1,3-benzene dicarboxylate, 1,3,5- benzene tricarboxylate and 3,3′,5,5′-biphenyl tetracarboxylate have been proven useful organic linkers for preparation of MOFs. In contrast, 3,3′,5,5′-azobenzene tetracarboxylate (H_4AzoBTC) and 5,5'-(hydrazine-1,2-diyl) diisophthalic acid (H_4 Reduced AzoBTC), one kind of tetra carboxylate linkers possess many advantages including azo groups presence, photo-chromatic in nature and rigidity in the structure inspired to study this work. The distinct difference in the studied structures were functional groups (N=N and HN-NH) and twisted nature presented in Reduced AzoBTC. The adsorption affinity of the HN-NH group over N=N and pore arrangement in the MOF can be understood through this study

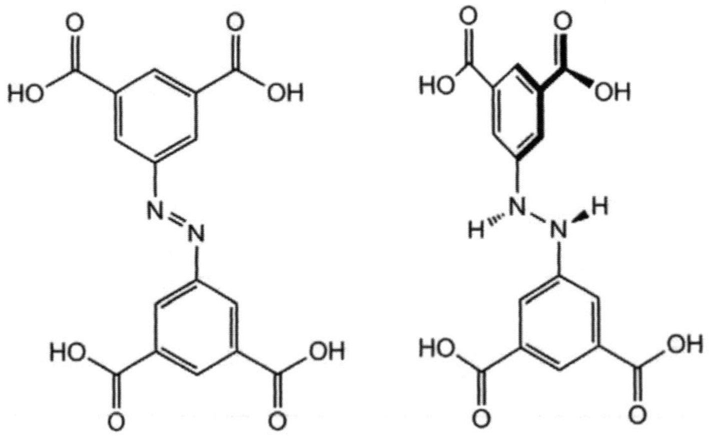

Scheme 1: Adsorption capacities of structures H_4AzoBTC (left) and H_4 Reduced AzoBTC (right) contained MOFs

compared in this work

The gas adsorption characteristics were evaluated by measuring adsorption isotherms over a wide range of polarity and polarizability *viz.* CO_2, CO, CH_4, N_2, C_2H_6, C_3H_8 and O_2. For all studied gases Type-I isotherm was observed. The isotherms were modeled using modified virial equation for polar gases (CO_2 and CO) and Langmuir model used for less polar gases (O_2, N_2, CH_4, C_2H_6 and C_3H_8). Model parameters used to calculate the enthalpies of adsorption. Ideal Adsorbed Solution Theory was used to predict selectivity of binary mixtures. The CO_2 Selectivity over N_2 increases significantly with increase in pressure. However, the CO_2 selectivity over N_2 for H_4 Reduced AzoBTC contained MOF is more than H_4 AzoBTC contained MOF due to strong affinity of functional group and framework-adsorbate interactions.

Keywords: metal-organic frameworks, adsorption

578839 Electrochemical Synthesis of Ammonia, and Development of Nanoporous Carbon-Composite Adsorbent Exhibiting High Ammonia Uptake

Recently, we synthesized nanoporous composite of an activated carbon and organic acids, which exhibited high amount of ammonia adsorption under very low pressure range (lower than 9.3 mmHg). The present nanoporous material was prepared by simple impregnation of organic acids (e.g., succinic acid, itaconic acid, and adipic acid) into the commercial activated carbon. A systematic investigation was performed to find the best choice of the organic acid, the optimized amount of the organic acid impregnation and the best impregnation condition. The activated carbon impregnating succinic acid showed 108 cm^3 g^{-1} (STP) under 9.3 mmHg of NH_3 at 298 K. The dramatic increase of the ammonia uptake could be explained by the strong interaction in the pair of the organic acid and basic ammonia molecule, and by the high dispersion of the organic acid owing to its strong affinity to the hydrophobic activated carbon support. We investigated also the ammonia-breakthrough property of the new adsorbent under very low concentration of ammonia flow (100 ppm in argon balance). At the beginning of the measurement, the concentration of the ammonia in product gas decreased to almost zero, and the breakthrough occurred at ten hours of the adsorption. The breakthrough amount was 1.89 mmol g^{-1}. This adsorbent could recover 90% of initial breakthrough amount after regeneration process at 473 K under nitrogen flow. The present adsorbent shows high potential in the application fields of purification of land-fill gas and bio-gas.

571548 Green and Sustainable Chromatographic Separation Strategy for Levulinic Acid from Biomass Hydrolysate

Levulinic acid (LA) is a promising and versatile biorefinery platform chemical, which can be used to synthesize a broad range of desirable chemicals and fuel additives. To date, an enormous amount of patents and research articles have been devoted to manufacturing LA from biomass by using mineral acids, solid acids, or metal chlorides as catalysts. Unfortunately, in these catalytic transformations, LA released from biomass hydrolysate is accompanied by formic acid (FA) and 5-hydroxymethylfurfural (5-HMF). How to economically, safely, and efficiently separation of LA remain difficult challenge from the viewpoint of industrial application.

In this work, a novel amide group modified mic/mesoporous hyper-cross-linked resin, SY-01, was prepared using suspension polymerization of 2-hydroxyethyl acrylate (2-HEA), styrene (St) and divinylbenzene (DVB), followed by a Friedel–Crafts reaction and an amination reaction. The physicochemical properties of the SY-01 resin were characterized by scanning electron microscope (SEM), Fourier transform infrared (FT-IR) spectroscopy, Element analysis (EA) and nitrogen adsorption-desorption isotherms. The results show that the specific surface area of the resin is as high as 1334 m^2/ g, and the microporous ratio and mesoporous ratio are 74.6 % and 9.48 %, respectively. Batch adsorption experiments showed that the affinity of SY-01 resin toward FA−LA−5-HMF were in the order of 5-HMF > LA > FA under noncompetitive and competitive systems. The highest adsorption capacity were 7.54 mg/g wet resin for FA, 103.51 mg/g wet resin for LA, and 107.73 mg/g wet resin for 5-HMF. The obtained thermodynamic parameters suggested that the adsorption of LA on SY-01 was spontaneous (ΔG^0<3.788 kJ/mol), and exothermic (ΔH^0=-11.764 kJ/mol). For kinetic study, the adsorption of LA onto SY-01 resin at various operating conditions follows the pore diffusion model and the intraparticle diffusion is the rate-limiting step for the adsorption of LA onto SY-01 resin. More importantly, mesoporous formation accelerates the adsorption / desorption diffusion of LA in the resin. The effective pore diffusivity was dependent upon temperature, but independent of initial LA concentration, and were 3.306 $\times 10^{-10}$, 5.274 $\times 10^{-10}$ and 7.707 $\times 10^{-10}$ m^2/s at 298, 318 and 338 K, respectively. In desorption process, the recovery efficiency of LA from SY-01 resin was 99.39% owing to the lower hydrogen bond energy as well as the reversibility of the hydrogen bond. Furthermore, a mathematical model based on the general rate model coupled with the noncompetitive single component and competitive multicomponent Langmuir isotherm was successfully developed to simulate the breakthrough curves of FA−LA−5-HMF from single, binary, as well as ternary-component mixtures. The proposed methodology for fixed-bed column multicomponent competitive adsorption model can be successfully implemented to completely design the separation unit of LA from aqueous solution or biomass hydrolysate. In summary, this work effectively solves the problems that the activated carbon can not be reused in the traditional adsorption and separation technology and that the ion exchange resin consumes a lot of acid, alkali and water in the process of elution and regeneration, reduces the environmental pollution and promotes the construction of beautiful countryside.

556744 Transport Kinetics and Selectivity for Carbon Dioxide and Water for Gas Phase Electrolysis

The effects of electrode and membrane chemistry and structure on the overall reaction kinetics for the electrochemical reduction of O_2, chemical reactions with CO_2, and the transport kinetics through a membrane electrode assembly (MEA) were characterized through physiochemical relationships. Gas phase electrolysis using an MEA provides an alternative to gas separation over adsorbents, selective membranes, or reactive systems. In the case of CO_2 separation, the MEAs eliminates the heat and humidity generated from exothermic adsorption beds and reduces the power demand associated with membrane systems. Electrochemical impedance spectroscopy (EIS) measurements were used to determine the balance between the control of the electrocatalyst formation chemistry, pore structure, membrane chemistry, and gas diffusion layer and electrode structure. The electrochemical system was scaled-up with a larger electrode area and demonstrated within a closed-circuit breathing loop to maintain near-zero CO_2 concentration in the effluent. A 5 mol% CO_2, high-humidity gas mixture was delivered to the MEA at a rate of 2.1–4.2 $mL/cm^2{\cdot}min$ while the current density was varied between 4–40 mA/cm^2. The MEAs maintained linear performance over a range of active area within a multi-cell system.

560401 Using Lithium-Aluminum Layered Hydroxide As an Efficient Adsorbent to Recover Lithium from Brine

Using lithium-aluminum layered hydroxide as an efficient adsorbent to recover lithium from brine

Jing Zhong[1], Sen Lin[1,2,], Jianguo Yu[1,2,*]*

1.National Engineering Research Center for Integrated Utilization of Salt Lake Resources, East China University of Science and Technology, Shanghai, China, 200237.

2.State Environmental Protection Key Laboratory of Environmental Risk Assessment and Control on Chemical Process, East China University of Science and Technology, Shanghai, China, 200237.

Abstract:

Lithium-aluminum layered double hydroxide (Li/Al-LDH) has been proved an efficient adsorbent for lithium recovery from high Mg/Li ratio brine[1,2]. Additionally, Li/Al-LDH after adsorption could be regenerated just by neutral water with no dissolution compared to other lithium adsorbents, including manganese ion-sieve and titanium ion-sieve.

In this study, Li/Al-LDH was prepared by co-precipitation and tested by various methods. The XRD spectrum showed that the prepared adsorbent was exactly the layered double structure where Li^+ entered the vacancies in the $Al(OH)_3$ layers and Cl^- and water molecules occupied the interlayer space[3]. The SEM images and BET results demonstrated the adsorbent was hexagonal platelet-shaped and had a large specific surface area of 161 m^2/g, respectively. The adsorption experimental results reflected that lithium adsorption capacity of Li/Al-LDH was more than 7.2 mg/g at 25°Cand the adsorption equilibrated after 2 h. Adsorption capacity was slightly affected by pH (3-11) during the lithium extraction process. Meanwhile, the lithium adsorption capacity rose by near two times when $MgCl_2$ concentration increased from 0 to 395.8 g/L, which indicated that the existence of Mg^{2+} would promote the lithium adsorption.

Besides, the lithium ions adsorbed on Li/Al-LDH could be eluted via pure water or low concentration lithium solutions and the desorption ratio was more than 90% without significant decrease in adsorption capacity after 30 adsorption/desorption cycles. Consequently, the as-prepared Li/Al-LDH had enormous potential for lithium recovery from high Mg/Li brine in China.

Keywords: Lithium recovery; Adsorption; Li/Al-LDH; Mg/Li separation.

References:

[1] Paranthaman M P , Li L , Luo J , et al. Recovery of Lithium from Geothermal Brine with Lithium–Aluminum Layered Double Hydroxide Chloride Sorbents[J]. Environmental Science & Technology, 2017, 51(22):13481-13486.

[2] Wu L , Li L , Evans S F , et al. Lithium Aluminum Layered Double Hydroxide Chlorides (, LDH,): Formation Enthalpies and Energetics of Lithium Ion Capture[J]. Journal of the American Ceramic Society, 2018.

[3] Britto S , Kamath P V . Polytypism in the Lithium–Aluminum Layered Double Hydroxides: The $[LiAl_2(OH)_6]^+$ Layer as a Structural Synthon[J]. Inorganic Chemistry, 2011, 50(12):5619-5627.

581339 MOF-GO Hybrid Nanocomposite Adsorbents for Methane Storage

In this study, the storage of methane in nanocomposite adsorbents comprising of metal-organic framework (MOF) and graphene oxide (GO) was investigated. Three different sets of MOF-GO nanocomposites comprising of HKUST-1 and pristine GO, reduced GO (rGO), and carboxyl-functionalized GO (fGO) were developed by solvothermal method and their methane storage characteristics were assessed through high-pressure methane adsorption measurements. The formation of MOF-GO nanocomposites was confirmed by XRD, FTIR, XPS, SEM, and TEM. All three types of nanocomposites exhibited higher surface area and porosity than the pristine MOF. Our results indicated that MOF@rGO nanocomposite with 10wt% rGO exhibited the best performance with a methane deliverable capacity of 193 cm^3(STP)/cm^3 in the pressure range of 5.8-65 bar and at room temperature which was approximately 30% higher than that of pristine HKUST-1 with capacity of 149 cm^3(STP)/cm^3. Moreover, the methane deliverable capacity of MOF@GO and MOF@fGO were found to be 181 and 162 cm^3(STP)/cm^3, respectively. The findings of this study demonstrate the synergistic effect of graphene oxide on methane storage performance of MOF-GO nanocomposites.

586140 Development of organic and inorganic desiccant materials for atmospheric water production

This work presents a study on the physical and chemical properties of a broad range of organic and inorganic materials having good affinity with water. These include polymers and minerals. The results demonstrate the possibility to efficiently exploit the hygroscopic properties of these materials to store water vapor under various relative humidity and temperature conditions. In a subsequent process, liquid water can be produced through an isothermal cycle. Furthermore, we implement a heat and mass exchanger that can regenerate the desiccant materials at reasonable temperatures below 50 °C, which is easily achievable with solar energy.

564613 Two-Dimensional-Material Mixed-Matrix Membranes for Gas Separation

Two-dimensional (2D) materials, with atomic thickness and micrometer lateral size, are emerging building blocks for separation membranes. Extraordinary molecular separation properties for purifying water and gases have been demonstrated by developing 2D-material membranes, which attract a huge surge of interest during last a few years. As a prominent characteristic of 2D-material membranes, inter-layer space has been proved to play a significantly important role in molecular transport.

We introduced mixed-matrix approach to develop 2D-material membranes for gas separation, with particular focus on tailoring sub-nanosized interlayer spaces between nanosheets for fast and selective transport of gases. Two typical 2D-materials will be demonstrated in this presentation: graphene oxide (GO)[1] and MXene[2]. First, membranes with fast and selective CO_2 transport channels of GO laminates were proposed based on the construction of hydrogen bonding between GO nanosheets and PEBAX chain. Notably, incorporating only 0.1 wt% GO could double the CO_2 permeability and CO_2/N_2 selectivity of pristine PEBAX membrane, transcending the 2008 Robeson upper-bound. Our explorations (e.g., TEM, positron annihilation and sorption) found that molecular-sieving of the interlayer spaces and CO_2-philic pathways of the GO laminates play critical roles in the enhancement of gas transport properties. Second, borate and amine intercalated MXene membranes with ultra-thin skin layer (~20 nm) were fabricated by a facile spin-casting assisted assembly technique. With finely tuned interlayer spacing and introduced CO_2-falicitate transport groups, the intercalated MXene membranes show excellent CO_2 separation performance that overcomes the CO_2/CH_4 upper-bound. The transport mechanism was also understood by analyzing the coefficients of sorption and diffusion. Our results demonstrate that 2D-material mixed-matrix membranes offer exciting opportunities for efficient gas separation.

References:

1. *Angew. Chem. Int. Ed.* **54**, 578-582 (2015).
2. *Adv. Funct. Mater.* 28, 1801511 (2018).

567387 Development of the Novel Manufacturing Process of High Purity $LiOH{\cdot}H_2O$ from Li_2CO_3 Using Acetic Acid

Recently, lithium carbonate has been applied as a lithium compound in various fields. As the trend in the future, it is conceivable that the use opportunity of lithium hydroxide-derived positive electrode material increases more than lithium carbonate-derived positive electrode material because lithium is mainly used as a battery. However, the current process using calcium hydroxide requires a large amount of heat of the process because it involves heating and concentration, and a large amount of pure water is required to use a slurry or the like. In other words, it is a process that requires high energy consumption and a large amount of water. This is the source of current environmental problems and causes contradictions.

Therefore, this laboratory aims to develop a synthesis process of lithium hydroxide monohydrate of the quality required for batteries, save energy and save resources of the process. We aimed to develop a crystallization process at room temperature that does not require latent heat, and at the same time focus on the immobilization of carbon dioxide that is synthesized as a by-product, as well as the process by thermal concentration up to now.

In this study, we applied the crystallization process and developed the synthesis process to synthesize high purity $LiOH \cdot H_2O$ from crude Li_2CO_3. There are various methods of crystallization, but in this study, we aimed to apply a crystallization process at room temperature without temperature change, in order to aim at development of an alternative process to heat concentration. In addition, as the crystallization operation, the possibility of high purification was studied from the crystallization operation of both reactive crystallization and antisolvent addition crystallization.

In order to develop a high purity process of LiOH · H2O, it was examined whether the process could be developed using major acids and alkalis. At this time, it has been confirmed that the synthesis process using CH_3COOH is applicable to the above-mentioned crystallization operation. Furthermore, it was confirmed from the XRD analysis result when the reaction was examined that other impurities could be sufficiently removed by using KOH. The amount of poor solvent addition crystallization using CH_3COOH and KOH and the amount of crystallization of the reaction were studied from the crystallization operation. As the crystallization conditions, the temperature is high, the stirring speed is slow, and the shorter the dropping time, the higher the crystals. It has been confirmed that purification can be performed. Finally, we report that it has become possible to increase the crystal purity to 99.95 wt%.

573908 Ionic Transport in Reverse Osmosis Membrane

Reverse osmosis (RO) has proven its efficiency to produce high quality water by reducing salinity of source from membrane treatments.

Transportation in a RO process can be well described by extended Nernst-Planck equation, where separation is a result of diffusion, convection and electro-migration of ionic species. Poisson-Boltzmann equation is used to describe potential profile. Ionic species are electro-neutrality everywhere in bulk of feed channel, and potential of solution is zero in this region. In electrical double layer region where is close to the charged membrane surface, negative static charged trends to repel co-ions and attracts counter-ions, and distinct ionic concentration distribution is formed. Co-ions and counter-ions would be balanced inside membrane, where potential change is assumed to be linear. We assume electro-migration effect in permeate site is minor. It is noticed from the numerical solution that the rejection of was lower than the rejection of throughout membrane. The rejection of and ions increased as the volumetric permeating flux based on membrane area () increased. An extension to multiple species shows that fixed charged membrane has a priority to reject divalent ions more than monovalent .

556462 Butene Oligomerization within Dividing Wall Column

The promotion of ethanol gasoline in China restricts the use of MTBE (methyl tert-butyl ether). This results in shortage of gasoline with high octane number and changes in the scheme of C4 resource processing. In butene oligomerization process, iso-octene, which is the ideal blend composition of gasoline with high octane number, is synthesized from iso-butene. The remnants which contain iso-butane and n-butene are favorable reactants for alkylation. Due to the high reactivity of the resin catalyst, inhibitor such as TBA (tert-butyl alcohol) is introduced to control the reaction temperature. Then TBA is separated from iso-octene in the inhibitor recovery column and recycled to the reactors. If C4 resource contains few pentane, which is common, a certain amount of distillates contain TBA, pentane, hexene and heptene must be discharged outside the battery limits to balance the concentration of pentane. After renovated to dividing wall column, the inhibitor recovery column manifests better separation performance. The weight concentration of TBA of the side product increases from 45% to 92%, which turns out to be more economic for cycle use. Total mass flow of hexene and heptene in the side product reduced 81%, which means less high boiling point olefin will be generated to ensure the gasoline endpoint is within the scope of the standard.

556671 Enhancement of Baffles with Different Sizes and Intervals on Mass Transfer in Microchannels

Generally, the mass transfer between gas and liquid two phases is a controlling step for a fast chemical reaction. However, the scaling-up through dimension enlargement of microscale devices usually deteriorate the mass transfer efficiency, thus the enhancement of gas-liquid mass transfer is an effective approach to improve mass transfer rate. In this work, a convenient strategy for intensifying gas-liquid mass transfer in the microchannel was proposed by embedding staggered rectangular baffles on both sides of the channel. The flow and mass transfer of gas-liquid two-phase for CO_2-water in all microchannels were visually investigated by a high-speed camera. The clean and unobstructed microchannels were utilized to compare and determine the enhancement of total volumetric mass transfer coefficient of the baffled microchannels. Based on Taylor flow and the Taylor-broken flow regimes in the baffled microchannels, the enhancement mechanism and pressure drop were explored by varying the operating conditions (the superficial flow velocity of gas and liquid phases), the baffle configuration (the blockage length of baffle entering the channel section and the spacing distance between two baffles) and the channel width. The results show that the mass transfer performance could be remarkably improved with only a slight increase in pressure drop in baffle-optimized microchannels, the difference of pressure drop between the clean and unobstructed channels is less than 0.5 kPa or 23%. The reduction of spacing distance between two baffles and the increment of the liquid flow velocity, baffle blockage length and channel width are beneficial to the mass transfer enhancement. Moreover, the mass transfer enhancement becomes more pronounced at higher gas flow velocities, especially in the Taylor-broken flow regime, the enhancement factor, which describes the relative ratio of total volumetric mass transfer coefficient in baffled microchannel to that in smooth microchannel, could reach up to 2.2. This study could be conducive to developing some novel gas-liquid microreactors.

564971 Solubility of L-Phosphinothricin and Its Hydrochloride in Water + (Methanol, Ethanol) Binary Solvent Mixtures from 278.15 K to 318.15 K

L-Phosphinothricin (L-PPT, (2S)-2-amino-4-(hydroxymethylphosphinyl)butanoic acid, CAS No. 35597-44-5) is the active ingredient of the broad-spectrum, nonselective herbicide DL-phosphinothricin ammonium salt. The separation and purification of L-PPT from aqueous solution containing organic and inorganic salts is one of the key technologies in the production of L-PPT. L-PPT can be solvated by water to form L-PPT monohydrate under certain conditions. L-PPT hydrochloride (CAS No. 73777-49-8) is usually obtained as an alternative product in the preparation process of L-PPT.

L-PPT and its hydrochloride can be separated from aqueous solution by using the method of anti-solvent crystallization. Methanol and ethanol are considered to be suitable anti-solvents for their good miscibility with the solution and high efficiency. The operational level of the anti-solvent crystallization process has a great effect on product quality; therefore, a reliable knowledge of the solubility of L-PPT and its hydrochloride in binary mixed solvents of water + (methanol, ethanol) is necessary for the design of a good crystallization process.

The solubility of L-PPT and its hydrochloride in water + (methanol, ethanol) binary solvent mixtures was experimentally measured from 278.15 K to 318.15 K by a static analytic method under atmospheric pressure. The experimental data showed that the solubility increased with the increasing temperature and decreased with the increase of alcohol concentration in the binary solvent mixtures. The experimental data were correlated by the modified Apelblat equation and CNIBS/R–K model. The correlation models were evaluated according to the relative average deviations. The results indicated the modified Apelblat equation showed better agreement with the experimental data. The local mole fractions of methanol in the local area around L-PPT were calculated, and the result showed that the preferential solvation of L-PPT was related to the solvent composition and temperature. The experimental data and the corresponding models can be used as essential data in the design of the crystallization process of L-PPT and its hydrochloride.

569260 H_2 Membrane Separation and C_3H_8 Vaporization Synergistic for Energy Saving in Oleflex Dehydrogenation Process

H_2 membrane separation and C_3H_8 vaporization synergistic for energy saving in Oleflex dehydrogenation process

Xuehua Ruan, Hongyan Xiao, Gaohong He*, Xiaobin Jiang

State Key Laboratory of Fine Chemicals, School of Petroleum and Chemical Engineering, Dalian University of Technology, Panjin, Liaoning, P.R. China, 124221

Corresponding Author Email: hgaohong@dlut.edu.cn

In Oleflex propane dehydrogenation technology, condensation for reaction effluent is extremely energy-intensive. On the opposite, C_3H_8 vaporization before the reaction, subject to the mismatch in temperature (vaporization is at about -25 ^oC, but condensation is sustaining from 5 to -100 ^oC), could not be exploited sufficiently. The accumulation in permanent gases along with condensation is the underlying factor. In this respect, H_2 membrane separation is embedded in the condensation system to ameliorate the cryogenic operation and the mismatch in temperature; meanwhile, H_2 can be yielded with the purity meeting the request for common hydro-treating processes. Our process design and optimization revealed that it is more suitable to arrange two shallow condensation units and two membrane units by turns and in series before the cryogenic condensation, and the suitable operation pressure is about 2.0 MPaG. For a 600-kt/a Oleflex plant with 221300 Nm^3/h reaction effluent, total membrane area is optimized to be about 42000 m^2, and the cold duty for cryogenic operation (below -25 ^oC) can be reduced by 90 %, the minimum temperature for condensation can be raising to -72 ^oC. Besides, the recycled H_2 and the exported H_2 can be concentrated from 92.3 to 96 and 99 vol%, respectively, with the total recovery ratio up to 95 %. According to our economic evaluation briefly, taking energy saving and H_2 purification together, the increment in annual gross profit is about $48.3*10^6$ USD. Overall, membrane separation is highly promising to save energy and enhance economic returns for Oleflex propane dehydrogenation process.

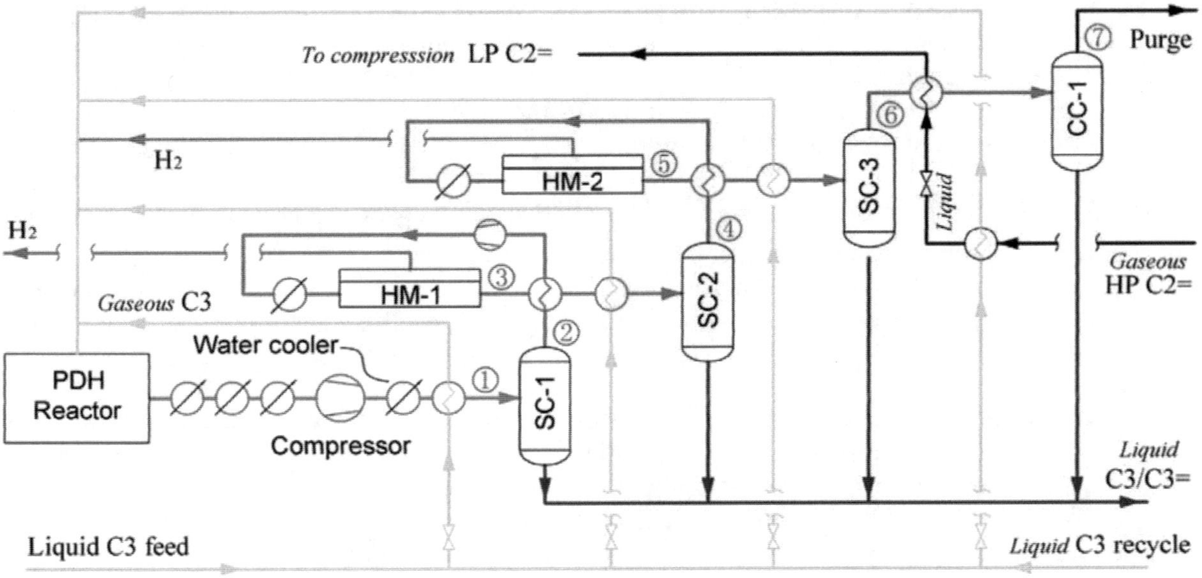

Figure 1 Retrofitted Oleflex condensation section with 2-step membrane-embedded system.

Acknowledgement:

The authors acknowledged the financial supports from the National Natural Science Foundation of China (Grant No. 21606035 and U1663223), Changjiang Scholars Program (T2012049), the Fundamental Research Funds for Central Universities (DUT18JC24), Liaoning Province S&T Department (Grant No.201601037) and Education Department (LT2015007).

568011 Development of Acoustic-Driven Packing Material a Novel Uky-Caer Technology for CO_2 Capture

Acoustic devices have shown to be beneficial in providing unique enhancements in the field of process engineering such as agitation, crystallization, cracking, de-caking, and de-foaming. A few of these devices take advantage of the acoustic streaming effect produced from traveling sound waves in an aqueous solution. A propagating surface wave in a liquid film will increase the film's surface area as it stretches and compresses the liquid, and promote internal mixing. In a gas-liquid absorption process an increase in solvent surface area and gas liquid mixing will increase absorption rate. A novel device is in development at UKy-CAER, acoustic-driven packing material, to scale this concept for use in a counter current CO_2 absorption process. Transmitting high frequency acoustic energy, at resonance, into an absorber column's packing material will cause that material to vibrate and become a transmitter of the incident acoustic energy. Solvent in contact with the vibrating packing material will oscillate at similar wave parameters to the source of acoustic energy, depending on the solvents physical properties.

This presentation will cover the development of acoustic-driven packing material starting from proof of concept, bench scale testing with random packing material, and ending on the first prototype structured packing test. Lessons learned for future development will also be discussed such as the importance of impedance matching and variable frequency emitters. At 20 and 28 kHz, acoustic-driven packing material has been shown to provide up to a 20% relative increase in solvent absorption rate with 30 wt% monoethanolamine. Additionally, adding 1 wt% fine solids, ≈ 40 µm, to the absorption solvent in combination with acoustic-driven packing material has been shown to further increase absorption rate by up to 40% over baseline testing. Acoustic amplitude was observed to be positively correlated with a minor increase in solvent absorption rate. This suggests that the smallest amount of power required to generate acoustic streaming in the solvent would be the most efficient choice in terms of power consumption optimization. Lastly, no decrease in performance has been observed when scaling from proof concept to prototype testing.

564497 Distillation Performance in Packed Columns with Structured Packing: Numerical Simulation Considering the Liquid Distribution Profiles

This paper provides a simulation tool with a cell-model to evaluate distillation performance considering liquid distribution profiles in packed columns with structured packing. A cell-model is one in which distillation performance is calculated by taking these condition into consideration that a packing element is considered as clusters of a cubic cell, and liquid descending from the top of columns is distributed into each cell according to a certain ratio which is called the 'splitting factor'. The splitting factor was obtained by referring to experimental results the author performed in advance.

Experiments for cryogenic distillation, in which some holes of the liquid distributor were intentionally plugged to cause initial liquid maldistribution, were performed under a total reflux condition for nitrogen-oxygen system to confirm the validity of the simulation tool. A column 0.3m in diameter and 1.5m high with structured packing of specific area 750 m^2/m^3 was used. Consequently, it is confirmed that the distillation performance was reduced significantly compared with homogeneous liquid distribution.

This paper presents the comparison between simulation and experimental results.

557077 Composition and Temperature Cascade Control with Model Predictive and PI Controllers for Dividing-Wall Distillation Columns

Despite dividing wall being a very effective technology for process intensification of distillation, the resultant dividing-wall distillation columns (DWDCs) exhibit strongly interactive and highly nonlinear behaviors, which may pose great challenges to process operations. Proportional-integral (PI) controllers can stabilize the operation of the DWDC, but their performances are frequently far from satisfactory. Model predictive control (MPC) can potentially improve performance sharply, but the stringent requirements on the qualities of the embedded model considerably lower its robustness. To suppress these two deficiencies, a cascaded control structure by combining the MPC and PI controllers is proposed for the DWDC in this work. The PI controllers regulate the stage temperature in the inner layer, serving to quickly stabilize the DWDC, while the MPC directly controls the product compositions in the outer layer, addressing the strongly interactive and highly nonlinear behaviors of the DWDC. Two typical cases i.e., a three-product Petlyuk DWDC and a four-product Kaibel DWDC, and five control schemes i.e., two impurity composition controls (PI-ICC and MPC-ICC), two purity composition controls (PI-PCC and MPC-PCC), and one pure temperature control PI-TC (which represents the case wherein MPC in the outer layer fails to work), are studied to evaluate the feasibility and effectiveness of the proposed control structure. It is found that the cascaded MPC/PI controllers (i.e., MPC-ICC and MPC-PCC) substantially outperform the pure PI controllers (i.e., PI-ICC and PI-PCC), with lower maximum and steady-state deviations, shortened settling times, and suppressed oscillations. In the case of failure of the MPC, the cascaded MPC/PI controller can still maintain stable operation of the DWDC. These outcomes highlight that the combination of MPC and PI controllers can be a potential method for the operation of highly interactive and complicated DWDCs.

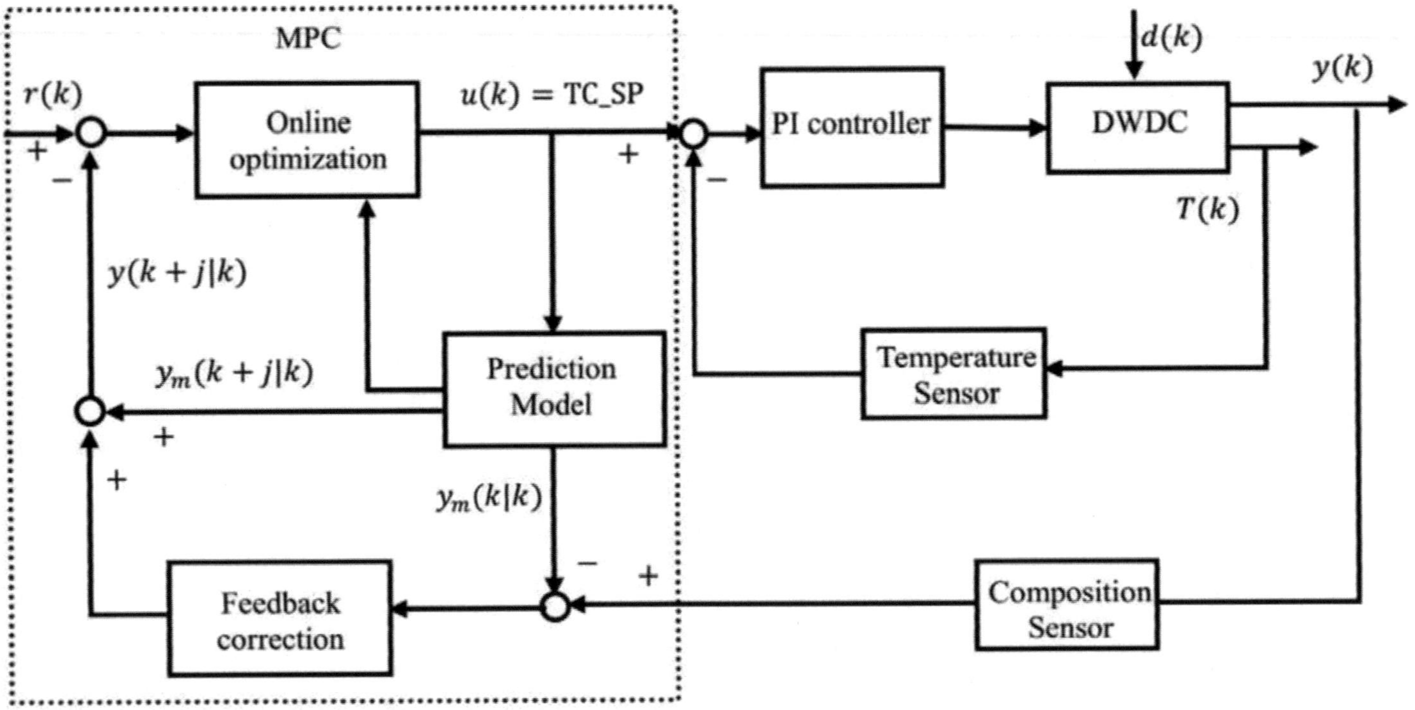

Figure 1 The novel composition/temperature cascade control scheme

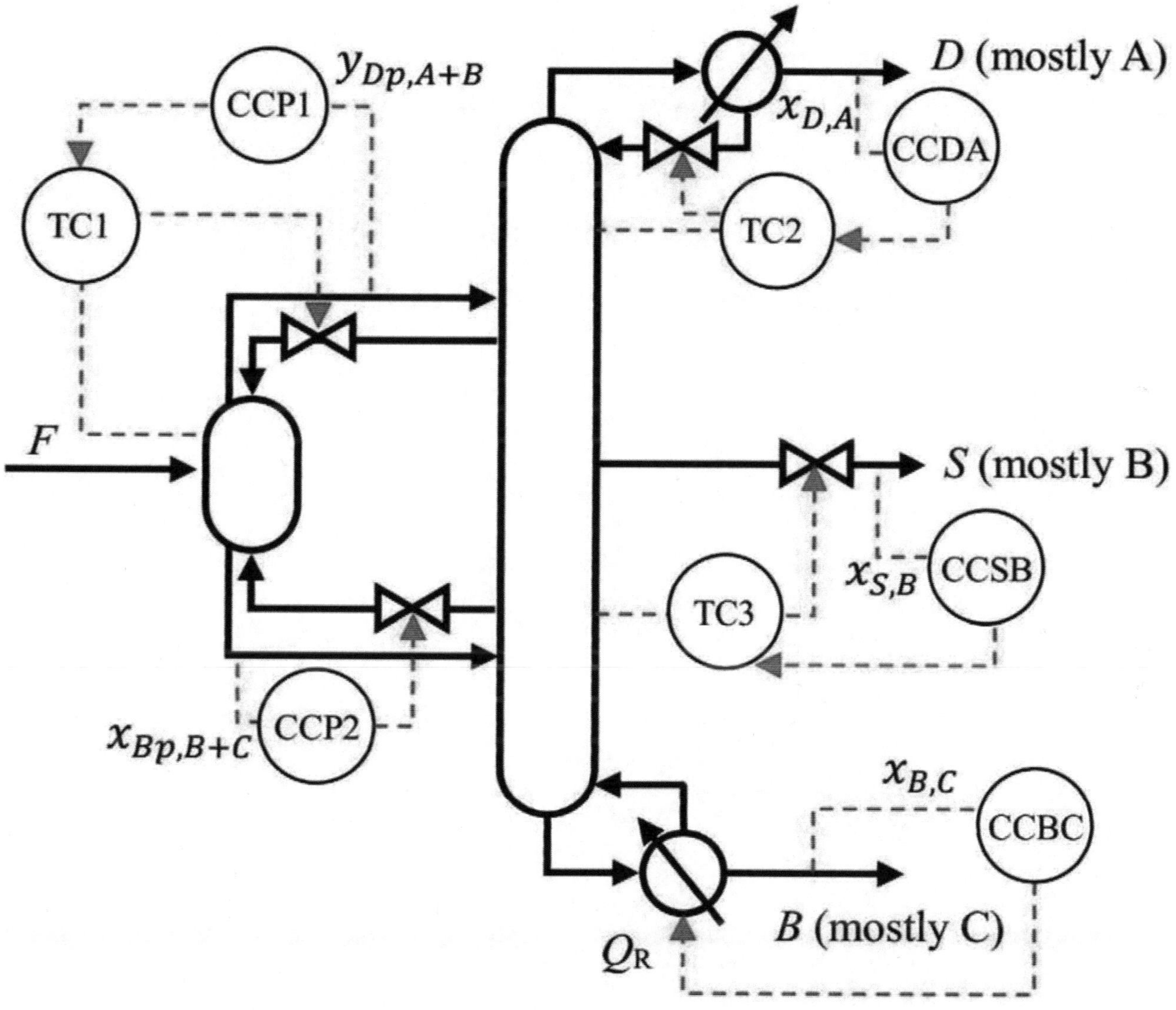

Figure 2 Composition/temperature cascade control of the three-product Petlyuk DWDC using pure PI control

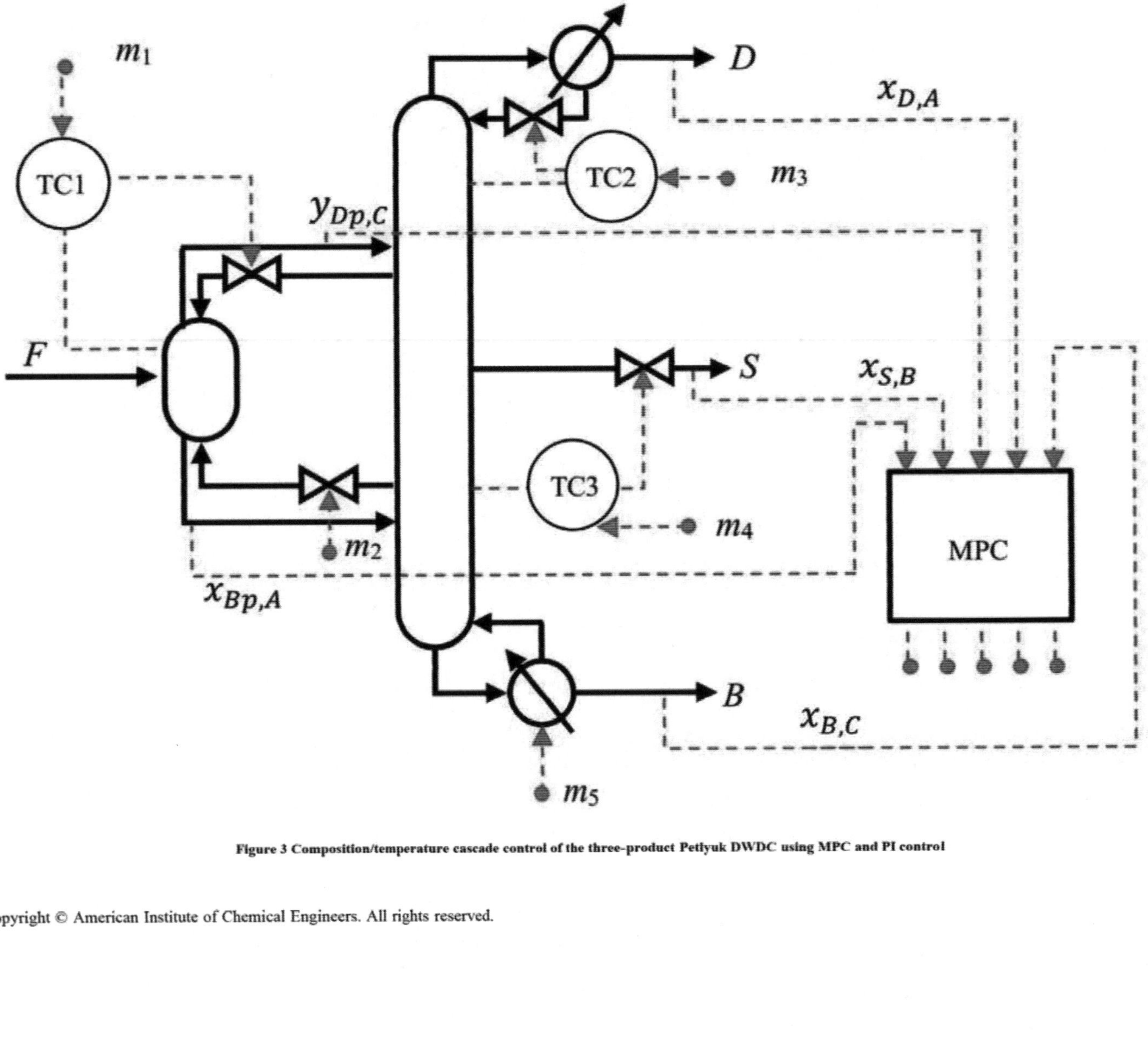

Figure 3 Composition/temperature cascade control of the three-product Petlyuk DWDC using MPC and PI control

556584 Minimum Reflux Calculation for Multicomponent Azeotropic Distillation Using Shortcut Method

Chemical, pharmaceutical, and agrochemical industries are frequently involved with separation of multicomponent mixtures exhibiting one or more azeotropes using distillation. To design a distillation column that is cost-effective to build and energy efficient to operate, key parameters such as the minimum reflux ratio are generally required. Therefore, design engineers need to understand and be able to determine the minimum reflux condition of a distillation column when separating a multicomponent azeotropic mixture.

Despite its practical significance, this problem is challenging to solve. Most existing approaches rely on tray-by-tray calculations or iterative guessing of reflux ratio, which can be too computationally expensive and complex to implement. Here, we present a simple and easy-to-use shortcut method to analytically calculate the minimum reflux ratio of a distillation column for multicomponent homogeneous azeotropic mixture separations. We treat each azeotrope as a pseudocomponent after exploiting the physical and mathematical properties of azeotropes and applying the proper linear transformation. Such transformation preserves the mathematical simplicity of the resulting system and allows us to derive the minimum reflux condition for multicomponent homogenous azeotropic distillation.

Through case studies, we demonstrate the accuracy and effectiveness of our new approach. We show that the classic Underwood's method used for ideal multicomponent distillation turns out to be a special case of this general approach. Compared to existing algorithmic approaches, our method does not require any tray-by-tray calculation and is iteration free. As a result, it can be easily incorporated into a global optimization framework that will allow industrial practitioners to, for the first time, quickly ranklist and identify a handful of attractive distillation configurations from the immense configuration search space.

558129 Biodiesel Separation Using Ultrafiltration Poly(ether sulfone) Hollow Fiber Membranes: Improving Biodiesel and Glycerol Rich Phases Settling

After alkali-catalyzed transesterification reaction for biodiesel production, the glycerol-rich phases have to be separated. These phases are traditionally separated by settling, requiring long residence time, especially when soaps and gels are formed. In this work, biodiesel, and glycerol rich phases separation was experimentally assessed using poly(ether sulfone) hollow fiber membranes (PES-HFM). Experimental data were obtained in a continuous bench scale system. The effect of pressure difference through the membrane (0 to 0.6 bar), feed composition and biodiesel-rich phase mass fraction (0 to 0.8) on permeability and permeate composition were studied. In addition, a mathematical model adopting the Hagen-Poiseuille transport equation for membrane ultrafiltration, where permeate fluxes and compositions depend on the Liquid-Liquid Equilibrium (LLE), was proposed, correlated and experimentally validated. Experimental results demonstrated only glycerol-rich phase permeated through the membrane following the LLE. Glycerol-rich phase flux increased when pressure difference through the membrane augmented, and decreased when permeate viscosity increased. The highest experimental permeability (33.2 kg $bar^{-1}h^{-1}m^{-2}$) was obtained at the highest methanol content in the feed stream, 66 %wt., equivalent to a molar ratio methanol to oil 18:1 fed to the reactor. The mathematical model predicts an increase in the glycerol-rich phase flux when temperature and methanol content augment. The transport mechanism coupled to the mathematical model explained accurately the membrane role in the separation of the compounds involved in transesterification of vegetable oils, as well as the membrane selectivity.

573645 Molecular Simulations of Carbon-Based Adsorbents/Membrane Materials for Sustainable Processes

According to the Brundtland Report: "Sustainable development is development that meets the needs of the present without compromising the ability of future generations to meet their own needs." Sustainable development implies the urgent needs to secure clean energy, food and water, known as the energy-food-water nexus. In this presentation, we will focus on some specific examples related to materials for CO_2 capture and for water treatment, as two specific examples to help in developing sustainable processes, either by avoiding the emission of greenhouse gases into the atmosphere or by providing quality water.

Commercial technologies used nowadays for carbon dioxide capture and water treatment face two major challenges: reduce capital costs and increase energy efficiency.[1,2] In the past few decades, porous materials have attracted considerable attention for separation purposes.[3] Common materials that have been extensively studied include zeolite, silica gel, metal-organic frameworks and carbon-based structures [4,5] Among them, carbonaceous materials have been well studied mainly due to their wide availability, low cost, high thermal stability and the possibility of synthesis from numerous carbons based naturally existing or spent materials.[6] Activated carbons, carbon nanotubes and membranes made of graphene or graphene-oxide structures are attractive and versatile materials. They are becoming vital for a wide range of industries, due to their unique structural and chemical properties. The ability to experimentally control adsorbent structural features at the molecular level, and recent progress made on modeling realistic carbon-structures and fluids force fields, enable building predictive models from molecular simulations for novel applications [7], [8]. Little is known about the effect of surface characteristics (*e.g.*, pore size, surface area, density and species of surface groups, etc.) and the topological nature of the connected pore structures on the phase changes for the confined fluids [9]. In this regard, molecular simulations provide a direct route from the properties of interacting molecules to the thermodynamic properties of their bulk phases, and hence, they provide an alternative source of property data, while allow gaining insights into the physical phenomena of the membrane/adsorption process.

Therefore, this contribution is devoted to molecular simulation on carbon-based materials, from building the model to the application in separating fluids. The final goal is to showcase how molecular simulations can be used as a robust, complementary tool to understand and optimize fluid separation processes of industrial relevance. We will present and discuss results from Molecular Dynamics (MD) and Monte Carlo simulations to study separation of multi-component mixtures with virtual activated carbons and multilayer graphene oxide membranes, using realistic operating conditions, also considering the presence of minor components. Improved understanding of the effects of confinement on the equilibrium composition of mixtures, and on adsorption and diffusion rates on complex surfaces, may lead to significant improvements in the industrial processes. We will focus on two specific applications: separation of CO_2 from a flue gas and water desalination, comparing results with experimental data.

This work is part of a collaborative bilateral project between the Korea Advanced Institute of Science and Technology and Khalifa University. Financial support for this work has been provided by Khalifa University through projects CIRA-103, RC2-2018-0024 and RC2-2019-007.

References

[1] M.M. Pendergast, E.M.V. Hoek. Energy Environ. Sci., 4, 1946 (2011).

[2] D. Bahamon, A. Díaz-Márquez, P. Gamallo, L.F. Vega. Chem. Eng. J., 342, 458 (2018).

[3] L. Wang, M.S.H. Boutilier, P.R. Kidambi, D. Jang, N.G. Hadjiconstantinou, R. Karnik. Nat. Nanotechnol., 12, 509 (2017).

[4] Y. Aristov. J. Chem. Eng. Jpn., 40(13), 1242 (2007).

[5] F. Perreault, A.F. De Faria, M. Elimelech. Chem. Soc. Rev., 44, 5861 (2015).

[6] R. Ben-Mansour, M.A. Habib, O.E. Bamidele, M. Basha, N.A.A. Qasem, A. Peedikakkal, M. Ali, M. Appl. Energy, 161, 225 (2016).

[7] L.F. Vega J. Theor. Comput. Chem., 18, 101 (2007).

[8] D. Bahamon, M. Abu Zahra, L.F. Vega, Fluid Phase Equilibria 492, 10-25 (2019)

[9] J.K. Brennan, T.J. Bandosz, K.T. Thomson, K.E. Gubbins. Colloids Surf. A, 187–188, 539 (2001).

564698 Model of the Multi-Channel Ceramic Membrane for so_2 Absorption

Multi-channel ceramic membranes have higher specific surface area and packing density than tubular membranes, making them more suitable for industrial applications. There is a broad application for ceramic membranes to absorb low concentrations of SO_2 in exhaust. the membrane phase resistance becomes the main resistance during the absorption of SO_2 by the hydrophobic membrane. Therefore, for the multi-channel ceramic membrane, the thickness of the central channels is too large, resulting in poor mass transfer performance. Wetting the membrane pores with NaOH accelerates the interfacial reaction rate which greatly reduces the membrane phase resistance. To find optimal NaOH concentration conditions and geometry of multi-channel membrane contactors, we employed a numerical 3D model to simulate SO_2 absorption process and verified the model by comparing its results with experimental data. Compared to conventional tubular ceramic membranes, multi-channel ceramic membranes have only a 25% decrease in flux under the same conditions. The effect of multi-channel membrane geometry was studied by this model. The gas concentration distribution in membrane contactor shows that channels have different desulfurization performance. Furthermore, the simulation reveals that the SO_2 removal performance is affected by the pore arrangement of the multi-channel membrane. The membrane tube of each configuration corresponds to an optimum NaOH concentration that achieves maximum flux. The multi-channel membrane contactor exhibit ed great application potential.

568337 A Predictive Model for Seawater Reverse Osmosis Module

A Predictive Model for Seawater Reverse Osmosis Module

Mingheng Li[1]

[1] *Califorina State Polytechnic University, Pomona, CA 91768, United States, minghengli@cpp.edu*

Optimal design and operation of membrane processes are important engineering problems arising at the water-energy nexus [1, 2]. To enhance process efficiency, it is essential to develop system-level models that can accurately predict of process behaviors including water flux, salt flux and pressure drops [3, 4]. While high-fidelity CFD models are available to explore local transport phenomena [5], enormous computational time is required for a whole membrane module. This presentation will focus on an efficient two-dimensional model that is capable of predicting the performance of commercial reverse osmosis modules with reasonable accuracy. The model explicitly accounts for longitudinal pressure drop in the brine channels and spiral pressure drop in the permeate channels as well as the effect of temperature on membrane permeability. Figure 1 shows that the model predictions of module performance from various feed conditions match fairly well with experimental data from a commercial Dow Filmtec™ module [6]. Model-based optimization of process design and operation will be also discussed.

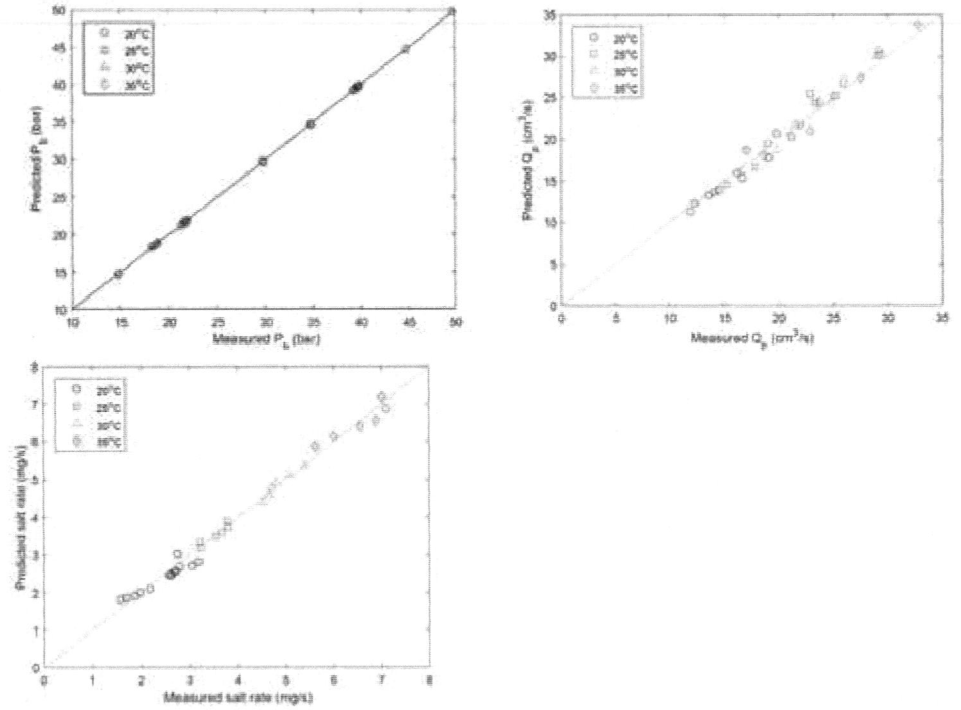

Fig. 1. Comparison of model predicted permeate outlet pressure (left), permeate flow rate (middle) and salt amount in permeate (right) and measured data.

References

[1] Li, M. *Chem. Eng. Res. Des.*, **2018**, 137, 1-9.

[2] Li, M. *AIChE J.*, **2018**, 64, 144-152.

[3] Li, M. *Desalination*, **2012**, 293, 61-68.

[4] Li, M.; Noh, B. Desalination, **2012**, 304, 20-24.
[5] Li, M.; Bui, T.; Chao, S. *Desalination*, **2016**, 81, 191-208.
[6] Avlomtis, S.; Hanbury, W.T.; Ben Boudmar M. *Desalination*, **1991**, 397, 194-204.

Acknowledgements

This work was supported by American Chemical Society Petroleum Research Fund.

573840 Modelling Oxygen Permeation across the Membrane of an Alternative Membrane Reactor for Oxidative Coupling of Methane

Oxidative coupling of methane to valuable fuels is currently investigated. Advance reactor concepts are currently been sought alongside high performing catalyst. One of such reactor is the Membrane contactor combining separation and reaction in a single unit. It has the potential of alleviating the kinetic and thermal problems of oxidative coupling of methane (OCM) reactions which is a typical problem of a conventional packed bed reactor. However, conventional membrane reactor still exhibits similar limitations of hotspot generation and trade-off between methane conversion and C_2 (ethane and ethylene) selectivity trade-off, hence, alternative membrane reactor concepts are emerging. One of such concepts is the varying thickness membrane reactor, hinged on the fact that, gradually increasing the membrane thickness along the length of the membrane reactor will minimize the amount of oxygen which can be detrimental for C_2 selectivity. This implies for such reactor, the right proportion and distribution of oxygen into the reaction zone is a key design aspect that can influence the performance of membrane reactors utilized for (OCM). In this study, a detailed kinetic model comprising of 15 catalytic reactions and 39 gas phase reaction is implemented in a membrane reactor. A 2D model taking into account radial dispersion is applied. An oxygen permeation model that distinguishes between the surface reactions at the two solid-gas interfaces of the membrane and the diffusion of oxygen across the membrane is implemented. Various oxygen permeation flux are considered by varying membrane thickness according to a linear, exponential and step change geometries. The importance of these flux profiles on both the performance of the membrane reactor and the oxygen transport limitations across the membrane are highlighted.

569359 Modeling the Ion Transport and Adsorption Behavior in a Membrane Capacitive Deionization (mCDI) Cell

In the arid west, the freshwater supply of many communities is limited leading to increased interest in tapping brackish water resources. Although reverse osmosis is the most common technology to upgrade saline waters, there is also interest in developing and improving alternative technologies. Here we focus on Capacitive Deionization (CDI) and specifically compare and contrast behavior between conventional CDI and membrane CDI (mCDI). The primary purpose of mCDI is to enhance the separation between salts and water. In this study, a comprehensive two-dimensional model describing ion transport and adsorption behavior was developed and solved numerically, taking advantage of Darcy's Law, Nernst-Planck equation, Donnan equilibrium and Nernst equation considering non-ideal membrane. The model behavior was supported by the experimentally observed variations in transient effluent concentration under different operating conditions. The advantages of mCDI over conventional CDI in adsorption amount and ion separation rates were shown in the simulation results. The simulated concentration, potential and ion flux distributions in the mCDI and CDI cells were used to better understand the device's behavior and optimal operating conditions. The model was used to examine the potential influences of cell geometry, membrane properties and operating conditions on the mCDI performance.

578222 A General Predictive Model for Comparative Study of Vacuum Membrane Distillation and Sweeping Gas Membrane Distillation

Membrane distillation is an emerging thermal membrane technology for the separation of salts and other non-volatile compounds from water streams. The process offers a unique solution for the treatment of concentrated solutions with high total dissolved solids, which are not viable for reverse osmosis. One of the major challenges associated with MD is the low energy efficiency, which is mainly due to the loss of the latent heat of evaporation. Therefore, multiple strategies are proposed to integrate MD with other conventional technologies for proper heat integration. We are designing a modular, integrated mechanical vapour compression-membrane distillation (MVC-MD) to improve the energy efficiency of the MD.

Two different MD configurations could be utilized for designing MVC-MD: vacuum membrane distillation (VMD) and sweeping gas membrane distillation (SGMD). Unfortunately, these two MD configurations have been remained as the least studied MD configurations. In this talk, we will present the basis for designing the first MVC-MD unit based on VMD and SGMD configuration. These two MD configurations represent the two extremes that MVC could be operational. We will discuss the optimized conditions of a sweeping gas-assisted vacuum membrane distillation unit that could be integrated with MVC to achieve respectable energy efficiencies. Our experimental work will be integrated with our process-level Aspen Custom Modeler simulations to further optimize the integrated MVC-MD unit, which will guide the future efforts to further optimizing MVC-MD process.

577861 Design and Fabrication of High-Performance Hollow Fiber Membrane Devices for Direct Contact Membrane Distillation

Membrane Distillate (MD) is an emerging water desalination technology that can allow for desalination of saline water using low-grade thermal energy. However, currently, there are no commercial membrane devices that are designed for water desalination through the membrane distillation process. This is mostly due to the lack of high-performance membranes and modules. Hollow fiber membrane modules are the most attractive candidates from the design perspective. In our recent modeling studies[1], we showed that the packing configuration of hollow fibers in a module plays an important role performance of modules. To validate our modeling studies, we fabricated Polyvinylidene Fluoride (PVDF) hollow fiber membranes using Nonsolvent Induced Phase Inversion (NIPS) process and packaged them into a module using custom-made spacers. We made spacers using additive manufacturing approach. Similar to the result of our modeling work, the experimental results showed that the close-packed configuration of membranes in the module provides about 13% more water flux compared to that of a randomly packed module. The average values for the permeate flux obtained from the experimental studies were 25±5 kg m^{-2} h^{-1} when the feed and distillate temperatures were set to 70 and 20 °C, respectively. Furthermore, we modified the outer surface of the hollow fiber membranes and deposited a thin coat of polytetrafluoroethylene (PTFE) through initiated chemical vapor deposition (iCVD) process. This coating imparted super-hydrophobicity to the surface of the membrane, rendering them self-cleaning. The result of our accelerated scaling experiment showed a significant reduction of salt scaling on the surface of the modified hollow fiber membranes.

References

(1) Mohammadi Ghaleni, M.; Bavarian, M.; Nejati, S. Model-Guided Design of High-Performance Membrane Distillation Modules for Water Desalination. *J. Memb. Sci.* **2018**, *563*, 794–803.

581340 Apply Nanocomposite Material for Hydrogen Separation and Purification

In this study, various adsorbents with tailored Nano porosities have recently been developed as potential separation media for CO_2/H_2 and CH_4/H_2. Nanocomposites were prepared by two different sets of MOF-GO nanocomposites comprising of HKUST-1 and pristine GO, and reduced GO were developed by solvothermal method and their gas separation characteristics were assessed through static test which is high-pressure adsorption measurements and dynamic breakthrough test. We found that the GO well-dispersed MOF-GO that has Nano sized and induced by the incorporation of GO significantly developed the CO_2, and CH_4 capacities and CO_2/H_2 and CH_4/H_2 the selectivity performance of the nanocomposites. The nanocomposites obtained exhibited about a (20~35) % increase in CO_2 and CH_4 capacities than the bare MOF at 25 ^{0}C and 40 bar, the breakthrough test for binary gas mixtures which are $40\%CO_2/60\%H_2$ and $20\%CH_4/80\%H_2$ were also considered at different pressure for the all adsorbents. Lastly, the CO_2/H_2, and CH_4/H_2 selectivity's were estimated according to single-component gas sorption experiment data. After knowing capacity, breakthrough time and selectivity will give us hints for the design and synthesis of Naonacomposite is promising candidate for industrial gas separation and purification processes.

566338 Isolation Development of Co-Crystal API

In the recent years, pharmaceutical cocrystal has attracted increasing attention to improve the physicochemical properties of APIs without modifying their chemical structure, such as the bioavailability, stability, and processability. How to isolate phase pure co-crystal API is an interest topic industrywide. This presentation shares a workflow for the co-crystal isolation process development with a case study. It provides an option for the scientists in the field to develop a robust co-crystal isolation process simply by following the workflow.

Polymer-Zeolite Dense Mixed-Matrix Membrane for Carbon-Dioxide Separation

Dragutin Nedeljkovic, American University of the Middle East, Kuwait

Introduction

The increased emission of the carbon dioxide has induced great concerns in the environmental engineering in the past decades. High concentration of carbon dioxide in atmosphere is responsible for the greenhouse effect, and can also contribute to the acid rains. The main emitters of the carbon dioxide are common combustion processes that include fossil fuels on the both industrial and communal scale (heating plants, power plants, different procedures in chemical industry, automobile engines)[1-3]. As demand for the energy increase every day, and the fossil fuels have no alternative on the world-scale, importance of carbon dioxide removal from the waste gases has attracted great efforts of the environmental engineering community [4-6].

One of the possible approaches to carbon dioxide capture is the implementation of membrane that would separate carbon dioxide from rest of the gases. Suitable membrane should exhibit high permeability for the carbon dioxide, and the lowest possible permeability for the other gases [7-9]. As the separation of gases is done on the molecular level, "standard" membranes with pores that use the difference in the eluent particle size are not suitable for this purpose. Therefore, so-called dense, non-porous membranes based on the solution-diffusion mechanism must be employed. In this type of membranes, effluent is adsorbed on the surface on the membrane, and then "dissolves" in the matrix of the membrane, and then diffuses through the thickness of the membrane layer. In that case, permeability of the membrane is defined as:

$$P = S \cdot D \qquad (1)$$

where P, S and D stands for permeability, solubility and diffusivity respectively. Selectivity of the dense membrane can be defined as:

$$\alpha_{i/j} = \frac{P_i}{P_j} \qquad (2)$$

$\alpha_{i/j}$ in equation 2 represents the selectivity of the substance i versus substance j, and P_i and P_j are the permeability of substances i and j respectively[10-13]. Combining equations 1 and 2, it is obvious that both solubility and diffusivity of the gases must be taken into account in order to define properties of the suitable membrane [14-16]. Approximate, theoretical comparison of molecular properties of carbon dioxide and other gasses (especially hydrogen) shows that carbon dioxide molecule is much bigger. Therefore, the diffusion coefficient for the small, non-polar hydrogen molecule is much higher compared to carbon dioxide, and thus the diffusion of the hydrogen is much faster. Therefore, the solubility of the carbon dioxide must be increased in order to obtain good selectivity for the carbon dioxide versus hydrogen. The situation is similar in the case of nitrogen and oxygen molecules, but those molecules are bigger, and thus the difference in diffusivity is not so high. However, the solubility of oxygen and nitrogen should be kept as low as possible [17-19].

In order to prepare proper film for the casting of the membrane, the zeolites have to be dispersed in the same solvent as the polymer. Important step is to provide good wetting of the zeolite particles by the polymer chain. Fulfilling this task can be tedious as the inorganic zeolite particles are highly polar and hydrophilic, and show very bad affinity to the highly hydrophobic polymer chains. Proper film must be free of any pin holes, with the homogenous distribution of zeolite particles in the polymer matrix. To fulfill this task, different additives can be added. The aim of the additive addition to the matrix is to act as a compatibilizer between the zeolite particles and polymer matrix [20-25].

Materials and Methods

For the research in this paper, possibility of application of polyether-*b*-amide with 60% of PEG was investigated. This polymer is available under the commercial name PEBAX 1657 from the supplier Arkema. The structure of this polymer is presented on Fig. 1.

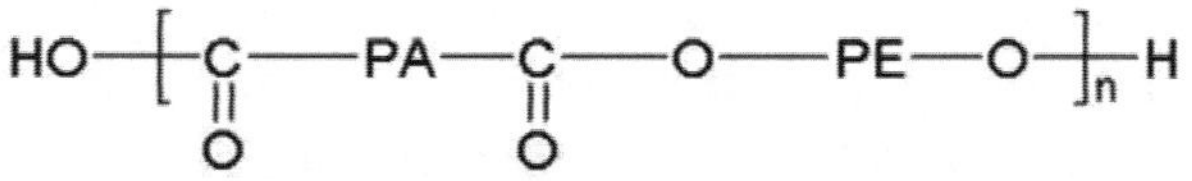

Fig. 1: Structure of the PEBAX 1657 used as a matrix

The second polymer used was poly(ethylene glycol)-poly(butylene terephthalate) copolymer with 77 wt% of poly(ethylene glycol) available under commercial name Polyactive (supplied by IsoTis OrthoBiologics).

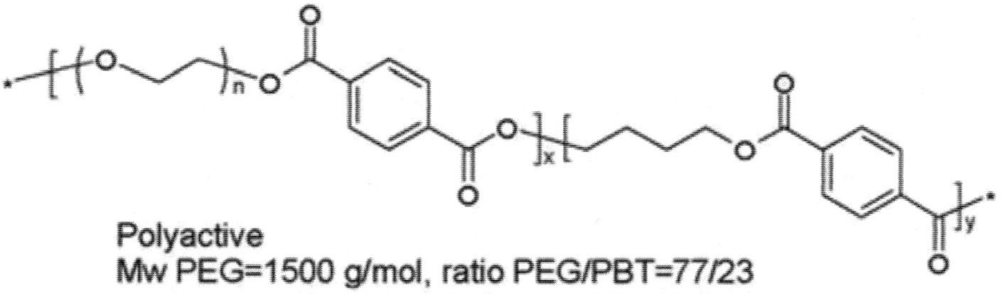

Fig 2: Structural formula of Polyactive co-polymer

In order to enhance the solubility of the carbon dioxide, inorganic nanopowder was dispersed in the polymer matrix. The properties of the zeolite can be described at the macro and micro level. At the macro level, the properties of the powder are determined by the specific surface of the particles. Particles used for the experiments presented in this paper have the specific surface of 300-600 m^2/g. For this experiment, five different types of the zeolites were used. One of the zeolites is with 1-dimensional pores (OSI), two are with two dimensional pores (IHW and TER) and 2 are with 3-dimensional pores (MFI and FAU). Properties of the used zeolite types are compiled in Table 1.

Table 1: Properties of different zeolite types used for this experiment

Framework type code	Channel system	Cell size, nm	Maximum radius of a sphere that can diffuse, nm
OSI	1d	1.85x1.85x0.53	0.63
IHW	2d	1.37x2.41x1.83	0.37
TER	2d	0.98x2.36x2.02	0.52
MFI	3d	2.01x1.97x1.31	0.47
FAU	3d	2.43x2.43x2.43	0.74

The zeolite powders were supplied by the company NanoScape, and characterization was done by the supplier. Two different additives were tested: n-tetradecane trimethyl ammonium bromide (n-C14-TMABr). It was supposed that the polar nitrogen – bromine bond would "attach" to the surface of the zeolite particle, and the long, normal, hydrocarbon tale would act as an "anchor" to the polymer matrix. By this mechanism, it was supposed that the compatibility between the zeolite and polymer would be achieved. The second additive that was used was dimethyl ammonium pyridine (DMAP). Although

DMPA does not contain long carbon chain, the mechanism for compatibilization should be similar as in the previous case. Beside this effect, due to the basic properties of the DMAP, solubility of carbon dioxide should be improved.

The procedure for the construction of the membrane was as following: Polymers were dissolve in suitable solvents (70/30 wt.% water/ethanol mixture for PEBAX, and chloroform for Polyactive) at appropriate temperature (80°C under reflux for Pebax and room temperature for Polyactive). The zeolite was dispersed in the small amount of the same mixture, and (eventually) additive was added. In order to get homogenous dispersion, ultrasound mixer was used. The titanium head was introduced in the solution, and the mixing was done by the application of full power for three minutes. The mixing time was short because during the mixing for a long period, nano particles of titanium are detached from the ultrasound head, and released in the solution. Then, this dispersion was poured in the polymer solution and stirred overnight at the 80°C under reflux (PEBAX based membranes) or room temperature (Polyactive based membranes). Viscous solution came as a result, and it was casted on the Teflon surface with the Teflon ring used as a border. This solution was covered with non-woven textile, and left overnight for drying at the room temperature and ambient pressure. The drying process had to be such slow, as in the case of the faster process, air bubbles are formed in the volume of the membrane, and membrane cannot be used for the gas separation. At this level it, is extremely important to obtain the polymer solution with the optimal viscosity. If the viscosity is too high, the surface tension dominates trough the casting of the membrane, and the result is the membrane with the uneven thickness, which negatively affects the overall permeation properties of the membrane. On the contrary, too low value for the viscosity causes the sedimentation of the zeolite particles, and the zeolite is concentrated on the one side of the membrane. If the latter is the case, the membrane self-rolls, and its application is negatively affected. After the drying at the ambient pressure, membrane was placed in the apparatus for the permeability measurements, and put at the high vacuum in order to remove any traces of the solvent.

The permeability measurements were done by the application of the gas at the one side of the membrane, while the high vacuum was applied from the opposite side. This difference in pressure acted as a driving force for the gas diffusion. Gases were measured one at the time, with high vacuum applied between different gases in order to remove any residual gas that was solved in the membrane. In the first series, the permeability of hydrogen, oxygen, nitrogen, methane and carbon dioxide were measured, and the selectivity were calculated versus carbon dioxide. The permeability of helium was also measured for the determination weather membrane contain pin hole or not. In the second series, prior to the measurement of the membranes' properties, all gases were saturated with the water vapor at the temperature of 40°C, and the permeability was measured. The obtained results of this "wet" series were compared with the first run of measurements ("dry" series) to check how the presence of the water vapor would influence the permeability and selectivity properties of the membrane.

Results and Discussion

The compositions and physical appearances of the constructed membranes are compiled in Table 2 and Table 3. All the presented percent were mass percent versus polymer. Three different membranes were synthesized for each polymer – zeolite – additive combination, in order to avoid samples made by chance. If there were discrepancies in the appearance between the membranes with the same composition, additional two samples were made as a control sample.

Table 2: The composition and optical appearances of the constructed membranes with PEBAX matrix

Sample number	Polymer	Filler	Filler, %	Additive	Additive, %	Appearance
1	PEBAX	-	-	-	-	transparent
2	PEBAX	-	-	n-C14-TMABr	-	transparent
3	PEBAX	-	-	DMAP	-	transparent
4	PEBAX	OSI	22.5		-	transparent
5	PEBAX	OSI	22	n-C14-TMABr	3.3	transparent
6	PEBAX	OSI	22.5	DMAP	8.5	white spots
7	PEBAX	IHW	22	-	-	white
8	PEBAX	IHW	22	n-C14-TMABr	3.3	transparent
9	PEBAX	IHW	21.1	DMAP	8.5	white
10	PEBAX	TER	22		-	white
11	PEBAX	TER	22.5	n-C14-TMABr	3.3	transparent
12	PEBAX	TER	23	DMAP	8.5	white spots
13	PEBAX	MFI	21.5		-	white
14	PEBAX	MFI	22	n-C14-TMABr	3.3	transparent
15	PEBAX	MFI	22	DMAP	8.5	white
16	PEBAX	FAU	22		-	white
17	PEBAX	FAU	23	n-C14-TMABr	3.3	transparent
18	PEBAX	FAU	22.5	DMAP	8.5	white spots
19	Polyactive	-	-	-	-	transparent
20	Polyactive	-	-	n-C14-TMABr	-	transparent
21	Polyactive	-	-	DMAP	-	transparent
22	Polyactive	OSI	22		-	white
23	Polyactive	OSI	21.5	n-C14-TMABr	3.3	transparent
24	Polyactive	OSI	22.5	DMAP	8.5	white spots
25	Polyactive	IHW	23	-	-	white spots
26	Polyactive	IHW	22.5	n-C14-TMABr	3.3	transparent
27	Polyactive	IHW	22	DMAP	8.5	white
28	Polyactive	TER	22		-	white
29	Polyactive	TER	21.5	n-C14-TMABr	3.3	transparent, rolled
30	Polyactive	TER	22.5	DMAP	8.5	white
31	Polyactive	MFI	22		-	white areas, rolled
32	Polyactive	MFI	22.5	n-C14-TMABr	3.3	transparent, rolled
33	Polyactive	MFI	22	DMAP	8.5	white spots, rolled
34	Polyactive	FAU	23		-	white, rolled
35	Polyactive	FAU	22.5	n-C14-TMABr	3.3	transparent
36	Polyactive	FAU	22.5	DMAP	8.5	white spots

Membranes that appeared white or contained white spots or were not measured. As it is stated in the introduction, white spots indicates aggregations of the zeolite particles, and thus implies non-uniform distribution of particles, and thus, bad permeation properties. Membranes that appeared white were not measured as well, as white color indicates bad contact between the surface of the zeolite and the surface of the polymer. Analyzing the data presented in table 2, it is obvious that DMAP is not suitable for the additive, as it does not provide good contact between the zeolite particle and polymer chains. Therefore, it might be concluded that only n-C14-TMABr was suitable for the compatibilizing additive. The permeability of the pure PEBAX 1657 was also measured, and compared to the literature data. All of the permeability were measured and presented in Barrer unit, which is usual unit in the membrane community. The permeability of 1 Barrer is the permeability of the membrane that allows permeation of 1 cm^3 of the oxygen at the standard temperature and pressure that diffuses trough the 1 cm thick membrane through the surface area of the membrane of 1 cm^2 and with the driving force of 1mmHg pressure difference in 1 second. (1 Barrer = 10^{-11} (cm^3 O_2) cm cm^{-2} s^{-1} $mmHg^{-1}$)

The first approximation of the membrane quality was obtained by the measurements of the permeability of helium. Helium atoms are very small, with perfect round shape and non-polar, and thus, have high diffusion coefficient, higher, even than the hydrogen. If the pin-holes or similar defects are present in the structure of the membrane, permeability of helium cannot be measured. Instead of the curve of nearly linear increase in pressure at the vacuum side of the membrane, rapid increase is observed which implies that helium passed through the pin-hole, rather than diffused trough the membrane.

Results of the permeability measurements for the gases performed at dry and wet conditions are presented in Table 3 and Table 4 respectively (all samples except sample 4 are prepared with n-C14-TMABr as an additive.

Table 3: The results of the permeability measurement of the membranes under dry conditions

Sample number	Polymer	Filler	Thickness, μm	P (CO_2), Barrer	α (CO_2/H_2)	α (CO_2/O_2)	α (CO_2/N_2)
4	PEBAX	OSI	191	95	8.9	25	52
5	PEBAX	OSI	217	125	9.3	27	57
8	PEBAX	IHW	185	137	9.9	22	59
11	PEBAX	TER	213	145	11.6	23	55
14	PEBAX	MFI	199	121	10.7	21	61
17	PEBAX	FAU	174	141	9.9	20	53
23	Polyactive	OSI	233	127	9.8	21	59
26	Polyactive	IHW	201	137	10.2	22	60
35	Polyactive	FAU	197	119	10.3	21	60.5

Table 4: The results of the permeability measurement of the membranes under wet conditions

Sample number	Polymer	Filler	Thickness, μm	P (CO_2), Barrer	α (CO_2/H_2)	α (CO_2/O_2)	α (CO_2/N_2)
4	PEBAX	OSI	191	80	7.2	22	50
5	PEBAX	OSI	217	98	8.1	26	54
8	PEBAX	IHW	185	101	7.7	20	57
11	PEBAX	TER	213	99	8.2	21	50
14	PEBAX	MFI	199	100	7.9	20	61
17	PEBAX	FAU	174	95	8.0	19	50
23	Polyactive	OSI	233	93	7.1	19	58
26	Polyactive	IHW	201	103	7.7	20	60
35	Polyactive	FAU	197	90	8.3	20	59

Analyzing the presented data, it is obvious that only n-C14-TMABr is suitable to serve as an additive. If the membranes were synthetized without it, good contact between polymer and zeolite particles could not be obtained. On the other hand, addition of DMAP induced agglomeration and/or sedimentation of zeolite powder. Comparing the properties of different polymer – zeolite combinations, it is obvious that the choice of the polymer plays no significant role in the obtain results, although PEBAX has shown slightly better properties in compatibility to zeolite particles. From the permeability data, it can also be observed that measurements presence of the water vapor decreases the permeability of carbon dioxide by approximately 20%, while the selectivity versus hydrogen is decreased by approximately 10%. Selectivity versus nitrogen and oxygen were changed at much smaller rate, and the decrease in selectivity was around 5%. Possible explanation might be that the water vapor competes in permeability with other gases, so the permeability of single gas is smaller, as two different gases should diffuse through the same thickness of the membrane with the equal driving force. The main challenge in the future will be downsizing the thickness and generation of higher fluxes of gases.

Conclusion

Comparing the permeability results between membranes of the different composition; it is obvious that zeolite particles need surface treatment in order to get good contact between with the polymer chains. Two different additives, n-tetradecane trimethyl ammonium bromide, and dimethyl ammonium pyridine were tested. From the observation of the membrane, it could be seen that zeolite particles without the additive cannot be evenly dispersed in the polymer matrix membrane. Therefore, additive should be included in order to obtain membrane that would fulfill requirements. In the case of DMAP, good results could not be obtained. On the other hand, n-tetradecane trimethyl ammonium bromide significantly improved the appearance of the membrane without negative influences to the permeability and selectivity. Permeability measurements have shown good results both in dry and wet measurements. The permeability of carbon dioxide was around ten times higher in comparison with permeability of the hydrogen in dry measurements (this ratio is usually called selectivity of carbon-dioxide versus hydrogen). In the wet measurements, lower selectivity was achieved, with average values between 7 and 8. However, as most of the separation processes should be done in the wet environment, it might be concluded that this is promising system for the gas treatment and carbon-dioxide separation.

References

1. Car, A., Stropnik, Č., Yave, W., Peinemann, K.-V. Sep. purif. technol., 2008, 33
2. R. Quadrelli and S. Peterson, The energy climate challenge: recent trends in CO2 emission from fuel combustion, Energy Policy 35 (2007), pp. 5938–5952
3. K. Thambimuthu, J. Davison and M. Gupta, CO2 capture and reuse, Proceedings of the IPCC Workshop on Carbon Dioxide and Storage Regina, Canada, November 18–21 (2002).
4. S. Loeb and S. Sourirajan, Sea-water demineralization by means of an osmotic membrane, Adv. Chem. Ser. **38** (1962), p. 117.
5. R.W. Baker, Future directions of membrane gas separation technology, Indian Eng. Chem. Res. **41** (2002), pp. 1393–1411.
6. H. Lin and B.D. Freeman, Gas solubility, diffusivity and permeability in poly(ethylene oxide), J. Membr. Sci. **239** (2004), pp. 105–117.
7. S.A. Stern, Polymers for gas separation. The next decade, J. Membr. Sci. **94** (1994), pp. 1–68.

8. L.M. Robeson, Polymer membranes for gas separation, Curr. Opin. Solid State Mater. Sci. **4** (1999), pp. 549–552.

9. . Car, Anja, Stropnik, Črtomir, Peinemann, Klaus-Viktor. Hybrid membrane materials with different metal-organic frameworks (MOFs) for gas separation. Desalination, 2006, vol. 200, iss. 1/3, str. 424-426.

10. I. Blume, I. Pinnau, Composite membrane, method of preparation and use, US Patent 4,963,165 (1990).

11. V. Barbi, S.S. Funari, R. Gehrke, N. Scharnagl and N. Stribeck, SAXS and gas transport in poly(ether-block-polyamide) copolymer membranes, Macromolecules **36** (2003), pp. 749–758.

12. L. Liu, A. Chakma and X. Feng, CO2/N2 separation by poly(ether block amide) thin film hollow fiber composite membranes, Ind. Eng. Res. **44** (2005), pp. 6874–6882.

13. A. Car, C. Stropnik, W. Yave and K.-V. Peinemann, PEG modified poly(amide-b-ethylene oxide) membranes for CO2 separation, J. Membr. Sci. **307** (2007), pp. 88–95.

14. J.M. Smith, H.C. Van Ness and M.M. Abbot, Introduction to Chemical Engineering Thermodynamics, McGraw Hill, New York (2001).

15. G. Soave, Equilibrium constants from a modified Redlich–Kwong equation of state, Chem. Eng. Sci. **27** (1972), pp. 1197–1203.

16. S. Kelman, H. Lin, E.S. Sanders and B.D. Freeman, CO2/C2H6 separation using solubility selective membranes, J. Membr. Sci. **305** (2007), pp. 57–68.

17. J.H. Kim and Y.M. Lee, Gas permeation properties of poly(amide-6-b-ethyleneoxide)–silica hybrid membranes, J. Membr. Sci. **193** (2001), pp. 209–225.

18. Stropnik, Č., Car, A. Some aspects of polymeric asymmetric porous membranes formation by phase separation method. Desalination., 2006, vol. 199, iss. 1/3, str. 130-132.

19. H. Lin, E. Wagner, B.D. Freeman, L.G. Toy and R.P. Gupta, Plasticization-enhanced hydrogen purification using polymeric membranes, Science **311** (2006), pp. 639–642.

20. L.N. Norman, Plasticizing effect of permeates on the selectivity of polymeric membranes, Engl. Chem. Prod. Res. Dev. **8** (3) (1969), pp. 281–286.

21. T. Visser, G.H. Koops and M. Wessling, On the subtle balance between competitive sorption and plasticization effects in asymmetric hollow fiber gas separation membranes, J. Membr. Sci. **252** (2005), pp. 265–277.

22. Car, Anja, Stropnik, Črtomir. Chemical modification of surface of wet phase separations membranes. Membrane news, Dec. 2004, no. 66, str. 41-43.

23. Klein, E., Mathematical Methods in Theoretical Economics, Academic Press, New York, 1973.

24. Raman, K., Chatterjee, R., "Optimal monopolist pricing under demand uncertainty in dynamic markets", Management Science, 45(1) (1995) 144-62.

25. Levin, R., Rubin, D., Stinson, J., E. Gardner, Quantitative Approaches to Management, McGraw-Hill, New York, 1989.

580616 Separation Applications of Reduced Graphene Oxide Membranes

Reduced graphene oxide (rGO) membranes have attracted increasing attention of the membrane community for aqueous and organic solvent nanofiltration applications. Although polymeric membranes can perform separations in this regime, same polymeric membrane platform cannot be extended to both aqueous and organic solvents. This research demonstrates the potential application of rGO membranes for treatment of water contaminated with emerging contaminants (perfluoroalkyl substance, PFAS) from water and separation of high-value lignin-derived oligomeric compounds from polar organic solvents. PFAS are negatively charged molecules and are expected to experience charge repulsion and steric hindrance from the rGO membranes. High removal of Perfluorooctanoic acid (PFOA) by rGO membranes was achieved, exhibiting 80% removal for 1 ppm feed at pH 7. Extensive retention studies conducted with PFOA suggested a charge-exclusion dominated mechanism for removal of PFOA. High rejection of PFOA at a high recovery of water observed in our study signifies the potential application of rGO membranes in the treatment of trace contaminants. Another application of rGO membranes investigated in this study was in organic solvent nanofiltration. The primary aim of this study was to examine the selective separation of lignin-derived oligomeric units from polar organic solvent media. This study addresses challenges in selective separation of the lignin depolymerization products, which is vital to fractionate and isolate high-value phenolic compounds. Impressive performance, with the rejection of over 70% for the model trimer, was achieved compared to only 20% rejection for dimer in the isopropanol-water mixture as a solvent. This corresponds to an encouraging selective separation with selective permeation of dimer 3.5 times compared to BMP trimer. Furthermore, selective separation of GO can be effectively modulated by controlling the extent of reduction of GO. Retention of trimeric compounds increased upon increasing the extent of reduction of GO. Remarkable performance of GO membranes could enable energy efficient fractionation of lignin oligomeric compounds from polar organic solvents. This research has been supported by NSF KY EPSCoR and by NIEHS-SRP.

555775 Crude Oil Distillation:General Heat Integration Considerations

Crude oil distillation, by itself, is one of the single largest industrial energy consumers. Even small improvements in energy efficiency have large benefits in crude distillation. The major energy recovery step in crude distillation is heat recovery from distillation towers to preheat the feed. Heat recovery mainly uses pumparounds on the tower or heat integration from the overhead condensers. How many heat integration levels and what types to use have significant consequences on unit investment and energy efficiency. Understanding the general issues involved is critical to maximizing unit performance.

Topics covered here include:

- General heat integration levels
- Constraints from desalter operating temperatures and desalter control
- Service order criteria
- Pump versus exchanger investment and product yields
- Split flow loops

557078 Determination of the Effective Impurity Distribution Coefficient for Stripping Crystallization of R-2-Amino-1-Phenylethanol from the Enantiomers

Stripping crystallization (SC) is a new separation technique which combines distillation and crystallization. Basically, SC is operated at a triple-point condition, in which the liquid is simultaneously vaporized and crystallized due to the three-phase equilibrium. In essence, the process is continued until the liquid phase is completely eliminated and only the pure solid crystals remain in the feed. Thus, crystal washing is not required since only pure solid crystals remain in the feed and no impurities are adhered on the crystal surfaces at the end of SC. In the previous research, SC has been successfully applied to separate p-xylene from the mixed xylenes by Shiau et al. (2008, 2013, 2018).

Stripping crystallization (SC) is introduced in this work for chiral purification of R-2-Amino-1-phenylethanol (R-APE) from the enantiomer mixture. As SC is operated in a series of the three-phase transformation conditions, SC combines melt crystallization and vaporization to produce crystalline product and mixture vapor from a mixture melt. The operating pressure for the three-phase transformation during the SC cooling process is determined based on the thermodynamic calculations. Sweating is applied at the end of SC to expel the mixture melt remained with the final R-APE crystalline product obtained from SC. A model based on the mass and energy balances is proposed to determine the enantiomeric purity and weight of the final product without the remaining mixture melt during SC. This model allows one to calculate the effective impurity distribution coefficient between the crystalline product and the mixture melt during SC.

References:

1. 1. Shiau, L. D. "Purification of styrene from a styrene/ethylbenzene mixture by stripping crystallization", *Eng. Chem. Res*, **57**, p6759-6765, 2018.

2. Shiau, L. D.; Liu, K. F. "Investigations into the effects of the cooling rate on stripping crystallization", Ind. Eng. Chem. Res, 52, p1716-1722, 2013.

3. Shiau, L. D., Wen, C. C. and Lin, B. S. "Separation of p-xylene from the multicomponent xylene system by stripping crystallization", AIChE Journal, 54(1), p337-342, 2008.

559194 Nineteen-Bore Hollow Fiber Membranes Fabrication and Applications in Oxygen Separation and Membrane Reactor

Hollow fiber membrane configuration owing to its unique, one-step formed asymmetric structure, thin separation layer and hence advanced oxygen permeation performance are intensively studied in recent years. There are large amount of reports can be found in regard to single bore hollow fibers. Although hollow fiber membranes have many advantages, their applications are limited by their brittleness when a force impingement or thermal shock occurred. Inspired by the nineteen channel monolithic membranes (in a diameter of more than 40 mm) which are wildly used in water filtration as their high flux and mechanical strength, we challenged BSCF nineteen-bore hollow fiber membranes in a small diameter of 4 mm. Existence of different paths for oxygen transport in the nineteen-bore hollow fiber were successfully demonstrated. The design of multi-bore hollow fiber will be easily exploited to build oxygen separator or membrane reactor, showing a prospect of commercialization.

561359 Ultrafiltration of Colloidal Silica

Treatment of colloidal wastewater is inevitable in many industries (paper mill, food, beverage, etc.) as it is used to recover valuable products through concentration, and/or to treat effluent to an acceptable discharge level in compliance with applicable regulations. In this work we studied the ultrafiltration of colloidal silica suspension and investigated the effects of membrane structure and operating parameters on permeation performance. The parameters studied were transmembrane pressure, pH, salinity, suspension concentration and turbulence. Flux and rejection were the parameters used to assess membrane performance. As expected, permeate flux decreased significantly with increase in feed concentration indicating that the reduced flux was likely due to concentration polarization. However, non-uniform variation of colloid rejection with feed concentration was observed and possible reasons for such outcome will be discussed. The results also showed that for relatively dilute colloidal suspensions (0.5 – 2.0 weight %), increasing the turbulence beyond 600 rpm has negligible effects on membrane performance. The structure of the membrane also provided valuable insights on the transport through the membrane.

564824 High-Performance Mixed-Matrix Membranes for Propylene/Propane Separation By *in-Situ* MOF Formation Using a Polymer-Modification Strategy

Despite the potentials of C_3H_6/C_3H_8 separation, there have been no industrial applications of ZIF-8 mixed-matrix membranes (MMMs) due to the moderate separation performances and several challenging processing issues. Herein, we present a new paradigm of MMM fabrication, named polymer-modification-enabled *in-situ* metal-organic framework formation (PMMOF), enabling *in-situ* formation of ZIF-8 nanocrystals inside 6FDA-DAM polymer. PMMOF consists of four steps including hydrolysis of polymer, ion-exchange, ligand treatment, and imidization. Each step was thoroughly analyzed and important processing parameters were identified, enabling the structural control of MMMs by PMMOF. The binary C_3H_6/C_3H_8 separation performance of the MMMs showed much higher separation factors than conventionally-prepared MMMs at similar filler loadings, satisfying the commercial C_3H_6/C_3H_8 separation performance criteria. PMMOF was successfully applied for other MOFs, demonstrating the process could be general. Finally, as a proof-of-concept, asymmetric mixed-matrix hollow fiber membranes with ultrathin selective skin layers were prepared by PMMOF, showing promising C_3H_6/C_3H_8 separation performances.

564868 Separation of NdFeB Scrap Powder from Hard Disk Drive By HD Process

Rare earth magnets based on NdFeB have been used in many clean energy and advanced applications, including Hard Disk Drive (HDD), electric vehicles, and electric generators in wind turbines. However, the rare earth elements such as neodymium and dysprosium are at highest supply risk of all metals for clean technologies. Therefore, the recycling of NdFeB magnets based on these materials can provide additional and securing supply. Hard Disk Drive (HDD) have been identified as one of the most abundant and readily available forms of Waste Electronics and Electrical Equipment (WEEE) scrap. However the separation of magnet is difficult because the magnet is glued into its functional position, fully magnetized, and coated with e.g. Ni for corrosion protection. It is well known that NdFeB material is hydrogenated by hydrogen gas during the Hydrogen Decrepitation (HD) process and subsequently crushed and pulverized due to a stress generated by volume expansion of the hydrogenated phases. Lately, a few studies have reported the possibility of HD process being an alternative method to effectively separate NdFeB magnets from waste HDD as anti-oxidant powders. However, the mechanistic and systematic study is still required to evaluate the several process factors and establish the optimal condition for selective and efficient extraction of NdFeB magnets from waste HDD. In this study, the effects of temperature, hydrogen pressure and magnet size on the pulverization of waste NdFeB magnets in HD process were investigated, and the separation process of magnets from HDD was established. More details will be discussed in the presentation.

565161 Molecular Transport of Aqueous Mixtures in Sulfonated Polyether-Based Ion Exchange Membrane

Understanding molecular transport in hydrated ion exchange membranes is crucial for promising energy technologies such as solar fuels devices. One of the major challenges with these applications is controlling molecular transport to prevent undesirable species crossover which impacts device performance. Polyethers are promising membrane materials due to their wide range of chemical functionalities and relatively facile fabrication technique, which involves UV-photopolymerization on diacrylate end groups. Sulfonates have been utilized as ion exchange functional groups in cation exchange membranes (i.e. Nafion 117). Membranes prepared from poly(ethylene glycol) diacrylate (PEGDA) and 2-acrylamido-2-methylpropanesulfonic acid (AMPS) have been used as a model system to investigate ion and molecular transport where permeability and selectivity are impacted by varying the PEGDA to AMPS ratio prior to photopolymerization allowing for variation in sulfonate and crosslink density. In previous work, our group has conducted direct permeation measurements using custom-built diffusion cells with in-situ ATR-FTIR spectroscopy to monitor concurrent multi-solute transport through ion exchange membranes. Here, we leverage this methodology to examine solution-diffusion transport of single and multiple solute through PEGDA/AMPS membranes. Finally, absorption-desorption experiments are conducted to determine solute solubilities to fully describe the transport behavior within these membranes.

565390 Surface Functionalization of Magnetite for Selective Recovery of Dysprosium (Dy^{+3}) from Permanent Magnet Waste

Dysprosium (Dy) belongs to the REEs in the lanthanide series, and its unique physical and chemical properties make it among the most indispensable heavy rare earth elements. Given its low concentration in ores, its recovery from secondary waste such as waste electrical and electronic equipment is highly needed. Herein, a magnetic nanoparticle (Fe_3O_4 NP)-based adsorbent for selective recovery of Dy is developed for its retrieval from permanent magnet waste. The Fe_3O_4 NPs was synthesized by co-precipitation and functionalized with diglycolic acid (DGA-NH_2-SiO_2@Fe_3O_4), a ligand with good affinity towards Dy^{3+}. Before diglycolic acid attachment, the surface of bare Fe_3O_4 NPs was coated with tetraethyl orthosilicate to ensure its stability in harsh condition. N-[(3-trimethoxysily) propyl] ethylene diamine was subsequently functionalized as a linker for DGA. The XRD analysis of the bare Fe_3O_4 NPs confirmed its structure, which remained unaltered after coating and functionalization of the linker and DGA ligand. The adsorbent was characterized thoroughly by SEM, BET, TGA, and VSM. The successful synthesis of DGA-NH_2-SiO_2@Fe_3O_4 was confirmed through FTIR and elemental analysis. The adsorption behavior of Dy^{+3} on DGA-NH_2-SiO_2@Fe_3O_4 was studied in batch equilibrium experiment, which was found optimum at pH 5. The isotherm equilibrium study shows that the material has a maximum adsorption capacity of 75 mg g^{-1} and fitted well with the Langmuir adsorption model. The model indicates that the adsorption is monolayer and that there is no interaction among the adsorbates. Kinetics study of the Dy^{+3} adsorption on DGA-NH_2-SiO_2@Fe_3O_4 is fitted well by Elovich model, which suggests the chemical interaction between the adsorbent and adsorbate with slow adsorption rate. Furthermore, the selectivity study shows that DGA-NH_2-SiO_2@Fe_3O_4 has good affinity towards Dy^{+3} ion over competing neodymium ion in permanent magnet waste. Overall results demonstrate the potential of DGA-NH_2-SiO_2@Fe_3O_4 for selective Dy^{3+} recovery from secondary waste.

This work was supported by Basic Science Research Program through the National Research Foundation of Korea (NRF) funded by the Ministry of Education (No. 2017R1D1A1B03028102 and No. 22A20130012051 (BK21Plus)) and by the Ministry of Science and ICT (No. 2017R1A2B2002109).

565630 Highly CO_2-Selective Membrane for Enhancing H_2 Utilization of Solid Oxide Fuel Cell

The recovery of H_2 from the anode exhaust of a high-temperature solid oxide fuel cell (SOFC) system is an emerging method to increase the H_2 utilization, and a polymeric membrane with a high CO_2/H_2 selectivity is the key for this separation task. However, one of the challenges is to maintain the high selectivity at a high temperature of 120 °C. The membrane reported here contained a polyelectrolyte with quarternaryammonium ions as the cations bound to the polymer backbone. Fluoride was chosen as the counterion as it can catalyze the reaction between CO_2 and water for enhancing CO_2 permeance. The composition also included a boric acid to further catalyze the reaction as well as a hydroxide-based mobile carrier for the facilitated transport of CO_2. The membrane showed a CO_2 permeance of 100 GPU and a CO_2/H_2 selectivity of 116 at 120 °C. Moreover, CO_2 partial pressures and membrane hydration were systematically studied to generate membrane performance data for the design of a membrane system.

Next, a high-level techno-economic analysis was used to explore the integration of the membrane process into the SOFC system. Owing to the high selectivity of the membrane, a single stage membrane system was able to achieve a 99% H_2 recovery with a CO_2 removal of 46.9%. Based on the CO_2 permeance of 100 GPU and the CO_2/H_2 selectivity of 116, the CO_2 removal cost was calculated to be $62.8/tonne for a set of optimized parameters. In order to improve the process further, a vacuum has been proposed to be pulled on the permeate side so that the removed CO_2 can be captured, which will decrease the carbon emission of the SOFC process.

Dr. Salim's current affiliation: Membrane Technology and Research, Inc. (MTR), Newark, CA

565648 CO_2-Selective Facilitated Transport Membrane for Carbon Capture: Module Fabrication and Scale-up

Carbon capture from flue gas using membrane technology has been studied extensively to combat global warming. It is often a challenge to scale up the membrane fabrication from lab scale to pilot scale and, subsequently, make the membranes into modules. Using a roll-to-roll fabrication machine integrated with a knife-coating method, a 165-nm thick facilitated transport membrane was successfully made. The scale-up membrane had a width of 14 inches and a total length of > 1200 feet, and it was used to fabricate spiral-wound (SW) elements and modules.

In order to increase the membrane packing density in an element, an improved multi-leaf rolling procedure was developed. Although there is extensive literature on the fabrication of a multi-leaf SW membrane module, most of the studies had all the membrane leaves directly attached to the central permeate tube. Hence, the number of leaves would be limited by the surface area of the central tube. For improving the fabrication, only a carrier layer is attached to the central tube, and the membrane leaves were stacked on the carrier layer. Moreover, with a thinner spacer at 6 mil in thickness, the membrane area was increased to 2.94 m^2 in the most recent modules. The transport results of the modules agreed well with those of the flat sheet membrane under various testing conditions. For instance, a CO_2 permeance of 1450 GPU and a CO_2/N_2 selectivity of 185 were demonstrated consistently at 67°C. Also, satisfactory pressure drops were achieved.

A scale-up factor of >70 has been attained since the start of this study on SW element fabrication process. By considering the limitations set by concentration polarization and pressure drop, a predictive model was used to relate the operating conditions to the separation performances of the SW module. The experience gained is helpful to design a full-sized element (8 inches in diameter and 40 inches in length) for industrial application.

Dr. Salim's current affiliation: Membrane Technology and Research, Inc. (MTR), Newark, CA

Dr. Wu's current affliation: Liquid Microcontamination Control (LMC) Membrane R&D Group, Entegris, Bedford, MA

565673 Silicene As a Novel Nanofiller for Thin Film Nanocomposite Forward Osmosis Membrane

Forward osmosis (FO) is an emerging membrane desalination process which gained popularity due to its energy-efficiency and lower propensity to membrane fouling. In FO, osmotic pressure gradient drives water molecules from the saline feed towards highly ionic draw solution. However, selection of appropriate draw solutes and FO membranes remains a challenge towards wide implementation of FO technology. Particularly, in thin film composite TFC-FO membranes, internal concentration polarization is a common problem. A phenomenon that reduces the effective osmotic pressure difference in the support layer and thus, the water flux. This study employs silicene nanoflowers as membrane support fillers to reduce the effect of internal concentration polarization by tuning the TFC-FO membrane structural (S) parameter.

Silicene was synthesized via magnesiothermic reduction of silica nanoparticles then characterized via XRD, TEM, N_2 adsorption-desorption test, and XPS. The nanoflowers were pulverized, sieved and then incorporated into polyethersulfone (PES) dope solution. Silicene/PES dope solutions with different filler loading were casted via phase inversion technique and used as support for TFC-FO membrane. On the silicene/PES, thin selective polyamide layer was deposited through interfacial polymerization of 2 wt% aqueous m-phenylenediamine and 0.15 w/v% trimesoyl chloride in n-hexane. The resulting silicene/PES TFC FO membranes were characterized using SEM, TEM, FTIR, porosity test and contact angle measurements.

Relative to the control PES TFC-FO, permeation tests reveal overall improvement in water flux for silicene/PES TFC FO membranes. This was attributed to decrease in S parameter, which was correlated to the increased porosity and reduced thickness of the support layer. This resulted in reduced water and draw solute tortuous paths which, in effect, enhanced the effective osmotic pressure difference across the FO membrane. This study demonstrates the potential of silicene, a silicon analogue of graphene as nanofiller, for TFC-FO membranes with improved water flux.

This research was supported by the National Research Foundation of Korea (NRF) under the Ministry of Science and ICT (No. 2016R1A2B1009221 and 2017R1A2B2002109) and the Ministry of Education (No. 22A20130012051(BK21Plus)).

565778 Co(II) Recovery from Electronic Waste Based on Ion-Imprinting of Reduced Schiff Base Ligand on Magnetite

The huge amount of generated electronic wastes (e-waste) worldwide is an attractive secondary source for the recovery of valuable metals. Cobalt in spent lithium-ion batteries (LIBs) is a critical metal that can be recycled into LIBs and other materials such as magnets, alloys, and printed circuit boards (PCBs). Thus, selective Co(II) extraction and recovery from secondary sources is of great importance. Schiff base ligands are known for their affinity towards Co(II) and have been used as extractants in liquid-liquid extraction (LLE). Schiff base-anchored support materials have also been reported, but none so far with magnetic particles as support. Schiff bases are typically synthesized via condensation of diamines or triamines with substituted benzaldehydes; their use as Co(II) ligands is relatively cheap and convenient. However, Schiff bases are susceptible to acid-catalyzed hydrolysis, especially during desorption of Co(II) prior to adsorbent recycling. The imine group of Schiff bases must be reduced to amine to render it acid-resistant while it retains its complexing ability with Co(II). In this study, Schiff bases as Co(II) ligand were prepared from the reaction of 4-allyloxy-2-hydroxybenzaldehyde (AHB) with o-phenylenediamine (PDA) or tris(2-aminoethyl)amine (TAEA) to form AHB-PDA and AHB-TAEA, respectively, which were further reduced by $NaBH_4$ (r-AHB-PDA and r-AHB-TAEA). All synthetic compounds were confirmed via infrared spectroscopy (FTIR), proton and carbon nuclear magnetic resonance (^{1}H-NMR and ^{13}C-NMR).

Preliminary results from LLE of Co(II) reveal q = 105.3 mg Co(II) g^{-1} for r-AHB-PDA and 12.2 mg Co(II) g^{-1} for r-AHB-TAEA. The ligand r-AHB-PDA was chosen for the synthesis of ion-imprinted polymers (IIPs) on a magnetic support material, a practical approach that would minimize reagent use and simplify the Co(II) recovery process. Ion-imprinting technique imparts selectivity towards Co(II) as the r-AHB-PDA binding sites are pre-arranged or templated specifically for the target Co(II). Meanwhile, the magnetic core renders the adsorbent useful for cycled adsorption and desorption of Co(II). For the synthesis of Co(II) selective magnetic-IIP using Schiff base as ligand, magnetite (Fe_3O_4) was initially prepared by co-precipitation of $FeCl_3/FeCl_2$ under basic conditions. Its surface was then coated with silica by tetraethylorthosilicate (Fe_3O_4@SiO_2), then functionalized with alkene by 3-methacryloyloxypropyltrimethoxysilane (Fe_3O_4@SiO_2-MATES) as co-monomers of r-AHB-PDA ligand. Radical polymerization between the Fe_3O_4@SiO_2-MATES and r-AHB-PDA was then carried out using azobisisobutyronitrile (AIBN) as radical initiator and ethylene glycol dimethacrylate (EGDMA) as cross-linker. The adsorbent was thoroughly characterized via FTIR, SEM, XRD, TGA and elemental analysis. Its adsorption properties (i.e. isotherms, kinetics and selectivity) and recyclability are currently being investigated. This work was supported by National Research Foundation of Korea (NRF) funded by the Ministry of Science and ICT (2017R1D1A1B03028102 and 2017R1A2B2002109) and Ministry of Education (2009-0093816 and 22A20130012051 (BK21Plus)).

566590 Application of Micropatterned and Chemically Modified PVDF Membranes for the Treatment of Produced Water Via Membrane Distillation

Produced water, specifically fracking wastewater, is considered as one of the principle waste streams from the oil and gas industries. Along with a high concentration of salts, fracking wastewater contains a wide range of harmful contaminants such as heavy metals, emulsified oil, and other organics, which pose serious environmental challenges. According to the Environment America Research & Policy Center, in United States, fracking wells located in seven states generated around 14 billion gallons of wastewater in 2014 and the amount is likely to increase in the future. Some of the techniques used to treat this water include filtration, cyclonic separation, flotation, evaporation, and deep well injection. However, most of these techniques require a large amount of energy for operation. Recently, membrane distillation (MD) has emerged as a promising candidate for treating fracking wastewater due to its energy efficiency. In contrast to other membrane processes where pressure is used as the driving force, MD is a thermal-driven separation process where water from the feed side (hot temperature side) transports to the permeate side (cold temperature side) through a hydrophobic membrane as vapor molecules. As fracking wastewater usually has a higher temperature than ambient temperature, the feed is not required to be heated which makes MD an energy efficient technique. Though MD is an effective way to treat fracking wastewater, flux reduction due to oil fouling of the hydrophobic membrane is a problematic issue. Because of oil fouling, the maintenance costs are high which makes the technique costly. As such, an efficient and cost-effective strategy is required to mitigate the fouling and will have significant economic and environmental impacts.

Previously, most of the strategies employed for mitigating oil fouling have been based on the surface chemistry modification of membranes. Recent studies showed that surface chemistry modification alone is not sufficient to tackle the problem. Membranes modified with a specific, ordered surface pattern exhibited reduction of fouling. The finding that patterning of membrane surfaces can enhance their resistivity towards fouling makes patterning a crucial parameter for membrane design. Based on the prior studies, we hypothesize that the combination of surface chemistry modification and surface patterning will make a membrane surface more effective in combatting oil fouling than either method alone.

In this work, a couple of micropatterns were imparted onto the surface of commercial polyvinylidene fluoride (PVDF) membranes by thermal embossing followed by the UV photopolymerization of a sulfobetaine zwitterionic polymer to the PVDF membrane surface as the antifouling coating. Profilometry and SEM were used for characterizing the pattern replication, whereas the surface chemistry modification was characterized by ATR-FTIR and contact angle. The modified membranes were tested in a custom membrane distillation system to compare their water flux performance and fouling resistance to the pristine membrane.

567363 Selective Separation of Rare Earth Elements from NdFeB Powders By Caustic Digestion

As the demand of NdFeB magnets increases, an abundance of waste magnets has been gradually released from the end-of-life devices and systems. Therefore, the recycling of permanent magnet has attracted much attention and several methods based on pyro/hydro-metallurgy have been tried and developed for recovering valuable metals from magnet. It is well known that NdFeB powders from permanent magnets are liable to be oxidized, so, the direct application of waste magnets is limited to use the production process of new and high-performance magnet. Therefore, recovery of pure REEs materials from waste magnet via hydrometallurgical route have been much studied and developed. However, the problems associated with the hydrometallurgical process (use of large amounts of strong acid and discharging of large amounts of acidic waste solution containing ferrous/ferric ions) has yet to be unsolved. In previous study, the selective separation of rare earth elements from non-oxidized magnets powders was reported to be possible through the caustic digestion and oxidation roasting of NdFeB metal powder. In this study, the caustic digestion and oxidation roasting of NdFeB metal powder was investigated for the recycling of waste NdFeB magnet powders obtained by Hydrogen Decrepitation (HD) process. The caustic digestion and oxidation roasting of NdFeB metal powder produced Nd_2O_3 and Fe_2O_3 except $NdFeO_3$. The selective separation of rare earth element from iron had been achieved by the dilute acid leaching of Nd_2O_3 and Fe_2O_3, The leaching yield of rare earth element was more than 95%, and that of iron was negligible.

567415 Experimental and Theoretical Characterization of Thiacrown Ether Epoxy-Polymer Adsorbents for the Selective Recovery of Pd and Pt from Spent Auto-Catalyst Leachate

Palladium (Pd) and Platinum (Pt) along with other platinum-group metals (PGMs) play an important role for technological advancement due to their high catalytic reduction and oxidation property, superior resistivity, high temperature and electrical stability, and resistance to tarnish. Hence, PGMs have been widely used in automotive emission control catalyst, industrial chemical catalyst, electronic devices, biomedical devices and jewelry. However, PGMs especially Pd and Pt are extremely expensive due to their low concentration on the earth's crust and the complex processes required for their extraction and refining. Currently, the separation and purification of Pd and Pt are carried out using liquid-liquid extraction (LLE) with commercial extractants (e.g. sulfur containing, oxime, thioglycoamides, ammonium and phosphonium functional groups) that require multiple stages of extraction process to achieve the desired purity. In terms of economic and technical view, adsorption is the most suitable method for the recovery of PGMs because of its high efficiency and low cost. Thus, many studies have attempted to incorporate thiourea or thioglycoamides in polymeric adsorbents but did not show similar LLE selectivities due to the non-selective multi-dendate coordination involved. Therefore, the development of a highly selective adsorbent will the improve sustainable supply PGMs especially if recovered from secondary sources such as spent automotive catalysts, industrial catalyst and electronic waste.

Herein, novel selective thiacrown ether ligands were engineered based on fine tuning the coordination specificity by changing the number of sulfur heteroatoms, structure of crown ether (CE) and cavity size ($Ø_{CE}$). The performance of each thiacrown ether was evaluated with the aid of experimental extraction and theoretical density functional theory (DFT) studies. Dihydroxy-dithia-, -trithia-, and -tetrathia-14-crown-4 ether derivatives (2g-2m) were effectively synthesized by bis-epoxide intermediate ring-opening cyclization reaction with 1,2-benzenedithiol using a mild base triethylamine and LiCl as template in water-tetrahydrofuran solvent. The resulting bis-epoxide cyclization reaction gave CEs with free di-hydroxyl groups necessary for further chemical modifications and subsequent polymerization.

Initial experimental liquid-liquid extraction in chloroform-water system and density functional theory (DFT) calculations using the *ab initio* Jaguar program B3LYP with LACVP** basis set reveal that the CE cavity size and denticity of the synthesized thiacrown ethers affected their selectivity (β) towards Pd and Pt. Among the thiacrown ether derivatives, bidentate dihydroxy-dibenzo-dioxadithia-14-crown ether (2i) with a DFT calculated cavity size of 1.58-1.65 Å was most selective towards Pd ($\beta_{Pd/Pt}=10^5$). Meanwhile, tetradentate dihydroxy-benzo-dimethyl-tetrathia-14-crown-4 ether (2m) with a cavity size of 1.31-1.60 Å was selective towards Pt ($\beta_{Pt/Pd}=10^3$). This is due to the CE cavity size-match relationship where the ionic size of Pd (1.56 Å) or Pt (1.48 Å) is suitable for the CE cavity. DFT also showed that the Pd and Pt are situated perfectly inside the CE cavity with an S_4 CE-M^{n+} complex coordination geometry and have the lowest calculated binding energies (ΔE_{CE}-M^{n+}). The complex 2i-Pd^{2+} showed the most stable conformer with ΔE_{CE}-M^{n+}=-69.10 kcal/mol whereas complex 2m-Pt^{2+} geometry with ΔE_{CE}-M^{n+}=-75.27 kcal/mol.

The synthesized dihydroxy-CEs 2i and 2m where further modified with epoxide groups by reacting with epichlorohydrin under strong basic conditions. The resulting CE bis-epoxide intermediates (2i-ep or 2m-ep) where cured with ethylenediamine in polyethylene glycol (PEG) at 150°C to afford porous thiacrown ether epoxy-polymerized adsorbents (EP-2i or EP-2m) that are selective towards Pd or Pt. The well-defined macroporous structure of the epoxy-polymer adsorbents were achieved by *in situ* polymerization in porogenic solvent PEG. Initially, the mixture of CE bis-epoxides and ethylenediamine in PEG is homogenous but with increasing cross-linking density, phase separation of the PEG occurs which leads to nucleation forming the porous structures. To evaluate the performance of each thiacrown ether epoxy-polymer adsorbent, the effects of HCl concentration was investigated to determine the optimum conditions for Pd and Pt separation as well as the adsorbent stability in very acidic solutions. Furthermore, the effects of the initial concentration of Pd or Pt, kinetics, and recyclability were also investigated to established optimum operating conditions. Lastly, the recovery of Pd and Pt was done using simulated auto-catalyst leachate containing high concentration of other competing metals (Al^{3+}, Ce^{3+}, Mg^{2+}, Pb^{2+}, Fe^{3+}, Ni^{2+}, Mn^{2+}, and Cr^{3+}) and showed excellent recovery of Pd and Pt.

This work was supported by Basic Science Research Program through the National Research Foundation of Korea (NRF) funded by the Ministry of Education (No. 2017R1D1A1B03028102) and by the Ministry of Science, ICT and Future Planning (No. 2017R1A2B2002109).

568339 Preparation and Application of Ceramic-Supported Metal-Organic Framework Membranes

Membrane-based separation outperforms other techniques for separation of matters at molecular level because of its low cost, energy saving potential, high capacity, continuous operation and easy scale-up, but raise high requirements for the membrane materials. Recently, metal-organic frameworks (MOFs) have attracted great attention in many fields because of their large surface areas, chemically functionalized cavities, flexible skeletons, and intriguing electronic properties. These unique features make them excellent candidates for applications in membrane separation fields. Herein, based on ceramic supports (ceramic disc and hollow fiber), a series of MOF membranes, including Zn-CD, ZIF-8, ZIF-71 and UiO-66 membranes, were successfully prepared and showed good separation performance in gas and liquid separation field.

571492 Research Concept for the Development of Novel Additively Manufactured Structured Packings

Additive manufacturing as a fundamental new way of manufacturing offers the opportunity to realise novel shapes of separation equipment and apparatuses. Optimizing the specific shape of equipment can, in turn, result in the intensification of the separation process, leading to environmental and economic advantages. Therefore, additive manufacturing offers great potential for the chemical industry, especially in applications where the shape of the equipment determines the process performance in a fundamental way. Such an application are structured packings which are widely applied in distillation columns.

Structured packings are used to intensify the mass transfer between the liquid and vapour phase within the distillation column. Besides maximized separation efficiency, minimal pressure drop and maximal capacity are essential packing parameters in leading to a successful distillation process. Improvements of these characteristics directly affect the performance of the distillation column. The mentioned parameters are directly correlated to the flow conditions of the gas and liquid phase, which are in turn strongly affected by the shape of the structured packing. For this reason, the shape of structured packings has been of research interest for decades, but previous developments were restricted by the limitations of conventional manufacturing methods. Additive manufacturing as an innovative manufacturing process now makes it possible to realise novel structures which are not feasible to manufacture with conventional methods due to their complexity. Accordingly, the main research goals of this work are the systematic identification of novel shapes of structured packings and the realisation of these packings by additive manufacturing.

The research concept incorporates a close interconnection of computer based methods and experiments. Different shapes of structured packings are designed and their key parameters, like pressure drop and liquid distribution, are determined via computational fluid dynamics. The packings are additively manufactured and examined in hydrodynamic and distillation experiments. The experimental results are in turn used for the validation of the simulations. In a first step this is done with conventional packing shapes to set up the models and experiments. The second step will be the design and examination of novel packing shapes.

The research concept for the systematic development of novel additively manufactured structured packings is presented in detail on the poster. Also, first results are shown for conventional structures, which are the base for further research.

572159 Polymer Composite Membranes Containing UiO-66 MOFs with Co-Continuously Segregated Structures for CO_2-Related Gas Separations

Gas separation membranes have been considered as one of the most promising techniques in replace of the conventional gas separations processes. Recently, UiO-66 metal-organic framework (MOF) containing polymer composite membranes have been demonstrated to achieve better gas separation performance, especially for CO_2-related separations. Until now, few studies used morphological control strategies for enhancing gas transport properties of composite membranes. In this work, we prepared polymer composites with co-continuously segregated structures, comprising UiO-66 porous nanoparticles and immiscible 6FDA-based polyimide blends. The gas permeability and gas sorption properties of pure gases (e.g., H_2, CO_2, N_2, O_2, CH_4) in the dense membranes and composite membranes were measured using the constant-volume variable-pressure method and the dual-volume pressure decay method. Based on the solution-diffusion model, the contributions of gas sorption and gas transport to the overall gas permeation properties were evaluated and studied. We found a considerable increase in CO_2 permeability by 635% (104.7 Barrer) along with the almost unchanged selectivities of CO_2/N_2 (16.5) and CO_2/CH_4 (32.8) for the composite membrane at 35 wt.% of the UiO-66 content. The overall gas separation results suggested that both the addition of UiO-66 porous nanoparticles and the morphology of co-continuously segregated structure would contribute to the dramatical enhancement in gas separation performance of the composite membranes.

574948 Design of Process Control Schemes and Experimental Investigation for a Transesterification in a Reactive Dividing Wall Column

Process integration is an important method to increase the process efficiency in the chemical industry. A highly integrated process is represented by the reactive dividing wall column (RDWC), which combines a chemical reaction and several separation steps in one column shell. Theoretical studies state that the energy demand of RDWCs can be reduced up to 15 % and investment costs up to 30 % in comparison to conventional separation sequences.[1] On the other hand the high degree of integration leads to very complex process behaviour, hence the chemical industry is reluctant to build and operate RDWCs. In order to assure constant product purities and a stable continuous operation, reliable process control strategies are required. To date various papers assess the design and steady state operation of the RDWC in simulation studies. However, there are only few papers about dynamic operation and process control and even less papers that show experimental results for RDWC process control.

Hence, our research aims to provide theoretical and experimental insights into the dynamic process behaviour of RDWCs. Therefore, we develop a systematic approach for the design of suitable RDWC process control schemes. Our research combines simulation studies with our validated dynamic RDWC model, developed at the Institute of Process and Plant Engineering, University of Technology Hamburg, with experimental tests of the process control schemes in a 12 m pilot plant RDWC. The employed reaction system, the transesterification of hexyl acetate with butanol, is well suited as a reference system for esterification, transesterification and hydrolases reaction systems.

The poster presentation will show the systematic development of process control schemes for the RDWC as well as the experimental investigation of these control schemes. First, the research concept, the reaction system and the pilot plant RDWC are shown. Our systematic approach for the development of control schemes suitable for an efficient and stable RDWC operation is presented. An exemplary control scheme is developed for the reference system and experimental test results are employed to assess its performance.

[1] Schröder, M.; Fieg, G. (2016): Influence of Reaction System Properties on the Energy Saving Potential of the Reactive Dividing-Wall Column: Separation Properties. In *Chem. Eng. Technol.* 39 (12), pp. 2265–2272. DOI: 10.1002/ceat.201600134.

575031 Physicochemical and Thermodynamic Property Determination & Modeling of Aqueous 3-Aminoproyl Triethoxysilane+Mdea/DEA within the Temperature Range of (298.15 to 333.15) K

Carbon dioxide is considered to be one of the major contributing agents for the global greenhouse gas emission. In this regard, absorption-regeneration technology aided with various improved solvents is an effective approach towards mitigation of this issue. The conventional solvents such as monoethanolamine (MEA), diethanolamine (DEA), methyldiethanolamine (MDEA) poses various drawbacks such as high solvent loss and degradation, equipment corrosion and higher regeneration energy requirement, making the process inefficient and economically unfavorable [1]. In order to cut down these limitations, physical solvents such as ionic liquids (IL) can be considered as a potential solvent for gas sweetening technology. Ionic liquid possesses many favorable features such as negligible vapor pressure which reduces the solvent losses to the environment. Other unique properties of IL includes its high thermal stability, higher conductivity, non-flammability as well as ease of separation since it require very less regeneration energy, and it possesses reversible and phase changing behaviors [2]. Despite its various advantages the applicability of ionic liquids as a solvent for acid gas separation is limited because of its high cost and lower CO_2 absorption rate. Hence, application of hybrid solvent such as mixture of (Alkanolamine + Ionic Liquid) or functionalized ionic liquid can be explored as an alternative solvent since it combines the advantages of both the system [3,4]. While functionalized ionic liquids have better capacities in comparison to the pure ionic liquids or mixture of ILs+ amine, but its application is limited because of advance purification steps required in the synthesis process which makes the overall process of CO_2 capture to be economically less favorable. Recently, reversible solvents have been explored as an alternative to these hybrid solvents. Since, they function both as physical and chemical absorbent for CO_2 capture. The advantage of using reversible ionic liquids includes the high capacity of CO_2 absorption over nitrogen (N_2) in dilute flue gas streams. In addition, the total quantity of solvent required can be drastically reduced compared to conventional alkanolamine solvents. Vittoria et al [5] has investigated the efficiency of 3-aminoproyl triethoxysilane (TESA) as a reversible solvent for CO_2 capture processes. Various thermo physical properties such as solvent densities, viscosities, thermal stability as well as total CO_2 capture capacities (14.76 mol of CO_2/kg of amine) at 35 ^{o}C and 62.5 bar has been reported. The present work reports the density, viscosity; sound velocity and refractive index data of two silane–amine blend systems viz. (1) 3-aminoproyl triethoxysilane (TESA) + methyldiethanolamine (MDEA) and (2) 3-aminoproyl triethoxysilane (TESA) + diethanolamine (DEA). The temperature range for which all the measurements were conducted is (298.15 to 333.15) K. For all two ternary aqueous mixture of (TESA+MDEA) and (TESA+DEA), TESA mass fraction was varied from (0.2 to 0.7). The density and viscosity results were further correlated as a function of temperature and concentration of ionic liquid with Redlich Kister and Grunberg - Nissan model, respectively. Two new models have been proposed for correlation of sound velocity and refractive index i.e. Model M_1 and Neural network model. Various other properties such as diffusivity of CO_2 in the solvents under study, enthalpy of activation, entropy of activation, Gibb's free energy of activation for the viscous flow and isentropic compressibility are evaluated using the available models in the literature.

References:

1. B. Sreenivasulu, D.V.Gayatri, I.Sreedhar, K.V.Raghavan Renewable and Sustainable Energy 41(2015)1324–1350

2. P.T. Anastas, J.C. Warner, Green Chemistry: Theory and practice, Oxford. University Press,New York,1998

3. A. Haghtalab , A.Shojaeian, J. Chem. Thermodynamics 81 (2015) 237–244

4. J. Aboudi, M. Vafaeezadeh, J. Adv. Res. 6 (2015) 571–577
5. B. Vittoria, R. Hart, V. L. Mestre, D. J. H., M. Burlager, H. Huttenhower, B. J. R. Thio, P. Pollet, C. L. Liotta, C. A. Eckert, Fuel 89 (2010) 1315-1319

575414 Microscopic Diffusion of Organic Molecules in Perfluorosulfonic Acid Polymer Membranes

Perfluorosulfonic acid (PSA) polymer membranes such as commercially available Nafion have high ion permselectivity and can be used as a solid fuel cell electrolyte and solid super-acid catalyst. PSA membranes are also among the most promising candidates for novel applications including optodes for chemical sensing and personal protective clothing. PSA membranes are composed of a hydrophobic backbone with hydrophilic acid head groups which form respective hydrophobic or hydrophilic domains. While the microscale transport of water and some small organic molecules such as methanol has been studied, the microscopic diffusion of larger organic molecules in PSA polymers is not well understood. In the present work, self-diffusion of organic molecules such as acetone and vanillic acid are studied in Nafion on micrometer and submicrometer length scale to address this gap in the current knowledge.

The diffusion studies were performed by multinuclear pulsed field gradient (PFG) NMR at high magnetic fields (14 T and 17.6 T) and large magnetic field gradients (up to 25 T/m). The combined application of C-13 and H-1 PFG NMR for diffusion studies of the same molecules allowed elucidating a possible influence of proton exchange and proton diffusion on the apparent self-diffusivities measured for organic molecules in Nafion by H-1 PFG NMR. The diffusion studies were performed for a broad range of diffusion times and at different temperatures and concentrations of water and organic molecules in Nafion. The diffusion properties of neat Nafion are compared with those of resorcinol- or vanillic acid-functionalized Nafion. These functionalized membranes have been shown to have potential as chemical sensors and improved acid catalysts. The role of the membrane functionalization in the diffusion process in Nafion will be discussed in detail.

575432 In Situ Measurement of Concentration Polarization in Pressure Driven Membrane Systems Using Laser-Based Spectrophotometric Method

Performance of membrane based separation process can be greatly influenced by scaling and fouling of membranes concentration polarization (i.e., increase in solute concentration near the membrane surface) is a phenomenon that has significant impact on mass transfer and formation of solids on the membrane surface. Previous studies focused on characterization of the concentration polarization layer in pressure driven system using semi-empirical models to estimate the extent of solute concentration increase compared to that in the bulk solution. However, actual concentration of solute near the membrane surface has not yet been experimentally characterized with requisite spatial resolution to allow comparison with semi-empirical models commonly used in practice.

This study presents a details of the spectrophotometric method specifically developed to measure the solute concentration profile near the membrane surface. To achieve this objective, a visible spectrophotometer using collimated laser beam as a light source and a CCD line camera as a detector were combined with specially designed membrane filtration cell to probe the solute concentration near the membrane surface. This experimental system was capable of probing the concentration polarization layer with spatial resolution of 4.5 μm. This system that was used to study concentration polarization layer in membrane distillation will be applied to UF and MF membranes using Direct Red 80 (dye) to study the effect of dye concentration, inlet pressure, feed flow rate and membrane characteristics on concentration profile. Results of these studies will be used to develop theoretical models for prediction of concentration polarization coefficient and boundary layer thickness that will be discussed in the presentation.

576295 Comparison of Methods of Incorporating Silver into Ultrafiltration Membranes to Prevent Biofouling

Biofouling on ultrafiltration membranes used for water purification can lower flux and shorten membrane lifespan. Prior work has shown that biofouling can be reduced by incorporating silver into the membranes, either as nanoparticles or in ionic form. This is due to the anti-microbial properties of silver. This project compares the effectiveness of various methods of incorporating silver into the membranes and how well each method prevents silver leaching. Ultrafiltration membranes are fabricated by phase inversion of a solution of cellulose acetate (CA) and polyvinyl alcohol (PVA) dissolved in n-methylpyrrolidone (NMP). Four methods of silver incorporation are tested. In two methods, 50 nm and 200 nm silver nanoparticles are added to the polymer solution prior to phase inversion. We expected the 50 nm silver nanoparticles to show a stronger initial anti-microbial effect than the 200 nm silver nanoparticles, but to dissolve and lose their anti-microbial effectiveness more rapidly. In the third method, silver nitrate is added to the polymer solution prior to phase inversion. We expected this to result in more rapid release of silver ions compared to when large nanoparticles are added. In the fourth method, a CA/PVA membrane is cast, then is functionalized with iminodiacetic acid (IDA), then is soaked in a 0.05M silver nitrate solution to allow silver ions to bind to the IDA. We expected the chelating action of IDA to slow the release of silver ions. Membranes are characterized with X-ray photoelectron spectroscopy (XPS) to confirm the presence of silver, infrared (IR) spectroscopy to determine functional groups present, and scanning electron microscopy (SEM) to examine the pore morphology of the membranes. To measure membrane performance, flux is calculated from dead-end filtration tests with deionized water and a 2 mg/mL solution of bovine serum albumin (BSA). Preliminary data show that IDA functionalization followed by silver nitrate addition improves pure water flux from 5.2 L/(m^2*h) to 35.0 L/(m^2*h), improves flux from 3.4 L/(m^2*h) to 24 L/(m^2*h) for filtration with BSA solution, and improves BSA rejection from 90.3% to 98.2% compared to an unfunctionalized CA/PVA membrane. The rate of silver leaching is determined by using atomic absorption spectrometry (AAS). Inductively coupled plasma mass spectrometry (ICP-MS) is used to measure the change in concentration of silver in the effluent over time during dead-end filtration. Antimicrobial performance is tested by the membrane's ability to kill E. coli over a period of time. Additionally, this project investigates predictive models of the phase inversion. The effects of PVA concentration and silver presence on pore morphology are modeled with molecular dynamics simulations and phase field simulations of the phase inversion process. The Flory-Huggins solution theory will be used to model the thermodynamics of the phase inversion process. We will investigate whether simulation results qualitatively agree with the experimental observations.

577460 Investigation of the Sulfur Adsorption Capability of Ion-Exchange Y Zeolites Using Density Functional Theory and IR Spectroscopy

Investigation of the Sulfur Adsorption Capability of Ion-Exchange Y Zeolites using Density Functional Theory and IR Spectroscopy

Kevin X. Lee, Stavros Caratzoulas, George Tsilomelekis and Julia A. Valla

Department of Chemical & Biomolecular Engineering, University of Connecticut, 191 Auditorium Road, Unit 3222, Storrs, CT 06269-3222, USA,

Phone: +1-860-486 4602, e-mail: ioulia.valla@.uconn.edu

The detrimental effect of sulfur emissions on human health and on the environment has been a growing concern. To overcome this problem, research has been focused on identifying adsorbents for the removal of sulfur molecules from hydrocarbons fuels. Specifically, there is a growing interest on fundamental studies of sulfur adsorption at the molecular level to obtain a better understanding of how sulfur species interact with metal cations in Y zeolites.[1,2] Experimental studies have shown that the incorporation of metals (e.g. Cu and Ce) plays an important role in enhanced adsorption capacity of refractory sulfur compounds such as dibenzothiophene (DBT) and alkylated-DBTs.[3] Cu^+ has an electronic configuration of $1s^2 2s^2 2p^6 3s^2 3d^{10} 4s^0$, which is highly desired for π-complexation, whereas Ce^{4+} is an f-block element with high positive charge and polarizability, which prefers to interact with sulfur via the direct S-M bond. While Cu^+ exhibits a higher sulfur capacity, aromatic compounds with similar shape and structure with sulfur molecules may also interact with the metal via the same adsorption mode. A better candidate for higher sulfur selectivity may be Ce^{4+}, but the trade-off is the low sulfur capacity. One way to overcome the individual limitations is to exploit the synergistic effects of both metals by ion-exchanging them into the Y zeolite. Spectroscopic studies have shown that CuCeY can adsorb sulfur compounds via both π-complexation and direct S-M adsorption modes, resulting in a higher sulfur capacity and selectivity.[4] The desorption temperature of CuCeY was higher compared to CuY and CeY, suggesting that the bond energies are stronger than the individual counterparts. To further understand the role of bond strength on the adsorption of sulfur, density functional theory (DFT) calculations were performed on sulfur and metal-modified Y zeolites. Here we will present the calculated binding/adsorption energies and the results of the natural orbital (NBO) analysis, which give further information on the adsorption modes of sulfur on the metal exchanged zeolites.

References

1. Hernández-Maldonado, A. J. & Yang, R. T. Desulfurization of Diesel Fuels by Adsorption via π-Complexation with Vapor-Phase Exchanged Cu(I)−Y Zeolites. *J. Am. Chem. Soc.* **126,** 992–993 (2004).
2. Hernández-Maldonado, A. J. & Yang, R. T. New Sorbents for Desulfurization of Diesel Fuels via π-Complexation. *AIChE J.* **50,** 791–801 (2004).
3. Lee, K. X. & Valla, J. A. Investigation of bifunctional zeolites for the adsorptive desulfurization of fuels. *Appl. Catal. B Environ.* **201,** 359–369 (2017).
4. Lee, K. X., Tsilomelekis, G. & Valla, J. A. Removal of benzothiophene and dibenzothiophene from hydrocarbon fuels using CuCe mesoporous Y zeolites in the presence of aromatics. *Appl. Catal. B Environ.* **234,** 130–142 (2018).

574643 Synthesis of Amine-Functionalized Zeolite for Carbon Dioxide Separation

With the growing concern about global warming placing greater demands on improving energy efficiency and reducing CO_2 emissions, the need for improving the energy intensive, separation processes involving CO_2 is well recognized. The US Department of Energy estimates that the separation of CO_2 represents 75% of the cost associated with its separation, storage, transport, and sequestration operations. Hence, energy efficient, CO_2 separation technologies with improved economics are needed for industrial processing and for future options to capture and concentrate CO_2 for reuse or sequestration. The overall goal of this research is to faster the development of zeolite adsorbent with amine functionalized to improve manufacturing efficiency and reduce CO_2 emissions. This research focuses on the power, petrochemical, and other CO_2 emitting industries, and provides a detailed review of the current commercial CO_2 separation technology, i.e. adsorption, an overview of the emerging adsorption technology for CO_2 separation, and both near and long term recommendations for future research on adsorption. Our research work aims at the synthesis of Y-zeolite and commercially available of Y60 zeolite with a modification by piperazine to study their characterizations and adsorption/desorption properties of carbon dioxide from gas streams. Further, the synthesis was carried out by hydrothermal method for large internal and external surface area with uniform particle size. By researching the characterization of Y-zeolite and amine functionalized Y-zeolite through X-ray diffraction pattern and Fourier Transform Infrared Spectrum. The diffraction peaks located at 2θ were clearly observed for both samples indicating evidence of a well ordered structure. The IR spectrum for both Y zeolite and amine functionalized zeolite shows significant bands which was related to O-H, H-O-H, and Si-O-Si stretching vibrations. The IR spectrum exhibit bands to $C\text{-}H_2$, $N\text{-}H_2$, N-H, and $-NH_3^{+}O\text{-}Si/\text{-}NH_2^{+}O\text{-}Si$ vibrations after the piperazine modification confirmed that piperazine has been grafted on Y-zeolite. Our research reflects that the both amine functionalized Y and Y60 zeolite can be potentially shows a promising and cost-effective adsorbents with high CO_2 adsorption capacity (<4.27mmol of CO_2 /g of sorbent) for the application in carbon dioxide separation for natural gas purification.

577845 Direct Air Capture of CO_2 Using Zeolites for the Production of High Purity CO_2

Direct Air Capture (DAC) of CO_2 has been an emerging field with studies originally investigating the use of NaOH solutions to absorb CO_2 as a means to reduce greenhouse gas emissions. However, in this field, there has yet to be a thorough investigation into using zeolites for DAC of CO_2 even though industry currently uses them in a similar process in air pre-purification. This presentation will include an experimental study of CO_2 capture from ambient air using vacuum temperature swing adsorption (VTSA) as a means to capture and effectively reduce the amount of CO_2 in the atmosphere in cold countries like Canada and concentrate it to 95% for sequestration. Another application of this process is for the production of high purity CO_2. Different grades of high purity CO_2 offers higher value as the purity of the CO_2 increases. Examples of different grades of CO_2 include medical, bone dry, food and beverage, anaerobic and laser, supercritical fluid, and research grade. An initial screening study was done on commercially available zeolites and breakthrough experiments have been used to determine the feasibility of the production of high purity CO_2 using the VTSA process.

562050 CO_2 Separation Process Based on Temperature Swing Adsorption Using Fluidized Bed

Carbon dioxide capture and storage technology has attracted recent attentionas as a technology to reduce greenhouse gas emissions and mitigate global warming; however, to facilitate its commercial implementation, its capture cost should be further decreased. This research proposes an innovative process using adsorption in a fluidized bed. In this process, the adsorption heat is supplied through a heat pump to desorption, leading to recirculation of all process heat. To determine energy consumption by simulation, fluidization experiments was conducted. Furthermore, process simulation using the experimental data was conducted with a commercial process simulator. The simulation results indicate that the proposed CO_2 separation process has large energy-saving potential.

562068 Enhanced CO_2 Adsorption Rate in Polyammonium Protic Ionic Liquids Functionalized SBA-15

Enhanced CO_2 adsorption rate in polyammonium protic ionic liquids functionalized SBA-15

Wei Zhang, Yi He*, Yao Shi*

Key Laboratory of Biomass Chemical Engineering of Ministry of Education, Institute of Industrial Ecology and Environment, College of Chemical and Biological Engineering, Zhejiang University, Hangzhou 310027, China

Abstract:

The development of cost-effective and high-efficiency adsorbents for CO_2 capture from flue gas deriving from the large-scale burning of fossil fuels represents one of the most significant scientific and technological challenges since the industrial revolution to face the global warming and anthropogenic climate change. Compared with conventional ionic liquid (IL) like imidazole ILs and amino acid-based ILs, polyammonium protic ionic liquids (PILs) possess the advantages of low-cost, simple-synthesis and high absorption capacity, but the same drawbacks of high viscosity and CO_2 diffusion limitation. To overcome these challenges, we develop mesoporous silica SBA-15 supported triethylenetetrammonium nitrate sorbents with excellent CO_2 uptake rate and great adsorption capacity. The CO_2 capture performance of the hybrid sorbents was evaluated under conditions mimicking a combustion flue gas (15% CO_2) at the temperature range of 298-348 K. The breakthrough experiments revealed that S15-66PIL (the mass ratio of PIL to carrier is 2:1) exhibited the highest CO_2 adsorption capacity of 2.11 mmol/g at 333 K. In order to better understand the adsorption kinetics and the rate-controlling step of the sorbent, the CO_2 adsorption capacity was plotted against $t^{0.5}$ using the intra-particle diffusion model (IPDM). The results demonstrated that S15-66PIL reached the fastest CO_2 adsorption rate and the adsorption rate of rate-controlling step was $131*10^{-3}$ mmol g^{-1} $s^{-0.5}$, approximately five times higher than bare support and two times higher than previously reported ILs-functionalized and amine-modified support systems. The high CO_2 adsorption rate of the adsorbent can be explained by the large surface area of the support which favors the diffusion of PIL and CO_2 molecules in its pores. In addition, the thermostability of S15-66PIL was comparable to other ILs-functional molecular sieve and was better than TEPA-modified porous materials. The long-term operating performance of the sorbent evaluated by 10 times cyclic adsorption/regeneration experiments and the results showed that 92% of original CO_2 capture capacity was reversible. FT-IR analysis coupled with isosteric heat and DFT calculation revealed that the adsorption nature of the samples was chemisorption and CO_2 preferentially interacted with the primary amine -N(3)H2 in [TETA][NO_3] to form carbamate due to the highest binding energy. Overall, [TETA][NO_3] impregnated SBA-15 sorbent provide a low-cost, high adsorption capacity alternative with fast adsorption kinetics, and desirable regenerability. Therefore, it may serve as a promising candidate for CO_2 capture in industrial applications.

Keywords: CO_2 adsorption; polyamine-based protic ionic liquids; SBA-15; Nanopore confinement; DFT calculation

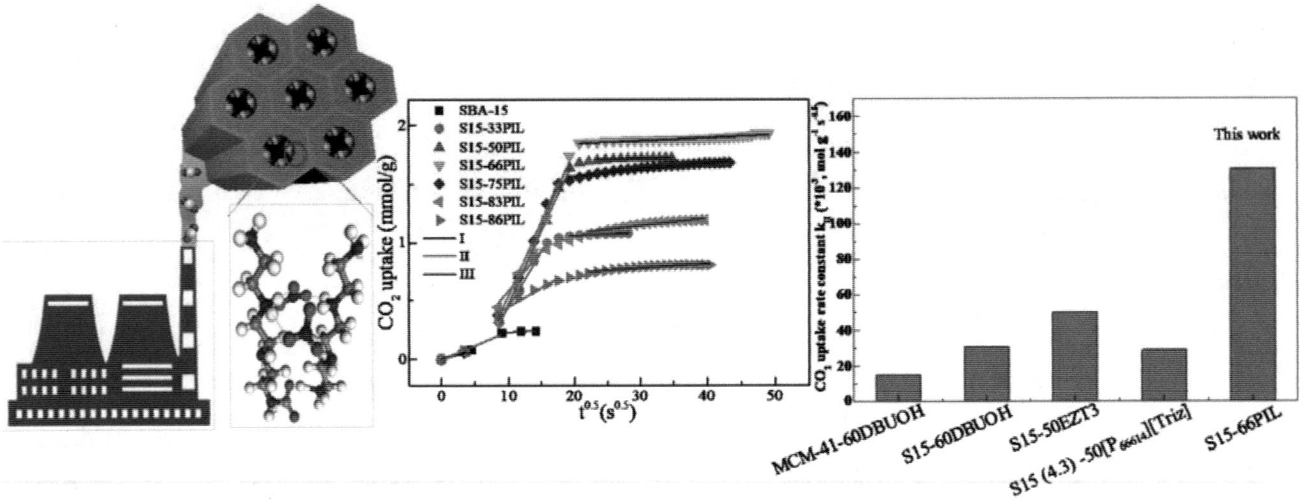

Fig. 1. CO_2 adsorption diagram, IPDM kinetic plots of S15-xPIL and comparison of diffusion rate constant of rate-controlling step for these sorbents

Enhanced Transport Characteristics for the Electrochemically Driven Highly Selective Separation of Gases

Philip Cox, Kayla O'Neill, Brinkley Artman, and Vincent Storhaug
Mainstream Engineering Corporation, Rockledge, FL

Abstract

Membrane-based gas separation represents an attractive route for the isolation of several important gases as well as an opportunity to improve the overall economic and environmental acceptability of a wide range of key industrial processes. In fact, high-performance membranes offer the potential to provide game-changing process energy advances when sufficiently high levels of selectivity and permeability are achieved. In particular, there is a significant push to increase membrane selectivity to drive the process enhancement of a wide range of chemical and combustion processes as well as reduce the generation and atmospheric accumulation of greenhouse gases. The effectiveness of thin, nonporous membranes or membrane coatings has been demonstrated with processes such as facilitated transport to enhance the selectivity for a range of important target gases. However, selectivity and permeability are typically inversely related, and increasing the selectivity typically comes with a penalty of decreased permeability, which is seen in both porous and nonporous membranes. Facilitated transport membranes incorporating targeted redox species have been shown to provide significant benefits in selectivity; however, significant permeability penalties persist even with extremely thin, nanometer-thick membrane layers. To address these selectivity and permeability limitations, Mainstream used targeted biomimetic redox molecules to provide controlled transport of target species across the membrane and membrane electrode assemblies (MEAs). To drive the permeability, we tailored the membrane/MEA with controlled porosity and surface chemistry. In this paper, we demonstrated the simultaneous enhancement of the gas separation selectivity and permeability for the targeted separation of gases from a range of mixed-gas-feed compositions using low-power, low-voltage, efficient electrochemically driven separation. We improved the permeability by a factor of up to 100 with a simultaneous improvement in the selectivity of 15 times using an electrochemical driving force for the separation of CO_2 from a range of mixed gas feed. We demonstrated permeate concentrations of 95 to 99% CO_2 at an 8000 Barrer permeation rate.

Introduction

Gas separation is an important step in a wide range of industrial and commercial applications. Examples include the separation of nitrogen and oxygen, removal of volatile organics from process streams removal of CO_2 from breathing air in recirculating respirators, removal of CO_2 from pre- and post-combustion processes, separation of oxygen from air in portable oxygen generators, and other applications. These separations have traditionally been performed using techniques such as cryogenic distillation, adsorbent beds, or pressure swing adsorption processes. However, over recent years with their lower capital costs, lower energy requirements, and simpler operation, membrane-based separations represent an increasingly attractive alternative for the separation of gases.

The separation of gases is a highly energy-intensive task and the demand for lower energy and cost separation processes such as membrane-based separations is high. The development of low-power membrane separations, which have both high flux (permeability) and high selectivity for the target gas, is critical to the commercial viability of the membrane-based separations. However, with conventional membranes, there is a tradeoff between the membrane permeability, or rate of transport, and the selectivity of the membrane, yielding the so called upper bound[1-2]. Thus, increased permeability is generally achieved at the cost of selectivity.

One method of separation is to incorporate redox-active molecules, such as transition metal porphyrins into the membrane, which have an affinity for the gas to be separated. Although this approach has been shown in the academic literature to be effective at increasing the selectivity, it does so at the cost of very low flux or permeability[3-4].

Mainstream's approach for electro-chemically driven membrane separation can potentially be applied to a wide range of gases by the correct selection of the active redox gas carrier species.

Experimental Methods

Membrane Electrode Assembly Fabrication

Mainstream fabricated MEAs to incorporate gel-based membranes with up to 60% electrolyte and a range of electrochemically active redox species and carriers. We coated the freestanding membranes onto either a glass substrate or a polymer-based carrier film using a highly scalable knife coating process as shown in Figure 1. We then incorporated the free-standing membranes into an MEA by laminating porous gas-diffusion-based electrodes on either side of the membrane.

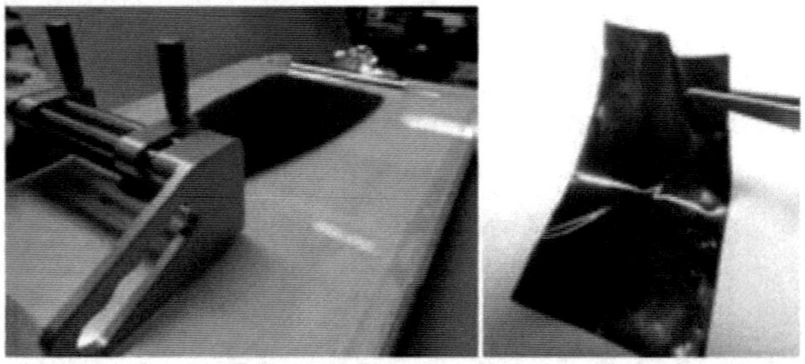

Figure 1. Gel-based membrane fabrication

To maximize membrane durability and performance, we fabricated the gas diffusion electrodes by coating carbon fiber paper with an optimized mixture of graphite and polymers to optimize the porosity and surface characteristics of the electrode structure. We used gas diffusion electrodes as both uncatalyzed (graphite-based electrode) and catalyzed with a range of electrocatalysts.

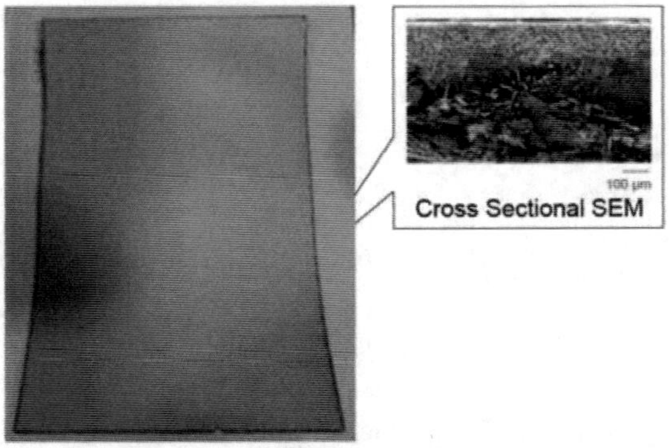

Figure 2. Gas diffusion electrode with optimized surface characteristics

Gas Permeability and Selectivity Measurements

We measured the permeability and selectivity of gas transport for the competitive separation of carbon dioxide from a range of gas mixtures including an exhaled breathing-air mixture (5% CO_2, 15% O_2, and 80% N_2), normal room air, and a methane CO_2 (50:50) mixture.

We custom designed an electrochemical cell with low pressure drop flow fields for the measurement shown in Figure 3.The test station is shown in Figure 4. We evaluated the gas-separating efficiency by using inline sampling to a gas chromatograph at regular intervals. For the initial screen test procedure, we used a helium sweep gas that could be flowed at a range of mass flow-controlled helium sweep rates. The gas chromatograph took a 1 cm^3 gas analysis sample at programmed intervals, and we used multiple columns and detectors to separate and analyze the gas composition. The testing was performed with no applied pressure differential across the cell.

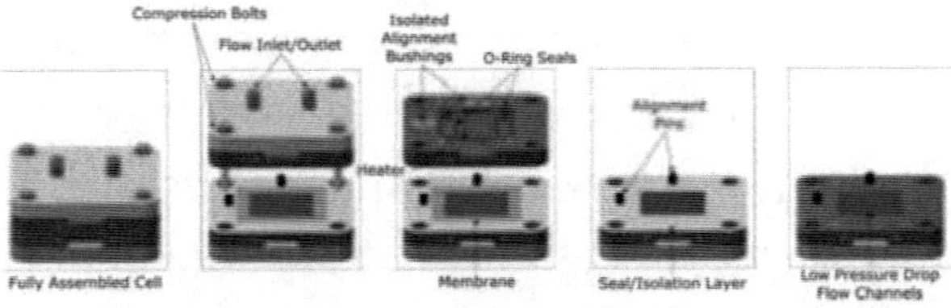

Figure 3. New cell design to lower the pressure drop and improve electrode sealing

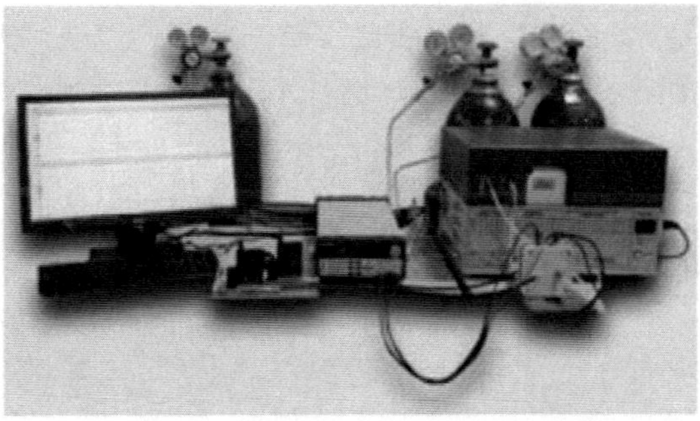

Figure 4. Mainstream's experimental setup for gas separation measurements

Results and Discussion

Cyclic Voltammetry of Electrode Coatings

To optimize redox reaction rates in our MEAs, we evaluated a range of electrocatalyst coatings on the carbon electrode surface. The tested catalysts were PtC, iridium oxide (IrO_2), and ruthenium oxide (RuO_2). After depositing catalysts on the standard carbon-coated gas diffusion layer (GDL) based electrodes, we used a Pt wire electrode for comparison. The catalysts were deposited by airbrush spraying of a catalyst ink onto the standard electrodes to give a catalyst loading of 1 mg/cm^2. We evaluated the catalyzed electrodes in CO_2-saturated solution by cyclic voltammetry scanning between 0.8 V to 1.6 V vs. saturated calomel electrode

(SCE). Representative voltammograms for the electrodes tested at a sweep rate of 100 mV/s are shown in Figure 5. The cyclic voltammograms show good reversible redox behavior for the redox carrier in the electrolyte solution. The peak shape and peak height response were consistent for all electrodes and typical of a solution mass-transport-controlled reversible redox reaction. Furthermore, similar behavior was obtained for a range of membrane and redox carrier combinations.

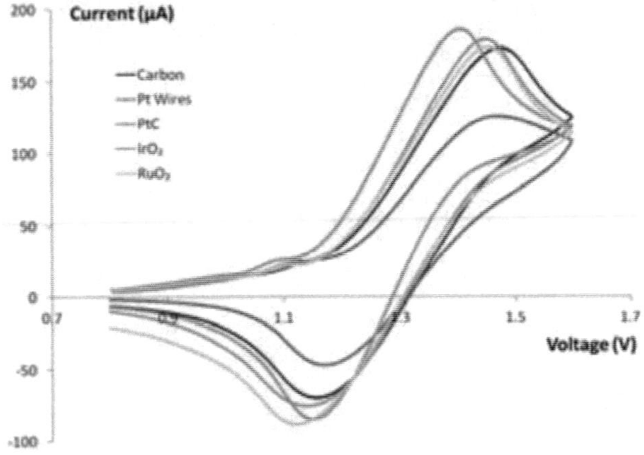

Figure 5. Voltammograms for electrodes tested

Applied Voltage Effects on Target Gas Permeability

To evaluate the performance of electrolyte and redox carrier pairings, we used a carrier/electrolyte-infused porous polyvinylidene difluoride (PVDF) membrane sandwiched between uncatalyzed, optimized gas-diffusion electrodes. We measured the initial performance with a breathing air mixture as the feed gas (5% CO_2, 15% O_2, and 80% N_2) both passively (no load) and with an applied voltage.

Figure 6 shows the permeability of O_2, N_2, and CO_2 through the electrolyte. Some selectivity for both O_2 and CO_2 is clear compared to N_2 from the redox carrier/electrolyte even without the applied field. However, when the load was applied, the CO_2 permeability increased by a factor of approximately 10, whereas the oxygen and nitrogen permeability remained constant.

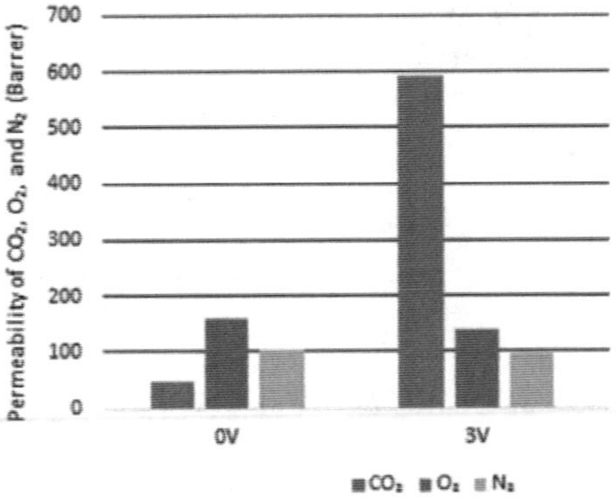

Figure 6. Permeability of CO_2, O_2, and N_2 with and without the applied load

Optimization of the Membrane Composition for Permeability and Selectivity

To improve membrane performance, durability, and manufacturability, we directly incorporated the redox carriers and electrolyte system into a range of freestanding, stable gel formulations. Figure 7 compares the transport of O_2, N_2, and CO_2 with an applied voltage through the gel membranes and the liquid film membrane in the porous PVDF matrix. As shown in Figure 7, with the applied load, the CO_2 permeability was significantly increased to as high as 8000 Barrer. This increase was achieved selectively without any associated increase in the O_2 or N_2 permeability. Figure 8 compares the selectivity of the CO_2 transport rate over those of O_2 and N_2 with the same three membranes under the applied load. Membranes A and B shared identical selectivity for CO_2/O_2 but were 20% more selective than the porous liquid film membrane.

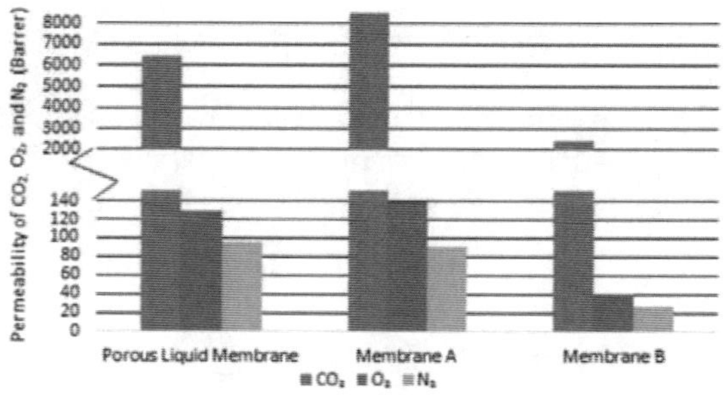

Figure 7. Permeability of CO_2, O_2, and N_2 from breathing air mixture with an applied voltage through aporous PVDF membrane and two gel-based membrane formulations

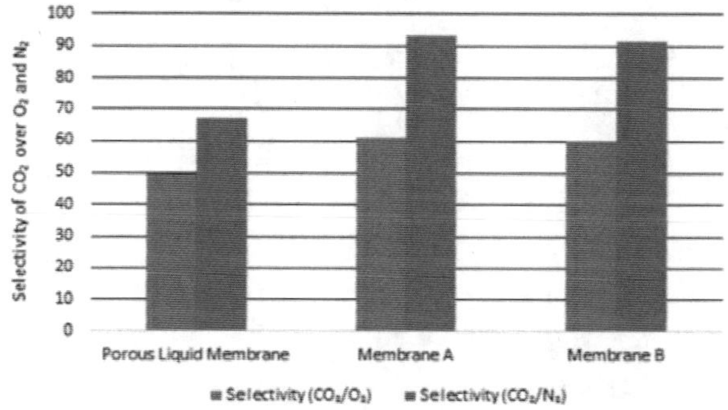

Figure 8. Perm selectivity of CO_2 over O_2 and N_2 from breathing air mixture with an applied voltage

Overall, this separation demonstrated the simultaneous enhancement of the permeability and selectivity of CO_2 without the associated tradeoff. Also, the permeability of CO_2 compared to the other gases was significantly increased relative to the porous membrane in the optimized gel membrane formulations.

Permeate Purity and CO_2 Selectivity

We measured gas permeation for feed gasses with a range of CO_2 including room air, breathing air, and CO_2: methane gas mixtures using a range of operating conditions. These conditions included the helium sweep gas measurements, the use of reduced pressure on the permeate side, and the direct electrochemical pumping with no applied pressure differential or sweep gas. In all cases, the permeate concentration of CO_2 was in the range of 95 to 99% CO_2 validating the high level of selectivity. Furthermore, this CO_2 purity in the permeate could be achieved with the electrochemical pumping of CO_2 from a 5% CO_2 feed to an over 95% permeate with the applied voltage.

Scale Up

Mainstream has scaled up the MEA fabrication and cell assembly to produce a multicell stack based on our existing fuel- and electrolysis-cell technologies as shown in Figure 9.

Figure 9. Scaled up separation stack

Conclusions

We demonstrated a scalable MEA-based technology can be used to simultaneously enhance the permeability and selectivity of CO_2 transport from a wide range of CO_2 gas mixtures and feed concentrations. Permeabilities as high as 8000 Barrier were achieved, which is a factor of up to 100 higher than the membrane without the applied voltage. We demonstrated high selectivities for the completive transport from a range of gas mixtures as high as 50 for CO_2/O_2 and 90 for CO_2/N_2. The results showed permeate concentrations as high as 99% for CO_2. Our prior experience with fuel and electrolysis cell technologies has allowed us to develop a membrane separation process that is effective and can be readily scaled to satisfy production-level demands.

Acknowledgements and Disclaimer

This material is based on work supported by the U.S. Department of Energy, Office of Science, Office of Basic Energy Sciences under Award Number DE-SC0015156

References

1. Robeson, L. M., The upper bound revisited. *Journal of Membrane Science* **2008,** *320* (1–2), 390-400.
2. Robeson, L. M.; Freeman, B. D.; Paul, D. R.; Rowe, B. W., An empirical correlation of gas permeability and permselectivity in polymers and its theoretical basis. *Journal of Membrane Science* **2009,** *341* (1-2), 178-185.
3. Bernardo, P.; Drioli, E.; Golemme, G., Membrane Gas Separation: A Review/State of the Art. *Industrial & Engineering Chemistry Research* **2009,** *48* (10), 4638-4663.
4. Preethi, N.; Shinohara, H.; Nishide, H., Reversible oxygen-binding and facilitated oxygen transport in membranes of polyvinylimidazole complexed with cobalt-phthalocyanine. *Reactive and Functional Polymers* **2006,** *66* (8), 851-855.

565189 Thermal Cross-Linking and Cyclodehydration Modifications of PI Membranes for Pervaporation Applications

Membrane is one of the critical factors for the separation efficiency of the pervaporation technology. A desirable pervaporation membrane should possess high flux, superior separation factor, and long-term chemical and mechanical stability. Common hydrophilic polymeric membrane materials have been widely used for pervaporation dehydration because of their high selectivity towards water, but they generally suffer severe swelling in the aqueous solution, leading to a poor operation stability. Alternatively, polyimide membranes have drawn more and more attentions for pervaporation applications in recent years, with their superior mechanical strength and chemical stability. Here we mainly introduce our research works on the polyimide membranes for the pervaporation dehydration in recent years, including the molecular design of membrane materials and various chemical modifications to improve their separation performance and operation stability, mainly including thermal cross-linking, chemical crosslinking and thermal cyclodehydration. Among then, thermal cyclodehydration is introduced into the hydrazide-crosslinked polyimide membranes for the first time, which incorporates aromatic oxadiazole rings into PI matrix and endows the polyimide membrane with larger d-spacing and improved solvent resistance, leading to the greatly improved separation performance and operation stability.

References

1. J. Membr. Sci. 2018, 549, 184–191.
2. Sep. Purif. Technol. 2017, 185, 215–226.
3. J. Membr. Sci. 2015, 496, 142–155.

560360 Application of High-Shear Wet Milling to an API Crystallization Process

High-shear wet milling was applied to the crystallization of a drug substance to reduce particle size. In this case study, ester hydrolysis under basic conditions followed by acidic pH adjustment crystallized the resulting API with large particle size. To generate API of smaller PSD for drug product evaluation, high-shear wet milling was investigated in conjunction with other crystallization factors such as seeding, temperature, concentration, and rate of pH adjustment. This application of in-situ particle size reduction avoided the dry milling of a potent compound, thus mitigating risk of exposure to plant operators. In developing the high-shear wet milling process, parameters such as shear frequency, tip speed, and turnover rate were investigated in the laboratory using scale-down calculations to simulate wet milling performance in a large-scale or production setting. These scale-down experiments comprised a series of multi-hundred gram batches to evaluate trends in API crystal growth and particle size distribution. During these studies, API particle size was monitored throughout the crystallization process from initial seeding to isolation using an in-situ FBRM probe in combination with PSD analysis of filtered slurry samples. In addition to reducing particle size, the solubility data showed that milling also increased the rate of API crystallization. These studies contributed to the development of a crystallization process with high-shear wet milling to provide API with the desired physical properties.

577487 Study on Impurity Control of 7-Adca Isomers

7-ADCA is an important intermediate in the production of cephalosporins. At present, the production of 7-ADCA in our country is mainly oxidized by penicillin K, ring-enlarged to obtain ring-enlarged acid, and then 7-ADCA is produced by enzymatic cleavage of ring-enlarged acid. However, there is a big gap between the final product quality level and foreign countries, mainly due to the high impurity content of 7-ADCA isomers. Therefore, improving the purity of 7-ADCA and reducing the content of isomer impurities are of great significance to the production market of 7-ADCA in China. This experiment optimizes the process of enzymatic cracking, crystallization and recrystallization in the production of 7-ADCA. Combining with single factor experiments, orthogonal experiments are carried out to study the effects of factors such as cracking temperature, cracking time, the use level of enzyme, crystallization temperature and crystallization time on the reduction of 7-ADCA isomer impurities. 7-ADCA with low isomer content is obtained, which provides reference for large-scale industrial production.

577798 Study on Recrystallization of Cephalosporanic Acid G

Abstract: The recrystallization of Cephalosporanic acid G in different solvents, such as methanol-water, ethanol-water, sodium bicarbonate-sulfuric acid, sodium bicarbonate-hydrochloric acid, ammonium bicarbonate-sulfuric acid, ammonium bicarbonate-hydrochloric acid, ammonia water-sulfuric acid, ammonia water-hydrochloric acid, was studied. The dynamic laser method was used to determined the solubility of Cephalosporanic acid G in ethanol, pure water and ethanol-water. In the mixed solvent of ethanol-water, the solubility of Cephalosporanic acid G was found to increase first and then decreased with the increase of the molar fraction of water, and reached the maximum when the molar fraction was 0.3.The mesostable region of Cephalosporanic acid G in ethanol-water system was determined. The thermodynamic model Apelblat equation, Van't Hoff equation, lh equation, Wilson equation and NRTL equation were used to fit data. The interaction energy between Cephalosporanic acid G and different solvents was simulated with Materials Studio software to discuss the difficulty of forming hydrogen bonds. Based on the thermodynamic properties of Cephalosporanic acid G, the crystallization process of ethanol-water was designed. The effects of temperature range, crystallization time, flux acceleration rate and agitation rate on product purity and yield were studied by univariate method. The optimum technological conditions for solvent crystallization of Cephalosporanic acid G were determined.
Abstract: The recrystallization of Cephalosporanic acid G in different solvents, such as methanol-water, ethanol-water, sodium bicarbonate-sulfuric acid, sodium bicarbonate-hydrochloric acid, ammonium bicarbonate-sulfuric acid, ammonium bicarbonate-hydrochloric acid, ammonia water-sulfuric acid, ammonia water-hydrochloric acid, was studied. The dynamic laser method was used to determined the solubility of Cephalosporanic acid G in ethanol, pure water and ethanol-water. In the mixed solvent of ethanol-water, the solubility of Cephalosporanic acid G was found to increase first and then decreased with the increase of the molar fraction of water, and reached the maximum when the molar fraction was 0.3.The mesostable region of Cephalosporanic acid G in ethanol-water system was determined. The thermodynamic model Apelblat equation, Van't Hoff equation, lh equation, Wilson equation and NRTL equation were used to fit data. The interaction energy between Cephalosporanic acid G and different solvents was simulated with Materials Studio software to discuss the difficulty of forming hydrogen bonds. Based on the thermodynamic properties of Cephalosporanic acid G, the crystallization process of ethanol-water was designed. The effects of temperature range, crystallization time, flux acceleration rate and agitation rate on product purity and yield were studied by univariate method. The optimum technological conditions for solvent crystallization of Cephalosporanic acid G were determined.

578306 Study on the Crystallization Process of Tazobactam Intermediate

Study on the Crystallization Process of Tazobactam Intermediate

Li Huimei,Sun Hua,Han Yiman,Liu Baoshu*

College of Chemical and Pharmaceutical Engineering, Hebei University of Science and Technology, Shijiazhuang, Hebei 050018, People's Republic of China

Abstract: Tazobactam is a ß-lactamase inhibitor. The synthesis step is up to ten steps, and the purification process is carried out during the preparation of various intermediates. The purification method used is recrystallization. For instance, in the preparation of intermediate 6a-bromopentanoic acid sulfoxide of tazobactam, the purity and yield of the product are improved by exploring the recrystallization process. The factors such as product dissolution temperature, cooling temperature, reverse solvent drop rate, primary crystallization and secondary crystallization were investigated systematically. Ultrasonic technology is also used to solve the solubility of drugs that are difficult to dissolve and the stability of heat-stable drugs, improve the dissolution rate and the number of dissolutions, and comprehensively study the quality, yield and process indicators of the products. For example, in the preparation of an intermediate pyrolysis ring-opening of tazobactam, the phenomenon of oil formation is solved by improving the solvent and using ultrasonic technology. This further increases the purity and yield of the product.

Key words: Tazobactam, recrystallization, ultrasound technique.

578391 Study on the Crystallization Process of 7-Adca Pivaloamide

Abstract: 7-ADCA pivaloamide is an impurity of the cephalexin. In order to control the quality of the drug, there is a clear requirement for the impurity content in the process of drug registration and declaration. The international counterparts research and verify the impurity reference substance in the drug to complete requirements. At present, compared with impurity reference substance, bulk produced 7-ADCA pivaloamide have some quality flaws such as low purity, small size,wide distribution of particle size. In addition, 7-ADCA pivaloamide in the crystallization process is likely to appear into the oil, coalescence and other phenomena. The crystallization process is extremely unstable. In order to improve the quality of 7-ADCA pivaloamide product, the crystallization process of 7-ADCA pivaloamide was systematically studied here.

By examining the effect of the reaction crystallization and the anti-solvent crystallization of the product on crystallinity and yield of the product, it was determined that the anti-solvent crystallization was used as the process to be optimized. Through the comprehensive study of the concentration of the solution, temperature, stirring and other operating parameters on the product performance and process indicators, the anti-solvent crystallization was determined.

577867 Separation and Crystallization of Solid-Liquid Mixtures

The solid-liquid mixtures studied in this study are o-chlorobromobenzene and p-chlorobromobenzene, as well as some small amounts of solvents and impurities. Currently, two substances are organic synthesis intermediates. There is no relevant literature report on the separation of the two substances.

Since the two substances are isomers, the physicochemical properties are very similar, but the melting points are different. Therefore, in this study, the crystalline product with higher purity can be obtained by melt crystallization. The separation is carried out by two methods: suspension melting crystallization and layer melting crystallization. The crystal growth process, growth mode, crystal morphology are obviously different, and two crystal products of different particle sizes and shapes can be obtained respectively. Therefore, by changing the crystallization process conditions, the morphology growth process and particle size distribution of the crystals were separately studied.

568159 Purification of Ethylene with Anion-Pillared Metal-Organic Frameworks As Adsorbents

Purification of ethylene with anion-pillared metal-organic frameworks as adsorbents

Qi Ding, Xili Cui, Zhaoqiang Zhang, Lifeng Yang, Huabin Xing*

College of Chemical and Biological Engineering, Zhejiang University, Hangzhou 310027

E-mail: xinghb@zju.edu.cn

As an important feedstock in petrochemical industries, ethylene (C_2H_4) is among the highest-yield chemicals in the world with a global production capacity of more than 170 MT in 2016. Separation of C_2H_4 from acetylene (C_2H_2) and ethane (C_2H_6) present as critical steps in preparing polymer-grade C_2H_4, which meanwhile exhibit great challenge due to the similar molecular size and physical properties of these light hydrocarbons and are generally realized by energy-extensive cryogenic distillation or solvent absorption, these processes accounting for nearly 85% of the total cost for C_2H_4 production. Adsorption based on porous materials is a commercially viable alternative technology for gas purification, while the ever reported adsorbents usually meet with the trade-off between uptake capacity and selectivity, confining their ultimate separation ability. In the presentation, we will discuss strategies addressing the problem and recent advances of our group in the context of ethylene purification. We reveal that finely control of pore chemistry and pore size in anion-pillared ultramicroporous metal-organic frameworks (MOFs) enables the materials brilliant performance for C_2H_2/C_2H_4 separation through exploiting supermolecular host-guest and/or guest-guest interactions. Furthermore, the tactic to give an overall consideration to adsorption thermodynamics and adsorption kinetics and take advantage of their synergetic effect leads to outstanding molecular recognition ability to C_2H_4 over C_2H_6. Impressive adsorptive selectivities setting new benchmarks were achieved in our work accompanied by high gas uptake.[1,2] The excellent performances of the anion-pillared ultramicroporous MOFs were confirmed by single-component static and kinetic adsorption, and multi-component break through experiments. The binding sites of C2 hydrocarbons on the materials were explored by DFT-D (dispersion-corrected density-functional theory) calculations to give insights into the interaction mechanisms.

Acknowledgement:

This research was supported by the National Natural Science Foundation of China (No. 21725603), Zhejiang Provincial Natural Science Foundation of China (No. LZ18B060001), the National Program for support of Top-notch Young Professionals (H. X.) and China National Petroleum Corporation.

References:

1. Cui XL, Chen KJ, Xing HB, et al. Pore chemistry and size control in hybrid porous materials for acetylene capture from ethylene. *Science*. 2016; 353(6295): 141-144.

2. Zhang ZQ, Cui XL, Yang LF, et al. Hexafluorogermanate (GeFSIX) anion-functionalized hybrid ultramicroporous materials for efficiently trapping acetylene from ethylene. *Ind. Eng. Chem. Res.* 2018; 57: 7266-7274.

556215 Oligo-Ethylene-Glycol Based Thin-Film Composite Nanofiltration Membranes for Effective Separation of Mono-/Di-Valent Anions

Oligo-ethylene-glycol based thin-film composite nanofiltration membranes for effective separation of mono-/di-valent anions

Dan Ren[1], Xiao-Tian Bi[1], Tian-Yin Liu[1,2*] and Xiaolin Wang[1*]

[1]Beijing Key Laboratory of Membrane Materials and Engineering, Department of Chemical Engineering, Tsinghua University, Beijing, China

[2]Barrer Centre, Department of Chemical Engineering, Imperial College London, South Kensington Campus, London, SW7 2AZ, UK

*email: *tianyin.liu@imperial.ac.uk *xl-wang@tsinghua.edu.cn

Separation of monovalent and divalent anions is highly required in water treatment technologies for recycling concentrated brine resources, upgrading industrial separation processes and reducing the release of hazardous compounds into the environment. For the selective separation of monovalent and divalent anions, the porous organic networks comprising oligo-ethylene-glycol units were finely tailored with pore structure, ion affinity and hydrophilicity. Enhanced cation affinity and tailored pore structures were achieved by using closely-related oligo-ethylene-glycol-containing building blocks, including 1,2-Bis(2-aminoethoxy)ethane (EDA), 4,7,10-trioxygen-1,13-tridecane diamine (DCA) and polyethylene glycol diamine (N-PEG). Thin-film composite (TFC) nanofiltration (NF) membranes were prepared by interfacial polymerization using the oligo-ethylene-glycol-based amines with trimesoyl chloride (TMC) on a polyethersulfone support (PA@DCA, PA@EDA and PA@N-PEG TFC membranes). The sub-nanometer pore size of the PA@DCA film was larger than the PA@EDA film in 2 angstroms, and the PA@DCA TFC membrane exhibited a NaCl permeation 3-fold higher than the PA@EDA TFC membrane. Furthermore, the cation permeances of the oligo-ethylene-glycol-based membranes increased in the same order of cation affinity based on the valent of ions. These results indicated a translation of the properties of the oligo-ethylene-glycol units to the ion-selective performances. Enhanced pure water permeance, faster NaCl permeance and high Na_2SO_4 rejection (more than 99%) were achieved at a reduced degree of cross-linking and a thinner thickness of the thin film layer. In terms of mono-/di-valent anion selectivity, the PA@DCA TFC membrane showed a performance 4-fold higher than the typically-used commercial NF membranes, such as DOW NF 270. This strategy paves the way to microporous organic networks for the fabrication of nanofiltration membranes.

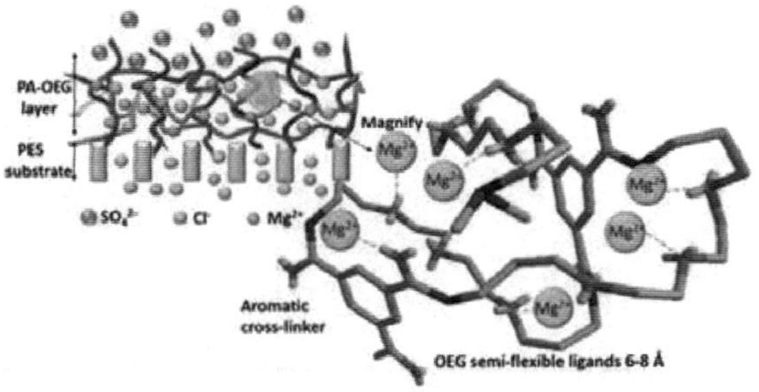

References

[1] M. A. Shannon, P. W. Bohn, M. Elimelech, J. G. Georgiadis, B. J. Marinas and A. M. Mayes, Nature, 2008, **452**, 301-310.
[2] O. Shyshov, K. A. Siewerth and M. von Delius, Chem. Commun., 2018, **54**, 4353-4355.

Keywords: Interfacial polymerization; nanofiltration membranes; oligo-ethylene-glycol; separation of mono-/di-valent anions.

570514 Tailoring Molecular Architecture of Heterogeneous Nanostructure for High CO_2 Separation Performance

Graphene Oxide (GO) has been extensively studied as a membrane materials for CO_2 separation due to its high CO_2-philic property. Although GO membrane is considered as one of promising membrane, it has a limitation to apply for practical application because of long diffusional pathway because of its two-dimensional structure.
Graphene Oxide (GO) has been extensively studied as a membrane materials for CO_2 separation due to its high CO_2-philic property. Although GO membrane is considered as one of promising membrane, it has a limitation to apply for practical application because of long diffusional pathway because of its two-dimensional structure. In general, gas molecules are penetrated along the GO layers, which lead to increase of diffusional pathway and resistance to oxygen functional groups on the basal plane of GO. As such, in this study, we modified GO to improve its performances by adjusting the architecture of GO using various methods as follows; 1) precise control of sp^2-graphitic regions and sp^3-oxidized regions without any chemicals, and 2) creating and decorating pores on GO sheets. By using these modification methods, we controlled interface resistance and diffusional pathways between GO layers. Both CO_2 permeability and CO_2/N_2 selectivity were increased by redesigning the structure. These results have the potential to open up new approaches for an improvement in the transport properties of two-dimensional membrane materials for use in next-generation membrane applications.

564944 Innovative Graphene Based Membranes for Gas Separation with Tunable Selective Properties

Atomic-thick 2D materials attract a great interest for the development of membrane materials due to their peculiar gas transport properties, combined to extremely high aspect ratios. In particular, graphene and its derivatives are very promising in this research area due to their very low permeability to atoms and molecules coupled with interesting selective properties towards smallest species, thus opening to new viable strategies for mass transport-related applications.

In this work, the layer-by-layer (LbL) technique is employed to fabricate thin coatings on top of polymeric substrates, using graphene oxide (GO) sheets as 2-D building blocks, piled up together with polyelectrolyte species in well-ordered structures. Electrostatic interactions drive the self-assembly of GO and polyelectrolytes, controlling both the growth and the coatings arrangement at the nanoscale.

The nanostructured architecture is achieved by alternated dip coating in highly diluted (typically 0.1 wt. %) aqueous polyelectrolyte solutions and GO dispersions, with a water rinsing step performed immediately after each dipping, in order to avoid inhomogeneity on sample surface and promote the required structural order. The whole process is repeated for a prescribed number of cycles to create long selective paths and, after being validated on several different polymeric substrates, both dense and microporous, is carried out on commercial Matrimid polyimide films, a benchmark membrane material for gas separation applications.

The obtained nanostructured coatings revealed an effective and homogeneous deposition of alternated layers, as observed by SEM analysis and by advanced scanning ellipsometry micromapping that allowed also the quantitative determination of the average thickness of each layer.

Gas permeation measurements showed a significant sieving ability of the coatings, which are able to discriminate the penetrant molecules on size basis, making these membranes very interesting for hydrogen purification (e.g. for H_2/CO_2 or H_2/CH_4 separation). Furthermore, due to their small thicknesses, the GO based layers provide a very limited effect on the gas transport of small penetrants such as He or H_2, eventually allowing to overcome the Robeson's technological limit for polymeric membranes.

Multiple strategies are explored in order to tune the permselectivities of the GO based membranes, considering different positive polyelectrolytes with varied steric properties, GO of different later sizes (as obtained by ultrasonication), and varying the ionic strength of the solutions during the LbL assembly (controlling the pH or using salts). The obtained materials are tested for He (used as safer H_2 surrogate) and CO_2, in order to evaluate the sieving ability of the membrane, and the results are compared with the state of the art data.

The best results are obtained considering a 10 bilayers coating (around 40 nm), whose values of permeability and selectivity (as high as 200) for the He/CO_2 gas pair lies well above the upper bound. Interestingly, the addition of small amount of NaCl is able to boost both selectivity and permeability, while the use of smaller GO flakes allow a more ordered packing of the layers thus leading to a decrease in the gas transport properties.

Therefore, this novel concept of membranes design demonstrated that by means of a simple and scalable method it is possible to fabricate innovative membrane able to overcome technological limits on this class of separation processes, by using innovative 2-D nanomaterial as graphene oxide.

572444 Polybenzimidazole-Derived Carbon Molecular Sieves with Microcavities and Ultra-Microporous Channels Achieving Superior Membrane H_2/CO_2 Separation Properties

Membrane technology is highly attractive for H_2/CO_2 separation for pre-combustion CO_2 capture, which requires membrane materials with both high H_2 permeability and H_2/CO_2 selectivity at 100 – 300 °C. Herein, we present carbon molecular sieve (CMS) membranes with superior H_2/CO_2 separation properties via the pyrolysis of polybenzimidazole (PBI, a leading polymer for H_2/CO_2 separation). We thoroughly investigated the effect of the pyrolysis temperature on physical properties (pore size, *d*-spacing, density, solubility, diffusivity) and H_2/CO_2 separation properties of the CMS films. PBI exhibits H_2 permeability of 27 Barrers and H_2/CO_2 selectivity of 14 at 150 °C, while the CMS films pyrolyzed at 600 °C, 850 °C, and 900 °C exhibit H_2 permeability of 370 Barrers, 190 Barrers, and 54 Barrers at 150 °C, respectively, and their corresponding H_2/CO_2 selectivity of 8.9, 16, and 80, respectively. The pyrolysis forms micro-cavities leading to high gas permeability and ultra-microporous channels leading to strong size-sieving ability, which is also confirmed by the Positron Annihilation Lifetime Spectroscopy (PALS). As the pyrolysis temperature increases from 600 °C to 850 °C and then 900 °C, the free volume element size increases from 5.08 Å to 5.46 Å first and then decreases to 4.90 Å. When tested with a mixture containing 50% H_2 and 50% CO_2 at 150°C, CMS pyrolyzed at 900 °C show H_2 permeability of 39 Barrers and H_2/CO_2 selectivity of 53, which surpasses the Robeson's upper bound for H_2/CO_2 separation at 150 °C. The film shows stable mixed-gas separation performance for 40 h. When water vapor of 0.31 mol% is introduced, the H_2 permeability slightly decreases to 37 Barrers while the H_2/CO_2 selectivity remains the same. The mixed-gas H_2 permeability increases to 39 after the water vapor is removed. The stability and robust H_2/CO_2 separation properties demonstrate the potential of PBI-derived CMS for practical H_2 purification and CO_2 capture.

557546 Pure and Mixed Gas Transport and Physical Aging in Glassy Polymers Exhibiting Configurational Free Volume

Plasticization and physical aging represent two major drawbacks for using glassy polymer membranes. These issues can be mitigated by using triptycene-based polymers. Triptycenes are formed by three aromatic rings arranged in a paddlewheel-like, 3-D structure. Once inserted into a polymer backbone, triptycene groups provide an internal free volume which greatly enhances gas permeability and selectivity. Unlike in conventional glassy polymers, the internal free volume provided by iptycene units is not related to the non-equilibrium, transient conformation, but to the molecular configuration. Such configurational free volume is intrinsic to the polymer structure and, as such, it is non collapsible, similar to the case of inorganic molecular sieves. Equally important, rigid triptycene units can accommodate penetrant molecules in their internal free volume without causing swelling and plasticization.

In this study, pure gas and vapor (He, N_2, CH_4, CO_2, C_2H_6, water vapor, hydrocarbons and alcohols) sorption and diffusion in a novel thermally rearranged co-polymer prepared from a co-polyimide precursor with controlled triptycene molar content, i.e., triptycene-dianhydride(0.25)-6FDA(0.75)-6FAP(1.0), was investigated in the temperature range 5-50°C and up to 32 atm. The experimental data were analyzed in the framework of the Dual Mode and Lattice Fluid models, and several meaningful structure/property correlations were identified. The dual mode parameters retrieved from the analysis of single penetrant sorption isotherms were used to predict the sorption behavior in mixed gas conditions. The mixed gas solubility-selectivity is significantly higher than ideal solubility-selectivity. The analysis of gas diffusion coefficients also disclosed precious information about the effect of the polymer structure on small molecule transport in this class of materials.

Finally, to accelerate physical aging, selected polymer samples were annealed at 220°C under vacuum for 10 days. Remarkably, transport properties were unaffected by thermal annealing, thus demonstrating the superior resistance to physical aging exhibited by these materials.

555871 A Thin Film Composite Membrane Prepared from Monomers of Guaiacol and Trimesoyl Chloride for Organic Solvent Nanofiltration

The main drawback of OSN membrane is low flux in weak polar solvent such as THF. In this study, a naturally occurring monomer, guaiacol (GA) was adopted to synthetize a hydrophobic polyarylester; OSN membrane via interfacial polymerization with trimesoyl chloride on the ethylenediamine cross-linked poly(ether imide) substrate. The prepared membrane was proved by chemical characterizations such as ATR-FTIR, XPS etc., it is illustrated that the interfacial polymerization occurs among TMC and GA on the C-PEI support to generate the polyarylester layers with crosslinked structure as well as enhanced diffusion property. GA-TMC NF membrane is composed of polyarylester, and the further contact angles measurements revealed that the hydrophobicity of the membranes increased with the GA loading. Besides, the study on the relationship between the membrane permeance and the membrane-solvent solubility parameter distance were performed by Hansen solubility parameter methods. The optimal NF-3GA membrane exhibited a higher chemical stability in THF and superior separation performance with the rejection of 96% for RB and a THF permeance of 76 L m^{-2} h^{-1} MPa^{-1}. These results suggest a novel strategy to to select potential monomers to develop advanced membranes applicable to organic solution separation for future sustainable usages.

583261 Crystallization Solvent Design Methodology for Pharmaceutical Processes Based on Crystal Morphology Quantitative Control

Solution crystallization is an important and efficient purification/separation unit operation in active pharmaceutical ingredient (API) production processes. The requirements of crystal products include not only purity and yield, but also crystal morphology. To obtain optimal crystal products, it is crucial to reveal the influence of solvents and mechanisms for the control of crystal growth. Nowadays, crystallization solvent selection methods are mostly experimental trial and error. Therefore, it is urgent to develop a systematic crystallization solvent design method from the interdisciplinary studies of supramolecular chemistry, condensed matter physics, and product design principles. In this paper, the organic small molecules gliflozin drugs are studied as an example for the crystallization solvent design. First, based on molecular dynamics, the crystal morphology in different solvents is predicted, and the quantitative relationship between crystal morphology and solvent descriptor is further established, which is the solvent evaluation index system to quantitatively regulate crystal morphology. Then, the Computer-Aided Molecular Design model, which integrates the established mechanism model, is developed for the design of crystallization solvent. Mathematical optimization model is established for the design of crystallization solvent to consider the requirements of crystal morphology, yield, green and safety simultaneously. Finally, experiments are implemented for verification. The results of the study are able to be expanded to various APIs for the efficient and clean production of pharmaceutical processes, and contribute to the pharmaceutical products upgraded and crystal engineering.

583263 Computer-Aided Design of Crystallization Solvents for the Recovery of High-Purity 2- Mercapotobenzothiazole

2-Mercapotobenzothiazole (MBT) is an important vulcanization accelerator, which is widely used in rubber industry. The production of high purity MBT mainly includes acid-base method and solvent-based method. In recent years, due to the restriction of environmental protection policies, the application of acid-base method has been greatly restricted. In solvent-based method, it is important to select a suitable crystallization solvent to ensure the refined MBT meets the industrial requirements. Therefore, a Computer-Aided Molecular Design (CAMD) model for designing crystallization solvents is proposed. The proposed CAMD problem is express as a Mixed-Integer Non-Linear Programming model, which contains objective functions, structural constraints, property constraints and process constraints. The objective functions are the potential yield and the purity of the MBT product. Other solvent-related properties are taken as property constraints, including molecular weight, normal melting point, normal boiling point, flash point, toxicity, solubility parameters, and SLE (solid-liquid equilibrium), etc. Among them, the activity coefficient related to the SLE is predicted by the COSMO-SAC model, and the other properties are predicted by the Group Contribution method. The model is then solved by the decomposition-based approach to obtain the optimal solvents, which are verified by experiments. Through the crystallization experiments, the effects of solvent ratio, initial crystallization temperature and the number of mother liquor cycles on the purity and yield of MBT product are investigated. Meanwhile, the experimental results are compared with the model results. Finally, optimal solvents are selected, which perform better results than the currently reported solvents for crystallization. The effectiveness of the model is verified, and it also provides guidance for the design of other crystallization solvents.

583373 Enriching Trace Bioactive Compound and Isolation of Pesticides from Natural Materials Using Scalable Supercritical CO_2 Continuous Countercurrent Extraction

Supercritical CO2 extraction is a green and sustainable technology and has been applied in the food and pharmaceutical industry. In this paper, we presented a study on enriching trace bioactive compound and isolation of pesticides from palm oil and spice oleoresins, respectively, using supercritical CO2 technology. Natural Beta-Carotene is an antioxidant and precursor of vitamin A, which has many benefits to human beings, such as it can reduce the risk of cancer, enhance immunity, and prevent night blindness, etc. Crude palm oil is the primary source of Natural Beta-Carotene, from which many methods have been reported to extract Beta-Carotene. At present, transesterification followed by distillation is a main method, which involves using a huge volume of organic solvent, and the chemical structure of palm oil was changed. In this work, a fractional pilot column with the height of 6 m and diameter of 50 mm, filled with packing, was manufactured and a continuous countercurrent supercritical CO2 extraction-distillation process was developed to enriching beta-carotene from crude palm oil directly. The influence of the process parameters (pressure, temperature, CO2, and feeding ratio) on the Beta-Carotene content was investigated. At the optimum conditions, the beta-carotene content could be enriched to 40 times and reached to 2% from 0.05% after two cycles. It is not necessary to make transesterification of palm oil, keeping its original form, and the process is simple. Also, we carried out a study on removing pesticides from capsicum oleoresin extracted from chilly. The results indicated the pesticides content could be reduced from the original PPM level to PPB level, which is expected to have wide application in removing toxic components from spice oleoresins. In conclusion, this approach can find wide applications in enriching and purifying compositions used in nutraceutical, pharmaceutical & cosmetic fields, etc.

Keywords: supercritical CO2 countercurrent extraction; enriching; isolation; beta-carotene; pesticides; spice oleoresins

576404 Modeling of Particle Retention and Penetration Depth in Deep Bed Filtration

A mathematical model that describes the transport and deposition of both nano- and micro-particles through porous media is important in industrial applications, such as oil exploration, water purification and pollutant treatment; however, existing experimental studies and mathematical models of deep depth filtration often fail to accurately quantify the particle retention and penetration depth. This is due, in part, to the fact that most existing models depend heavily on empirical parameters; inaccurately describe pore structure of the filter medium and particle size distribution of contaminants; and neglect the probability of deposition when $d_{particle} < d_{pore}$.

We develop a deep bed filtration model that is able to predict the migration of micro- and sub- micro-particles in porous media. The model couples the transport and deposition characteristics of particles. A transport probability function P(i) is proposed to predict the probability of flow through a pore based on the analysis of pore resistance. The model also includes a deposit probability function $P^*(i)$ that captures the likelihood of particle deposition within a pore. The P*(i) is measured experimentally using the change of concentrations during filtration tests with different pore structures and particle sizes, including cases where $d_{particle} < d_{pore}$. The empirical P*(i) shows the probability of particle deposit/ pore capture is highly related with the size ratio ($d_{particle}/d_{pore}$) and flow velocity (i.e., P*(i) is positively correlated with the size ratio and is negatively correlated with the flow velocity). This model can predict both the particle retention concentration in a pore and the approximate penetration depth of contaminants within a filter bed. By combining this new model with our earlier macroscopic description of filter flow, we can predict the filtrate concentration, flow dynamics and the penetration-scale of the filter medium from first principles.

559518 Methods to Study Influence of Surfaces in Crystallization Processes: Scaling and Supersaturation Depletion

Methods to study influence of surfaces in crystallization processes: scaling and supersaturation depletion

Jacob Hutfles[1], Sankaranarayanan A. Ravichandran[1], and John Pellegrino*

Department of Mechanical Engineering, University of Colorado-Boulder, Boulder, CO 80309-0427

**corresponding author:* john.pellegrino@colorado.edu

Abstract

We present methods to study the effects of various polymeric surfaces on the heterogeneous crystallization of calcium carbonate from supersaturated aqueous solutions using shaken jars. The methodology allows for bench scale evaluation of different materials, the extent of mixing and the interaction of these parameters on crystallization kinetics. The crystallization process was quantified using turbidity, crystal collection, and depletion on Ca ions in solution using ICP-MS. From measuring the aforementioned parameters, we identified surfaces that were able to promote heterogeneous crystallization while avoiding scale build up. Our experimental variables were the type of polymer, time, and mixing energy input. An unanticipated observation was that the polyethylene and polymers contributed to similar levels of overall supersaturation depletion, but then poly (ethylene-co-vinyl alcohol) was found to resist surface build-up i.e. was resistant to calcium carbonate scaling. The methodology we have developed can aid in improving crystallizer designs to increase efficient water recovery from brine waste in terms of the material of construction and mixing devices.

[1] *both authors contributed equally to the research and manuscript preparation*

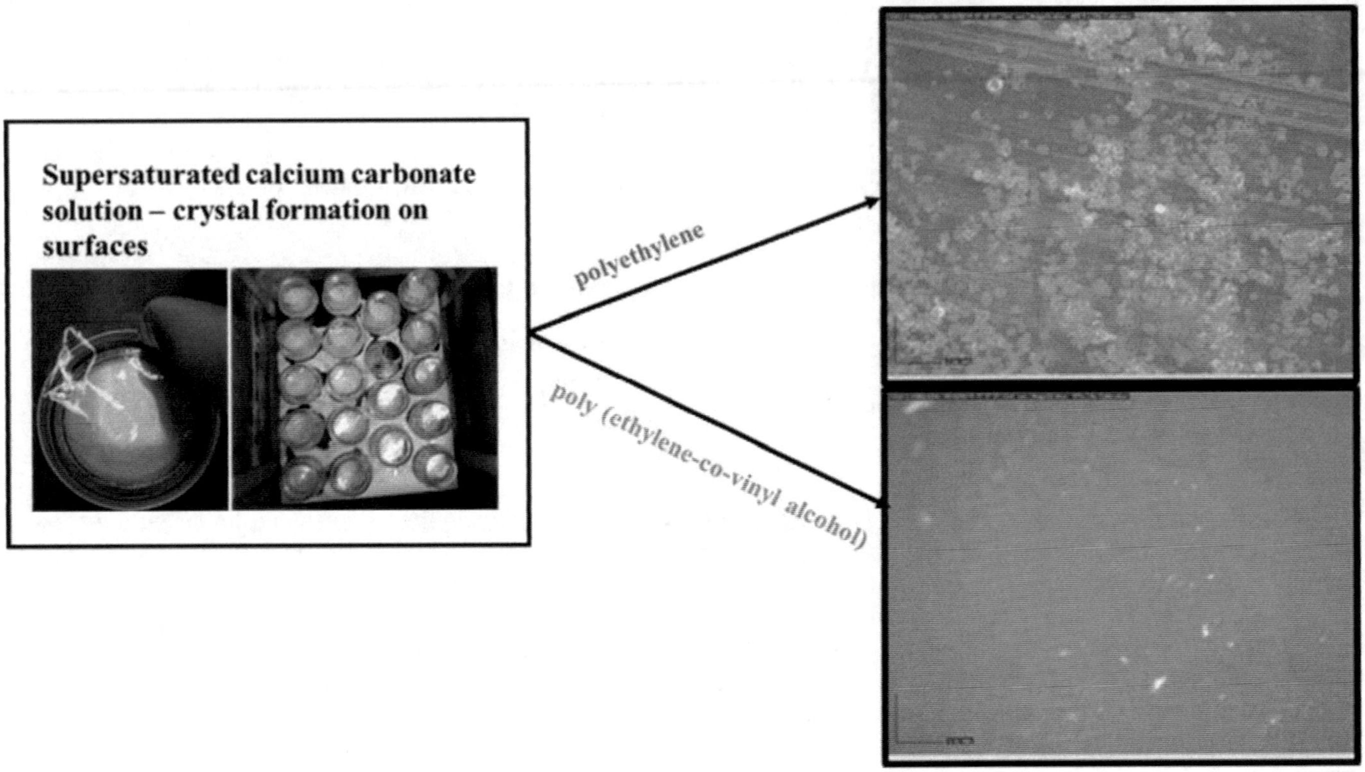

Figure 1. Observe the that polyethylene shows significantly more scaling than poly (ethylene-co-vinyl alcohol)

556475 Effect of Glycine Pre-Leaching of Copper- Gold Bearing Low Grade Ore Prior to Cyanidation

The aim of this work was to explore the opportunities of increasing the gold recovery and reducing the cyanide consumption in the leaching of a copper gold ore. The effects of glycine in recovering copper prior to cyanidation from local mine ore was investigated. Several test works were carried out to come up with the optimum parameters for excellent copper and gold dissolution. Direct cyanide and ammonia assisted leaching of the ore were also carried out as to evaluate the responsiveness of the ore to the reagents. The data reveals that glycine pre-leaching of copper can be done to treat the low grade ore Cu-Au and still save in terms of reagent costs whilst optimizing recovery of gold and copper metals.

556547 Main Operating Parameters of the Spouting Regenerator for the Collecting Particles of the Moving Bed Filter

The combustion of fossil fuel inevitably results in serious environmental pollutions such as discharging flue gas containing SO_2, NO_2 and CO_2. Thus efficient flue gas purification techniques, especially for the high-temperature gas purification process, are required to protect the environment as well as the downstream units. The moving bed filter is undoubtedly one of the most effective ways to high-temperature dedust. The collecting particles in moving bed filter are usually anticipated to be used recycling, thereby an effective regenerating method for the collecting particles is particularly needed. That is to say, in a circulating system, the dusts in gas are caught in the moving bed filter and they were separated from the collecting particles in a regenerator sequentially.

In this paper, the spouting regenerating for the collecting particles of the moving bed filter is investigated. The fundamental of spouting regenerator is that the operating lift gas velocity was considerably higher than the delivery velocity of dust while fairly lower than that of the collecting particles.

Therefore, it is necessary to determine the main operating parameters of the spouting regenerator for the collecting particles of the moving bed filter. The experimental research on the regenerating efficiency of the regenerator with the collecting particles containing dust was carried in a large plexiglass cold model experimental setup. The dust-collecting particle mixture was sampled both in the spent collector inclining tube and regenerating inclining tube under different operating conditions. These effects of the operating parameters including the dust-collecting particle ratio, the gas velocity and the circulating particle flux on the regenerating efficiency were investigated, respectively. Besides, the abrasion of collecting particle in the spouting bed was measured. On the other hand, the particle size variation during the regenerating process was also analyzed. By measuring the particle size distribution and the mass frequency of the sampled mixture, the corresponding regenerating efficiency under different operating conditions was obtained. It was found that, the dust-collecting particle ratio, the gas velocity and the circulation flux were all the key operating parameters for regenerating. Increasing the dust-collecting particle ratio and the gas velocity is beneficial to the regenerating efficiency; while increasing the circulating particles flux results in an opposite effect. Moreover, the spouting regenerator can achieve the high separation efficiency under the condition of high lifting gas flow velocity. If the gas velocity too high, however, the collecting particles would also be brought away with the dust and sequentially, the magnitude of the collecting particles decreases. The spouting regenerating process inevitably leads to the collecting particles abrasion and thus a suitable gas velocity is quite vital for the whole regenerating system. Based in the experimental results, the recommended gas velocity is generally 1.1-1.3 times that of the collecting particles.

574209 Diffusion Measurements Using a Volumetric Differential Pressure Apparatus

In this work we present the first adsorption kinetic experiments carried out on an innovative differential volumetric system we have recently designed and built. The system is equipped with a differential and an absolute pressure transducer as well as thermocouples to measure the temperature of the dosing and the uptake cells, including the direct measurement of the sample temperature. Differently from conventional single-branch volumetric systems, in the case of differential systems, the methodology is based on following the differential pressure between two symmetric branches (namely, sample and reference) as the gas expands from the dosing to the uptake volumes.

Double-branch volumetric systems present some key advantages compared to conventional single-branch ones. The use of the differential pressure (as opposed to absolute) significantly improves the quality of the transient pressure signal allowing high accuracy kinetic measurements over the entire range of absolute pressure available. Despite these advantages, these type of volumetric apparatuses are not very common and limited literature is currently available. This is particularly true regarding the use of volumetric differential apparatuses for the measurements of adsorption kinetics for which, at the knowledge of the authors, no kinetic data are available in the literature.

Here we present the methodology to use the apparatus for accurate kinetic measurements over a wide range of pressures. The kinetic model developed for the conventional volumetric systems has been adapted to the new configuration to take into account the use of the differential pressure and possible asymmetries of the system.

To validate both the system and model we have carried out experiments on commercial adsorbents used for air separation using different gases in a range of experimental conditions to determine the capabilities (and the limits) of the technique for the measurements of adsorption transport kinetics.

572719 Anomalous Relationship between Molecular Size and Diffusivity in ZIF-11 By a Combined Application of Diffusion NMR, Volumetric Adsorption Method, and IR Microscopy

Zeolitic imidazolate frameworks (ZIFs), a sub-class of metal organic frameworks (MOFs), have been considered for gas separation due to their great thermal and chemical stability, large porosity, and tunable pore apertures comparable with sizes of gas molecules. ZIFs are known to exhibit significant framework flexibility that allows molecules with sizes larger than the nominal pore aperture size to enter and diffuse inside ZIF crystals. Clearly, framework flexibility can play a large role in diffusion of guest molecules inside ZIF pore systems.

In this work, ^{1}H and ^{13}C pulsed field gradient (PFG) NMR was used to study intracrystalline self-diffusion of ethane and ethylene in batches of ZIF-11 crystals with different mean crystal sizes between around 3 and 100 microns in a broad range of temperatures and sorbate concentrations. The observed agreement between the PFG NMR diffusion data measured with different nuclei types for the same molecules and samples allowed ruling out any measurement artifacts. For ZIF-11 crystals loaded with single sorbates, despite the larger size of ethane compared to ethylene, a larger intracrystalline self-diffusivity was observed for ethane than ethylene at large and intermediate sorbate loadings. It was also observed that replacing a fraction of ethane molecules by ethylene molecules reduces intracrystalline self-diffusivity of ethane. In order to further investigate this anomalous relationship between ethane and ethylene intracrystalline diffusivities in ZIF-11, uptake curves were measured using IR microscopy and volumetric gas adsorption method for single ZIF-11 crystals and thin ZIF-11 crystal beds, respectively. The corrected diffusivities obtained from these complementary measurements under the same or comparable conditions were found to be in agreement with the self-diffusion data from the PFG NMR measurements. The observed anomalous relationship between the diffusivities of ethane and ethylene in ZIF-11 is attributed to a reduction in the framework flexibility of ZIF-11 in the presence of ethylene molecules. This flexibility reduction manifests itself in a decrease in the effective ZIF-11 aperture size.

569341 Impact of MOF Pore Size on the Diffusion of CWA Simulants

Metal-organic frameworks (MOFs) have been explored for the capture and destruction of chemical warfare agents (CWAs) in protective equipment. Due to the large number of possible MOFs, a set of general predictive parameters is desired to help determine which MOFs to pursue. One such parameter is mass transfer, which can hinder a materials adsorption or catalytic performance. This study aims to provide general guidance on minimum required pore sizes for the diffusion of toxic chemicals in MOFs. The UiO family is a highly stable Zr-based MOF series, where varying linker lengths creates an isoreticular topology with a variation in pore size. This type of system parameterization will eliminate the impact of different intermolecular interactions on diffusion as a variable. Select TICs and simulants with different sizes will be tested. When comparing data sets of each adsorbate on the UiO series of MOFs, we hypothesize that there will be a step increase in diffusion rate as a function of pore size which indicates the minimum pore size necessary for effective diffusion. This could lead to generic design rules for necessary pore sizes of MOFs used to filter toxic chemicals.

573281 Integrating Material and Process Design for Kinetic Separation

Although equilibrium adsorptive separation has become one of the commonly used industrial techniques for gas and liquid separation, kinetic adsorptive separation has not received equal attention. The first known commercial application of kinetic separation is oxygen and nitrogen using Carbon Molecular Sieve [1]. Several recent studies have focused on biogas purification (CH_4/N_2) using kinetic separation [2, 3]. Recently with the development of highly tunable MOFs and ZIFs there is a great potential to exploit kinetic separation for difficult mixtures such as propane and propylene [4].

Kinetic separation exploits the differences in diffusion coefficients of the different adsorbents in the mixture irrespective of their equilibrium loading amounts. This may even result in the selective adsorption of the weakly adsorbing component for a certain duration of time[5]. Using the traditional Fickian model for gas diffusion is however not adequate for predicting this transient overshoot. We have used the Maxwell-Stefan diffusion equation which has been proposed and extensively validated with crystal diffusion experiments by Krishna et al [5]. A few past studies have attempted to screen materials and subsequently develop PSA cycles for kinetic separations incorporating this more accurate description of crystal diffusion.

There is, however, no straight forward method of screening adsorbents for kinetic separation from diffusion coefficients and isotherm data alone. This is mostly due to the complex coupling between the kinetics and equilibrium. This talk will present a parametric study of the influence of isotherm shape and diffusion coefficient on the uptake within a single crystal and light component transient overshoot. This will be used to screen materials which have better potential for kinetic separation. The selected adsorbents will then be incorporated into a PSA cycle to determine the effectiveness of the primary screening using crystal level data. Our goal is a systematic method to co-design materials and adsorption cycles for kinetic adsorption processes.

References

1. Kapoor, A., & Yang, R. T. (1989). *Chem. Eng. Sci., 44*(8), 1723-1733.
2. Erden, L., Ebner, A. D., & Ritter, J. A. (2018). *Energy Fuels*, *32*(3), 3488-3498
3. Effendy, S., Xu, C., & Farooq, S. (2017). *Ind. Eng. Chem. Res.*, *56*(18), 5417-5431.
4. Pimentel, B. R., & Lively, R. P. (2016). *Ind. Eng. Chem. Res.*, *55*(48), 12467-12476.
5. Krishna, R., & van Baten, J. M. (2018). *Microporous Mesoporous Mater.*, *258*, 151-169.

564920 Characterization of Sorption Kinetics of CO_2 from N_2-Rich Gas Mixtures Studied By Breakthrough Experiments on Zeolites

The removal of CO_2 from CO_2-containing gas mixtures is essential in many purification processes. This purification step plays a major role for cleaning of natural gas, for upgrade of bio- or synthesis gas as well as for the purification of hydrogen after a steam reforming process and removal of CO_2 from air before cryogenic air separation [1,2,3]. For dry gas mixtures this purpose can be achieved by adsorption of CO_2 on zeolites. Along with the knowledge of isotherms the comprehension of the kinetic behavior of such materials is of high interest for an optimal process design.

In this study we investigated the shape and steepness of breakthrough curves for CO_2 from mixtures with N_2 on zeolite 3A, 4A, 5A and 13X. All materials were purchased from the same supplier with same particle sizes of 1.6 mm to 2.5 mm. Superficial velocities up to 0.8 m s^{-1} were investigated. The breakthrough of CO_2 on zeolite 3A results in a spontaneous breakthrough curve due too narrow pore size in this material and therefore by kinetic-steric exclusion of CO_2. All other zeolites exhibit considerable adsorption behavior with different shapes of the breakthrough curves and some differences in the corresponding temperature profiles in the adsorber bed. Zeolites 5A and 4A have similar breakthroughs in the initial part of the curves. As we expect, one can observe the steepest breakthrough curve for the zeolite 13X material. Although the pure component isotherm of CO_2 in the pressure range of 50 mbar results in a higher sorption capacity for CO_2 on 5A zeolite, the initial breakthrough for CO_2 on 13X starts latest. This observation can be explained by the highest sorption kinetics for 13X.

Based on triples of pure component isotherms for CO_2 and N_2, mixed equilibria were predicted with simple multi component approaches and compared to the experimental results. Afterwards mass- and energy balances were used to calculate breakthrough curves and to estimate Linear Driving Force constants by fitting the model on experimental breakthrough curves.

[1] S. Cavenati, C.A. Grande, A.E. Rodrigues, *Energy Fuels,* ***20** (6),* 2648–26591 (2006)

[2] A. Möller, R. Eschrich, C. Reichenbach, J. Guderian, M. Lange, J. Möllmer, *Adsorption.,* ***23** (2-3),* 197-209 (2017)

[3] C.A. Grande, R. Blom, A. Möller, J. Möllmer, *Chem. Eng. Sci.,* ***89**,* 10-20 (2013)

579569 Studies on the Potential of Graphene-Oxide Based Membranes for Waste Water Treatment

Membranes constructed by stacking atomic layers of graphene-oxide into a large self-assembled permeable film has gathered considerable attention in the last few years. Despite many outstanding properties regarding the ability to sieve small molecules and ions in combination with high water permeance, a generic consensus is that these membranes are difficult to upscale and produce in a commercially viable manner. Over the past few years, we have pioneered a novel approach to manufacture these membranes by harnessing the liquid crystalline nature and viscoelastic properties of colloidal dispersions of graphene oxide. Using conventional printing processes such as gravure, we have demonstrated feasibility for translation to a roll-to-roll approach. In this talk, I will discuss the performance of the membranes towards removal of natural organic matter such as tannic acid, and perfluoro-octanoic acid (PFOA) - a typical compound associated with degradation of fluorinated organics. The performance of these membranes were investigated by undertaking mass balance of species, and ability to concentrate these species in a dead-end filtration mode showing 60-80 % rejection of PFOA and >95% rejection of tannic acid. Results from cross-flow filtration of tannic acid indicates excellent long term stability (> 200 h) and high rejection of tannic acid (>95%). The membranes can be cleaned by hypochlorite and are resistant to significant damage at dosage of bleach treatment (upto 1000s ppm) typically much larger than what is used in typical membrane cleaning operations. Finally, we will discuss some preliminary results on the use of the membranes in the form of 1812 membrane modules for treatment of secondary waste water. These membranes are in various stages of development, so performances discussed here are indicative of samples used & ongoing investigations for enhancing the permeance and flux as a result of various post-fabrication membrane modification steps are in progress.

564970 Hydrophobic Thin-Film Composite Membranes Coated with UV-Curable Fluoropolymer for Concentrating Pharmaceuticals and Food Additives in Organic Solvents

A UV-curable fluorinated polymer, perfluoropolyether (PFPE, Fluorolink® AD1700), is coated and polymerized onto a cross-linked polyimide substrate to modify its surface properties for organic solvent nanofiltration (OSN). Different coating concentrations, ranging from 1 wt% to 10 wt%, are applied. The PFPE coating is found to convert the hydrophilic substrate to a hydrophobic membrane and narrow the pore size of the membrane to make it suitable for OSN applications. The prepared composite membranes are tested using various organic solvents (acetonitrile, ethanol, isopropanol and hexane) and the permeances of these solvents are found to be inversely proportional to the products of their viscosities and molar volumes. The separation performance of the composite membrane coated by 10 wt% PFPE exhibits rejections of >90% to orange II sodium salt (MW= 350.32 $g\cdot mol^{-1}$) and remazol brilliant blue (MW= 626.54 $g\cdot mol^{-1}$) in isopropanol. 7-day tests are also conducted to (1) separate tetracycline from ethanol/tetracycline solutions and (2) recover hexane from hexane/-carotene solutions. The fluxes and rejections of both membranes do not fluctuate significantly during 7-day continuous tests with performance comparable with and superior to most literature data. Therefore, the PFPE coated membranes may have great potential to be the next-generation OSN membranes for industrial applications.

568012 Novel Spirocyclic Polymers for Membrane-Based Organic Solvent Separations

Solution-processable microporous polymers have the potential to outperform traditional polymeric membranes in the area of organic solvent separations. In this work, novel spirocyclic polymers were incorporated into thin-film composite membranes with the aim of improving solute rejection when compared against current spirocylic polymers such as PIM-1. The separation system of interest is a complex mixture of hydrocarbon solvents that are typically found in the light fractions of crude oil. These compounds were utilized not only to create a molecular weight cut-off (MWCO) curve, but also to assess the performance of this class of membranes in the fractionation (as opposed to purification) of complex mixtures. We find that the novel spirocyclic polymers have higher rejections than PIM-1 membranes, although the permeances are lower using our non-optimized thin film composite fabrication techniques. A MWCO of ~ 250 Da can be obtained, which is accompanied by high selectivities amongst different classes of hydrocarbons of a similar size. Moreover, high selectivities are also observed amongst molecules within the same class with only sub-nm difference in size. Such size- and shape-selective separation enables the use of these polymeric membranes in a wide range of applications.

573806 Membrane Separations with High Selectivity

We have recently discovered that a particular variety of a type of perfluoropolymer membrane, perfluoro-2, 2-dimethyl-1, 3-dioxole copolymerized with tetrafluoroethylene, (PDD-TFE), has extraordinary selectivity in separating organic solvent mixtures by reverse osmosis. For example, from a mixture of an aromatic solvent, toluene, with a variety of polar aprotic solvents such as NMP, DMSO, DMF, pure toluene can be obtained as permeate. We have been making preliminary explorations of a variety of systems using such a glassy copolymer with a very high fractional free volume. The types of systems studied include aromatic-polar aprotic, polar protic - polar aprotic, aromatic - polar protic1/2 among others. We will describe our results for these and other types of systems and seek potential explanations for such behavior.

574018 Optimization-Based Computational Tools for Precise Engineering of Channel Proteins and Organic Cages to Perform Angstrom-Scale Separations

Monodispersed angstrom-size pores embedded in a suitable matrix are promising for highly selective membrane-based separations. They can provide substantial energy savings in water treatment and small molecule bioseparations. Such membrane proteins (primarily aquaporins) are commonplace in biological membranes but difficult to implement in synthetic industrial membranes due to their modest and non-tunable selectivity. Here we describe PoreDesigner, a computational design workflow for the redesign of the robust beta-barrel Outer Membrane Protein F as a scaffold targeting of any specified pore diameter (spanning 3–10 Å), internal geometry and chemistry. PoreDesigner uses a mixed-integer linear program to optimally place long side -chain hydrophobic amino acids at the pore constriction region that yield a smaller and more hydrophobic pore by maximizing the interaction energy between the pore wall and the permeating water wire. We appended a design assessment step in each iteration by accepting only those designs that fit the user-fed pore dimensions. We first ran PoreDesigner to obtain pore sizes lesser than 4 Å that would exhibit aquaporin-like single file water transport yet maintaining high water permeation rates. 40 accepted OmpF redesigns were obtained and were classified as off-center (OCD), uniform closure (UCD), and cork-screw designs (CSD) dictated by their internal pore architecture. The narrowest pore design from each category was chosen and set in a membrane-patch and an all-atom 200ns molecular dynamics forward-osmosis simulation was performed to corroborate the PoreDesigner-predicted pore sizes. Subsequently, stopped-flow light scattering experiments on these three designs revealed complete salt rejection by the UCD mutant and an order of magnitude higher single-channel water permeabilities than any reported aquaporin till date (for all three designs). Follow-up efforts are being made to tune the membrane-pore interactions for various biomimetic membrane materials, by systematic alteration of the hydrophobicity of the membrane-facing residues without altering their pore size. This would enable easier incorporation of these redesigned proteins in 2D planar membrane sheets and serve as viable filtration assemblies for performing precise angstrom-scale separations. PoreDesigner has been made freely downloadable from http://www.maranasgroup.com/software.htm. Additionally, a similar computational architecture has been used to design complex organic cage-like molecules and their various oligomeric states have been explored to yield gyroid single-file water transport networks. These when incorporated in biomimetic membrane assemblies have been experimentally seen to result in extremely effective membrane materials that allow high water permeation rates without leaking any ions whatsoever.

577976 Pi Electron Cloud Mediated Separation of Aromatics Using Supported Ionic Liquid (SIL) Membrane Having Antibacterial Activity

Supported ionic liquid (SIL) membranes using imidazolium based ionic liquids and hydrophobic PTFE base membrane were used for the π electron mediated separation of aromatic compounds including dyes in aqueous and non-aqueous medium. The FTIR of SIL membranes showed the signature of the functionalities for both base membrane and the imidazolium moiety. A significant reduction in contact angle, especially the dynamic contact angle, was observed for SIL membranes compared to the virgin membrane indicating the enhancement in surface hydrophilicity of the membrane, while considerable changes zeta potential was seen at different pH indicating the modification in surface charge. The SEM image confirmed the pore filling of base membrane with ionic liquids. The stability of the SIL membranes were studied using hexane, water and DMF as solvents. Tuning the anionic moieties, ionic liquids were utilized for aqueous and non-aqueous application through SIL membranes. For non-aqueous applications, the trend in separation efficiency was found to be divinylbenzene > styrene > toluene, which was attributed to the more π electron cloud density on extended conjugation leading to favorable interaction with ionic liquid influencing its transfer through SIL membranes. Preferential separation of Congo red (CR) and Ramazol Brilliant Blue R (RBBR) was achieved in water based application of the SIL membranes.

The ionic liquids were having affinity for the π electron cloud. This π electron cloud density for toluene is lower than styrene followed by divinyl benzene, due to the presence of additional double bonds in conjugations with the aromatic π electron cloud. Due to this extended conjugation, divinyl benzene interacted strongly with the ionic liquid sitting inside pore. This might be responsible for bringing the selectivity in the separation for the SIL membranes. The lower transport in case of allyl ionic liquid might be attributed to the presence of electron withdrawing sp^2 hybridized carbon of the ally group resulting reduction in the electron density over imidazolium ring. On the other hand, for choloro and bromo ionic liquids, butyl and hexyl groups were present in the side chain of the imidazolium ring. The (+) inductive effect of the alkyl group containing sp^3 hybridized carbon atom pushed the electron density towards imidazolium ring. More electron cloud on the imidazolium ring, it will be more favorable for the π electron interaction resulting enhancement in the separation efficiency.

Similar investigation was also carried out to understand the selectivity in separation of different dyes using SIL membranes with water insoluble ionic liquids. In this case three different dyes CR, EBT and RBBR were used. The SIL membranes were found to be more selective for CR. In case of hexyl imidazolium ionic liquid, almost 60 % CR was transferred to the receiver side within one and half hour, while for allyl imidazolium ionic liquid % transport of CR reached around 50 % after two hours. For the other dyes RBBR and EBT, the % transport through both the SIL membranes were found to be less than 5 %. Though all these dyes have aromatic moieties, their complicated structures with different functionalities and stereo-chemical arrangement might lead to different overall electron cloud density on them resulting different selectivity through SIL membranes.

The stability of the supported liquid membranes is of major concerns. Therefore, it is interesting to investigate the stability of the SIL membranes. In view of this SIL membranes were allowed to be equilibrated in presence of water, DMF and hexane and at different time interval the conductivity of the medium were measured. For SIL membranes with water soluble ionic liquid, the conductivity of the medium was found to be modified in presence of water and DMF. In presence of water, the water soluble ionic liquids came out of the membrane pore resulting changes in water conductivity. DMF being polar in nature, certainly interacts with the ionic liquid moieties in the membrane due to electrostatic interaction. As result, ionic liquid might get dissolved in DMF. This also led to changes in conductivity in DMF, while in case of hexane, upto 60 days, there is no change in the conductivity of the medium indicating the stability of the SIL membranes. Using these equilibrated membranes (30 days' equilibration), the separation of toluene, styrene and divinyl benzenes were carried out. There modification in separation efficiency were found to be within 10 – 15 %. These investigations revealed the high stability of SIL membranes with water soluble ionic liquid in hexane medium.

The anti-microbial activity of these SIL membranes were investigated using gram-positive bacteria (**staphylococcus aureus** and **pseudomonas aeruginosa**). More the number of carbon atom on the side chain of the imidazolium ring, more antimicrobial activity was observed. For ***staphylococcus aureus***, the L/D ratio for the virgin membrane was kept 52. Over the time frame of 24 hours, the virgin membrane did not show any appreciable modification in the ratio, though some fluctuation is noticed. In case of SIL membrane a drastic decrease in L/D values were observed for all the ionic liquids upto a definite time (3 hours for bromohexane and chlorobutane ionic liquids; 5 hours for allylbromide ionic liquid), followed by almost constant values for L/D upto 24 hours. This investigating clearly demonstrates the antibacterial activity of SIL membranes compared to the virgin membrane. The antibacterial activity of allylbromide SIL membrane was found to be slightly lower than that of chlorobutane or bromohexane SIL membranes. The end point was taken as 24 hours, because beyond that, there might be some appreciable modification in the L/D ratio for the original bacteria due to their life cycle.

Similar study was also carried out with another gram positive bacteria, ***pseudomonas aeruginosa*** starting with L/D ratio ~ 56 by monitoring the L/D ratio using luminescent spectra for 24 hours with a regular time interval. The trends were found to be similar to ***staphylococcus aureus***. Though, this method is very convincing for signaturing the antibacterial activity, still it has some limitations. The bacteria having compromised membranes may be able to recover and reproduce but this method identify them as 'dead' bacteria. Similarly, some bacteria may have intact cell membrane but unable to reproduce in nutrient medium, are assigned as 'live' bacteria. Moreover, since imidazolium moiety is also able to give fluorescence, it may interfere in the determination of fluorescent intensity of the dyes indicating live and dead bacteria. Therefore, additional non fluorescent based technique should also be used to confirm the antimicrobial activity of SIL membranes.

The quantitative antibacterial analysis was carried out in terms of colony forming units per mL for bacteria. Large CFU/mL values, 2.82 E+09 and 3.5 E+09 were taken as the starting point for ***staphylococcus aureus*** and ***pseudomonas aeruginosa***, respectively. Within 24 hours, there was almost no change in the CFU/mL values for both the bacteria for virgin membrane, while significant reduction in CFU values were observed for SIL membranes. After 5 hours of contact with the membranes the CFU/mL values become 2.82E+09, 4.4E+06, 1.10E+03 and 1.2E+04 for virgin membrane, SIL membranes with allylbromide, bromohexane and chlorobutane ionic liquids, respectively; whereas after 24 hours CFU/mL became 2.80 E+09, 9.01E+04, 2.10E+0 and 2.04E+01. The investigation clearly demonstrates, not only the antibacterial activity of SIL membranes, but also the trend of allylbromide < cholorobutane < bromohexane. This trend was found to be similar than that of earlier investigation. More the number of the carbon atom on the side chain of the imidazolium ring more pronounced antibacterial activity was observed. The quantitative antibacterial analysis was carried out in terms of colony forming units per mL for bacteria. Large CFU/mL values, 2.82 E+09 and 3.5 E+09 were taken as the starting point for ***staphylococcus aureus*** and ***pseudomonas aeruginosa***, respectively. Within 24 hours, there was almost no change in the CFU/mL values for both the bacteria for virgin membrane, while significant reduction in CFU values were observed for SIL membranes. After 5 hours of contact with the membranes the CFU/mL values become 2.82E+09, 4.4E+06, 1.10E+03 and 1.2E+04 for virgin membrane, SIL membranes with allylbromide, bromohexane and chlorobutane ionic liquids, respectively; whereas after 24 hours CFU/mL became 2.80 E+09, 9.01E+04, 2.10E+0 and 2.04E+01. The investigation clearly demonstrates, not only the antibacterial activity of SIL membranes, but also the trend of allylbromide < cholorobutane < bromohexane. This trend was found to be similar than that of earlier investigation.

572368 Continuous Zeolite MFI Membranes from 2D MFI Nanosheets on Ceramic Hollow Fibers: Fabrication Processes and Hydrocarbon Separation Properties

The advent of 2D zeolite nanosheet materials has created new possibilities of accelerating the large-scale application of zeolitic membranes. In this regard, an important issue is the development of economical and scalable membrane fabrication processes that incorporate 2D zeolite nanosheets into thin membranes with high flux and selectivity. Such processes should be amenable to the use of low-cost, scalable supports that do not require any specialty materials, surface modifications, or microstructural engineering. This talk will focus on the fabrication of thin (submicron), *b*-oriented, selective MFI-type zeolitic membranes on low-cost ceramic hollow fibers (< 800 micron diameter), starting from 2D MFI nanosheet materials. First, 2D MFI nanosheet coatings are prepared on a-alumina hollow fiber supports via application of vacuum filtration. As-prepared 2D MFI nanosheet coatings are uniform and preserve the desirable (*0k0*) out-of-plane orientation; under optimized conditions no significant voids (such as delaminated areas, or uncovered areas) are found via microscopy. However, this 2D MFI coating contains uncovered regions between the stacked nanosheets where gas molecules can freely diffuse, which must be selectively filled to render the film capable of molecular sieving, while preserving the out-of-plane orientation of the molecular sieving channels. This is achieved by two-step hydrothermal treatments; the first involving a tetrapropylammonium fluoride (TPAF) structure-directing agent, and the second involving a tetraethylammonium hydroxide (TEAOH) structure directing agent. The two-step hydrothermal treatments with an appropriate choice of structure directing agents are effective in improving adhesion of the membrane on the support, closing macroscopic voids/nanoscale gaps between the 2D nanosheets, as well as highly suppressing the overgrowth/twinning of crystals and maintaining a high degree of out-of-plane *b*-orientation. This microstructurally optimized 2D MFI-based hollow fiber membranes are demonstrated to have excellent performance in a number of binary and multicomponent hydrocarbon separations, including for the separation of butane isomers and removal of natural gas liquid (NGL) components from methane. Overall, our findings have significant implications for utilizing zeolite nanosheets to accelerate scale-up and applications of thin and selective membranes.

Reference

B. Min, S. Yang, A. Korde, Y. H. Kwon, C. W. Jones, and S. Nair, "Continuous Zeolite MFI Membranes Fabricated from 2D MFI Nanosheets on Ceramic Hollow Fibers", Angew. Chem. Int. Ed., 2019, Accepted

560425 Factors Controlling Protein Transmission in Continuous Dynamic Filtration

Solids removal by membrane filtration promises several advantages over centrifugation, such as a cleaner permeate fraction, lower energy demand, and more gentle process conditions. Dynamic filtration, by rotation of the membrane, can facilitate continuous solid separation in the bio-chemical industry.

When designing membrane processes the transmission of the product through the membrane is crucial for the yield, but difficult to predict. A better understanding of the transmission can potentially lead to design of robust processes with a consistent high yield.

This project aims to describe how different process conditions affect the filtration performance and the transmission of specific proteins in a rotating filtration system. In order to separate the complexity of dynamic filtration from the variability of biological feeds, a synthetic particle system was used for the initial investigations. The system contains chalk ($CaCO_3$) particles, purified proteins (BSA), and salts.

During the experiments, the transmission in response to different operational conditions, such as particle concentration, differential pressure and membrane rotational speed, was monitored by UV. A model for the mechanisms controlling transmission was established, linking cake formation, fouling, and operating conditions to transmission. The model can be used for the design of processes with persistent high transmission.

570830 Understanding Membrane Fouling during a Combined Tangential-Flow and Depth Filtration Process

The rapid advancement in upstream cell culture operations has led to a significant increase in product titers. However, this high-level of productivity is accompanied by a rather high cell density which places a significant burden on the traditional downstream clarification and purification operations, especially during the initial clarification and capture steps. Depth filtration, centrifugation and tangential-flow filtration (TFF) have been used to remove cells, cell debris and other impurities. Depth filter consists of a thick porous bed that can trap particles within the filter in contrast to screen type filters that largely reject particulate matter based on surface filtration. On the other hand, the tangential flow of the feed to the membrane surface leads to the suppression of cake growth during TFF operation. BioOptimal MF-SL from Asahi KASEI operates as a hybrid TFF-depth filter. However, the plugging mechanism(s) for BioOptimal MF-SL and other hybrid filters are poorly understood. Here a combined pore blockage and cake filtration model has been able to describe well the fouling mechanism of this hybrid filter in the normal flow mode whereas a resistance in series model has been able to predict performance of the filter operated in the TFF mode. Significant insights have been obtained in understanding the fouling mechanism and optimal operation condition of this hybrid filter.

571742 New Multimodal Anion-Exchange Membranes for Purification of Therapeutic Proteins

Multimodal membranes bind proteins and bioprocess impurities through combinations of interaction modes that include Coulombic interactions, hydrophobic interactions, and hydrogen bonding. Because of the available modes of binding, these membranes can operate over a wider range of conditions than conventional membranes, enabling their use in chromatographic purification steps with minimal or no pretreatment of the feed.

In this study, we are evaluating the performance of Purilogics' new multimodal anion exchange (MM-AEX) membrane in bind-and-elute and flow-through bioprocess operations. We begin by presenting performance data from dynamic binding capacity measurements for human immunoglobulin G, and salmon sperm DNA for a range of conductivities (1-30 mS/cm) and pH values (5-9). After showing that these membranes have high capacities at short residence times and from low to high conductivity, we will demonstrate their unmatched performance for aggregate removal during polishing step antibody purification. Protein aggregation is a growing problem in the manufacturing of biologics due to increasing solution titers. Antibody aggregates and other impurities often are removed in a three or four-step process involving ion-exchange and hydrophobic interaction chromatography columns after an initial Protein A capture step. We present case studies in which aggregates and host cell proteins (HCP) are removed in a simplified two-step monoclonal antibody (mAb) purification scheme. In this scheme, the mAb is isolated from supernatant using a Protein A chromatography step followed by a polishing step purification using Purilogics' MM-AEX membranes operating in flow-through mode. Several mAbs were purified using the two-step purification strategy. Different buffer types (e.g., tris base, phosphate buffers) and buffer conductivities (1-15 mS/cm) were explored. HCP level and fractions of monomer, dimer and higher-order aggregates were compared before and after the polishing step. The results were compared with studies using two other leading commercial membrane products.

567496 A Membrane Process for the Purification of Nanosized Vesicles

Extracellular Vesicles (EVs) and exosomes have significant physiological properties that make their use particularly interesting in the diagnosis of different diseases, including cancer, in drug delivery and for the regeneration of tissues or wound healing, as specific therapeutic agents.

EVs can transport different biomolecules, such as lipids, proteins, nucleic acids and polysaccharides, thus playing a clear key role in intercellular communication.

The recent discovery of plant based EVs made possible the application of edible plant vesicles as carriers of many different molecules in the fields of diagnostics and advanced therapy.

The purification of EVs from biological and vegetal complex fluids has so far been based on the use of ultracentrifugation, sucrose gradient centrifugation and immunoaffinity chromatography: despite the high yield of product recovery achieved with these techniques, their application on large industrial scale is not feasible. Aim of this work is to present a more simple, less expensive and much faster membrane based process for the purification of EVs. Nanosized vesicles from *Citrus Limon* were chosen as a simple and economic system to develop the purification process. Indeed, *Citrus Limon* juice extracts represent an innovative and still poorly studied source of EVs that have very promising therapeutic use as they inhibit cancer cell proliferation on different cell lines.

The process starts from the fruits to obtain clear lemon juice, the pre-treatment consists in an extraction with a cold press juicer machine and a centrifugation to remove the heavy fibrous parts. The supernatant is recovered and treated with a cascade of membrane processes to purify and concentrate the EVs. All solutions, feed, retentate and permeate are analysed using gel filtration chromatography, technique that allows to completely isolate the EVs from the other components, demonstrating that no exosome-like vesicles are present in the permeate of the final microfiltration step. However, gel filtration chromatography is a time consuming technique, therefore a diafiltration process, using a 0.45 μm regenerated cellulose membrane, was used to replace this stage obtaining an average recovery of 95%, while the contaminants are reduced to a maximum of 4% with respect to their initial concentration.

The purified EVs were characterized using Transmission Electron Microscopy (TEM), Atomic Force Microscopy (AFM) and Dynamic Light Scattering (DLS) to determine size, size distribution, integrity, mechanical properties. Current work is dealing with the quantification of EVs in all process stages.

557535 Membrane Developments for Decellularization of Blood Via Gravity in Dry Plasma Spot Applications

Membrane developments for decellularization of blood via gravity in dry plasma spot applications

Jie Gao, Kwee Hiang Jackson Low, Chester Lee Drum, Tai E. Shyong and Tai-Shung Chung*

Department of Chemical & Biomolecular Engineering,

National University of Singapore, Singapore 117585

*Corresponding author

Tel: +65-65166645; fax: +65-67791936; Email: chencts@nus.edu.sg

We describe the use of advanced materials to cheaply fabricate a pre-analytical sample stabilization matrix. There is a strong need for a robust blood & plasma-collecting device, engineered to stabilize small molecules, antigens, antibodies, DNA or other biomarkers of interest, at the point of collection, at an affordable price, utilized for surveillance and mass screening of diseases or to monitor outbreaks. Thus, we proposed to decellularize and dehydrate plasma with a combined system made from a piece of the thin membrane and a water-drawing adsorbent layer. When specimen blood is dropped onto the device surface, blood cells will be retained and prevented from entering the membrane matrix due to size exclusion. Driven by gravity, the liquid plasma will penetrate into the plasma drawing layer with all the biomarkers being retained in the layer.

Commercialized membranes designed to remove blood cells from blood include the Noviplex card from Shimadzu and Vivid® membranes from Pall Corporation. However, the systems and membranes are relatively expensive. Development of high-efficiency membranes for decellularization via gravity at affordable prices is very meaningful, especially for dry plasma spot (DPS) applications. In the study, we have developed polyacrylonitrile (PAN) membranes which are able to recover plasma as high as 28.05 % with zero retention to glutamine by using gravity as the driving force. The developed membranes have a higher plasma recovery as compared to the commercialized membranes, which have plasma recoveries of 17.04 % or less. Novel membranes that can provide point of care separation of blood components may have great potential in microvolume sample collection applications.

578284 Macromolecular Modulation of Calcium Oxalate Nucleation and Growth

The crystal nucleation, growth, and agglomeration in solution is strongly influenced by the presence of soluble molecules in minute quantities. In this work, we study the crystallization kinetics of calcium oxalate, the major constituent of kidney stones, by employing a constant composition potentiostatic technique. We evaluate the effects of supersaturation, temperature, and additives on the nucleation and growth of calcium oxalate in aqueous solutions. We have demonstrated that certain anionic polyelectrolytes can be very effective toward kidney stone inhibition by reducing the growth rate, altering the crystal morphology, and favoring the formation of metastable forms due to adsorption on active growth sites. We compare our results to citrate, a naturally occurring inhibitor in the urinary system and currently the standard pharmacological treatment for stones. Our findings offer a better understanding of the calcium oxalate crystallization kinetics and provide information on the viability of synthetic polymers as potential calcium oxalate stone therapies.

564064 Direct Measurement of Growth Rate Dispersion in the Reactive Crystallization of Ampicillin Trihydrate

Direct measurement of growth rate dispersion in the reactive crystallization of ampicillin trihydrate

Matthew A. McDonald*, Andreas S. Bommarius*, Martha A. Grover*, and Ronald W. Rousseau*

School of Chemical & Biomolecular Engineering, Georgia Institute of Technology, Atlanta, USA

Keywords: Growth rate dispersion, ß-lactam antibiotics, penicillin G acylase

In this study, the growth rates of individual ampicillin trihydrate crystals in isothermal and quiescent conditions were measured by optical microscopy. Supersaturation was generated by the penicillin G acylase (PGA) catalyzed reaction to synthesize ampicillin from 6-aminopenicillanic acid (6-APA) and phenylglycine methyl ester (PGME) [1]. The growth rate of ampicillin trihydrate crystals under the same conditions showed considerable variation, an example of growth rate dispersion (GRD). Quiescent conditions and isothermal generation of supersaturation allowed nucleation, breakage, and heat transfer to be eliminated as sources of GRD. Curiously, the crystals showed zero growth at low supersaturation; growth was activated only beyond a threshold supersaturation, possibly due to the presence of other species such as the reactants 6-APA and PGME. The growth rate and threshold supersaturation are correlated, and both also correlate with seed crystal size. Changes in solubility of ampicillin due to 6-APA or PGME were also ruled out, leading to the conclusion that each crystal has an intrinsic growth rate dependence on the supersaturation. [2]

[1] McDonald, M. A., Bommarius, A. S., and Rousseau, R. W. (2017) Enzymatic reactive crystallization for improving ampicillin synthesis, *Chemical Engineering Science 165*, 81-88.

[2] McDonald, M. A., Bommarius, A. S., and Rousseau, R. W. (2019) Direct measurement of growth rate dispersion during reactive crystallization of ampicillin trihydrate, *Processes* IN PREPARATION

572686 Dynamics of Struvite Formation: Insights from the Macroscopic to Molecular Scale

The formation of so-called *infection stones* arising from urinary tract infections is one of the most difficult and dangerous stone diseases due to its large size, rapid growth, and high recurrence. Infection by urease positive bacteria leads to a cascade of reactions that elevates the pH, as well as concentrations of NH_4^+ and PO_4^{3-} ions, resulting in precipitation of struvite ($MgNH_4PO_4 \cdot 6H_2O$). These crystals have clinical implications as approximately 50% of all long-term catheterized patients experience catheter blockage that leads to more serious complications due to reflux to kidneys by urine flowing back[1].

Research efforts have focused on understanding the effects of different parameters (e.g., supersaturation, pH, modifiers) on crystallization of struvite; however, conventional batch experiments are subject to limitations, which include their ability to simulate the flow conditions in urinary tract system, notably under conditions that permit the evaluation of crystal growth in real time. To this end, we have established a new platform to evaluate struvite growth at both macroscopic and microscopic length scales in the absence and presence of various modifiers.

Here, we will present the results of integrated microfluidics and *in situ* atomic force microscopy (AFM) measurements. Our findings demonstrate how these techniques serve as a useful platform to capture the time-resolved dynamics of crystal growth. A unique microfluidics platform creates an environment that mimics the flow conditions of a urinary tract system, while offering distinct advantages of bulk assays under quiescent conditions. Seeded growth decouples nucleation and growth, while the continuous supply of growth solution maintains constant composition without the depletion of solutes or the reduction in pH. We compare struvite growth under static and flowing conditions and evaluate growth kinetics and morphology transformation as a function of supersaturation. In parallel, we conduct *in situ* AFM measurements to elucidate the molecular mechanisms of struvite growth in the presence and absence of modifiers. As will be discussed, these integrated approaches have allowed us to identify modifiers that inhibit struvite formation and rationalize their molecular mechanisms.

1. Stickler, D. J. (2014). Clinical complications of urinary catheters caused by crystalline biofilms: something needs to be done. *Journal of Internal Medicine*, *276*(2), 120-129.
2.

tetrasodium pyrophosphate in artificial urine: chemical and physical aspects of nucleation and growth. *Crystal Growth & Design*, *16*(6), 3519-3529.

567190 Comparative Screening of Polymorphs of o-Aminobenzoic Acid Using Microtiter Plates and Novel Continuous-Flow, Microfluidic Devices

Screening of polymorphs, morphologies, and crystallization kinetics is crucial for the development of crystallization processes for the synthesis of materials such as pharmaceuticals, agrochemicals, catalysts, semiconductors, and metal-organic frameworks. The existing screening techniques use either microtiter plates or droplet-based microfluidic devices consisting of small compartments in which the crystals nucleate and grow while supersaturation depletes continuously affecting polymorphs and morphologies. Setbacks are often encountered during the scale-up of crystallization process due to the effect of varying supersaturation in screening evaluations. Therefore, a screening technique where supersaturation can be controlled is much needed for the effective design of crystallization processes. Additionally, such constant-supersaturation screening technique is an indispensable tool to boost innovation in continuous manufacturing, also supported by the U.S. Food & Drug Administration. Here we analyze five different types of continuous-flow microfluidic devices including conventional T-junction, cross-flow, and cell sorter devices; and novel H-shaped, and cyclone mixer designs for their ability to trap crystals that are nucleating and growing in a constant supersaturation. These continuous-flow microfluidic devices have three connected sections- thermalizer, mixer, and diffuse, whose optimal dimensions and operating conditions are determined using Multiphysics simulations. A facile technique is also shown to fabricate such microfluidic devices using a stereolithographic3D printer. Although T-junction, cross-flow, and cell sorter devices are able to operate at constant supersaturation near stagnation point, the crystals are constantly dragged away from the trap zone and hence such devices cannot be used for screening. The H-shaped design relies on countercurrent diffusion of solvents to create stable supersaturation gradient along the mixer, which can be used to screen proteins or metal-organic framework with longer induction time and slower growth kinetics. The cyclone mixer design with multiple inlets is demonstrated to not only operate at constant supersaturation but also trap crystals for *in situ*analysis of polymorphs and morphologies. The polymorphs and morphologies of o-aminobenzoic acid (o-ABA) at two different supersaturations are analyzed using cyclone mixer design and a traditionalmicrotiter plate. While the polymorphs and morphologies of o-ABA are affected by depleting supersaturation in a microtiterplate, the cyclone mixer design shows consistently stable and metastable forms of o-ABA. We expect the H-shaped and cyclone mixer designs will advance innovations in crystalline materials research in various applications.

575601 Environmentally Friendly Molecular Modifiers Reveal Unique Inhibition and Dissolution Mechanisms in Barite Scale

Adsorption and subsequent precipitation of sparingly soluble compounds during oil and gas production, water treatment, and manufacturing processes may severely damage equipment and cause reduction or loss of finished product, thus posing serious challenges for these industries. Designing effective biodegradable chemical treatments to reduce or eliminate scale formation requires an understanding of the molecular-scale interactions of chemical additives in supersaturated and undersaturated environments. Here we apply a unique approach to design and test environmentally friendly chemical treatments for scale formation using a combination of surface science techniques, microfluidic technology, and molecular modeling. Crystallization can be inhibited in highly supersaturated solutions and dissolution can be promoted in undersaturated media using dilute quantities of modifiers. The precise effect of hydrodynamics, which alters modifier-crystal interactions, on inhibitor and dissolver efficacy remains elusive. Our studies focus on barium sulfate (barite), a highly insoluble and common scaling mineral. In this presentation we will describe a microfluidic platform that is used to perform multiple bulk crystallization assays simultaneously and capture real time crystal inhibition and dissolution in the presence of molecular additives. Using *in situ* scanning probe microscopy, we explore the interfacial interactions between the modifiers and different surfaces of barite over a range of undersaturated and supersaturated conditions to elucidate the role of these putative inhibitors/dissolvers during the growth and dissolution phases. Concurrently, we use molecular modeling of inhibitor adsorption of barite surfaces to probe the interactions that govern modifier-crystal specificity and efficacy. Collectively, these studies generate new understanding of the mechanisms of crystal growth modifiers that can be applied to the design of improved scale treatments.

564661 Nucleation Behavior of Ethyl Vanillin: The Partial Dependence of Kinetic Parameters on Interfacial Energy

The experimental metastable zone width (MSZWs) of ethyl vanillin (EVA) in ethanol, n-propanol and n-butanol were measured by the polythermal method to describe the nucleation behavior in its unseeded cooling crystallization. In accordance with the classical nucleation theory (CNT), the dependences of MSZWs on saturation temperature T_0 and cooling rate R were mainly discussed via Nývlt's semiempirical model and modified Sangwal's theory, respectively. Meanwhile, the pre-exponential factor A, crystal-liquid interfacial energy γ and the corresponding critical parameters r_{crit} (critical nucleus size), G_{crit} (critical Gibbs free energy) can be obtained. It is interesting to find that γ increases linearly with decreasing T_0 and increasing R, while the value changes of A do not have regularity against thermodynamics and kinetics. As the result, the increase or decrease of nucleation rate J mainly depends on the increments of γ more than the slight variation of A. Furthermore, it suggests that the thermodynamic dependence of γ is greater for the system with higher supersaturated concentration of solute.

579823 Scaling up PSA for Renewable Energy Applications

This presentation will address scale-up of pressure swing adsorption (PSA) systems. It focuses on systems used for upgrading biogas and landfill gas to produce renewal fuels. First, the nature of those applications is reviewed: raw gas compositions, adsorbents, operating conditions, and generic PSA cycles. Next, a step-by-step engineering procedure will be explained, and the steps will be illustrated by a case study. The case is a two-stage PSA system for splitting CO_2, H_2O, VOCs, siloxanes, N_2, and O_2 from landfill gas to obtain pipeline grade gas, CNG, or even LNG. It has been possible to scale-up from 1 inch diameter to 9 ft diameter adsorbers in one step. The corresponding volumetric scale-up factor has typically been more than 10,000 in one step.

This presentation will cover adsorbent types, properties and selection criteria, process flowsheets, design principles, and performance. There will be time for questions and answers.

580103 23 Years of Adsorption and Ion Exchange: Lessons Learned

The intent of this talk is to impart useful knowledge, tips, tricks, and identify potential pitfalls that have been assembled after spending the last 23 years as a corporate specialist in the area of sorptive separations. These are lessons learned developing, troubleshooting, optimizing and scaling up a myriad of adsorption processes, both liquid phase and vapor phase, across a very wide range of chemistries. While this presentation will focus primarily on liquid phase separations, often the principles and learnings apply to both liquid and vapor phase systems. Some (sanitized) real life examples will be presented.

574919 Accurately Modeling Liquid Phase Adsorption for Industrial Applications with Bench Scale Testing

Granular activated carbon (GAC) has been proven to remove contaminants in vapor and liquid phase applications for decades. Vapor phase adsorption data has been well documented in literature and modeling exists for the design of vapor phase applications. Liquid phase applications, however, are more difficult to model due to the complex interactions present in liquid chemistry. The contaminants of interest are in a matrix of organic and inorganic compounds that interfere with adsorption. Therefore, laboratory testing is suggested to determine the effectiveness of carbon for specific applications, the size necessary for an industrial scale treatment system, and the carbon usage rates in full-scale systems to optimize removal and cost.

Lab scale testing can be divided into two categories; feasibility and process simulation. Feasibility testing is typically accomplished with an isotherm test. An isotherm is a batch test designed to demonstrate the degree to which a particular dissolved organic compound (adsorbate) is adsorbed on activated carbon (adsorbent). For a single adsorbate, a Freundlich equation can be used to estimate the carbon usage rate at equilibrium loading conditions. Continuous flow tests such as small-scale column tests and pilot studies simulate not just the thermodynamic adsorptive properties demonstrated in the isotherm, but they also allow for an accurate representation of the kinetic factors which effect adsorption in a dynamic system. The adsorption breakthrough curve obtained is scaled to accurately represent treating an aqueous stream in a large adsorption system.

The results from this testing provide sufficient data to determine the most economical approach for installing and operating GAC for a wide range of liquid phase applications.

573899 Development of Adsorption Processes for the Separation of Close-Boiling Gas Mixtures

Distillation is the most widely used process for the separation of homogeneous mixtures in industry. However, it is very expensive when the components to be separated have similar volatilities. This issue is even more pronounced for highly volatile gas mixtures which do not condense at ambient conditions. For such mixtures, very high pressures or low temperatures, or a combination of both, are required in distillation systems, thus significantly increasing the cost of equipment as well as operating costs.

Efforts to increase the energy efficiency and reduce costs associated with these types of processes have primarily focused on improving or optimising existing technologies, leading to highly heat-integrated and complex flowsheets. This approach, although leading to improvements in process performance, is inherently limited in its potential to bring about energy savings, as the core technology is the same. Novel approaches are required to truly revolutionise the performance of such processes.

One possible alternative is the use of pressure or temperature swing adsorption to achieve the required separation. As separation by adsorption is based on differences in chemical interactions between the components in the mixture and a solid adsorbent, rather than volatility differences, adsorption-based flow sheets could potentially bypass bottlenecks seen in distillation processes.

In this work, we investigate the potential of adsorption processes as alternatives to cryogenic distillation-based processes for the separation of close boiling gas mixtures using a systematic two-stage approach. In the first stage, simple shortcut adsorption process models are used for quick screening and ranking of candidate adsorbents, and to identify the most appropriate cyclic process (i.e. PSA or TSA) for a given separation. Detailed dynamic adsorption column models are then used for more in-depth evaluation of promising adsorbents and separation schemes identified in the first stage. This approach allows for suitable adsorbents and separation schemes to be identified with minimal computational cost, and without recourse to cyclic adsorption experiments, which could be very costly if novel adsorbents are being investigated, and are time-consuming in any case.

The separation of CO – a major feedstock in the production of acetic acid – from syngas containing N_2 is the case study in this work. We investigate the potential of metal-organic frameworks (MOFs) – a class of highly-tuneable adsorbent materials – for high purity CO recovery from N_2 containing mixtures. Conventional cryogenic distillation-based processes for CO production are restricted to low N_2 feeds as energy costs are prohibitive for high N_2 feeds: it is difficult to separate CO and N_2 by distillation as they have very similar physical properties. Adsorption-based flowsheets could potentially result in lower energy costs for traditional feedstock, as well as potentially allowing the use of cheaper feedstock with higher N_2 content, such as the off-gas from steel production, for the production of CO.

Results from the benchmarking of existing cryogenic distillation-based technology for CO production, systematic screening of several MOFs for CO/N_2 separation, and dynamic modelling of adsorption processes for CO/N_2 separation will be presented. The effect of the selected process scheme, choice of adsorbent material and operating conditions on the energy demand and capital cost of the adsorption-based flowsheets are investigated. The performance of the adsorption-based flowsheets is compared with conventional technology.

557563 Advancements in Scaling up Metal Organic Framework (MOFs) Materials

Metal Organic Frameworks (MOFs) are a new class of highly versatile and adaptable adsorbent nanomaterials. Since their discovery in the 1990s there has been an explosion of research efforts to create and characterize new structures and develop new applications in areas including storage, separation, filtration, capture, and controlled release of gases and liquids. MOFs have unsurpassed surface areas and their pore size and functionality can be tailored with unprecedented precision. The potential of selecting from hundreds of metal or metal organic nodes and thousands of organic linkers creates possibility to architect nearly an infinite number of theoretical MOF structures. To date, thousands of structures have been realized by research institutions many showing properties that outperform traditional adsorbents. Unlike the early structures, the new generation of MOFs exhibit the thermal, mechanical, and chemical stability necessary for practical use in industry. As the interests in using these advanced materials in real-world applications grows, there are pressing needs to scale-up promising MOFs from few grams produced in academic environments to high volumes. In this paper, challenges and lessons experienced by NuMat Technologies in efforts to successfully scale up and manufacture several MOFs will be discussed. In addition, advancements in the formation techniques required to densify and compress small MOF particles in to workable granules, pellets, and extrudates will also be reported.

557927 An Adsorbent Screening Tool with Integrated Process Economics for Carbon Capture By PVSA

There is significant interest in the development of step-change technologies for CO_2 capture, with adsorption processes in general and MOF-based systems in particular attracting a lot of attention. Although there are several screening models available, none attempt to quantify the process economics[1-4]. In order to prioritise research efforts in this important area, the development of screening tools which link fundamental scientific insights with engineering considerations is key.

In this work, we use a reduced order pressure/vacuum swing adsorption model[5] which models a one bed, three step cycle made up of feed, desorption and feed re-pressurisation steps. This allows for the non-isothermal nature of adsorption to be accounted for, yielding more reliable results than simple isotherm analysis, while still allowing a rapid solution.

The cycle performance results based on one kilogram of adsorbent are then taken as the input to the process design tool. The number and size of the process equipment is then determined for a user specified amount of feed gas, and the capital and operating costs are estimated to obtain the cost of capture per tonne of carbon dioxide.

Having the cost of capture as a screening metric is beneficial on several fronts. It allows an adsorbent to be selected based on overall process cost in addition to traditional parameters such as purity, recovery, working capacity and specific work. The relationship between adsorbent characteristics, separation performance, and capture cost can also be uncovered which allows additional insights. For example, in the case of a 150 kPa_a feed of 4 $\%_{mol}$ CO_2, reducing the desorption pressure to 1 kPa_a from 15 kPa_a for ZIF-8 results in a 44 % and 57 % increase in CO_2 purity and recovery respectively, however the capture cost increases by 3.8 times. In the case of Ni-MOF-74 which displays good isothermal performance, the cyclic performance is poor which results in CO_2 purities and recoveries 5.7 and 3.0 times lower than 13X and a capture cost 3.9 times higher at a desorption pressure of 3 kPa_a.

[1] G.D. Pirngruber *et al.*, *ChemSusChem*, **5**, 762-776, 2012

[2] L. Joss *et al.*, *Ind. Eng. Chem. Res.*, **54**, 3027-3038, 2015

[3] M. Khurana, S. Farooq, *Ind. Eng. Chem. Res.*, **55**, 2447-2460, 2016

[4] V.S. Balashankar *et al.*, *Ind. Eng. Chem. Res.*, **58**, 3314-3328, 2019

[5] B.J. Maring, P.A. Webley, *Int. J. Greenh. Gas. Con.*, **15**, 16-31, 2013

564946 Bi-Functionalized Ionic Liquid/MOF Composite Material for Efficient Carbon Dioxide Capture and Conversion

To alleviate environment issues derived from CO_2 emission, it is significant to develop an economical and effective means of CO_2 capture and conversion. Ionic liquids (ILs) possess remarkable properties such as low volatility, excellent thermal stability and widely tunable structure, which have been used as CO_2 absorbents and conversion catalysts. However, most of ionic liquids have the characteristic of high viscosity, which limits the mass transfer of guest molecules. Metal-organic framworks (MOFs) have recently attracted significant attention because of their large surface area, high porosity and inherent functionality. Therefore, combination of ILs and MOFs is a meaningful strategy to enhance capture ability and catalytic performance.

In this study, bi-functionalized ionic liquid immobilized on MOF material was developed for CO_2 capture and conversion. First, bi-functionalized ionic liquid 1-(2-carboxyl-ethyl)-3-methoxyethyl-imidazolium chloride ([CeMOEim][Cl]) and metal-organic framework MIL-53(Al) were respectively synthesized. Furthermore, [CeMOEim][Cl] was immobilized on as-synthesized MIL-53(Al) *via* ultrasonic-assisted impregnation method. Chemical structure, microstructure, and thermal stability of as-synthesized materials were characterized by NMR, FT-IR, XRD, SEM, TEM, N_2 adsorption-desorption measurement and TG analysis. Finally, [CeMOEim]/MIL-53(Al) composite materials with different loading rates were used as CO_2 sorbent and catalyst. For comparision, capture capacity and catalytic performance of the functionalized ionic liquid and MIL-53(Al) were also studied.

-

569641 Capture and Methanation of CO_2 Using Sodium Promoted MgO Based Catalytic Sorbent

A combined cycle of capture and conversion of CO_2 performed under isothermal conditions may be an advantageous process intensification approach compared to traditional CO_2 capture and sequestration in that the combined cycle can remove energy intensive temperature or pressure swing steps and minimize process units.[1] Furthermore, the produced fuel source from conversion of CO_2 can be directly used again as an energy source for power generation. A first generation of catalytic sorbents, such as $Ru+CaO/Al_2O_3$, for such an application has previously been reported.[1,2] In this work, sodium promoted MgO and loaded with ruthenium metal was used for capture and methanation of CO_2 in a combined cycle. The MgO support with various $NaNO_3$ and $NaNO_2$ promoter loadings was tested for CO_2 capture at 250 °C. MgO_$NaNO_3$ (1:0.1 molar ratio) showed adsorption capacity of 7.6 mmol CO_2/g in 12 h. MgO_$NaNO_3$_$NaNO_2$ (1:0.1:0.1 molar ratio) showed adsorption capacity of 8.6 mmol CO_2/g in 12 h. To evaluate performance of these materials in a capture and methanation cycle, the methanation capacity, which is equivalent to number of moles of methane produced per mass of catalytic sorbents in a cycle of capture and methanation, was measured for the prepared catalytic sorbents. It was found that a physical mixture of MgO_$NaNO_3$ and Ru supported MgO showed the highest methanation capacity at 250 °C with 1.1 mmol CH_4/g. This methanation capacity is comparable to the highest methanation capacity reported up to date, making sodium promoted MgO sorbent promising materials for such applications.

Reference:

(1) Duyar, M. S.; Treviño, M. A. A.; Farrauto, R. J. Dual Function Materials for CO_2 Capture and Conversion Using Renewable H_2. *Appl. Catal. B Environ.* **2015**, *168–169*, 370–376.

(2) Duyar, M. S.; Wang, S.; Arellano-Treviño, M. A.; Farrauto, R. J. CO_2 Utilization with a Novel Dual Function Material (DFM) for Capture and Catalytic Conversion to Synthetic Natural Gas: An Update. *J. CO_2 Util.* **2016**, *15*, 65–71.

570090 Amine-Based Latex Coatings for Indoor Air CO_2 Control in Commercial Buildings

Amine-Based Latex Coatings for Indoor Air CO_2 Control in Commercial Buildings

Anirudh Krishnamurthy, Ali Rownaghi, Thomas Schuman, Fateme Rezaei

Department of Chemistry, Missouri University of Science and Technology, 1101 N State Street, Rolla, MO, 65409, United States

Abstract

High levels of indoor air CO_2 in commercial buildings can lead to various health effects, commonly known as sick building syndrome. Passive control of indoor air CO_2 through solid adsorbents incorporated into the paint offers a high potential to handle CO_2 without utilizing much energy. This study focuses on incorporating silica supported aminopolymers into a polyacrylic based latex that could be used as a buffer material for passive control of CO_2 in enclosed environments. To maximize the effect of the pigment (adsorbent), paints were all prepared at critical pigment volume concentration (CPVC) levels. CO_2 at 800 ppm and 3000 ppm were used to asses both low and high level contamination. The removal efficiency of the surface coatings was evaluated within typical time frames (10 h for adsorption and desorption). Our lab-scale chamber results indicated that silica-TEPA based paint with 70 wt% loading exhibits the best adsorption performance, comparable to that of powder-based sorbent, with only a ~20% decrease in adsorption efficiency. Our results also revealed that optimization of paint formulation is critical in passively controlling indoor air CO_2. The findings of this study highlight the potential of amine-based adsorbents as pigments in high PVC paints for indoor CO_2 control in commercial buildings.

571116 Combined Capture and Utilization of CO_2 in Ethylene Production over Dual-Function Materials

Finding cost-effective approaches to capture and utilize the emitted CO_2 could open new avenues to the sustainable creation of energy with less environmental footprint. Combined capture -utilization approach offers a promising route to use waste CO_2 as a feedstock for producing valuable commodity chemicals and fuels. In this investigation, we report *in-situ* capture and utilization of CO_2 in ethylene production through oxidative dehydrogenation of ethane over dual function materials (DFMs). The DFMs considered for this application consisted of double salts (K-Ca and Na-Ca) as adsorbent, Cr nanoparticles as catalyst, and γ-Al_2O_3 as support. The structural and physiochemical characteristics of the materials were assessed by N_2 physisorption, XRD, XPS, TPD, and H_2-TPR. The adsorption step was conducted at 650 ºC using a feed gas mixture of 10%CO_2/Ar followed by the reaction step at the same temperature using ethane/Ar at WHSV of 5000 mL/g.hr. The effects of CO_2/C_2H_6 mole ratio and bed temperature on catalytic activity of the DFMs were systematically investigated to determine the optimum process conditions. In addition, the properties and compositions of the DFMs were correlated with their adsorption and catalytic performance to determine the best DFM that yields the highest CO_2 removal efficiency, ethane conversion and ethylene selectivity.

575115 Batch Adsorber Analogue Model Based Screening of Large Adsorbent Databases for Post-Combustion CO_2 Capture

Advancements in molecular modeling have made it possible to synthesize hypothetical materials and predict the adsorption behaviour of gases. Hence some large databases containing hundreds of thousands of such materials have been reported. Screening such large databases to find suitable candidates can be a challenge. Recent studies have shown that adsorption metrics, such as selectivity, working capacity are not good predictors of the ability of an adsorbent's performance when subjected to a process [1,2]. In this work, a simplified proxy-model called Batch adsorber analogue model (BAAM) is introduced [3]. In this model, the adsorption column is represented as a well mixed batch adsorber with the adsorbent and the fluid phase in equilibrium. This assumption reduces the system of partial differential equations to a system of two ordinary differential equations, thereby reducing the computational complexity. A 4-step cycle with LPP, which is shown to provide a low parasitic energy is mimicked using this simplified model. Owing to the assumptions made in the model development, the BAAM does not provide an accurate representation of the key performance indicators, e.g. Purity, recovery and energy consumption. However, this restriction was overcome by calibrating the the model's output against results from detailed models. This procedure allows computing the Purity, recovery and energy consumption for the concentration of CO2 from a mixture of CO2 and N2. It is shown that the BAAM can predict the ability of an adsorbent to satisfy US-DOE capture requirements (95% CO2 purity with 90% recovery) for 83% of the adsorbents and the energy consumption can be estimated within 15% accuracy. Finally, the BAAM also results in a simple graphical approach

Using the BAAM, two databases were screened. The first one, the NIST/ARPA-E database containing experimental isotherms and the Carbon capture material database (CCMDB) containing over 100,000 hypothetical Zeolites and ZIFs. The top performing materials from each of these database were further studied using detailed process models combined with optimization algorithms which show an excellent reliability of the BAAM. The screening of such a large database also provides key insight about the ranges of properties for effective adsorbents.

1. Rajagopalan, A. K.; Avila, A. M.; Rajendran, A. Do adsorbent screening metrics predict process performance? A process optimization based study for post-combustion capture of CO2. Int. J. Greenhouse Gas Control 2016, 46, 76-85.

2. Farmahini, A.H., Krishnamurthy, S., Friedrich, D., Brandani, S., Sark- isov, L.: From crystal to adsorption column: challenges in multi- scale computational screening of materials for adsorption separation processes. Ind. Eng. Chem. Res. 57, 15491–15511 (2018)

3. Subramanian Balashankar, V., Rajagopalan, A. K., de Pauw, R., Avila, A. M., & Rajendran, A. (2019). Analysis of a Batch Adsorber Analogue for Rapid Screening of Adsorbents for Postcombustion CO2 Capture. Ind. Eng. Chem. Res. 58, 3314–3328 (2019)

562894 Liquid-Liquid Based *in Situ* Recovery Systems Enhance Bioproduct Production and Enable New Bioprocesses

The economics of a bioprocess are generally driven by the expense of the downstream separation of product from dilute aqueous streams. Continuous separation techniques, such as *in situ* product recovery (ISPR), are attractive in that they can simultaneously remove and concentrate products from bioreactors, thereby minimizing product inhibition and downstream separation cost. Although there are multiple benefits of ISPR, there are few reports of its integration with further downstream processes to recover highly pure bioproducts. This seminar will cover recent work in developing a robust ISPR system to recover carboxylic acid products from sterile monocultures, and from non-sterile anaerobic digestors. Results will be presented showing the enhanced effects the system has on microorganism titers, rates, and yields. In one application the system was able to reduce fermenter size by a factor of ~10 compared to a standard fed batch fermentation. Additionally, the system's application to an anaerobic digestor is found to effectively arrest methanogenesis producing higher value volatile fatty acid products that can be used as platform chemicals in lieu of methane.

567698 Design and Performance of Amine Based Solvents

Liquid-liquid extraction of biobased products such as acetic acid, hydroxypropionic acid etc. with tertiary amines is seemingly on a rise again. Reactive solvents based on tertiary amines have not established in acid extraction with distillative recovery due to strong chemical interaction between acid and amine. However, applications in liquid-liquid extraction with caustic recovery may gain attention in industrial applications. The design and performance of triisooctylamine based solvent mixtures in liquid-liquid extraction of the bulk chemical acetic acid was investigated.

Amine based solvents need admixture of modifiers and diluents to correct the liquid-liquid partition characteristic of aqueous acids with the pure extractant, and to correct the physical properties of the laden solvent by adjusting density, interfacial tension and viscosity, and to avoid third phase formation.

To gain data about the effect of solvent composition on liquid-liquid partition and mass transfer of acetic acid with tertiary amine based solvents liquid-liquid extraction of acetic acid was investigated with solvent mixtures comprising triisooctylamine (TiOA), dissolved in the industrial diluent Shellsol T (SST), and modified with either the Lewis basic modifier tributyl phosphate (TBP) or the Lewis acidic modifier isodecylalcohol (ISO). The modifiers were investigated for their potential of adjusting the association of acetic acid and triisooctylamine. Investigations considered the determination of liquid-liquid equilibrium data and the effect of solvent composition on mass transfer. The liquid-liquid equilibrium data were obtained from temperature controlled shaking funnel experiments. The mass transfer data were obtained from single droplet mass transfer experiments. Phase transfer of acetic acid starts with protonation of TiOA at the interface, followed by salt formation with the acetate anion and transfer of the triisooctylammonium acetate salt into the solvent phase. In principle the salt may stabilize the interface and retard mass transfer of the constituent acetic acid. On the other hand enhancement of phase transfer may be expected from neutralisation of acetic acid with triisooctylamine. To account for mass transfer phenomena, different solvent mixtures as well as different acid concentration in the aqueous phase and the effect of preloaded solvent on mass transfer were investigated. By preloading the solvent with the solute the experimental validation of extended droplet residence time was simulated. Due to salt formation enhancement of mass transfer might be expected. However high viscosity of the solvent and the low diffusion coefficient of the amine in the solvent exclude any enhancement. Data analysis of mass transfer experiments resulted in the conclusion that mass transfer of aqueous acetic acid in amine based solvent mixtures is simply controlled by diffusion of the constituent. Solvent viscosity, adjusted by mixing of appropriate amounts of solvent constituents controls mass transfer of acetic acid in the amine based solvent droplet. Whether stabilizing effects, provided by the "surfactant" TiOA and the reaction product triisooctylammonium acetate, or just viscosity or both effects lead to a significant difference between the dynamically obtained solvent load from single droplet mass transfer experiments and the liquid-liquid equilibrium partition, cannot be concluded from experiments.

559226 Direct Numerical Simulation of Liquid-Liquid Droplet Interactions

Due to its lower costs regarding money and time compared to performing experiments, computer simulations grow in importance when developing industrial separation processes. On plant scales, process simulators are widely applied and on particular scales, Computational Fluid Dynamics (CFD) calculations can support design. However, those CFD simulations often do not incorporate information about the thermodynamic equilibrium that constitutes the achievable efficiency of any separation process. Regarding extraction processes, it is usual to use experimentally adjusted droplet population balances [1] or phase field models based on non-thermodynamic expressions for the free energy (e.g. $f = (x-1/2)^2(x+1/2)^2$, [2]) in multi-phase simulations. On the one hand, droplet population balance based models do not consider the microscopic interfacial behavior and on the other hand, phase-field models that rely on an arbitrary two-minimum potential will not achieve the thermodynamic equilibrium.

Regarding mixtures, applying a phase building component's concentration as phase field [3] together with a so-called influence parameter coming from Density Gradient Theory (DGT) [4] leads to a full physical representation of the interfacial behavior. To adjust the influence parameter, the model requires only one experimental interfacial tension. The DGT accounts for the impact of concentration gradients on chemical potentials and thereby defines the value of chemical potentials in the non-stable phase diagram region. The non-steady DGT [5] together with the momentum balance equations form the system of Cahn-Hilliard/Navier-Stokes equations. However, the fact that CFD requires a closure in form of a relation between density and pressure still holds. When considering liquid-liquid extraction, it is convenient to assume a mixture of two viscous, incompressible liquids, where the density just depends on concentration but not on pressure. Hence, the component balance determines also the overall density. The task then is, to calculate the pressure field in a way that mass conservation holds. Such a mixture is usually referred to as "quasi-incompressible".

In this work, we apply this model in two dimensions and additionally consider the energy balance. Here we account for the enthalpy of mixing that is an outcome of the thermodynamic model and is calculated from chemical potentials. This procedure results in the necessity to solve the energy balance in terms of specific energy, where the solution of a non-linear equation determines temperature from the specific energy. This equation must be solved at each grid node at each time step. As thermodynamic model for equilibrium, we use the Non-Random-Two-Liquid (NRTL) model. Due to the DGT, the interfacial tensions are inherent in this model.

Having this framework implemented, we are able to investigate influences of transport properties on decomposition behavior as well as influences of temperature and convection. The model allows calculating droplet formation, coalescence and even break-up in a thermodynamic consistent approach without any simplifying assumptions. In our model calculations, we consider phase separation of binary and ternary mixtures with various mixing gaps and we can induce temperature gradients or boundary heat flows (temperature quenching) or mechanical disturbances.

In our presentation, we will show a variety of dynamic simulations of binary and ternary mixtures that have a single mixing gap. Besides investigating the influence of temperature (and therefore the influence of tie line length) on coalescence we also show, how interfacial enrichment affects mass transport across the interface. With wall boundary conditions on the bottom and top of the simulation domain, we are also able to include gravitation and study its effects on e.g. falling droplets in combination with various density and viscosity ratios.

Those and further researches may help to build large-scale fluid dynamic models that incorporate more physics rather than empiric findings. For example, in our model it is possible to measure coalescence time parallel to interfacial tensions, which then can be utilized as parameters in population balance based models.

[1] V. Alopaeus, J. Koskinen, K.I. Keskinen, J. Majander, Chem. Eng. Sci., 57, 2002, 1815.

[2] D. Jacqmin, J. Comput. Phys., 155 (1999), 96.

[3] E. B. Nauman, D. Q. He, Chem. Eng. Sci., 56 (2001), 1999

[4] J. W. Cahn, J. W. J. Chem. Phys., 42, 1965, 93.

[5] K. Kruber, M. Krapoth, T. Zeiner, Fluid Phase Equilib., 440, 2017, 54.

580532 Effects of Different Additives on Fluid Dynamics and Mass Transfer in Single Drop Extraction

Extraction processes in liquid/liquid systems rely strongly on the characteristics of the interface and the occurring transport phenomena. Additives, such as surface-active substances or nanoparticles, which are added intentionally or occur as impurities, may influence fluid dynamics and mass transfer substantially. Residence times and hold-up in liquid/liquid extractors may increase while the additives impose an additional resistance against mass transfer or promote the same by inducing interfacial effects like Marangoni convection. Since the interfacial phenomena can have contrary effects on the mass transfer in multiphase systems, the prediction of their occurrence and process variables linked to them based on theoretical or empirical basis is a challenging task.

The present work focuses on the characterization of fluid dynamics and mass transfer in the presence of different surface-active substances. Due to the interaction between the transport processes, fluid dynamic measurements of single droplets can act as an indicator for the prediction of interfacial phenomena and mass transfer rates. Measurements of single drops are done in a drop rising cell of one meter length. Droplet paths and velocities were detected optically with a high-speed camera. Mass transfer was determined by collecting drops at different heights and analyzing the concentration of the transfer component.

Acetic acid was used as a transfer component and fluid dynamics and mass transfer from the disperse 1-octanol phase to the continuous water phase were measured for different drop sizes and initial concentrations of the transfer component. As additives, the surfactant Butyldiglycol (C4E2) and the partly hydrophobic silica nanoparticles HDK H20 were chosen. Drop rising velocity, droplet path and concentration for different contact times were measured and compared between the ternary reference system and the two systems with additives.

At high concentrations of acetic acid the pure system showed lateral displacement resulting in a break out from linear droplet paths, which could indicate Marangoni effects. For low concentration no Marangoni effect occurred. Surfactant and nanoparticle systems were, therefore, compared for low concentrations of the transfer component.

Upon addition of C4E2 without transfer component the rising paths of the drops likewise indicated Marangoni convection. Stunningly, the dependencies of concentration and drop size were reversed compared to the standard reference system toluene/water/acetone known for Marangoni effects. With acetic acid as transfer component the behavior became even more chaotic, which indicates a strong influence on the mass transfer.

When adding nanoparticles, no influence on the fluid dynamics is observed in the nanoparticle concentration range investigated. Although the fluid dynamics stayed the same a decrease of mass transfer was measured with increasing nanoparticle concentration.

559255 Unifac Correlated Parameters for Liquid-Liquid Equilibrium Prediction of Ternary Systems Related to Biodiesel Production Process

Two phases liquid systems composed by vegetable oil, biodiesel, fatty acids (FAs), glycerol, low molecular weight alcohols, and water are present in most of the biodiesel production process stages, including oil refining, free FAs esterification, transesterification, biodiesel and glycerol-rich phases separation and biodiesel refining by water washing. For that reason, the correct prediction of the liquid-liquid equilibrium (LLE) is a key-factor for an accurate process understanding. In this work, a new set of group interaction parameters (GIP) for the UNIFAC model that describes the LLE for all the systems involved in biodiesel production was correlated and validated. A database including 1414 LLE tie lines from 97 references was constructed and employed to correlate the GIP using a genetic algorithm. These parameters describe the LLE involved in the correlation process with an average deviation up to 2.07%. Additionally, a sensitivity analysis using the proposed GIP was performed to understand the effect of the molecular structure on the LLE behavior for the studied systems. Results showed that composition of the distributed component (alcohol, fatty acid or biodiesel) in both phases depends on the number of carbon atoms, unsaturated bonds and hydroxyl groups in the oleochemicals conforming the system. The variable with the highest incidence in the sensibility analysis was the distributed component overall mass fraction, followed by the alcohol length chain.

566587 Influence of Nanoparticle Wettability on Absorption and Desorption Mass Transfer Rates at Elevated Pressure in a Simple Hydrocarbon System

In the petroleum industry, the time required for degassing, or the rate of gas evolution within a gas-liquid separator at elevated pressure, is usually determined by assuming thermodynamic equilibrium between the two phases and estimating the time required for entrained gas bubbles to rise out of the super-saturated liquid. Gas-liquid separator design using this approach does not fully account for the dissolved solution gas present within the super-saturated solution. Further, crude oil naturally contains several oil-wet and water-wet particles such as inorganic clay, sand, rust produced from corrosion, and precipitated asphaltenes. The influence of such nanoparticles on the rate of gas evolution from super-saturated systems has not been investigated, so it is critical to investigate this phenomenon in order to improve our understanding of separator design and performance.

In this study, we measured the rate of absorption and desorption of methane in n-dodecane with 1 wt.% solid (silica) nanoparticles of varying wettability. These measurements were conducted in a batch stirred tank system at a constant pressure of 1,000 psia, a temperature of 45 ^{o}C, and a mixing speed of 250 rpm. Three different silica nanoparticles were selected based on their wettability characteristics: hydrophobic (R812S), partially hydrophobic (R711), and hydrophilic (A200). The rates of absorption and desorption of methane in pure n-dodecane were found to be within experimental uncertainty. The rate of gas absorption and desorption decreased with the addition of partially hydrophobic nanoparticles (R711) and hydrophilic nanoparticles (A200). Hydrophobic nanoparticles (R812S) did not have any noticeable effect on the rate of gas evolution. In addition, the wettability of the nanoparticles influenced the bulk viscosity of n-dodecane.

Keywords: absorption, desorption, gas evolution, gas-liquid mass transfer, particles, silica, wettability, gas-carry under, solution gas

559508 Measuring Crystallization Kinetics of Sparingly Soluble Salts from Supersaturated Waters Using CSTRs-in-Series: Studies with $CaCO_3$

Measuring crystallization kinetics of sparingly soluble salts from supersaturated waters using CSTRs-in-series: studies with $CaCO_3$

Sankaranarayanan A. Ravichandran[a], Jordan Krist[b], Dakota Edwards[a], Saied Delagah[c], John Pellegrino[a,*]

[1]Department of Mechanical Engineering, University of Colorado, Boulder, CO USA

[2]Department of Chemical and Biological Engineering, University of Colorado, Boulder, CO USA

[3]Bureau of Reclamation, Denver, CO USA

[*]Corresponding author: john.pellegrino@colorado.edu

Abstract

We present an approach for studying crystallization kinetics of sparingly soluble salts, particularly supersaturated reject streams from reverse osmosis (RO) membrane processes. Our focus was to study crystallization kinetics to develop heuristics to improve the design of crystallizers used in water treatment applications. We used a system of six continuously stirred tank reactors in series (CSTRs-in-series), the fluid between CSTRs is transported using peristaltic pumps. We used turbidity and pH changes to track crystallization. Our steady-state approach provides a better signal to noise ratio than extant batch mode methods to study the kinetics of crystallization processes. Moreover, steady-state bench scale approaches are more realistic in describing large scale continuous processes. We used a realistic calcium carbonate supersaturated stream modeled on an RO reject stream found at a brackish water treatment facility at Brighton, Colorado. The residence times in each CSTR ranged between ~3 to 11 min. The maximum hard water inlet feed was ~0.6 L/min for the shortest residence time. For six CSTRs-in-series the total system residence time of ~18 to ~ 66 min and showed supersaturation relief in the range of ~ 25 to over 50 percent respectively without added chemicals. We used a metric of 5 NTU turbidity to describe a point of discernible crystal crystallization where crystal nucleation is significant. We studied the effects of impeller mixing and the mixing in the peristaltic tubing using the Kolmogorov mixing theory. Interestingly, for all residence times studied, a peristaltic recirculation loop in the 1st CSTR where mixing takes place in a smaller surface to volume ratio was required to induce discernible crystallization with turbidity values >> 5 NTU. Moreover, increased impeller mixing corresponding to mixing energy inputs higher than that of peristaltic recirculation loops could not induce discernible crystallization in the 1st CSTR. We developed a parametric semi-empirical model to co-relate mixing and, turbidity in the 1st CSTR using Kolmogorov mixing theory and, laws of mass action. Our studies show that optimizing the surface to volume ratio along with mixing energy input is key to controlling crystallization kinetics. We were able to extend the scope of this study for a more complex supersaturated system involving co-precipitation of calcium carbonate and calcium sulfate.

Figure 1. CSTR-in-series setup, observe the evolution of turbidity in successive CSTRs

561630 Assessment of Oil Fouling By Oil-Membrane Interaction Energy

Membrane-filtration is promising for treating the voluminous oily wastewater, especially when the oil emulsions are smaller than 20 μm. However, studies on the inevitable membrane fouling phenomenon by oil are rather scarce. In particular, a question that remained to be addressed was whether the DLVO or XDLVO model provides better predictions of the oil-membrane interfacial interactions and thereby the extent of fouling. Accordingly, this study investigated four oil types (namely, hexadecane, soybean oil, fish oil and crude oil) that were stabilized by the same non-ionic surfactant (namely, Tween 20) and had similar mean droplet diameters. The direct observation through the membrane (DOTM) technique was used to quantify the critical flux of the different emulsions, and both the XDLVO and DLVO models were used to quantify the foulant-membrane and foulant-foulant interactions. DOTM results indicated that the critical flux values were similar for all oils under the conditions tested, except crude oil. Although the XDLVO model appears to be more comprehensive than the DLVO in terms of accounting for the additional Lewis acid-base polar (AB) interaction that is acknowledged to be important in membrane-filtration, results indicate that the dominance of the AB component drowns out the other interactions like that of electrostatics (EL) in this case, which impedes accurate prediction of the different fouling tendencies by the different oil types.

574809 Adsorption-Induced Transformations in Flexible Microporous Frameworks: Insight from Phonon Spectra Analysis

Adsorption induced transformations of the pore structure have received recently broad attention in view of discovery of novel flexible adsorbents. Low frequency lattice vibrations (phonons) can be used as indicators of a variety of structural transformations. In particular, conformational changes in flexible metal-organic frameworks (MOFs) can be successfully explained by analyzing materials phonon spectrum in frequency range below 200 cm^{-1} [1,2]. It has been shown that such lattice vibrations are responsible for pressure- and adsorption-induced gate-opening in zeolitic imidazolate frameworks (ZIFs), and pore breathing in materials from MIL-53 family.

In this work, the density functional theory (DFT), Grand Canonical Monte Carlo (GCMC) and adsorption stress model (ASM) [3] were used to analyze low-frequency phonons and their impact on adsorption-related structural transformations.

We analyze adsorption-induced transformations in ZIF-8 and MIL-53. The step-like adsorption isotherm observed in this ZIF-8 was interpreted in the literature as a signature of gate-opening (enlargement of the pore free volume) induced by the increasing number of guest molecules entering the pore. To analyze this phenomenon at the microscopic level, we first calculated and visualized normal modes of the lattice vibrations. For that, density functional theory was used. Only the modes which fulfill two criteria: (i) their frequency is below 200 cm^{-1}, and (ii) the related lattice deformation leads to increase of the pore free volume, have been further analyzed in GMCM adsorption simulations. For MIL-53 we used ASM to analyze softening of the vibrational modes related to breathing transformation.

Studies of low-frequency vibrations in MOFs provide important, qualitative insight into the phenomena of transformations in MOFs. These dynamical fluctuations may become a precursor to a static deformation under any external stimuli such as temperature (thermal energy), pressure (mechanical energy) or adsorption (intermolecular interactions) which overcomes the energy barrier of the transition. The main focus is made on particular vibrations which are related to breathing and gate-opening transitions in studied MOFs.

[1] MR Ryder, et al., Physical Review Letters, 2014 113, 215502
[2] F Formalik, et al., Microporous and Mesoporous Materials, 2019
[3] A Neimark, et al., Journal of Physical Chemistry C, 2010, 114, 22237

570555 Material Discovery By Model Guided Targeted Synthesis for Natural Gas Storage Application

The increased availability of natural gas (NG) has revolutionized the energy industry. However, penetration of NG in transportation is limited due to lower energy density compared to liquid fuels. We are focused on improving state of the art sorbents by model guided targeted synthesis to achieve higher dynamic capacity for a given adsorption-desorption cycle.

We will present an automated modeling workflow to construct and simulate methane adsorption performance for hypothetical MOFs using state of the art force fields. By limiting the design space to specific combinations of linkers and topologies, we have come up with a combination of synthetically amenable new MOFs which can offer higher methane storage performance. Our targeted synthesis of MOF lead to discovery of new materials with better understanding of the unique structure-property relationships.

576923 On the Use of Single, Dual and Three Process Langmuir Models for Gas Mixture Components That Exhibit Different Combinations of These Processes

There are several interesting cases that arise when one of the components in a gas mixture is described by either the single process Langmuir (SPL) or dual process Langmuir (DPL) (LI) model and the other component is described by the three process Langmuir (TPL) model. These cases are very subtle and have never been addressed in the literature.

Methodologies will be presented based on three approaches to predict the mixed gas isotherms of a binary system, wherein one of the components is described by either the SPL or DPL model (usually the lighter component) and the other one is described by the TPL model (usually the heavier component). These approaches differ on whether the individual processes of the isotherms correlate in a perfect positive, perfect negative or unselective fashion. For these cases, a consistent set of single and binary isotherms from the literature that behave these ways are used as the basis for comparison to demonstrate these concepts.

562351 Determination of Optimal Hot Purge Gas Temperature in Regenerating Thermally an Air-Drying Zeolite 13X Column By Equilibrium Theory Analysis

Air-drying TSA processes have been widely commercialized for reducing air humidity down to ppb levels, producing a dry air suitable for production of engineering plastics and food. In designing such a cyclic adsorption process driven by temperature swing, it is crucial to find optimal regeneration temperature, so that the adsorbents can be regenerated fully for the next adsorption run and simultaneously the time taken for regeneration has to be short enough for good bed productivity.

In this study, the regeneration temperature was optimized by Equilibrium Theory. The theoretical breakthrough curves and their temperature profiles were constructed by the hodography in which a Toth isotherm was taken as equilibrium isotherm [1]. Interestingly, the breakthrough curves constructed by Equilibrium Theory exhibited different trends at varying regeneration temperatures. Its initial part always consists of two plateaus connected by a shock. The second plateau is followed by a simple wave at a low regeneration temperature. With increasing regeneration temperature, however, a shock wave appears and evolves in the middle of the trailing wave. This interesting results indicates that the regeneration temperature must be sufficiently high so that the shock wave could be large enough to reach below the targeted dew point of the dry air.

It was also investigated the effect of capillary condensation on the breakthrough curve. To this end, Aranovich-Donohue isotherm [2] was taken for equilibrium isotherm of water vapour on zeolite 13X. Incorporating the Type II isotherm into the governing equations made the second plateau elongated in time, but did not affect much the rest of the trailing curve.

Full numerical simulations of the thermal regeneration breakthroughs were also carried out. Their results were coherent with the experimental data measured earlier by the author at three different regeneration temperatures (423K, 473K and 523K) [3]

[1] Wang, Y. and LeVan, M.D., Adsorption Equilibrium of Binary Mixtures of Carbon Dioxide and Water Vapor on Zeolite 5A and 13X, *J. Chem. Eng. Data*, **2010**, 55, 3189.

[2] Kim, K.-M., Oh, H.-T., Lim, S.-J., Ho., K., Park, Y., Lee, C.-H., Adsorption Equilibria of Water Vapor on Zeolite 3A, Zeolite 13X, and Dealuminated Y Zeolite, *J. Chem. Eng. Data,* **2016**, 61, 1547, 2016

[2] Ahn, H. and Lee, C.-H., Effects of Capillary Condensation on Adsorption and Thermal Desorption Dynamics of Water in Zeolite 13X and Layered Beds, *Chem. Eng. Sci.*, **2004**, 59(13), 2727.

577665 Thermodynamic Modeling of Ion Removal Processes

Removal of ionic contaminants from industrial wastewater is a widely practiced water purification process. From thermodynamic modeling perspective, the ion removal process can be considered as either liquid phase adsorption [1, 2] or ion exchange equilibrium problem [3-5]. In this study, we investigate the ion removal process with three different approaches. First, treating the ion removal process as an adsorption process, we explore the applicability of a recently developed adsorption non-random two-liquid (aNRTL) model [6, 7] to describe the ion adsorption behavior on selected adsorbents. Taking into consideration both the adsorbate-adsorbent interactions and the lateral adsorbate-adsorbate interactions, the model successfully correlates the highly complex competitive ion adsorption systems with only three binary interaction parameters. Treating the ion removal process as an ion exchange equilibrium problem, the second approach requires information for the equilibrium constant K and the activity coefficients of ionic species in both the bulk phase and the resin phase. The electrolyte NRTL model [8] and the aNRTL model are used to calculate the activity coefficients of ionic species in the bulk phase and the resin phase respectively. Both approaches correlate the experimental data successfully.

The third approach considers the ion removal process as a phase equilibrium problem for the ions between the bulk phase and the resin phase. Again the electrolyte NRTL model is used to describe the ionic activity coefficients in the bulk phase while the resin phase is considered as a polyelectrolyte system modeled with the polyelectrolyte NRTL model [9]. By establishing equality of chemical potentials of electrolytes in both the bulk phase and the resin phase, this novel approach provides a complete account on the physical nature of both the resin phase and the bulk phase. It uniquely allows computation of concentrations and activity coefficients of both counter-ions and co-ions in the resin phase.

Keywords: ion removal, adsorption NRTL model, electrolyte NRTL model, polyelectrolyte NRTL model

References

[1] V.C. Srivastava, *et al.*, Modelling Individual and Competitive Adsorption of Cadmium(II) and Zinc(II) Metal Ions from Aqueous Solution onto Bagasse Fly Ash, Sep. Sci. Technol., 41 (2006) 2685-2710.

[2] A. Erto, *et al.*, Modeling of single and competitive adsorption of cadmium and zinc onto activated carbon, Adsorption, 21 (2015) 611-621.

[3] R.P. Smith, E.T. Woodburn, Prediction of multicomponent ion exchange equilibria for the ternary system SO-NO-Cl from data of binary systems, AICHE J., 24 (1978) 577-587.

[4] D.C. Shallcross, *et al.*, An improved model for the prediction of multicomponent ion exchange equilibria, Chem. Eng. Sci., 43 (1988) 279-288.

[5] B.S. Vo, D.C. Shallcross, Modeling Solution Phase Behavior in Multicomponent Ion Exchange Equilibria Involving H+, Na+, K+, Mg2+, and Ca2+ Ions, J. Chem. Eng. Data, 50 (2005) 1995-2002.

[6] H. Kaur, H. Tun, Local Composition Activity Coefficient Model for Mixed-Gas Adsorption Equilibria, (In progress).

[7] H. Kaur, Investigating Liquid Mixture Adsorption Equilibria using modified aNRTL model, (In progress).

[8] Y. Song, C.-C. Chen, Symmetric electrolyte nonrandom two-liquid activity coefficient model, Ind. Eng. Chem. Res., 48 (2009) 7788-7797.

[9] Y. Li, *et al.*, Modeling of Divalent and Multivalent Polyelectrolytes System with polyelectrolyte NRTL (In progress).

572729 Recycling of Nimh Batteries and Recovery of Rare Earth Elements Using ETS-10 Titanosilicate

Recovery of rare earth elements (REEs) is necessary in order to maintain a sustainable supply for future green energy applications. Magnets, lamps and NiMH batteries are some of the key sources containing a large amount of recyclable REEs. In the current work we explore the adsorption of critical REE cations via ion exchange on ETS-10 titanosilicate. ETS-10 is a three-dimensional solid framework material containing ordered -O-Ti-O-Ti- chains which generate negative charge necessary for cation exchange. Batch experiments in acidic aqueous solutions prove that ETS-10 is a promising material for this application and that it has high equilibrium adsorption capacity for REE cations. We further also explore the competitive separation of divalent Ni^{2+} cations from trivalent Nd^{3+} cations usually found in aqueous solutions generated during recycling of NiMH batteries. Based on these preliminary results we believe ETS-10 can be used to fabricate inorganic ion exchange membranes for continuous separation and recycling of REE cations.

565449 Spiral-Wound Module with Facilitated Transport Membrane for CO_2 Capture from Flue Gas

Carbon capture from flue gas using membrane technology has been studied extensively to combat global warming. It is often a challenge to scale up the membrane fabrication from lab scale to pilot scale and, subsequently, make the membranes into modules. Using a roll-to-roll fabrication machine integrated with a knife-coating method, a 165-nm thick facilitated transport membrane was successfully made. The scale-up membrane had a width of 14 inches and a total length of > 1200 feet, and it was used to fabricate spiral-wound (SW) elements and modules.

In order to increase the membrane packing density in an element, an improved multi-leaf rolling procedure was developed. Although there is extensive literature on the fabrication of a multi-leaf SW membrane module, most of the studies had all the membrane leaves directly attached to the central permeate tube. Hence, the number of leaves would be limited by the surface area of the central tube. For improving the fabrication, only a carrier layer is attached to the central tube, and the membrane leaves were stacked on the carrier layer. Moreover, with a thinner spacer at 6 mil in thickness, the membrane area was increased to 2.94 m^2 in the most recent modules. The transport results of the modules agreed well with those of the flat sheet membrane under various testing conditions. For instance, a CO_2 permeance of 1450 GPU and a CO_2/N_2 selectivity of 185 were demonstrated consistently at 67°C. Also, satisfactory pressure drops were achieved.

A scale-up factor of >70 has been attained since the start of this study on SW element fabrication process. By considering the limitations set by concentration polarization and pressure drop, a predictive model was used to relate the operating conditions to the separation performances of the SW module. The experience gained is helpful to design a full-sized element (8 inches in diameter and 40 inches in length) for industrial application.

Dr. Salim's current affiliation: Membrane Technology and Research, Inc. (MTR), Newark, CA

Dr. Wu's current affliation: Liquid Microcontamination Control (LMC) Membrane R&D Group, Entegris, Bedford, MA

569306 Highly Selective Holey Graphene Oxide Membranes for Gas Separation

Graphene oxide (GO) nanosheets stacked in parallel with sub-nanochannels have been demonstrated with an excellent size-sieving ability for membrane-based gas separation. However, pristine GO sheets are almost impermeable to all gases, so gas molecules can only diffuse through these tortuous nanochannels, leading to low gas permeability. Drilling nanopores on GO sheets will reduce the tortuosity of the gas transport path and thus increase gas permeability while retaining the sharp size-sieving ability. Herein, nanopores were created on GO sheets by etching with hydrogen peroxide to prepare holey GO (HGO) sheets. The pore size of HGO can be tuned from several nanometers to ~100 nm by increasing the etching time. The as-prepared HGO sheets are highly dispersible in water and can be made into membranes by vacuum filtration. Increasing the etching time increases gas permeability. For example, the HGO membrane with an etching time of 4 h (HGO-4h) shows He permeability of 5.8 Barrers at 35 °C, which is much higher than the pristine GO membranes (0.97 Barrer), while both membranes show similar He/N_2 selectivity of ~200. Moreover, the microstructure of HGO membranes can be controlled by tuning the pH of HGO solution. For example, when the pH value decreases to 2, the obtained HGO-4h membrane shows the He permeability of 57 Barrers, which is ten times that of membranes obtained at pH value of ~6. The HGO-4h membranes (pH = 2) exhibit He/N_2 selectivity of 1300 and He/CH_4 selectivity of 1800 at 35 °C, which surpasses the 2008 Robeson's upper bound for He/N_2 and He/CH_4 separation.

565857 Hydrogen-Bonded Polyimide/Metal-Organic Framework Hybrid Membranes for Ultra-Fast Separations of Multiple Gas Pairs

Membranes can effectively reduce the enormous amount of energyrequired for gas separations, compared to conventional separation processes, such as cryogenic distillation. Currently, polymers dominate gas separation membrane markets due to their appealing manufacturability and performance. However, polymeric membranes suffer from the productivity-selectivity tradeoff, i. e., the Robeson upper bound. To overcome this challenge, growing attention has been shifting towards hybrid materials synergistically utilizing advantages of each component in the hybrid membranes, i.e., easy processing of polymers and molecular sieving of fillers, such as metal-organic framework (MOFs). However, only a few MOF-based hybrid membranes perform above current Robeson upper bounds, predominantly relying upon rigid Polymers of Intrinsic Microporosity or aggressive thermal annealing as high as 240 °C, adding complications for scaling-up. Moreover, most hybrid membranes are specialized for a particular gas pair (e.g. CO_2/N_2), lacking a general approach to separating multi-component gas mixtures.

This work demonstrates new classes of polyimide/MOFs hybrid membranes residing above current Robeson upper bounds via a generalizable approach. The hydrophilic amino ($-NH_2$) groups in MOFsintensively interact with carboxylic moieties in a 6FDA-based polyimide via hydrogen bonds to foster a firm interphase adhesion. The polyimide/MOFshybrid membranes demonstrate a $CO_2$and H_2permeability of 2494 and 2932 Barrers, respectively, with a CO_2/CH_4, $H_2/CH_4$and H_2/N_2selectivity of 29.3, 34.4, and 23.8 respectively, surpassing current Robeson upper bounds considerable. The permeability was enhanced by record-high 16 times comparing with the neat polymer. With a processing temperature as low as 80 °C and performance continuously exceeding the upper bounds for over 5300 hr, the membrane is readily compatible with state-of-the-art membrane manufacture processes.

575073 Enhancing H_2-Permselectivity of High-Flux Hollow Fiber Membrane Via in-Situ Layer-By-Layer Surface Treatment

A primary challenge for realizing a hydrogen economy is the production of inexpensive and high purity hydrogen. Current techniques for the purification of hydrogen such as Pressure Swing Adsorption or cryogenic distillation are costly, energy intensive and complex in operation. Polymeric gas membranes with low production cost, improved mechanical properties, and high selectivity have the potential to replace current state of art technologies. Membranes have already found widespread application at industrial scale for the separation of nitrogen, air enrichment and the removal of CO_2 from synthesis gas. However, these size selective membranes have a tradeoff between permeability and selectivity.

Our previous work demonstrated layer-by-layer (LbL) deposition of thin functional films as a robust, low-cost method to introduce size-selective functionality to a membrane substrate [1]. Current industrial polymeric membranes generally employ high-flux, low-cost and low-selectivity Hollow Fiber membranes for light gas (H_2, CO_2, CH_4) separations. In this work, we demonstrate that LbL deposition of sub-micron conformal coating on a highly selective membrane with competitive permeability on hollow fibers (HF) for enhanced light gas separation. For the first time, we show increase in the H_2 selectivity of high flux HF membranes by addition of Polyethylenimine (PEI)/polyacrylic acid (PAA) bi-layers.

Polyamide imide (Torlon) Hollow fibers were provided by collaborators at NETL. These HFs were first potted in a tube and shell assembly followed by passing PEI and PAA solution alternatively. The optical microscopic images of these membranes indicated a thickness of 5 microns. These membranes showed high degree of 'ionic-crosslinking' achieved high selectivities of H2: CO2 (>200:1) and H2: N2 (>2000:1) at 40 Barrers of H_2 permeability exceeding the Robeson upper bound for homogenous polymer films [2]. The deposition of a low-cost, water-based polymers and easy to fabricate LbL membranes on conventional HF has potential use at industrial scale for hydrogen separation.

1. Kim, D., et al., *Highly size-selective ionically crosslinked multilayer polymer films for light gas separation.* Adv Mater, 2014. **26**(5): p. 746-51.
2. Robeson, L.M., *The upper bound revisited.* Journal of Membrane Science, 2008. **320**(1-2): p. 390-400.

559330 Polymer Gas Separation Membranes Containing High Free Volume Elements and Ionic Liquids

This presentation will detail our development of new polymer architectures which contain high free volume elements (e.g., 6-FDA, spirobisindane, Tröger's Base, etc.) which are covalently bond to ionic groups forming gas separation membranes which not only can absorb ionic liquids (ILs) into their structures, but also align/rearrange in the presence of ILs. This "bottom-up" approach allows for high degrees of membrane tunability and influence/control over polymer nanostructure and associated performance with respect to gas separations involving CO_2, N_2, CH_4, H_2 and other species.

557054 Vapor Phase Infiltration of Metal Oxides into Nanoporous Polymers for Solvent Stable Nanofiltration Membranes

Membrane-based organic solvent separations promise a low-energy alternative to traditional thermal separations but require materials that operate reliably in chemically aggressive environments. While inorganic membranes can withstand demanding conditions, they are costly and difficult to scale. Polymeric membranes, such as polymers of intrinsic microporosity 1 (PIM-1), are easily manufactured into forms consistent with large-scale separations (e.g., hollow fibers), but perform poorly in aggressive solvents. Here, a new post-fabrication membrane modification technique, vapor phase infiltration (VPI) is reported that infuses PIM-1 with inorganic constituents to improve stability while maintaining the polymer's macroscale form and microporous internal structure. The atomic-scale metal oxide networks within these hybrid membranes protect PIM-1 from swelling or dissolving in solvents. This stability translates to improved separation performance (90 % molecular weight cutoff is 204 Da) in a variety of solvents, including solvents capable of dissolving PIM-1. The infiltrated inorganic phase also appears to give new control over solute sorption in OSN. These hybrid membranes further show promising performance for organic solvent reverse osmosis (OSRO) separations in challenging solvents, even at small molecular weight differentials (14 Da). Because the VPI process can be integrated with state-of-the-art membrane modules, this treatment could be readily adopted into the large-scale manufacturing of advanced membranes.

564267 Bio-Inspired Immobilization of Casein-Coated Silver Nanoparticles on Cellulose Acetate Ultrafiltration Membranes: From Laboratory to Scale-up

Casein-coated silver nanoparticles (casein-AgNP) had been previously loaded on cellulose acetate (CA) ultrafiltration membranes for biofouling control[1]. However, when physically added to the membrane casting solution, AgNPs tend to leach off the membranes. Therefore, there is strong interest in developing methods to chemically bind the AgNPs to the membranes without losing their activity, and in making these methods scalable. Inspired by the affinity of silver to the thiol groups of cysteine proteins in bacteria, CA membranes were cast and imbedded with AgNPs via attachment with functionalized thiol groups using polymerized glycidyl methacrylate (GMA) and cysteamine. However, the chemical reactions required three steps after casting the membranes, and each step needs a large amount of solution bath. To minimize the materials and energy needed in the chemical attachment process and maximize the reaction efficiency, this study integrated all the chemical reactions in the stage of preparing dope solutions. The recipe and procedures changed due to the difference of dope solution and membranes. Membranes immobilized with AgNPs were successfully cast for laboratory-scale studies as well as for scale up studies. This method successfully prevented leaching of AgNPs during cross-flow and dead-end studies. Pseudomonas fluorescens Migula in brackish water was used for biofouling test and the membranes displayed a significant reduction in the accumulation of bacterial cells. This method provided a promising approach of casting CA membranes immobilized with AgNPs.

1. Sprick, C.; Chede, S.; Oyanedel-Craver, V.; Escobar, I.C. Bio-inspired immobilization of casein-coated silver nanoparticles on cellulose acetate membranes for biofouling control. *Journal of Environmental Chemical Engineering* **2018**, *6*, 2480-2491.

572821 Fouling Resistant Ultrafiltration Membranes with Ultrathin Hydrogel Selective Layers Containing Zwitterionic Moieties

Membrane separation processes are of paramount importance in water and wastewater treatment. However, membrane fouling by organic compounds is still one of the main obstacles to more extensive use of membrane separation processes. Hydrogels are materials that can absorb large quantities of water and swell. This high degree of hydration makes them inherently fouling resistant. Hydrogels also have tunable mesh size – enabling control of their permeation selectivity – and are functionalizable. These features make them excellent membrane materials. However, the applicability of hydrogels as fouling-resistant selective layers of thin film composite membranes has, so far, been hindered by the lack of a fabrication process that allows for the formation of a thin, defect-free hydrogel layer on a porous support. We have recently demonstrated a new method, namely Interfacially Initiated Free Radical Polymerization (IIFRP), to address this challenge. In IIFRP, an ultrathin hydrogel selective layer, as thin as ~80 nm, is synthesized on top of a porous support by separating the monomer and photoinitiator into two immiscible phases. The hydrogel is formed at the interface of these two phases. The goal of the present study was to explore the incorporation of other hydrophilic functional groups, particularly zwitterionic (ZI) groups, into hydrogel selective layers by IIFRP. We have shown that the incorporation of ZI moieties can further increase membrane fouling resistance and improve membrane properties. For instance, the partial substitution of poly(ethylene glycol) diacrylate (PEGDA) units in the hydrogel layer with the zwitterionic monomer sulfobetaine methacrylate (SBMA) led to a permeance increase of almost an order of magnitude (from 3.9 ± 0.4 L/h.m^2.bar to 34.7 ± 0.8 L/h.m^2.bar). Although the addition of SBMA also decreased the crosslinking density of the hydrogel selective layer, the rejection properties of these membranes remained essentially constant (99.2 ± 0.6% and 98.7 ± 0.6% myoglobin rejection, respectively). Moreover, the zwitterionic hydrogel membranes have shown higher fouling resistance to oil emulsions compared to both commercial membranes and the zwitterion-free hydrogel membranes. Consequently, zwitterionic containing hydrogel selective layer obtained by IIFRP are very promising membrane materials.

564930 Zwitterion-Ag Complexes That Simultaneously Enhance Biofouling Resistance and Silver Binding Capability of Thin Film Composite Membranes

Biofouling can be overcome with zwitterion grafts and anti-microbial, metallic nanoparticles. However, the mechanism underpinning this effective approach remains unclear. To elucidate the role of each component in this system whilst maximizing membrane anti-fouling and anti-microbial properties, here we performed a comparative study to investigate the impact of zwitterion type and their interactions with Ag of various states. Two different zwitterions (SO_3^--based and COO^--based) were employed to modify polyamide (PA) thin film composite (TFC) membranes, and the metallized and mineralized membranes were developed via *in-situ* formation of silver (Ag) nanoparticles and deposition of silver chloride (AgCl) particles on the zwitterion-modified TFC membranes. The presence of zwitterions was key to enhancing Ag content, resulting in significantly improved anti-microbial and anti-fouling properties without compromising the nanofiltration separation performance. COO^--based zwitterions were found more favourable towards Ag metallization and mineralization compared to SO_3^--based zwitterions. The underlying mechanisms underpinning this discovery were further revealed using the Density Functional Theory (DFT) to reveal Gibbs free energy of the binding between zwitterions and Ag^+ ions. This fundamental knowledge is crucial for designing next-generation anti-biofouling strategies.

564491 Fabrication of Fouling-Resistant Electro-Conductive Thin Film Composite Membranes By Laminating Polyaniline (PANI) and Reduced Graphene Oxide (rGO) on Polyethersulfone

Fabrication of Fouling-resistant Electro-conductive Thin Film Composite Membranes by Laminating Polyaniline (PANI) and reduced Graphene Oxide (rGO) on Polyethersulfone

Amin Karkooti[a], Masoud Rastgar[b], Mohtada Sadrzadeh[b], Neda Nazemifard[a]

[a]Department of Chemical and Material Engineering, University of Alberta

[b]Department of Mechanical Engineering, University of Alberta

Abstract:

The adsorption of organic matter onto the membrane surface is the most significant obstacle that restrict the sustainable application of membranes for residential and industrial wastewater treatment. Fouling on membrane decreases water flux, shortens membrane lifespan, and consequently leads to a higher operating cost. Development of electro-conductive membranes (ECMs) can potentially provide an effective solution to enhance fouling resistance characteristics of membranes. The application of electrical potentials to the ECMs offers a promising enhancement to existing membrane-based water treatment processes. Membrane surface charge is one of the effective parameters that influence fouling at the initial stage of filtration through electrostatic interactions. By applying voltage to a conductive membrane, membrane's surface becomes more negative and it can effectively inhibit the adsorption of organic contaminants, which are predominately negative charged. In addition, in electro-conductive membrane, filtration process is combined with electro-chemical oxidation (EO) technique, which is a powerful technique in removal different harmful pollutants from water. Simultaneous to filtration, electrical potential causes organic contaminants in water to be eliminated by either direct or indirect oxidation mechanisms. Depending on operational time, anodic material, and the applied electrical potential, the oxidation could be progressed entirely which clean water and carbon dioxide would be produced through overall degradation of organic compounds. In this work, we fabricated a novel and mechanically-stable electro-conductive membrane by pressure-assisted laminating of reduced graphene oxide-polyaniline (rGO-PANI) suspension on commercial polyethersulfone (PES) substrate. PANI (50 ppm) and rGO (20 ppm) were respectively dispersed in N-Methyl-2-pyrrolidone (NMP) and deionized water via sonication and were mixed together in different weight ratios. Afterward, Using 20 psi air pressure, the obtained mixtures were forced to pass through PES substrate having surface area of 41.8 cm^2. In this coating layer, two neighboring rGO sheets can create special interconnected channels that is selectively able to permeate water, while rejecting all other solutes. Three different membranes including M1 (PANI (0.5 mg)-rGO (0 mg)), M2 (PANI (0.25 mg)-rGO (2 mg)), and M3 (PANI (0.5 mg)-rGO (2 mg)) were provided and used for organic fouling experiments. Prior to laminating the thin film layer, graphene oxide (GO) nanosheets were synthesized based on modified Hummers' method. Next, reduced GO (rGO) nanosheets were produced by heating GO at 1000 °C. The reduction procedure partially removes functional groups, such as hydroxyl (R-OH), carboxyl (R-COOH) and carbonyl (R-O-R), giving rGO high electrical conductivity because of its none or low oxidation state. Both rGO and PANI are electrically conductive and we observed that the obtained composite membranes were mechanically stable in aqueous environments. In addition, It was found that rGO-PANI ECMs fabricated using sulfuric acid as the dopant showed the best transport, electrical, and stability characteristics, making them ideal for water treatment applications. In the next step, the surface morphology, chemical composition and permeation properties of the fabricated rGO-PANI membranes were characterized by transmission electron microscopy (TEM), field emission scanning electron microscopy (FESEM), attenuated total reflectance-fourier transform infrared (ATR-FTIR) spectroscopy, zeta potential measurements, contact angle analysis. The filtration tests were performed by both pure water and feed solution containing 50 ppm sodium alginate (SA) as an organic model foulant and 5 mM NaCl as the background electrolyte. The combination of rGO with PANI were found to have significant advantages such as increasing surface hydrophilicity and enhancing attachment and stability of coating under anodic oxidation and natural pH conditions. The top view FESEM images revealed that linear PANI chains made very porous regions on top layer, while rGO deposited parts were almost defect-free. Through cross-sectional TEM imaging, the average thickness of top rGO-PANI layer was determined to be 120 nm. Furthermore, the water contact angle measurements showed that by increasing the ratio of PANI to rGO, membranes demonstrated more hydrophilic characteristics. Thus, these membrane materials are promising candidates for electro-conductive membrane materials suitable to participate in electro-oxidation reactions. For flux measurement, the membranes were placed in a custom-made cross-flow non-conductive chamber in which membrane top layers were connected to a DC power supply (anode). A circular piece of stainless steel (Diameter = 2 cm) was used as a counter electrode (cathode). The distance between anode and cathode was fixed in 5 mm. The water flux and organic fouling behavior of three fabricated electro-conductive membrane (M1, M2, and M3) containing different ratio of PANI to rGO were compared by applying either no or 2 V anodic potential and the total flux decline ratio (R_t) values were assessed for each membrane. The result is presented in Fig. 1. The fouling resistance propensity of all membranes were enhanced by applying 2 V potential compared with no electrical potential. The highest improvement was achieved by M3 that the total flux decline (R_t) due to organic fouling was 33.2% less in comparison with the same membrane without applying potential. The total flux decline ratio in M1 membrane was 18.2% and 10.9% in M2 membrane after applying 2 V anodic potential. As it mentioned before, electro-chemical oxidation offers a broad range of possible reactions to control organic fouling. The SA macromolecules could be directly oxidized over anodic electrode (membrane surface) without involving of any other substances. This reaction is kinetically slow and may decline catalytic performance of membrane during long-term applications. Besides, fouling may be mitigated through indirect mechanism. In this mechanism, super-active intermediates like hydroxyls (OH^-), chlorine (Cl^-), hypochlorous acid (HClO) and hypochlorite (ClO^-) may be generated in close distance to the ECMs membranes. These super active oxidizer are able to rapidly oxidize organic materials and eliminate fouling layer in advance. Oxygen evaluation reaction (OER) is usually driven over carbon-based electrodes in pretty low electrical potentials. Therefore, oxygen bubbling can also help membranes to overcome fouling in some extents. The experimental results showed more than 30% improvement in flux recovery ratio (FRR) when 2 V DC potential is applied on the surface of the rGO-PANI ECMs membrane compared with rGO-PANI membrane without applying potential. Fouling experiments conducted with organic foulant sodium alginate demonstrated the capacity of the rGO-PANI ECMs for in situ oxidative cleaning.

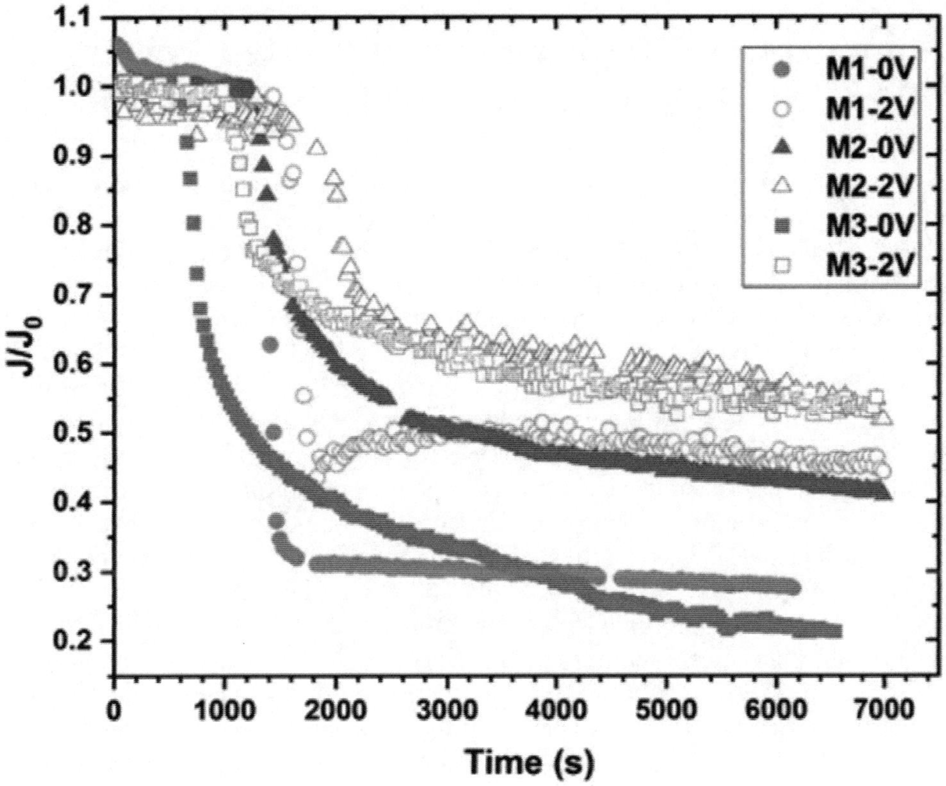

Figure 1 Fouling behaviors of different fabricated membranes.

571851 Development of Graphene Oxide-Coated Polyethersulfone Hollow Fiber Membranes with Improved Antifouling Property

Membrane technology has been used widely in many separation applications. However, membrane fouling, which is the adsorption of particulates or compounds on the membrane surface forming a layer or the blockage of membrane pores, is the biggest issue with this technique. For fouling remediation, cleaning is required which accounts for additional operational cost and reduces the membrane durability as well. Although, there are many methods for the surface modification of membrane such as blending, grafting, cross linking for fouling mitigation. Coating is an emerging membrane modification method. Membrane coating is the formation of a layer on the surface of membrane. It is one of the cheapest methods of membrane surface modification because it does not involve any expensive equipment, and additional treatment steps after membrane fabrication. Graphene oxide (GO) is one of the promising nanomaterials, which provides good mechanical strength, antifouling property, outstanding antimicrobial property against different bacterial species, hydrophilic in nature because of presence of carboxyl, hydroxyl and epoxy functional groups. In this study, GO was synthesized by the modified Hummer's method. The prepared GO was characterized for surface morphology, functionality and elemental analysis. GO solution was used as a bore solution to coat the lumen of the hollow fiber membranes (HFMs) during their fabrication. GO-coated HFMs were characterized by different imaging and spectroscopy techniques, which confirmed the coating of GO layer on the membrane surface. Minor decrease in pure water permeability of modified membrane was measured, however, a considerable improvement in the antifouling property was measured. Thus, this study presents a facile method to develop improved fouling resistant GO-coated HFMs for separation applications.

576061 Nanostructured Layer-By-Layer (LbL) Surface Modification of Nanofiltration Membranes for Effluent Treatment

Layer-by-layer (LbL) is a versatile surface modification tool, which has found applications in diverse fields like sensors, gas barrier films and drug delivery. This tool has also been extensively employed to improve the performances of existing membranes; yet the application of such membranes has been primarily limited to ion removal. Moreover, these surface-modified membranes have not been evaluated in the presence of real and complex feed streams. This talk will highlight our work focused on developing LbL-modified nanofiltration membranes for *real effluents*. By careful tuning of process parameters, such membranes showed impressive performances in terms of flux, rejection and very importantly in terms of fouling resistance. The potential to further improve the anti-fouling properties *via* introduction of nanomaterials (*e.g.* clay nanoplatelets) will also be briefly discussed. The results shown by LbL-modified membranes for such complex feed streams make them suitable for wastewater reuse and reclamation purposes.

556378 Molecular Drilling and Bonding for Highly Permeable and Fouling-Resistant Reverse Osmosis

Higher permeability and lower fouling propensity have been the persistent pursuit for reverse osmosis (RO) membranes over the past decades. Though it is argued that the benefits in energy saving for seawater desalination brought by further increment in the permeability of current RO membranes is only marginal, recent research found that membranes with a three-times-higher permeability can reduce the energy consumption by 46% in brackish water desalination. In addition, fouling is one of the major concerns in the membrane desalination process. Attachment of inorganic and organic foulants from the feed solution onto the membrane surface can cause significant flux decline, which leads to increased operating pressure and frequent physical/chemical cleaning. To reduce the operational cost and increase energy efficiency of the current system, proper antifouling strategies are critical. Existing methods reported in literature to modify reverse osmosis (RO) membranes for enhanced anti-fouling properties usually involve two or more complicated reactions on membrane surface and toxic chemicals, which are difficult to scale up. One-step and environmentally friendly approaches are strategically important for the upsize and commercialization of any new types of antifouling RO membranes.

This paper reports a fast, cost-effective and environmental-friendly method to engineer the 3D membrane structure, and simultaneously enhance the permeance and fouling resistance in RO. By free radical polymerization in water, copolymers containing fouling-resistant zwitterionic and anchoring groups are synthesized and quickly bonded to the polyamide membrane surface. The copolymer also functions as the molecular engine to reduce the thickness of the polyamide layer and drill nanopores, thereby re-shaping its 3D structure for higher permeability. It is revealed that by design of the size and functionality of the copolymer, the water permeability and salt rejection may be simultaneously increased. Our hollow fiber membrane demonstrates a pure water permeability of 10.5 ± 0.5 L m^{-2} h^{-1} bar^{-1} which outperforms currently available membranes by 2 to 3 times, NaCl rejection of 98.1 ± 0.2 %, and superior fouling resistance towards various foulants and realistic feed from the local RO plants. This paper provides a new perspective for the development of highly permeable, fouling-resistance membranes not only for RO, but also nanofiltration.

565606 The Taylor-Couette Disc Contactor, a Novel and Flexible Apparatus for High Throughput Liquid-Liquid Extraction

The Taylor-Couette Disc Contactor (TCDC) [1, 2] shows high potential for application in liquid-liquid extraction. This agitated counter currently operated extraction column makes use of the hydrodynamic advantages of the rotating disc contactor (RDC) and the Taylor-Couette reactor (TCR). The TCDC design is based on the design of the rotating disc contactor (RDC) but without stator rings and with increased shaft and disc diameter. The shaft diameter is increased to an optimum ratio of $D_{Column}/D_{Shaft}=0.5$ to prevent formation of hydrodynamic dead zones at the area of the shaft. The shaft and the discs induce banded two-phase flow similar to banded flow in the Taylor-Couette reactor. The simple design makes the TCDC insensitive to any kind of crud accumulation, and thus it becomes highly tolerant to varying quality of the feed streams. The simple design of internals allows high hydraulic load beyond $B = 30$ [m^3 m^{-2} h^{-1}]. For appropriate column design, the hydrodynamic design specification is required. Therefore, the effect of varying hydraulic load, rotational speed and phase ratio on the dispersed phase holdup, residence time distribution and drop size distribution was investigated in a 0.1 m and 0.3 m TCDC column. The total hydraulic load has a minor influence on the DSD. The effect of varying phase ratio on the dispersed phase holdup is insignificant at low hydraulic load. The effect of the phase ratio at high hydraulic load has a significant impact on dispersed phase holdup. Comparison of operation in 0.1 m and 0.3 m scale has proven CFD based simulation results of scale up rules, including the effect of varying rotor disc diameter on operation performance. Axial dispersion was determined by measuring the residence time distribution. The effect of varying rotor disc diameter on dispersed phase holdup, residence time distribution and drop size distribution was investigated to test the flexibility of this column design. The rotor disc diameter of the TCDC can be varied from design without discs (Taylor-Couette reactor) to a maximum rotor disc diameter of 90% of the column diameter. The total hydraulic load is hardly affected by the ratio of rotor disc diameter to column diameter. This makes the TCDC column easily adjustable to specific separation tasks.

[1] E. Aksamija, C. Weinländer, R. Sarzio, M. Siebenhofer, *Sep. Sci. Technol.* **2015**, 50 (18), 2844–2852. DOI: 10.1080/01496395.2015.1085406

[2] A. Grafschafter, M. Siebenhofer, *Chemie-Ingenieur-Technik* **2017**, 89 (4), DOI: 10.1002/cite.201600142

567464 Heterogeneously Catalyzed Reactions with Solvent Extraction

Isolation of carboxylic acid from dilute pulping effluents has a huge energy demand. Target constituents do not show up in broths in appropriate concentration and composition for application of simple separation technologies. Via combination of chemical conversion (e.g. esterification) with simultaneous solvent extraction, economic isolation may become feasible. In dilute aqueous systems, the high water content shifts the chemical equilibrium composition of esterification reactions towards reactants. By extracting the reaction product, liquid-liquid separation of the product shifts conversion in the aqueous carrier. Slow chemical reactions, such as esterification reactions, require catalytic acceleration. For continuous operation, heterogeneous catalysts are very suitable since they can be separated from the reaction broth via sedimentation, simplifying downstream processing. Heterogeneously catalyzed esterification combined with solvent extraction requires adequate apparatus design. The Taylor-Couette Disc Contactor (TCDC) [1], a rotating liquid-liquid extraction column, is very suitable for multiphase flow since it has a simple design of internals and can withstand harsh operation conditions. The internals of the TCDC are similar to the Rotating Disc Contactor (RDC), but without stator rings and with increased shaft and rotor disc diameter.

For investigation of the hydraulics of multiphase flow as well as conversion efficiency, continuous liquid-liquid-solid flow has been implemented in a TCDC with 50 mm column diameter and 700 mm active mixing height. Acetic acid with equimolar amount of methanol was fed on top of the TCDC. Biodiesel was used as solvent for methyl acetate extraction, and it was fed at the bottom of the column. The esterification was accelerated with the solid cation exchanger Amberlyst15®, which was fed on top of the column, separated after esterification from the continuous phase at the bottom, and recycled with a peristaltic pump. Due to the hydrophilic properties of the catalyst, it remains dispersed in the aqueous phase. Operation temperature was T = 65°C. For comparison, batch experiments were performed with the same operation conditions.

Beyond a critical rotational speed, the vorticity of the continuous phase overcomes sedimentation force of the dispersed phase and the solid phase and ensures intensive phase contact in the single compartment of the TCDC. The experimental results confirm applicability of the multiphase reactor TCDC for intensifying heterogeneously catalyzed chemical reactions such as esterification combined with liquid-liquid extraction.

[1] E. Aksamija, C. Weinländer, R. Sarzio, M. Siebenhofer, *Sep. Sci. Technol.* **2015**, 50 (18), 2844–2852. DOI: 10.1080/01496395.2015.1085406

571661 Cloud Point Temperature Purification of Butyric Acid from Aqueous Solutions Using Non-Ionic Surfactant

Butyric acid is a volatile organic acid that can be used as starting material for a wide range of applications. Butyric acid can be produced through fermentation and is typically recovered by precipitation, distillation, adsorption, reverse osmosis and membrane filtration. Previous phase equilibrium study demonstrated the ability of a non-ionic surfactant (N23E2) to extract butyric acid from aqueous solutions. After extraction, however, a surfactant-rich phase that contains most of the butyric acid is generated, which requires additional downstream process(es) to produce high purity butyric acid. The present study was conducted to evaluate the cloud point temperature of N23E2 as the main parameter for the aforementioned downstream process(es). For non-ionic surfactants, the cloud point is the temperature when the surfactant tends to separate from a solution to form an almost pure surfactant phase. Typically, cloud point temperatures are measured using a specific solvent, and thus, literature information might not reflect the true cloud point temperature of a given surfactant. Experiments were performed to produce a butyric acid-laden surfactant-rich phase (BLS phase) based on the pre-established phase diagram of butyric acid + N23E2 + water at 35°C. The BLS phase was recovered and incubated at different temperatures from 55 to 70°C for 24 hours. This temperature range was decided based on preliminary observations in the laboratory as well as information provided by the manufacturer of N23E2. After incubation, the resulting phases were sampled and analyzed for butyric acid and surfactant concentrations using liquid chromatography. Initial results indicated phase separation at various temperatures tested. It is hypothesized that the butyric acid will separate out of the surfactant and will constitute majority of the phase generated. Ongoing experiments are designed to determine the purity of the butyric acid produced using the cloud point temperature of the surfactant as the main parameter. The non-toxic and biodegradable characteristics of this non-ionic surfactant as solvent would be useful in developing a purification process for producing butyric acid as a food ingredient or component in cosmetic formulations.

588482 Solvent Design for Acid Extraction

Liquid-liquid extraction of carboxylic acids has been widely investigated. Several solvents, including low boiling esters ethers and ketones as well as high boiling reactive solvents have been proposed. The extraction capacity of low boiling solvents is poor, and economics of downstream processing of laden low boiling solvents fails because of disastrous energy consumption. Trialkylphosphane oxides and trialkylamines provide excellent partition properties. Trialkylphosphane oxides may be applied without dilution. They may also be mixed with hydrocarbon based diluents to adjust the viscosity, density and interface tension. Interaction of trialkylphosphane oxide based solvents with carboxylic acids is of Lewis acid – Lewis base character. Distillative downstream processing of laden trialkylphosphane oxide based solvents is simple, and it does not need excessive energy. Lately, trialkylamine based solvents seem to become fashionable again. Trialkylamine based solvents interact with carboxylic acids by ammonium salt formation. The laden solvent forms greasy salts. The physical properties of the basic reactive constituent must therefore be severely corrected by admixing modifiers and diluents. Distillative separation of carboxylic acids from laden trialkylamine based solvents demands high temperature (to break up the salt) and it needs an entrainer for separation. Carboxylic acids form heterogeneous acetotropes with hydrocarbons. In distillative separation of carboxylic acids from trialkylamine based solvents we may therefore make use of the diluent. However, these technological limitations do not really suggest to make trialkylamine based solvents the first choice in solvent selection for carboxylic acid separation. In conclusion, trialkylphosphane oxide based solvents offer several advantages in carboxylic acid isolation from aqueous feed over low boilers and over trialkylamine based solvents. Trialkylamine based solvents may show interesting potential in acid extraction with downstream processing by reextraction.

570430 Evaluation of the Solvent Polarity on Extraction of Castor Oil By Soxhlet Method

Conversion of castor oil into biodiesel can be an alternative technological to cover the current demand of biofuels, this vegetable oil feedstock does not challenge with the food industry and avoiding the use of the oil palm. Soxhlet extraction method was used to evaluate the polarity of three extracting solvents (hexane, diethyl ether and ethanol) on the oil extraction from castor oil seeds. A factorial design 2^3 experiments were used to study the effect of the size particle (1.981, 2,80 and 3.35 mm) and extraction time (1, 2 and 3 h) on the oil yield for each extracting solvent employed. Peroxide value, iodine value, saponification value, acid value, refractive index, viscosity index and density were the physicochemical properties used to characterize the castor oil extracted. The analysis of these properties shows that oil is suitable for the production of biodiesel. The solvent polarity has a significant effect on the oil yield, indicating that a greater oil yield was obtained when the solvent polarity is lower. Hexane showed to be the best solvent for the castor oil extraction, in terms of high oil yield (48%). However, diethyl ether reached an oil yield of 45% under the same extraction conditions used for hexane, which had a particle size of 2.80 mm and an extraction time of 3 h. Based on oil yield, the castor oil seeds have a higher affinity by non-polar solvents.

574692 Solvent Extraction for Nanoparticle Purification

Colloidal nanoparticles have been used as catalysts for many liquid-based reactions. They are typically synthesized via chemical methods with excess reducing agents and stabilization molecules. Conventionally, they need to be purified before use. There are several purification methods, including washing and centrifugation, dialysis, electrophoresis, and filtration. In this work, we explored the use of solvent extraction to remove impurities from colloidal nanoparticles in solution. The effect of purification methods on catalytic performance were studied.

567834 Digital Adsorption: 3D Imaging of Adsorption Equilibria and Dynamics in Technical Microporous Solids

Gas adsorption is one of the key physical processes that underlie the numerous industrial uses of porous solids. Typical applications include separation processes, catalytic reactions, energy storage, and gas recovery from unconventional plays. Key to maximise the efficiency of these processes is our ability to understand the microsctructure of the material at the pores scale (1–10 Å) and its interaction with the operating fluids. Yet, to be useful for industrial implementation, the powders obtained in the laboratory need to be shaped into so-called technical solids, i.e. millimetre-sized bodies with sufficient porosity, with improved chemical and mechanical stability. However, these densified solids also reveal variations in terms of the structure and composition over length-scales much larger than the pore size, from a single pellet (μm–mm) up to the packed-bed (cm – dm)[1]. Understanding the scale and extent of heterogeneity in technical adsorbents by means of advanced experimentation is a precursor to the rational scale-up of tailor-made materials, from powders to the sub-kilogram scale, and to the robust design of separation processes. Multi-dimensional imaging techniques represent an essential tool to characterise these systems over the range of relevant length scales. However, the application of these methods to systems whose function is based on the presence of nanometre sized pores remains a challenge due to the trade-off between sample size and image resolution.

We show here how X-ray transmission Computed Tomography (XCT) can be applied to obtain a three-dimensional spatial characterisation of an adsorbent bed (50 cm^3) in terms of its microstructural properties, non-invasively. We report the results of two case studies with commercially available samples of activated carbon and zeolite 13X, demonstrating the deployment of digital adsorption workflow[2,3] to produce for the first time an integrated and quantitative three-dimensional visualisation of a packed-bed adsorber *operando*. Results are presented that include both static and dynamic (breakthrough) experiments using Helium and CO_2as inert and adsorbing gas, respectively. From the static experiments, spatially distributed adsorption isotherms are measured at mm resolution and pressures 0.1–3 MPa; when combined with conventional analyses of physisorption data, we determine three-dimensional maps of the specific surface area, pore volume and other process performance metrics. We further use a machine learning approach to identify and locate different materials within the packed bed. The breakthrough experiments have been carried out at a total pressure of 0.1 MPa and at room temperature, while varying inlet composition, flow rates and tracking the adsorption fronts by XCT. All measured outputs (outlet *and*internal profiles) are described by a detailed model of the column that solves the partial differential equations consisting of mass, energy and momentum balances coupled with the appropriate initial and boundary conditions. Notably, while 1D and average properties are well captured by the model, significant variability is observed in the local adsorbed amount. The latter is associated with packing variability at the mm-scale and, possibly, inter-pellet heterogeneity

This novel ability to measure the adsorbed amount in-situ with both spatial and temporal resolution paves the way towards the characterisation of adsorption processes over the continuum of relevant length scales. This is key towards reconciling fundamental studies on the scale of a single crystal or pellet with those performed at the column scale. The gained insight is expected to contribute in bridging the gap between materials research and process design.

References

[1] R. Pini, L. Joss, *Curr. Opin. Chem. Eng.* **24**, 37–44, 2019; [2] L. Joss, R. Pini, *J. Phys. Chem. C***121**, 26903–26915, 2017; [3] L. Joss, R. Pini *ChemPhysChem***20**, 524–528, 2019

571172 Direct Characterisation of Gas Adsorption with Raman Spectroscopy

Characterising the sorption properties of materials is necessary for many industrial applications, as well as the development of novel materials with improved capacity and advanced functionality. However, depending on the characterisation technique, measurements are typically limited by a combination of relatively long acquisition time, pressure range, and required sample size. To overcome such constraints, we demonstrate a direct characterisation of gas adsorption using high-resolution Raman spectroscopy. Raman spectroscopy allows the physical and chemical features of optically compatible materials to be directly investigated by detecting vibrational modes that identify the composition and phase of the constituent molecules. In conjunction with a custom built high-pressure, temperature controlled visual cell, this approach enables absolute adsorption capacity to be rapidly measured over a wide range of temperature and pressure. In this work, we validate this novel technique by reporting results for the CO_2 adsorption capacity of a commercial silica gel, measured at 30 °C and at pressures up to 3 MPa. The results from Raman spectroscopy are compared to independent measurements using commercial gravimetric sorption and physisorption apparatuses at high and low pressures respectively, with good agreement over the entire pressure range within the combined uncertainties of each technique. Work is currently underway to extend the functionality of this technique to include adsorption selectivity. Moreover, Raman imaging opens the possibility for heterogeneous adsorption to be spatially resolved and studied within individual particles.

573661 A New Technique to Measure Multicomponent Adsorption Equilibrium: Integral Mass Balance (IMB)

Majority of adsorption applications involve gas mixtures. Accurate equilibrium data is necessary to design and operate these applications. Data is also important to test adsorption equilibrium models. Although models which can predict binary/mixture equilibrium for a class of systems, certainly any industrial entity will seek to check the actual data before investing large resources. On the other hand, Collection of binary and higher mixture data with existing approaches are tedious and very time consuming, as slow as one data point per day. There is a need to collect quality multicomponent adsorption data quickly and accurately. It is also preferable to collect data with full control of the independent variables { T, P, yi } as the gas phase properties. The data measured are the partial amount adsorbed of each component, Ni.

A new flow-through system is developed and tested to quickly measure binary adsorption equilibrium data at any desired T,P and yi. More importantly the experiments are very fast in the order of an hour per data point, compared to one point per day with systems/techniques with comparable accuracy. Unlike other similar systems based on breakthrough integration, the incorporation of an electronic balance eliminates the need to measure the outlet flowrate. Hence, the technique does not have an exit flowmeter which is usually the source for large uncertainties in measured partial amount adsorbed. The composition of the outlet gas is measured as usual in all chromatographic techniques. A quadrupole-Mass Spectrometer is used at outlet to closely monitor composition over time.

In this presentation, we will explain the concept, show the experimental system, explain data collection details, and examine the limitations of the approach. We will also show the test results comparing new measurements on O_2/N_2 mixtures in 5A-zeolite to those published by two different labs.

574033 A Combined Volumetric and FTIR Apparatus for the Measurement of Adsorption Kinetics

The measurement of adsorption kinetics in zeolites is challenging due to the combined heat and mass transfer mechanisms involved in the process. In the case of a volumetric apparatus it is possible to reduce thermal effects by modifying the thermal mass of the sample, but for strongly adsorbed and rapid uptake systems this may not achieve near isothermal behaviour.

In this contribution we present an approach based on performing a volumetric experiment with an uptake cell that can be inserted in an FTIR instrument. In the first instance a sample of ZSM-5 powder is pressed into a thin pellet and the FTIR is used in transmission mode. With this configuration both the external pressure and the average adsorbed phase concentration are measured simultaneously. The results obtained using n-hexane and toluene indicated that heat effects were not negligible. A thermal imaging camera was added in order to measure also the surface temperature of the pellet.

The system in this configuration provides a direct measurement of total pressure, average adsorbed phase concentration and surface temperature of the solid. These experimental signals are matched to the theoretical non-isothermal volumetric model and provide information that can be used to determine both heat transfer and mass transfer coefficients.

573921 Use of a Novel Volumetric Differential Pressure Apparatus for Adsorption Equilibrium Measurements

In this work we present the design of a new volumetric differential pressure apparatus for the measurement of adsorption equilibrium and kinetics. Conventional volumetric systems are based on the measurement of the absolute pressure to calculate the amount adsorbed in a calibrated volume. On the other hand, differential volumetric systems rely on the measurement of the differential pressure between two symmetric branches (namely, sample and reference) as the gas expands from the dosing to the uptake volumes. The key advantage of the use of the differential pressure (as opposed to absolute) is that it ensures the same accuracy in the entire pressure range available. This is particularly beneficial when adsorption isotherms up to high pressure need to be measured.

The system is designed to measure high pressure adsorption isotherms up to 270 bar. It is equipped with a differential and an absolute pressure transducer as well as thermocouples to measure the temperature of the dosing and the uptake volumes, including the direct measurement of the sample temperature. The system is designed with a focus on minimising the volumes, allowing accurate equilibrium and kinetic measurements with less than 100 mg of sample. In addition, the apparatus is designed to be completely modular. This allows to easily modify the system to adapt the volumes (enlarging the volumes if necessary, for example) or to change the setup in different configurations.

We present the detailed design of the experimental apparatus as well as the methodology to accurately calibrate the volumes and carry out reliable adsorption equilibrium measurements.

To validate the system we have carried out experiments on reference adsorbents using different gases in a range of experimental conditions. We also highlight the advantages and disadvantages of using the system in two different configurations.

560451 Commercializing a Membrane Technology in the Oil and Gas Industry

By some accounts, chemical separations account for roughly half of the industrial energy use in the US and membranes have been heralded for many years as a way to reduce energy as well as capital needed for those separations. Compact Membrane Systems has been focused on one of the most energy-intensive separations, namely that of separating light olefins and paraffins. Our efforts have resulted in a successful field demonstration of our Optiperm^{TM} fluoropolymer membranes on a stream of mixed C3 olefins and paraffins. We are now scaling up the membrane cartridges and developing larger commercial demonstrations.

This talk will focus on some of the highlights of our product development path thus far as well as lessons learned along the way. We have ruthlessly focused our energies on suitability for commercial adoption. This approach has led to an economically attractive, technologically capable, user friendly, and readily manufacturable product In addition to separation performance (permeance and selectivity) the technical characteristics most valued by users are stability, reliability and ease of operation. We have opted to work with potential customers as early as possible in an effort to learn at the beginning of the product development process what is feasible and what is not. We believe this has led to our success thus far, although there is still a long road to travel before widespread commercial adoption. This talk will cover what we see as the path forward and where we believe strategic partnerships will be necessary or important to ease the way.

579696 Major Applications of Osn Membrane Technology

Abstract

For more than 60 years, membranes have been used for molecular scale separations in aqueous solutions with applications in reverse osmosis and nanofiltration – e.g. waste water treatment, water desalination, dairy industry - but the new frontier in liquid filtrations is molecular separation in organic solvents [1]. Organic solvent nanofiltration (OSN) has emerged in the last 5-10 years as an effective technique for product purification in a wide range of process chemical industries (e.g. Pharmaceuticals, Specialty, Chemicals, Petrochemicals, Natural Oils etc.) where molecules in the 150 to 1,500 Da range need to be separated or concentrated in presence of an organic solvent. The breakthrough in OSN technology has occurred with the commercialization of Evonik's DuraMem® and PuraMem® range of membranes. This has opened up possibilities for these OSN applications in a variety of organic solvents ranging from non-polar through polar to polar aprotic.

In majority of process chemistries, up to 90% of the capital investment involves separation processes [2]. OSN can be applied as a unit operation on its own or in combination with other unit operation(s) to achieve various separation tasks. OSN membrane technology is particularly advantageous when replacing or minimizing a high temperature (e.g. evaporation) or a low temperature (e.g. chilled extraction / de-waxing) unit operations thus leading to substantial energy savings. With separation of molecules of different molecular weights in an organic solvent, OSN offers a particularly energy efficient way to achieve the purification and thus enhancement in product quality and ease of manufacturing.

The relevance and importance of these applications will be demonstrated through several major case studies on applying OSN. Benefits such as mass efficiency via increase of product yields, reduction of process time, and reduced use of high-energy consuming technologies will be presented.

References:

[1] Vandezande, P., Gevers, L.E.M., Vankelecom, I.F.J., "Solvent resistant nanofiltration: Separating on a molecular level", Chemical Society Reviews, 37 (2), 365-405 (2008).

[2] Cuperus P., Ebert K., "Nanofiltration- principles and applications", Elsevier Advanced Technology, 521-536, 2002.

564542 Membrane Based Acid Gas Separation: Past, Present and Future

Past:

Membrane based CO_2 separation was discovered and practiced as early as 1980s. The overall membrane adoption rate for these type of separation was lower due to several challenges related to pretreatment and unknown process design expertise for membrane systems. The overall experience in early adoption of membrane was mixed bag with some installations were successful while others did not make mark due to lack of process design and inadequate pretreatment. The early adoption of CO_2 and H_2S removal membrane started for Enhanced Oil Recovery (EOR) applications where high CO_2 feed gas streams were uneconomical to treat using conventional technologies like Hot potassium carbonate and Amine treating. Gas membrane system provided an economical solution to remove CO_2 and recover maximum hydrocarbon in the product gas. The higher CO_2 recovery with higher purity improved overall CO_2 EOR processing economics.

Present:

Gas separation membrane adoption has increased over last 30 years and has shown significant adoption in offshore CO_2 removal projects. The high CO_2 natural gas producing assets around the world require more compact and light weight separation technology which membrane technology is a clear winner. The offshore CO_2 removal membrane systems must be more agile to address offshore processing challenges with higher up time availability. The new FPSOs and fixed platforms requires bulk CO_2 to be removed offshore followed by onshore treatment of gas for LNG or midstream gas processing. The current challenge for offshore is that membranes are still little less competitive for low $CO_2 < 10\%$ streams compared to amine system. This is due to pretreatment and recycle compression for higher hydrocarbon recoveries. However for large CO_2 removal applications, membrane is still a preferred option.

Future:

The future for gas processing CO_2 removal membranes is around applications which are usually not economically feasible for gas membranes, like low $CO_2 < 10\%$ removal applications, membranes with limited pretreatment and remove CO_2 and H_2S more effectively with maximum hydrocarbon retention. The approach of current membranes in combination with other technologies, hybrid approach, could address some of these applications more cost effectively than standalone membranes solution. However new age of nano composite with highly cross linked polymer membranes can limit the size of pretreatment and provide long term benefits. Also, digital integration of membrane performance and real time advisory tool can provide higher visibility of membrane performance and optimization making membranes more viable with higher adoption rate for onshore and offshore gas processing applications.

580036 Onboard Fuel Separation

ExxonMobil, Corning and Toyota have collaborated on an Onboard Separation System (OBS) to improve gasoline engine efficiency and performance [1]. OBS is a membrane based process that separates gasoline into higher and lower octane fractions, allowing optimal use of fuel components based on engine requirements. The novel polymer -ceramic composite monolith membrane has been demonstrated to be stable to E10 gasoline, while typically providing 20% yield of high octane product when using regular unleaded gasoline. The OBS system makes use of wasted exhaust energy to effect the fuel separation and provides a means for managing the separated fuels. This presentation illustrates how the principles of Process Intensification (PI) were applied in the OBS project.

556413 Gas Separation Membrane Innovation at Air Liquide

Currently, Air Liquide's gas separation membranes are utilized on land, air, and sea. On land, Air Liquide's membrane products produce purified nitrogen, hydrogen, natural gas and biomethane. At sea, Air Liquide's membranes provide nitrogen blanketing, on-site fuel generation and natural gas purification. In the air, Air Liquide's OBIGGS (On Board Inert Gas Generating Systems) membranes provide enhanced safety by inerting fuel tanks for commercial and military aircrafts.

Although tremendous progress has been made for industrial applications, polymeric gas separation membranes with robust performance towards real gas feed streams and improved gas selectivities are still critically needed. In the last 10 years, Air Liquide has rolled out two innovative gas separation membrane platforms: PEEK-Sep, poly(ether ether ketone) based membranes, and PIX, polyimide based membranes. PEEK-Sep membranes are robust for purifying feed gases from almost all contaminants, and PIX membranes demonstrate exceptional high gas separation selectivities. The platform generated by combining PEEK-Sep and PIX technologies has significantly simplified the gas separation membrane process scheme and made it much more competitive towards cryogenic, and adsorption separations. This talk will briefly review the development of the two membrane platforms and discuss their applications for natural gas and biogas purifications. We emphasize the importance of achieving contaminant resistance and high gas separation selectivity as two key elements for next generation membrane development.

576124 Nuclear Quantum Effects on Adsorption in Nanoporous Structures

Nuclear quantum effects are typically not important for most adsorption and separation processes. Therefore, almost all molecular simulations of adsorption, diffusion, and separation in porous materials use a classical treatment of the nuclei. However, there are notable cases where differences in nuclear quantum effects provide the driving force for a separation process. These are rare, but important problems, primarily dealing with isotope separations. In this work, we have employed path-integral molecular dynamics simulations to rigorously account for quantum motion of the nuclei. We have used this method to study the selectivities of isotopologues of oxygen molecules in MFI zeolites and carbon nanotubes. We also use these methods to study structural changes in Calcium-based squarate frameworks brought about by adsorption of H_2O and D_2O. We use a combination of high-order splitting techniques and appropriate thermostats to achieve convergence of quantum mechanical properties at a much reduced computational cost. Our results indicate that a difference of nuclear quantum kinetic energy results in the preferential adsorption of heavier isotope of oxygen in MFI zeolites and a negative thermal expansion as a result of entrapping of heavier water in squarate-based frameworks.

576680 Toward Sustainable Metal-Organic Frameworks for Post-Combustion Carbon Capture – Identifying Improvement Opportunities By Molecular Simulation and Life Cycle Assessment

Anthropogenic CO_2 emitted from the combustion of fossil fuels has been a significant contribution to the global warming effect, and fossil fuels will continue to be used for electricity generation in the coming years. To sustain the usage of fossil fuels and decrease CO_2 emission, carbon capture and sequestration (CCS) has been regarded as a near-term viable solution, but implementing such a process using the amine technology that uses monoethanolamine (MEA) to capture CO_2 involves a large energy penalty which decreases the efficiency of power generation. To achieve a more energy-efficient CCS process, exploring novel materials to efficiently capture CO_2 has drawn significant attention. Specifically, metal-organic frameworks (MOFs) have been identified as promising candidates for carbon capture, owing to many of their favorable properties (e.g., highly tunable nature, large adsorption capacity, and selective adsorption). Identifying promising MOF candidates has been an important research area, and most of the studies reported to date mainly attempted to evaluate MOFs based on their adsorption properties (e.g., selectivity) or their performance using a process model. Although MOFs have been demonstrated to possess potentially better performance (i.e., less energy intensity) than MEA, the overall impact by this emerging class of materials remains unknown. To this end, to facilitate the development of a new technology based on MOF adsorbents, the impacts of implementing MOF-based carbon capture, including the environmental impacts and resource depletion imposed from MOF synthesis process, should be considered. In this study, we carry out a comprehensive lifecycle assessment (LCA), coupled with molecular simulations, to investigate a selected set of 50 MOFs with diverse characteristics for their impacts when used in carbon capture and compared with that by using MEA. A variety of indicators such as global warming potential, eutrophication potential, and resource depletion are quantified from this study to shed light on different aspects of environmental impacts by using MOFs in carbon capture. In this study, extensive experimental data reported in the literature is used as a reference in the LCA to estimate the impact by the MOF synthesis from, e.g., solvent usage and synthesis energy. Molecular simulations are also conducted to calculate the adsorption properties of each studied MOF, which are subsequently used to determine the energy requirement of the MOF to separate a given amount CO_2 as well as the amount of adsorbent materials needed. Our results again show the great promise of MOFs in carbon capture, from the life cycle point of view. Moreover, our study highlights the significant environmental impact due to the usage and the type of the solvents, which suggests the importance of optimizing the synthesis conditions as well as seeking for the use of green solvents in the synthesis. The key role of MOF stability is also identified in determining the impact of MOFs in carbon capture. Overall, the insights obtained in this study can help guide the development of MOF-based CCS, and the analysis framework established herein can also be used to facilitate the development of other applications using MOFs.

576538 DFT Study for Developing Novel and Specific B5-Type Step Site Scaling Relations and BEP Analysis for Stepped HCP Catalyst Surfaces

Introduction

In the literature recently there have been reports that show that HCP metal catalyst nanoparticles and their activity may be highly sensitive to the exact nature and number of 'B5' step and edge sites present in the faceting and facet-intersections of the nanoparticle.1-3 Unfortunately to date, theory and modeling treatments in the literature for 'stepped' HCP surfaces largely rely on using ad-hoc deletion of atoms (or atomic rows) from ideal HCP(0001) surfaces to create representative surface models. Our work in this submission highlights the possible failings of such an approach; specifically we examine for differences in adsorption energies at various B5 step edge types, and any possible trends across the same type of B5 sites on various HCP catalyst species that have not yet been thoroughly characterized. Our work in this manuscript uses the low energy (1016) Miller Index surface of Co, Os, and Ru which exposes 2 distinct and strongly adsorbing step edge sites, the B5-B and B-5A step edge which have been reported as relevant in the literature for Cobalt nanoparticle catalysis applications. Results from this study should be used to help further understand atomistic processes on the stepped surfaces of catalytically active HCP elements.

Materials and Methods

The work described and presented was performed using modern, state-of the art plane-wave DFT calculations as implemented in the VASP 5.4x code deployed in the Materials Design MEDEA environment 2.21. Calculations and supercells for surface-mediate adsorption and catalytic phenomena involved dipole corrections parallel to the surface normal, vacuum spacing of at least 65% of the overall calculation cell height, and slab models of at least 6-8 Å or equivalence to 4 crystalline layers of the material(s) studied. Typical electronic convergence of the SCF was performed to at least 10-6 eV, and atomic force relaxations of the uppermost slab model atoms and adsorbed species was performed to less than 0.035 eV/ Å . Specific functional(s) used for the calculations in this study was rPBE.

Results and Discussion

Our results have determined that there are commonalities and trends in binding for various mono-atomic and diatomic adsorbate molecules on the HCP(1016) surface(s) of Co, Os, and Ru. Though there are some exceptions, we have used these results to examine adsorption energies and transition state energies on this and other stepped HCP surface to determine step-edge dependent scaling and BEP relations related to their potential catalytic activity. Our results show it is very important to understand that not all HCP "steps" are equivalent, and that previous literature attempts to model and develop scaling relations using the simple approach of ad-hoc deletion of atoms from a model HCP(0001) surface slab may be insufficient to truly model atomistic processes on the surface of HCP metal catalyst nanoparticles.

Significance

We present results from DFT studies on the effects local geometric and electronic structure on the relative adsorption and catalytic properties of differing B5-type steps (B5-A, B5-B, etc) of actually stepped HCP model surfaces such as the HCP(10-16) surface. Trends on various metals such as Co,Os, and Ru are discussed. Specific scaling and BEP relations for trends in catalytic activity on these different active sites have not been reported to date; we hope our new work helps fill this gap and leads to enhanced understanding of atomistic processes on HCP metal nanoparticle catalysts.

References

1. Petersen, M. A.; van den Berg, J.-A.; Ciobîcă, I. M.; van Helden, P., Revisiting CO Activation on Co Catalysts: Impact of Step and Kink Sites from DFT. ACS Catalysis 2017, 7 (3), 1984-1992.
2. García-García, F. R.; Guerrero-Ruiz, A.; Rodríguez-Ramos, I., Role of B5-Type Sites in Ru Catalysts

used for the NH3 Decomposition Reaction. Topics in Catalysis 2009, 52 (6), 758-764.
3. Agrawal, R.; Phatak, P.; Spanu, L., Effect of phase and size on surface sites in cobalt nanoparticles. Catalysis Today 2018, 312, 174-180.

576001 Impact of H_2O and CO_2 on Methane Storage in Metal-Organic Frameworks

We investigated eight representative metal-organic frameworks for methane storage using molecular simulation. Validated force fields were used to calculate the amount adsorbed for pure methane and its mixtures with CO_2 and H_2O at 5.8 and 65 bar at 298 K within the composition limits specified for natural gas. Within the analyzed concentrations, the MOFs without open metal sites were minimally influenced by the presence of CO_2 and H_2O. However, for the MOFs with open metal sites, the presence of these species proved to be harmful. We found that concentrations as low as 25 ppm of water can reduce the delivered volume of methane by more than 20%. A detailed analysis of the adsorption mechanisms leading to this poisoning is presented. These results highlight the limitations of MOFs with open metal sites for use in natural gas storage.

574413 Predictive Models of Azeotropic Adsorption from Molecular Simulations and Macroscopic Theories

The knowledge of multicomponent adsorption equilibria is of crucial relevance in the design of adsorption separation processes[1]. Prediction of multicomponent equilibria from single component measurements mainly relies on the assumption that the adsorbed mixture can be treated as an ideal solution[2]. However, some adsorption systems deal with heterogeneity of the solid sorbent and fluid mixtures composed of species different in size and polarity. Azeotropic behaviour can also occur in such cases, and its analysis and macroscopic modelling still remains a major challenge[34].

This works has two objectives: to understand behaviour of complex adsorbed mixtures, capable of forming azeotropes, on a molecular level; using insights from a molecular-level picture, develop a predictive macroscopic model of adsorption of these mixtures, suitable for design of adsorption processes.

For this, we employ Grand Canonical Monte Carlo (GCMC) simulations to explore how the heterogeneity of the solid, and mixture of gases of different size and polarity, can lead to azeotropic adsorption behaviour. As a case study we consider binary isobutane-ethylene mixture adsorbed in 13X zeolite[5]. Within molecular simulations adsorptive behaviour of different regions of the porous space can be explored in isolation from the rest of the system[6]. This allows us to define the adsorption sites within the system of interest and obtain specific information about properties of these sites, such as: heat of adsorption, Henry's law constants, full single and multicomponent isotherms of both isobutene and ethylene on each site. This analysis elucidates a microscopic origins of the azeotropic behaviour of the system.

The results from GCMC simulations and the analysis above provide a foundation for the parametrisation of the thermodynamic "Rigid Adsorbent Lattice Fluid" (RALF) model, recently developed by Brandani[7]. The novel heterogeneous version of RALF is presented here to account for different adsorbing sites. Indeed, RALF model has shown comparable performance for the prediction of single component isotherms with respect to empirical equations, and good *a priori* predictive capability for complex multicomponent systems. Therefore, using a combination of molecular simulations and advanced RALF theory, we achieve the second objective of this study.

References

1. Ruthven, D. M. *Principles of Adsorption and Adsorption Processes*. (1984).
2. Myers, A. L. & Prausnitz, J. M., *AIChE J.* **11,** 121–127 (1965).
3. Do, D. D. & Do, H. D., *Adsorption* **5,** 319–329 (1999).
4. Lucena, S., Snurr, R. Q. & Cavalcante, C. L. J., *Adsorption* **13,** 477–484 (2007).
5. Hyun, S. H. & Danner, R. P., *J. Chem. Eng. Data* **27,** 196–200 (1982).
6. Krishna, R. & Paschek, D., *Phys. Chem.* **1,** 453–462 (2001).
7. Brandani, S., *AIChE J.* **65,** 1304–1314 (2019).

570436 Density Functional Theory Study of Biomolecule Adsorption to Graphene and Modified Graphene: Molecular Insights into Biofilm Formation and Adhesion

Despite their prevalence in natural and engineered systems, the exact mechanisms and forces controlling biofilm formation and adhesion are relatively unknown. Moreover, most modeling studies to date have focused more on continuum or macroscopic film behavior. While those studies are important and necessary for biofilm engineering, investigations at the molecular level are needed, too.

Recent work has hypothesized that biofilm formation and adhesion may be related to the adsorption of key, early molecules including exopolysaccharides (EPS) to surfaces.[1] To explore this hypothesis, we have studied the adsorption of mannose and glucuronic acid to graphene and copper-modified graphene surfaces using molecular modeling and simulation.

Specifically, the Adsorption Locator module within Biovia Materials Studio was used to sample adsorption configurational space through Monte Carlo simulation, using the COMPASS and Universal classical force fields.[2-4] After identifying the most energetically stable physisorption configurations, these same configurations were used as starting structures for subsequent density functional theory (DFT) quantum chemical calculations. DFT calculations are conducted using the DMOL3 module within Materials Studio.[5] The generalized gradient approximation with PBE functional, Grimme's DFT-D dispersion correction, and a DNP basis set are used for all calculations.[6-8] In addition to vacuum calculations, we have also studied the effect of liquid water solvent on adsorption by implementing the COSMO continuum solvation model.[9]

[1] Wang, J., Exopolysaccharides from a thermophilic bacterium *Geobacillus* sp. WSUCF1 – production, characterization, bioactivities and mechanisms, Ph.D. Thesis, South Dakota School of Mines and Technology, 2019.

[2] https://www.3dsbiovia.com/products/datasheets/adsorption-locator.pdf

[3] Sun, H., *J. Phys. Chem. B*, **102**, 7338 (1998).

[4] Rappé, A. K., Casewit, C. J., Colwell, K. S., Goddard III, W. A., and Skiff, W. M., *J. Am. Chem. Soc.*, **114**, 10024 (1992).

[5] Delley, B., *J. Chem. Phys.,* **113,** 7756 (2000).

[6] Perdew, J. P., Burke, K., and Ernzerhof, M., *Phys. Rev. Lett.*, **77**, 3865 (1996).

[7] Grimme, S., Antony, J., Ehrlich, S., and Krieg, H., *J. Chem. Phys.,* **132,** 154104 (2010).

[8] Delley, B., *J. Chem. Phys.*, **92**, 508 (1990).

[9] Eckert, F. and Klamt, A., *AIChE J.*, **48**, 369 (2002).

573442 Understanding the Adsorption of 4th Generation Refrigerants on Metal-Organic Frameworks By Molecular Simulations

To reduce the overall emission of greenhouse gases, the Kigali Amendment to the Montreal's Protocol is forcing to phase-out hydrofluorocarbons (HFC) with high Global Warming Potential (GWP) and replacing them by refrigerants with low GWP.[1] In this regard, the development of refrigerants has evolved to the fourth generation, with sustainable and environmental friendly substances mainly known as hydrofluoroolefins (HFOs) [2], while part of the challenge remains on finding the best refrigerants and blends for the selected applications. For instance, taking into account that adsorption cycles can be integrated into typical refrigeration systems to enhance the cycle efficiency -by reducing the condenser temperature-, such fluorocarbon-based refrigerants have been examined recently with traditional adsorbent materials such as activated carbons and zeolites.[3]

Metal-Organic Frameworks (MOFs) comprise an important class of novel solid-state materials. MOFs are highly crystallized structures composed by the combination of metal clusters and organic ligands through the coordination bond or intermolecular interactions, possessing good thermal stability, high specific surface area and strong adsorption affinity for adsorbates, and making them promising for energy storage and gas separation.[4] Nevertheless, studies related to their adsorptive performance for HFO refrigerants is rather scarce.[5] Understanding the adsorption and phase behavior by accurately predicting the thermophysical properties of these systems with low GWP refrigerants is essential for designing and evaluating refrigeration cycle performances and determining optimal conditions and possible compositions.[6]

In recent years, molecular simulation studies have been gaining acceptance as an important complement to experiments in order to obtain reliable thermodynamic and transport properties of these new class of refrigerants.[7] Pioneering work on the use of these techniques for the adsorption of refrigerants in MOFs has been very recently published by Hu et al. [8]. In this contribution, we will present and discuss results concerning the adsorption of fourth generation refrigerant molecules (HFOs and blends) on different MOFs families, searching for the best conditions for cold thermal energy storage. Simulations were performed using the Grand Canonical Monte Carlo (GCMC) technique, to gain a molecular-level understanding of the adsorption mechanism of gases in these porous materials, and were also validated with measurements obtained experimentally in our laboratories. Focusing on the behavior at operating conditions, the effects of temperature and pressure on the adsorption and desorption (i.e., regeneration) steps were evaluated, as well as the compositions of the blends. Interfacial properties and wetting phenomena were also investigated through the molecular simulations.

We acknowledge support for this work from Khalifa University of Science and Technology (project CIRA 121)

[1] United Nations Treaty Collection. *Kigali*, (2016).

[2] Mota-Babiloni, J. Navarro-Esbrí, A. Barragán-Cervera, F. Molés, B. Peris, G. Verdú. *Int. J. Refrigeration*, 57, 186-96 (2015).

[3] D. Banker, M. Prasad, P.Dutta, K.Srinivasan. *Appl.Therm.Eng*., 29, 2257–64 (2009).

[4] K. Henninger, H. A. Habib, C. Janiak, *J. Am. Chem. Soc*., 131, 2776–77 (2009).

[5] B. Getman, Y. S. Bae, C. E. Wilmer, R. Q. Snurr. *Chem. Rev*., 112, 703–23 (2012).

[6] A. Fouad, L. F. Vega. *AIChE J*., 64, 250–62 (2018).

[7] Raabe, E. J. Maginn. *J. Phys. Chem. B*, 114, 10133-42 (2010).

[8] Hu, C. Liu, Q. Li, X. Shi. *Int. J. Heat & Mass Transfer* 125 1345–1348 (2018).

571670 Modeling the Adsorption of Isobutane/Butene on Hydrophobicity/Hydrophilicity Modified Solid Acid Surface

Modeling the Adsorption of Isobutane/Butene on Hydrophobicity/Hydrophilicity Modified Solid Acid Surface

Piao Cao, Weizhong Zheng, Weizhen Sun, Ling Zhao

State Key Laboratory of Chemical Engineering, East China University of Science and Technology, Shanghai 200237, China

The alkylate produced by the C4 alkylation reaction with a high-octane number and cleaning burning, is an ideal blending component for gas upgrading. Compared with the traditional concentrated sulfuric acid process, the solid acid catalyst has the advantages of environmental friendliness and easy separation, but the easy deactivation characteristics is the dilemma faced by the industrialization of solid acid process. The key to inhibiting the inactivation of solid acid is to increase the ratio of isobutane/olefins (I/O) in the microenvironment around the active site of solid acid, and to avoid oligomerization of the initially formed carbocation with butene molecules. Researches have shown that the surface modification by grafting functional groups is an effective way to modify the textural, chemical and morohological properties of solid acids, thus influence the adsorption selectivity and catalytic properties. However, the interaction mechanism between surface groups and C4 molecules and the adsorption and diffusion behavior of C4 molecules are still unclear.

In this work, the adsorption behaviors of isobutane and 2-butene on silica surface with a series of functionalized surface groups, i.e. [-OH], [$-C_3H_7$], and [$-C_3H_6-SO_3H$], were studied using molecular dynamics (MD) simulation. The transport diffusion coefficients of C4 molecules were obtained in order to assess the effect of surface functionalization with organic groups.

MD simulations reveal a strong interaction of isobutane molecules and butene molecules with the silica surfaces, which induces significant concentration gradients between two silica walls in all systems. In addition, it was found that the hydrophobic modification of the surface by grafting propyl groups could inhibit the adsorption of butane on the surface, thus improving the ratio of I/O which is beneficial to inhibit the polymerization reaction of butene and obtain a higher quality of alkylate. However, the hydrophilic modification of the surface by grafting propanesulfonic acid groups results in a higher ratio of I/O compared to unmodified surface with silanol groups, which can be attributed to the stronger interaction between propanesulfonic acid groups with butene than isobutane. In addition, the butene molecules diffuse faster than isobutane molecules in all simulation systems while the difference in diffusion coefficient between isobutane and butene is minimal in propyl functional system, which indicates the ability of the hydrophobic groups to tune the ratio of I/O in the interface between C4 reactant phase and solid acid catalyst. This simulation study provides a microscopic insight into the adsorption and diffusion behavior of C4 molecules on hydrophobicity or hydrophilicity modified solid acid surface, which can further guide the design and improvement of the solid acid catalyst of isobutane alkylation.

574734 Prenucleation Motifs of Various Dimensionality in Entropic Crystallization

In multi-step crystallization pathways, a single or multiple prenucleation state appears during the crystallization process. Such multi-step crystallization pathways occur in many chemical and physical systems, for instance, calcium carbonate ($CaCO_3$) solutions favor compact pre-nucleation clusters at the early stage of nucleation, and in the crystallization of a gas hydrate, water molecules form amorphous frameworks around guest molecules before clathrate nucleation occurs. The driving force for the appearance of the precursors has been explained by unique characteristics of the system, but universal characteristics shared by the multi-step crystallization process are not well understood. Here, we report three hard particle systems (truncated tetrahedra, pentagonal bipyramids and triangular bipyramids) exhibiting two-step crystallization via a metastable fluid-fluid phase transition. Monte Carlo simulations reveal that these systems first form a high-density fluid with prenucleation motifs in the form of clusters, fibers, and networks from a low-density fluid. Subsequently, complex crystal structures (*cF*432, *oF*244 and *cP*92) nucleate from the high-density fluid. Our observations demonstrate the existence of diverse crystallization pathways in entropic systems and reveal various dimensionalities of prenucleation motifs.

576276 A Multidimensional Free Energy Approach for Polymorph Selection in Crystallization of Glycine from Solution

Being able to predict the polymorph that is likely to crystallize from solution under given conditions is of utmost importance to many industries, as the resulting polymorph will determine key physical properties of the end-product. In the pharmaceutical industry, for example, the bioavailability of a drug molecule depends on its crystal structure, which directly affects the solubility and dissolution rate of the drug. Moreover, obtaining appropriate models of industrial crystallization units requires insight into the mechanism of nucleation and knowledge of polymorph specific kinetic properties such as nucleation rates. However, polymorph prediction remains a challenge both experimentally, due to the small time and length scales associated with formation of critical nuclei and computationally, due to the rare event nature of nucleation. Classical nucleation theory (CNT) has been the most prominent theoretical model to study and understand nucleation of a new thermodynamic phase. While it has been proven to work well for simple systems, the one-dimensional free energy description in CNT makes it less applicable to complex systems such as crystallization from solution. CNT assumes that the free energy change associated with formation of a nucleus only depends on the size of the nucleus and fails to provide any information about the structure of the critical nucleus. However, recent studies have revealed that the changes in the structure of the nucleus also play an important role in the nucleation mechanism. Especially in the presence of multiple possible polymorphs, structural information as well as the nucleus size should be present in the free energy description.

In this work, we describe a methodology to predict the polymorph of glycine that will crystallize from aqueous solution by using molecular simulations. Based on the discussion above, we aim to increase the dimensionality of free energy by introducing polymorph specific nucleus size coordinates. A template matching algorithm is used to differentiate between the three known polymorphs of glycine and to calculate polymorph specific nucleus sizes. These collective variables provide information about both the size and the structure of the nucleus. The "string method in collective variables" can be used to obtain the minimum free energy paths on this multidimensional free energy surface connecting the solution phase to each of glycine's polymorphs in the crystallized phase. This makes it possible to determine the free energy barrier for nucleation of each polymorph. We describe how hybrid MD/MC simulations can be used to evaluate the gradient of free energy at each point along the "strings", allowing the use of discrete collective variables. The described methodology allows polymorph selection for nucleation from solution, without assuming the nucleus shape or structure.

576991 Towards Digital Design of Crystals: Predicting Absolute Chemical Potentials for Solid, Solution and Gas Phases

Currently, more than 90% of small molecule drugs are delivered in crystalline form. Solubility of these crystalline molecules plays a crucial role in drug design as it directly affects the bioavailability of a drug. Also, the solubility as predicted by the force field is crucial in all efforts to predict nucleation and growth rates. The thermodynamic definition of solubility is, the concentration of the solute in the solvent which makes its (solute) chemical potential equal in both the phases (solid & solution).

Solubility prediction using atomistic simulations is a challenging task. The solubility prediction of NaCl took nearly a decade of research efforts. The solubility prediction of polyatomic molecules is an ongoing effort and we have developed a new approach that employs independent predictions of absolute chemical potentials in the solid and solution phases without having a common starting reference system. For the solid phase we apply the Einstein crystal method in its unaltered form although we have extended it differently from some of the previous works to compute the free energy of the molecular crystal (solid). In order to compute the absolute chemical potential in the solution phase, we have developed a new gas phase reference system to compute the absolute chemical potential of an isolated polyatomic gas molecule. To which we add the solvation free energy of the molecule at a given solute concentration to give the absolute chemical potential in the solution phase.

We leverage and demonstrate the computational tools developed so far to predict the solid-vapor equilibrium curve for succinic acid (our model compound), i.e., predict its sublimation vapor pressure as a function of temperature. We find excellent agreement between our predicted results and experiments in the literature. Thus, having tested our methodology we are currently applying it to the solubility problem.

576552 Irreversible Inhibition of Hematin Crystallization By Cooperative Phase Behaviors Induced By Antimalarials

Crystallization is the central process of materials synthesis in biological, geological, and extraterrestrial systems. Nature achieves remarkable diversity of shapes, patterns, compositions, and functions of the arising crystalline structures by combining simple strategies to control the number of nucleated crystal and the sizes to which they grow. To promote or inhibit crystallization in both natural and engineered environments, soluble foreign substances are deployed that interact with the solute or the crystal-solution interface. For solution grown crystals, two classes of inhibition of step propagation are discussed. In the first mechanism, known as step-pinning, inhibitors bind to flat terraces and arrest step growth over broad areas of the crystal surface. Alternatively, inhibitors may associate and block kinks, the sites where solute molecules incorporate into steps. We examine the growth of beta-hematin crystals, a component of the physiology of malaria parasites, in the presence of artemisinin and quinoline derivatives, drugs that represent the current front line of antimalarial defense. We demonstrate two novel mechanisms of growth inhibition. At moderate hematin concentrations, a heme-artemisinin adduct, H-ART, promotes copious nucleation of hematin nanocrystals in the solution bulk, but impedes their growth above ca. 100 nm. The nanocrystals associate to the surface of larger hematin crystals and completely arrest their growth. The strain due to the imperfect incorporation of the nanocrystals constrains growth even after H-ART is removed, leading to irreversible inhibition. Pyronaridine, PYR, a quinoline of relatively large size, suppresses nucleation of nanocrystals in the solution bulk and exhibits an entirely different mechanism of growth suppression. PYR stabilizes pairs of steps and leads to abundant step bunching. At elevated hematin concentrations, the step bunches engender solute occlusions that resolve as screw dislocations outcropping on the crystal interface. The high density of steps originating at the screw dislocations strongly delays step propagation owing to competition for supply between adjacent steps. Removal of PYR does not restore the initial growth rate value since the high step density is preserved. The two examples of nonclassical irreversible and partially reversible inhibition may provide guidance in the search for suitable inhibitors control crystallization of pathological, biomimetic, and synthetic materials. In a broader context, our results highlight modifier interactions mediated by the dynamics and structures in the solution and on the crystal interface as a primary element of the regulation of the shapes and patterns of crystalline structures in nature and industry.

568311 Identification of Polymorph Specific Molecular Interactions during the Process of Crystallization

Mechanism of molecular attachment during crystallization is known to be impacted based on the degree of supersaturation. Rapid crystallization at high supersaturations is known to produce metastable polymorphic forms and slow crystallization at moderate to low supersaturations produces stable forms of the molecule being crystallized. Experimental approaches of crystallization often rely on maintaining the operating conditions in the regime of desired crystal structure. However, the underlying mechanism of this phenomena has not yet been studied. In this study, Molecular Dynamics simulations were used to show that the dynamics of solvation shell around the molecule being crystallized plays a critical role in the formation of desired crystal structure and the effect of supersaturation can also be quantified in terms of activation barrier of crystallization. The identification of these dynamics further allows reducing complexity of modeling molecular interactions using stochastic simulation approaches. The specific molecular interactions during stochastic simulations can be reliably associated to a certain crystal structure to identify the nucleation pathway of the molecules to crystallize in certain structure. The multiscale approach highlights the importance of identifying molecular-scale kinetics of crystallization processes to successfully predict the outcome of crystallization.

577445 Crystallization of MIL-53 Metal Organic Frameworks (MOFs) through Changes in the Hydrothermal Process

Metal organic frameworks are a class of porous materials with highly ordered 1-,2-, or 3-dimensional networks which are formed by the coordination of metal clusters with organic ligands. Because of their large surface areas and high pore volumes and long-range order, these materials can capture molecular species and are thus may be applied to a variety of functions such as molecular storage, catalysis and separation. MIL-53 crystals are well-studied materials which belong to a subclass of MOFs with rod-packing motifs and thus 1-dimensional channels, which makes them particularly interesting for their applications to passive separation processes as size-selective separation materials. In order to demonstrate their feasibility as separation materials some challenges must be addressed. Among these are the control of crystal size and number of defects and crystal intergrowths through the hydrothermal synthesis method which is perhaps the most popular process used to produce the most robust thermodynamic MOF products. Though typically MIL-53 crystals are made as large as a micrometer in diameter, we apply principles of crystallization to make large MIL-53 single crystals, some of which may be visible to the naked eye, through the hydrothermal method. We also explain under which conditions these approaches may lead to the formation of lesser defects.

566023 Avoiding Distribution Confusion: Using the Information in Particle Size Distributions

Particle size distributions of solid active pharmaceutical ingredients (APIs) govern the material properties important to bioavailability such as dissolution. To characterize API properties, rich data sets such as Morphologi G3 or surrogate measures such as laser light scattering are used to measure the full population of particles within a sample. However communicating this distribution often relies on histograms of the distribution or simple summary measures of volume weighted quantiles, e.g D10, D50, D90. These measures cannot adequately capture multi modal distributions and can be insufficient for setting material specifications.

By using concepts from nonparametric statistics and information theory, the information loss incurred by using these summary metrics and histogram binning can be quantified. A method is proposed for compressing a high resolution distribution into a summary measure that best captures the underlying particle characteristics. In addition to a theoretical study of information, an example of different summarization approaches applied to a PBPK model for a model API will be presented. The resulting approach of maximizing the value of particle characterization data is a useful methodology for communicating and modeling the behavior of particulate pharmaceutical compounds.

574784 Combining Liquid-Liquid Phase Separation and Mesoporous Carrier to Deliver Drug Product Intermediate

Combining Liquid-Liquid Phase Separation and Mesoporous Carrier to deliver Drug Product Intermediate

Moussa Boukerche, Sean Garner, and Nandkishor K. Nere

Center of Excellence for Isolation and Separation Technologies (CoExIST), Process Research and Development, AbbVie Inc

Liquid-Liquid Phase Separation (LLPS), typically known as oiling out is an undesired phenomenon when developing a robust and scaleable antisolvent crystallization process. LLPS can occur in ternary phase system composed of the molecule, a solvent and an anti-solvent. The solvent and anti-solvent alone are fully miscible but the presence of the solute can force the system to phase split and deliver a binary solution made of a continuous (solute rich) and a dispersed phase (solute lean). This phenomenon will impact the purity and physical properties of the molecule of interest and typically delivers a sticky amorphous phase not suitable for drug product development. In order to circumvent this issue and take advantage of LLPS, a phase diagram determination combined with the use of Mesoporous carrier (Neusilin US2 in this case) allows the preparation of Drug Product Intermediates (DPI) with tunable drug load from 0.1 to 65wt%. The DPI has the physical properties of the carrier and is not impacted by the physical properties of the Drug Substance (DS). This DPI has the potential to be readily suitable for direct compression, reduce the risk of variability of content uniformity, sticking behavior, polymorphic phase transition during DP processing. This approach is different than the Incipient Wetness Impregnation technique that is traditionally used to load mesoporous carrier. This presentation will describe the workflow of this new approach to quickly enable DPI by triggering LLPS purposefully to load mesoporous carrier with a defined drug load.

All authors are employees of AbbVie and may own AbbVie stock. AbbVie sponsored and funded the study; contributed to the design; participated in the collection, analysis, and interpretation of data, and in writing, reviewing, and approval of the final publication.

565588 Optimising the Nucleation of Polypeptides Using Nanotemplates: The Case Study of Insulin

Optimising the nucleation of polypeptides using nanotemplates: The case study of insulin

Frederik J. Link*[a], Jerry Y.Y. Heng[a]

[a]Department of Chemical Engineering, Imperial College London, South Kensington Campus, London SW7 2AZ, UK

*f.link18@imperial.ac.uk

Keywords: insulin, nanotemplates, crystallisation

Bio-active polypeptides have attracted interest in research because of their significant use in medical, biological and pharmaceutical industries. Thereby, the crystalline form offers significant pharmacokinetic advantages with regards to achieving higher bioavailability and better release control [1]. In the case of insulin, which is widely used for treating diabetes, the most common route of administration is by subcutaneous injections of either pure microcrystals or a mixture of microcrystals and amorphous insulin. Due to the slow dissolution of the crystals, the crystalline administration leads to a slow release of insulin into the blood stream [2]. However, one of the main limitation is the lack of controllability of the nucleation mechanism. However, nanotemplates are remarkable in controlling nucleation. Recent studies showed that by using small inorganic particle with a highly porous surface the induction time of nucleation of polypeptides can be significantly accelerated even if the supersaturation is low [3]. Furthermore, it is believed that nucleation is enhanced if the pore diameter has the same order of magnitude as the radius of gyration of the polypeptide [4]. This stems from the suggestion that nucleation in porous surfaces gets enhanced due to local increases in supersaturation which is controlled by the diffusibility of the protein into the pores. However, little is known about this mechanism and the impact of the chemical properties of the surface of the nanotemplates on controlling nucleation. It is believed that porous surfaces trigger nucleation due to the local increase in supersaturation. Therefore, investigating the non-covalent interaction between the surface patches of the nanotemplates and insulin will open up new routes for controlling the crystallisation of insulin.

Insulin has a molecular weight of 5.8 kDa and consist of 2 polypeptide-chains, A and B, with combined 51 amino acids. Furthermore, the two chains are inter-linked by two disulfide bonds and, additionally, chain A is intra-linked with an additional disulfide bond. Citrate buffer with a pH near the isoelectric point of insulin was used with zinc-sulphate as precipitant. Different concentrations were tested to identify the most promising ones. Since insulin is stored in a rhombohedral crystal configuration in combination with zinc-sulphate in the pancreas, zinc-sulphate was selected as the most appropriate precipitate. The impact of nanotemplates on crystallisability is investigated by mean of measuring induction time and crystal yield. Therefore, batch crystallisation experiments will be carried out in which the crystal yield and induction time will be tracked via measuring the concentration of insulin within the solution over time. To achieve this, the concentration of insulin is determined via UV-vis absorption in combination with Bradford's method [5]. Different types of nanotemplates will be investigated which differ in their surface chemistry and porosity, amount and size.

By using nanotemplates, the induction time of crystallising a complex molecule such like insulin can be reduced significantly even at low supersaturation. Although the nanotemplates do not have a notable influence on the crystal quality, multiple configurations can be observed which is due to the local availability of dissolved zinc-sulphate. When in the vicinity of zinc-sulphate insulin crystallises in its rhombohedral form, whereas otherwise it crystallises in a cubic shape.

Further work will be done to characterise the impact of the nanotemplates with regards to understanding the impact of the chemistry of the surface patches and whether the crystallisation can be enhanced solely due to local supersaturation or also because of specific non-covalent interaction between the surface patches and insulin. Additionally, the impact of the seeds on the solubility of insulin has to be investigated to evaluate the impact of the nanotemplates on the overall yield. All these findings should ultimately result in the ability of engineering appropriate nanotemplates to enhance the crystallisability of insulin. Further work will also be done in investigating the potential of scaling up from 1 ml to 10 and 100 ml scale.

References

1. Pechenov, S., et al., *Injectable controlled release formulations incorporating protein crystals.* J Control Release, 2004. **96**(1): p. 149-58.
2. Norrman, M. and G. Schluckebier, *Crystallographic characterization of two novel crystal forms of human insulin induced by chaotropic agents and a shift in pH.* BMC Struct Biol, 2007. **7**: p. 83.
3. Shah, U.V., et al., *Crystallization of Proteins at Ultralow Supersaturations Using Novel Three-Dimensional Nanotemplates.* Crystal Growth & Design, 2012. **12**(4): p. 1772-1777.
4. Shah, U.V., D.R. Williams, and J.Y.Y. Heng, *Selective Crystallization of Proteins Using Engineered Nanonucleants.* Crystal Growth & Design, 2012. **12**(3): p. 1362-1369.
5. M., B.M., *A Rapid and Sensitive Method for the Quantitation of Microgram Quantities of Protein Utilizing the Prinicple of Protein-Dye Binding.* Analytical Biochemistry, 1976. **72**: p. 248-254.

600d - SURFACE-INDUCED NUCLEATION STRATEGIES: SEEKING SYMMETRIES BETWEEN SELF-ASSEMBLY OF HETERONUCLEANTS AND CRYSTALS

Fiora Artusio[1], *Giacomo Ceccone*[2], *Roberto Pisano*[1*]

[1]*Politecnico di Torino, Torino, Italy*
[2]*European Commission, Joint Research Centre, Ispra, Italy*

*Corresponding author, e-mail: roberto.pisano@polito.it

Biography

Fiora Artusio is a Ph.D. candidate in Chemical Engineering at Politecnico di Torino. She is working at the Molecular Engineering (MolE) Lab under the supervision of Prof. Pisano. She got her MSc Degree in 2016 studying photo-activated polymerization for the synthesis of nanocapsules for drug delivery. Her research is now focused on the development of experimental methodologies for the control of pharmaceuticals and biopharmaceuticals crystallization. She has carried out research activity in Paris (LSI, Ecole Polytechnique) and at JRC – European Commission, concentrating on electron microscopy and surface analysis techniques respectively. She is currently working on protein crystallization at Laboratorio de Estudios Cristalograficòs (LEC) in Granada.

Giacomo Ceccone is a Staff Scientist at Joint Research Centre (JRC) – European Commission in Ispra. He received his Degree in Physics from University of Pavia (Italy) in 1984. After post-doctoral fellow at the University of Minnesota (USA) and TASC Laboratory, (Trieste, ITALY) he joined the JRC in 1990 where he is working in the Directorate F – Health, Consumers and Reference Materials. Dr Ceccone's main research interests concern the application of surface analysis techniques (e.g. XPS, ToF-SIMS, Synchrotron Radiation-based spectro-microscopies) to the physico-chemical characterization of surfaces with particular emphasis on nanomaterials and micro/nano-plastics. Dr Ceccone has published more than 150 papers in peer-reviewed journals, 4 book chapters and delivered more than 80 presentations at international conferences.

Roberto Pisano is a Professor of Chemical Engineering at Politecnico di Torino (Italy), where he received his Ph.D. in 2009. Professor Pisano's research focuses on the application of both computational and experimental methods to engineering chemical products and processes, with particular emphasis to pharmaceutical processing and formulation of both small molecules and biologics. He has worked with many pharmaceutical companies in research or consulting. He has published more than 80 papers, 11 book chapters, 1 edited book on freeze-drying of pharmaceuticals and currently has 4 patents issued or pending.

Abstract

The present paper presents the application of surfaces having defined and controlled attributes as heteronucleants for the crystallization of a model pharmaceutical molecule. The synthesis of the substrate was optimized in order to relate surface features to the crystallization outcome. Extremely flat and topographically uniform glass supports bearing amino and thiol head groups were successfully synthesized and characterized by means of contact angle, AFM and XPS analyses. Such surfaces were then used as supports for aspirin (ASA) crystallization in order to investigate their influence on nucleation kinetics. Compared to untreated glass, amino-functionalized glass was dramatically nucleation-active, whereas thiol-functionalized supports strongly repressed ASA heterogeneous nucleation. The promoting or inhibiting action towards the stabilization of ASA nuclei on a functionalized surface and their successive growth into crystals was therefore related to the chemistry of exposed head groups.

Introduction

Nucleation represents the core of a variety of processes occurring in nature, such as ice or cloud formation [1], but it is also responsible of amyloid aggregation leading to Alzhemeir's disease [2] or even creation of virus capsids [3]. The formation of new phase in defined environment as a consequence of self-assembly and ordering processes represents a *fil-rouge* for all these phenomena. More specifically, nucleation can occur homogeneously or heterogeneously, depending on whether aggregation mechanism involves the presence of single or multiple phases. In the latter case, stable nuclei are formed taking advantage of favorable interactions with foreign surfaces which lower the energetic cost for new interface area formation. However, heterogeneous nucleation often occurs in uncontrolled manner because of impurities in the system, leading to significantly different crystallization outcomes. This aspect is pivotal to pharmaceutical and biopharmaceutical industries as nucleation not only affects process kinetics, but also crystalline form, habit and size selection, as well as downstream processes [4].

Recently, the role of external surface features on crystallization of Active Pharmaceutical Ingredients (APIs) is being investigated by relating surface chemistry and morphology to nucleation kinetics and polymorph selection [5–7]. For example, surfaces carrying pores were found to strongly accelerate nucleation kinetics, provided that these were not spherical [8]. Selecting one specific crystalline form of a given substance can also be achieved by simply acting on the heteronucleants, keeping constant all the other crystallization conditions [9–11]. However, it is difficult to isolate and distinguish between chemical and morphological interactions between a surface and a to-be-crystallized molecule.

Spatial confinement of chemical reactions represents a useful strategy for the synthesis of a variety of devices, such as nanocapsules and nanohydrogels [12,13]. Here, by confining the functionalization reaction to an active solid-liquid interface, Self-Assembled Monolayers (SAMs) on glass were synthesized. We present an organic approach to the study of heteronucleants quality attributes in order to relate them to crystallization of small molecule drugs. The complete and diversified characterization of surface properties is vital, as the interplay of complex and different aspects of a surface is a key-aspect when promoting or inhibiting nucleation. By carefully tailoring synthesis conditions, we synthesized high-quality SAMs on amorphous substrates and successively applied them to the crystallization of a pharmaceutical model molecule.

Materials & Methods

Chemicals

Glass coverslips were purchased from Neuvitro (Vancouver, USA). Hydrogen peroxide (30 wt % in water, ACS reagent), sulfuric acid (ACS reagent, 95.0 – 98.9%), anhydrous toluene (99.8%), toluene (ACS reagent, ≥ 99.5%), ethanol (puriss. p.a., ≥ 99.8%), 3-aminopropyltrimethoxysilane (APTMS, 97%), 3-mercaptopropyltrimethoxysilane (MPTMS, 95%) and water (HPLC grade) were purchased from Sigma-Aldrich (Cesano Maderno, Italy).

Synthesis of self-assembled monolayers

Glass coverslips were pre-activated and cleaned in 3:1 *piranha* solution for 1 h and thoroughly rinsed with water afterwards. Then, they were incubated in 5.4×10^{-2} M solution of silane in anhydrous toluene. Reaction time was 30 min for APTMS and 15 h for MPMTS. Final rinsing with toluene, toluene-ethanol mixture and ethanol served as last synthesis step.

Surface characterization

Contact angle (CA) analyses were firstly carried out in order to assess the effectiveness of functionalization step and surface uniformity. A MSE DigiDrop (GBX, France) was used to dispense 3 µL HPLC water drops on substrates and calculate sessile drop contact angles. Surface topography was evaluated by means of Atomic Force Microscopy (AFM, Solver NANO, NT-MDT Spectrum Instruments, Russia). Cantilever frequency was 1 Hz, scanned area was 1 x 1 µm and 256 lines per scan were acquired. Finally, chemical analyses were performed with X-Ray Photoelectron Spectroscopy (XPS, AXIS ULTRA, DLD Kratos Analytical, UK). Pass energy was set at 80 keV. Wide spectra were collected from 1100 to 0 eV values of binding energy. The angle between sample surface and analyzer was set at 90°. Calibration for all the spectra involved setting hydrocarbon C1s at 285.0 eV. Data were processed with CasaXPS software (version 2.3.20).

Crystallization studies

We designed and implemented a high-throughput crystallization platform in order to perform a large number of crystallization experiments simultaneously. Aspirin was dissolved in 38% ethanol in water and crystallization was carried out in wells containing SAMs. Supersaturation was set at 1.8 and crystallization temperature was 15 °C. Crystallization outcome was studied according to kinetic and thermodynamic aspects, involving nucleation induction time and crystalline form and habit selection. The first aspect was studied by means of optical microscopy coupled to motorized X-Y stage (M125C, Leica Microsystems, Germany) and *in-house* cooling system, enabling time-lapse multiple-positions acquisition and ensuring maintenance of precise supersaturation conditions until the nuclei formation. Nucleation kinetics was evaluated by calculation of detectable induction time, corresponding to the time when detectable crystals could be evinced inside each crystallizer.

Results

SAMs were built starting from methoxy-silane molecules carrying a propyl chain with a terminal functionality. The latter represented the only difference among our monolayers, being thiol or amino-terminated. The portfolio of elected end groups permits the deep investigation of secondary interactions between surface and solute molecules and their impact on nuclei formation. As a matter of fact, our final goal is to design and develop a versatile benchmark for crystallization studies assessing the extent of specific and non-specific interactions. This is achieved by only changing substrate, while all crystallization conditions, *e.g.* supersaturation, solvent, temperature, are not to be modified.

Proving surface functionalization

Trimethoxysilanes were chosen because of their high reactivity and degree of packing when organized in monolayers. These molecules have been anchored to pre-activated glass substrates by means of condensation reactions between the surface hydroxyl groups and the hydrolyzed methoxy groups of silanes. The synthesis conditions have been optimized in terms of reaction solvent, functionalizing agent, reaction time and temperature in order to ensure the formation of reproducible single layers of organic matter on glass. As a matter of fact, for crystallization studies devoted to the mechanistic understanding of surface effects on nucleation, it is crucial to deal with extremely uniform substrates both as concerns the sample itself, but also *intra* and *inter*-batch homogeneity. This need arises from the intrinsic stochastic nature of nucleation, thus requiring a high number of experiments involving the same experimental conditions to get statistically significant data.

Thorough characterization of the crystallization substrates has been carried out in order to confirm functionalization effectiveness and gain a structured overview of surface properties. First, contact angle analyses confirmed surface functionalization: as regards MPTMS, glass surface turned out to be more hydrophobic after monolayer assembly, whereas, for APTMS, a more hydrophilic surface was obtained, as reported in Table 1. Absence of local deposits of polymerized silanes was confirmed by AFM, which also confirmed the achievement of an extremely regular functionalization. Moreover, some interesting insights into surface roughness were collected. It turned out that the presence of a monolayer on the surface did not alter surface mean roughness, whereas, when uncontrolled functionalization was carried out, major effects on this parameter were detected. More specifically, untreated glass represented an extremely flat surface, having R_q well below 1 nm, *i.e.* 130 pm. Then, covalent bonding of an organic monolayer on top led to comparable average roughness values for all the functionalities. This feature represents a particular appealing aspect of the application of high-quality SAMs to heterogeneous nucleation studies since, as topography is not altered by the functionalization process, the effect of different superficial chemistries may be isolated from morphological interactions between substrate and solute molecules.

The effectivity of functionalization was corroborated by means of XPS. Analyses were carried out for different purposes: the main goal was the confirmation of the presence of characteristic elements of SAMs on glass. Compared to untreated glass, MPTMS and APTMS-functionalized supports led to sulfur and nitrogen characteristic peaks in the XPS spectra, respectively. Further confirmation derived from the comparison of characteristic elements ratios between the ideal and real case. C/S or C/N theoretical ratios for SAMs are equal to 3. We measured 6.6 and 4.4 respectively. Slightly higher elemental ratios were attributed to superficial contamination of the samples, but also denoted that functionalization was likely to occur in monolayers.

The characterization of SAMs led us to confirm the suitability of these surfaces to be used as heteronucleants in crystallization studies.

Table 1. Contact angle, roughness and elemental composition of untreated glass, thiol- and amino-terminated SAMs.

Sample	**CA, °**	**%O**	**%Si**	**%C**	**%S**	**%N**
Untreated glass	49.1 (+/- 1.9)	54.6	21.6	20.4	-	-
Thiol-SAM	65.7 (+/- 1.7)	55.3	21.3	19.1	2.9	-
Amino-SAM	37.1 (+/- 0.6)	30.3	15.0	43.1	-	9.8

Addressing the effect of surface chemistry on crystallization of pharmaceuticals

In this study, self-assembled monolayers (SAMs) have been selected as ideal candidates to probe superficial chemical effects on nuclei formation and successive crystal growth. An amorphous substrate was selected for supporting monolayer assembly in order to exclude epitaxial interaction between the final heteronucleant and solute molecules. In this way, we were able to exclude complex ordering phenomena rising from lattice matching between surface and growing nuclei. In addition, such an approach enables the development of crystallization supports which are structurally inert, but chemically active towards nucleation. Extremely flat surfaces were selected in order to deconvolute, as much as physically possible, the contribution of surface chemistry and topography to nucleation phenomena.

Starting from experimental observation, cumulative probability distribution curves for ASA crystallization were calculated and reported in Figure 1a. Untreated glass was considered as reference for comparing drug nucleation kinetics.

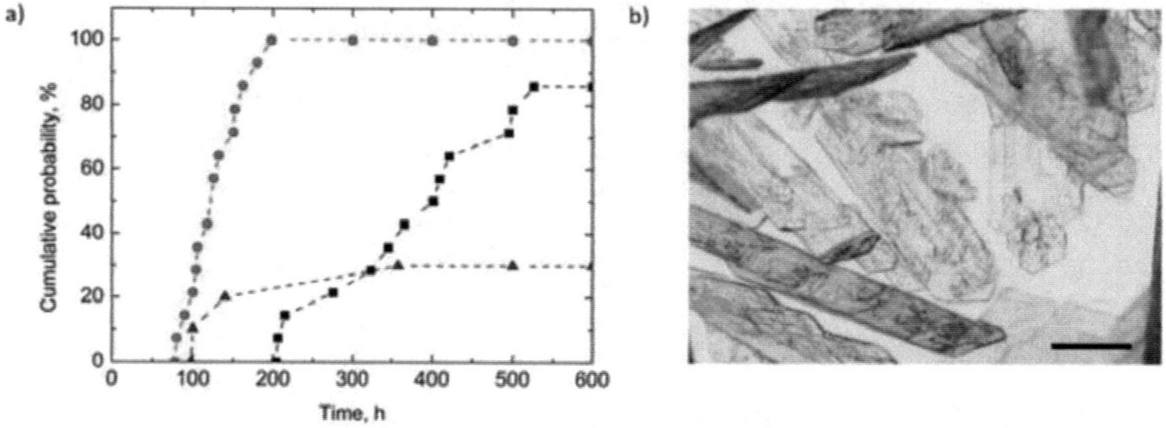

Figure 1. a) Cumulative probability of observing nucleation phenomena for untreated (square, black curve), thiol-SAMs (triangle, blue curve) and amino-SAMs (circle, red curve). b) Optical microscope image of ASA crystals grown on SAMs, scale bare is 100 µm.

Both SAMs were responsible for halving nucleation onset time compared to the untreated substrates. However, the amino-SAMs led to a dramatic boost in nucleation kinetics, as well as the concentration of nucleation events in a shorter time interval. These observations corroborated the nucleation activity of amino surface groups towards ASA. Moreover, crystals were observed in all the wells during the test time, leading to 100% probability of observing nucleation in the system. We related this ability to favorable acid-base interactions between surface and ASA. Conversely, the thiol-SAMs strongly inhibited ASA nucleation, leading to only 30% crystallized wells. This property was attributed to electrostatic repulsion between solute molecules and surface, impeding stable cluster formation in contact with the surface. ASA crystallized according to a plate-like configuration on all the investigated surfaces, as depicted in Figure 1b.

Conclusion

An optimized method for the synthesis of APTMS and MPTMS self-assembled monolayers is here presented, along with a set of characterization techniques for the evaluation of surface quality and features. Such a reference platform is to be linked to the crystallization behavior of heteronucleants. Self-assembly phenomena were therefore exploited at two different levels: firstly, for the synthesis of heteronucleants with defined properties and, then, for the study of the interplay between surfaces and self-assembly of solute molecules in solution leading to crystals. Such surfaces could act as ASA nucleation promotors or inhibitors, thus enabling the tuning of nucleation kinetics.

Abbreviations

AFM, Atomic Force Microscopy; API, Active Pharmaceutical Ingredient; APTMS, 3-aminopropyltrimethoxysilane; ASA; Aspirin; CA, Contact Angle; MPMTS, 3-mercaptopropyltrimethoxysilane; SAM, Self-Assembled Monolayer; XPS, X-Ray Photoelectron Spectroscopy.

Acknowledgements

Alexandre Yuri Kitamukai is kindly acknowledged for the experimental activity on ASA crystallization. Part of this work was carried out thanks to the access to the Nanobiotechnology Laboratory under the Framework for Open Access to the Joint Research Centre Research Infrastructures of the European Commission.

References

[1] J.M. Campbell, F.C. Meldrum, H.K. Christenson, Observing the formation of ice and organic crystals in active sites, Proc. Natl. Acad. Sci. 114 (2017) 810–815. doi:10.1073/pnas.1617717114.

[2] P. Friedhoff, M. Von Bergen, E.M. Mandelkow, P. Davies, E. Mandelkow, A nucleated assembly mechanism of Alzheimer paired helical filaments, Proc. Natl. Acad. Sci. U. S. A. 95 (1998) 15712–15717. doi:10.1073/pnas.95.26.15712.

[3] R. Zandi, P. Van Der Schoot, D. Reguera, W. Kegel, H. Reiss, Classical nucleation theory of virus capsids, Biophys. J. 90 (2006) 1939–1948. doi:10.1529/biophysj.105.072975.

[4] A.S. Myerson, Handbook of Industrial Crystallization, 2nd ed., Butterworth-Heinemann, 2002.

[5] F. Artusio, R. Pisano, Surface-induced crystallization of pharmaceuticals and biopharmaceuticals: A review, Int. J. Pharm. 547 (2018) 190–208. doi:10.1016/j.ijpharm.2018.05.069.

[6] Y. Diao, A.S. Myerson, T.A. Hatton, B.L. Trout, Surface design for controlled crystallization: The role of surface chemistry and nanoscale pores in heterogeneous nucleation, Langmuir. 27 (2011) 5324–5334. doi:10.1021/la104351k.

[7] J.M. Ha, J.H. Wolf, M.A. Hillmyer, M.D. Ward, Polymorph selectivity under nanoscopic confinement, J. Am. Chem. Soc. 126 (2004) 3382–3383. doi:10.1021/ja049724r.

[8] Y. Diao, T. Harada, A.S. Myerson, T. Alan Hatton, B.L. Trout, The role of nanopore shape in surface-induced crystallization, Nat. Mater. 10 (2011) 867–871. doi:10.1038/nmat3117.

[9] J. V. Parambil, S.K. Poornachary, R.B.H. Tan, J.Y.Y. Heng, Template-induced polymorphic selectivity: The effects of surface chemistry and solute concentration on carbamazepine crystallisation, CrystEngComm. 16 (2014) 4927–4930. doi:10.1039/C3CE42622J.

[10] S. Roy, N.R. Goud, A.J. Matzger, Polymorphism in phenobarbital: discovery of a new polymorph and crystal structure of elusive form V, Chem. Commun. 52 (2016) 4389–

4392. doi:10.1039/C6CC00959J.
[11] C.P. Price, A.L. Grzesiak, A.J. Matzger, Crystalline polymorph selection and discovery with polymer heteronuclei, J. Am. Chem. Soc. 127 (2005) 5512–5517. doi:10.1021/ja042561m.
[12] F. Artusio, M. Bazzano, R. Pisano, P.E. Coulon, G. Rizza, T. Schiller, M. Sangermano, Polymeric nanocapsules via interfacial cationic photopolymerization in miniemulsion, Polym. (United Kingdom). 139 (2018) 155–162. doi:10.1016/j.polymer.2018.02.019.
[13] F. Artusio, A. Ferri, V. Gigante, D. Massella, I. Mazzarino, M. Sangermano, A. Barresi, R. Pisano, Synthesis of high payload nanohydrogels for the ecapsulation of hydrophilic molecules via inverse miniemulsion polymerization: caffeine as a case study, Drug Dev. Ind. Pharm. (2019) (in press). doi:10.1080/03639045.2019.1672714.

572187 Olanzapine Crystal Symmetry Originates in Preformed Centrosymmetric Solute Dimers

The symmetry of crystals is a central quality that determines their function and allure. Given the uncorrelated symmetries of a crystal and its constituent molecules, the symmetry breaking that begets a crystal phase has been thought to ensue during nucleation. Here we demonstrate that the one of the two symmetry element of olanzapine crystals, an inversion center, emerges in solute dimers extant in the solution prior to crystallization. Time resolved *in situ* atomic force microscopy monitoring of growth steps reveals that crystals grow nonclassically, exclusively by incorporation of dimers; this growth mode has been speculated and inferred from crystal and solute structures, but never experimentally observed. The growth rate of crystal layers exhibits a quadratic dependence on the solute concentration, characteristic of the second-order kinetics of dimer incorporation. Olanzapine dimers constitute a minority solution component. We show that growth by dimers is preferred not owing to greater activation entropy or faster dissociation of attached solvent molecules. The monomer and dimer adsorption energies on the crystal surface, evaluated in all-atom molecular dynamics simulations, suggest overwhelming accumulation of adsorbed dimers, which expedites their incorporation into growth sites. The proposed outlook on the cooperative creation of crystal symmetry illuminates analyses of structure selection, polymorph transformations, chiral separation, and crystal properties of natural and engineered crystals.

572060 Development of a Hybrid Quality-By-Control Framework for Optimal Crystallization Process Design

Strict regulatory standards are dictated by the FDA, EMEA and WHO for the active pharmaceutical ingredients (API) to meet high level quality criteria such as purity, size and shape of the API crystals or polymorphic structure. One of the main criteria is to achieve size uniformity as well as desired yield in industry. Despite of the efforts put in pharmaceutical industry, intensive fines formation and agglomeration of crystals are still common problems, often having a negative impact on the product crystal size distribution (CSD). A general practice to eliminate fines and agglomerates is to apply temperature cycles for internal fines dissolution and de-agglomeration.[1] For this purpose, quality-by-design (QbD) approaches can be applied for the design of the suitable temperature profile, but this requires intensive experimental investigation to explore the whole design space. Quality-by-Control (QbC) has been introduced more recently as a new novel alternative to QbD, which can be applied to control critical quality attributes of a certain process, whereby the desired operating curve like temperature profile is automatically determined by the feedback control algorithm rather than by excessive open-loop experimentation. In this way QbC can significantly reduce required process development time and material used.[2]

The work proposes a novel hybrid framework combining the model-free and model-based quality-by-control (mfQbC and mbQbC) paradigms for the optimal crystallization design with the aim of minimization of fines and agglomerates in the product. This developed framework describes the mfQbC based rapid design used in conjunction with direct nucleation control (DNC) and supersaturation control (SSC), along with (2) the mbQbC framework using a novel population balance model-based platform for the robust optimization of the operating trajectory of the crystallizer to achieve crystals with desired quality attributes.[3] This framework provides an improved alternative to the current industrial practice, by the combined use of feedback control and mathematical models for rapid design of robust crystallization processes. The implementation of the proposed framework enhanced the prediction of crystallization events and was able to achieve better control of the crystallization processes by providing controlled temperature cycles, for de-agglomeration and fines destruction. The information for parameter estimation and optimization calculations was collected from the experiments and the system diagnostics was done in mfQbC part, which also provided a rapid design of robust operating conditions, that were improved to achieve optimal operation subsequently in the mbQbC approach.

The advantage of the mfQbC is that it requires minimal preliminary system information, hence, it enables quick and efficient crystallization process design. On the other hand, the mf-QbC operation policy might be sub-optimal, which can be overcome by exploiting the predictive capabilities of the process models used by the mbQbC. In the mfQbC part, DNC experiments were run for two purposes: to determine the primary and secondary nucleation metastable zone widths (MSZW) and to obtain the temperature cycle profile that is required for producing the crystals with desired properties. DNC is an efficient approach to eliminate fines due to the application of temperature cycles. SSC was applied after DNC generates the MSZW information to obtain the temperature profile for only growth mechanism. The resulting temperature profile from SSC can also be easily implemented in industrial distributed control systems (DSCs). After mfQbC, using the mbQbC framework, a population balance model (PBM) was developed to simulate primary and secondary nucleation, growth, agglomeration and de-agglomeration mechanisms as well as dissolution. Model parameters were identified using experimental data for a model API crystallization with high agglomeration and secondary nucleation tendency. Using a model-based robust optimization, in which parametric uncertainty obtained from the model-identification step was also included, different temperature profiles were determined considering the dual objective of fine minimization and maximal de-agglomeration. The optimal temperature profiles obtained was validated experimentally to show the higher quality product that can be obtained using the this QbC framework.

References:

1. Wu, Z., Yang, S. & Wu, W. Application of temperature cycling for crystal quality control during

crystallization. *CrystEngComm* **18**, 2222–2238 (2016).
2. Yang, Y., Song, L. & Nagy, Z. K. Automated Direct Nucleation Control in Continuous Mixed Suspension Mixed Product Removal Cooling Crystallization. *Cryst. Growth Des.* **15**, 5839–5848 (2015).
3. Szilágyi, B., Borsos, Á., Pal, K. & Nagy, Z. K. Experimental implementation of a Quality-by-Control (QbC) framework using a mechanistic PBM-based nonlinear model predictive control involving chord length distribution measurement for the batch cooling crystallization of L-ascorbic acid. *Chem. Eng. Sci.* **195**, 335–346 (2019).

580441 Advances in Gas-to-Liquid Solid-Liquid Separation Systems

Advances in Gas-to-Liquid Solid-liquid separation systems

The advances in the design and operation of the solid liquid separation (SLS) systems for a conventional GTL plant has been investigated. The focus is more on the GTL technologies with the slurry type FT reactors, since these technologies encounter more challenging catalyst/wax separation issues than the fixed bed FT reactor types. The review covers advances in solid-liquid separation systems, which are internal and external to the FT reactors. The selection and efficacies of various filter elements and their features when used for FT catalyst/wax separation are highlighted. Operational modes of the filters and potential operation issues are also reviewed. The use of settlers and other solid-liquid separation systems is also discussed.

557469 Separation Characteristics in a Novel Gas-Liquid Vortex Separator

A novel multi-swirling arms gas-liquid vortex separator (GLVS) was designed in order to improve the gas-liquid centrifugal separation process and realize the high-efficiency separation. It can provide a new idea to design for large-scale gas-liquid separator. The pressure drop characteristics are investigated in a large scale cold-separator model, the mainly diameter of this separator is 500 mm, as shown in Figure 1. The static pressure drop between the inlet and outlet was measured at different velocities of the swirling arm under pure air flow conditions, as shown in Figure 2. The velocities of the swirling arm ranged from 5.65 to 16.95 m/s, which can cover the operating conditions of the industry gas liquid separator. The experimental results show that the pressure drop and the velocity of the swirling arm presence a square relationship, as shown in Figure 3. The dimensionless standard deviation of the static pressure drop is maintained within 2%. Furthermore, the total pressure drop (ΔP) includes three parts, i.e., loss in inlet friction (P_1), loss in the separation space (P_2), and loss in outlet pipeline (P_3). We found that the pressure drop occur in the separation space is the largest. In addition, the three parts of pressure drop and the velocity head of the swirling arm presence a linear relationship, as shown in Figure 4. A model between the pressure drop of each part and the velocity head of the swirling arm was then given. The resistance coefficient of this rig is 16, it has no significant increase compared to common cyclone separators. The forecast equation of the resistance coefficient was obtained based on the experimental of four similar structures with different sizes. Compared to real resistance coefficient, the predicted results error is less than 0.5%. Then we proposed the separation characteristic of GLVS under the gas-liquid mixture condition. We discovered the gas-liquid condition, especially increase the liquid, has a little effect on pressure drop, as shown in Figure 5. Figure 6 illustrates the results of separation efficiency performed on different swirling arm velocities. We obtained the optimal velocity of swirling arm is 12.4 m/s.

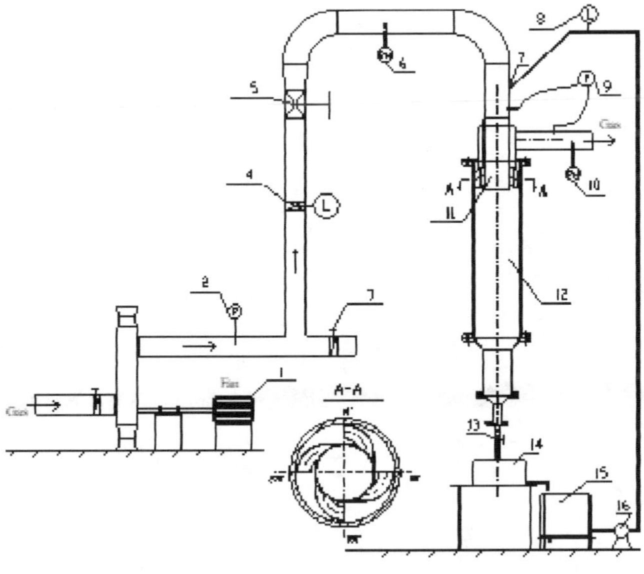

1 air blower; 2 pressure gage; 3 atmospheric valve; 4 vortex flowmeter; 5 control valve; 6 hygrometer in inlet; 7 ultrasonic atomizing nozzle; 8 liquid flowmeter; 9 digital manometer; 10 hygrometer in outlet; 11 swirling head; 12 separation space; 13 drain valve; 14 measuring tank; 15 water wank; 16 pump

Fig.1 Schematic diagram of the experimental apparatus

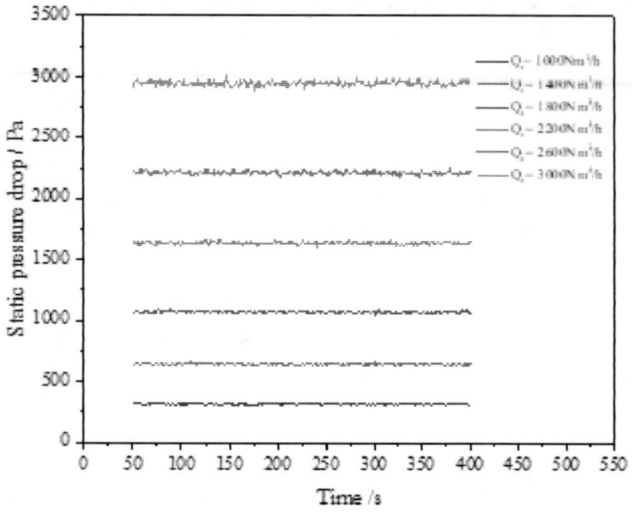

Fig.2 Fluctuation of static pressure drop vs. recording time at different air flow rates

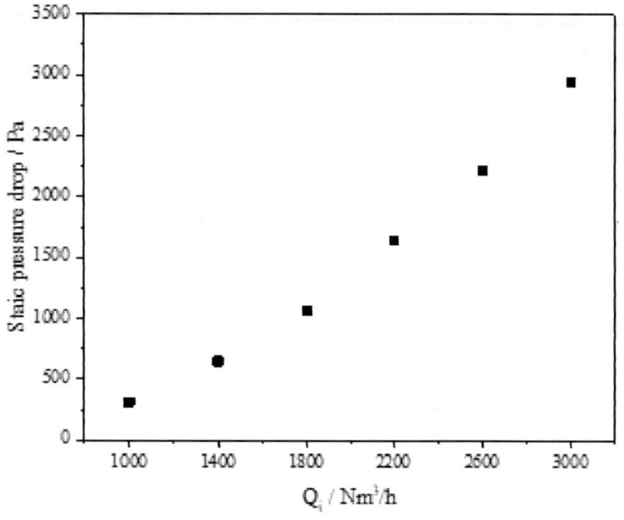

Fig.3 The static pressure drop vs. inlet gas flow rates

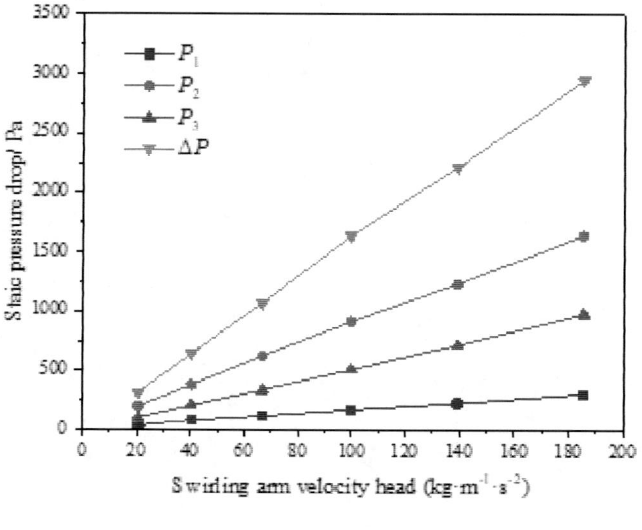

Fig.4 The relationship between segmental pressure drop and velocity head of swirling arm

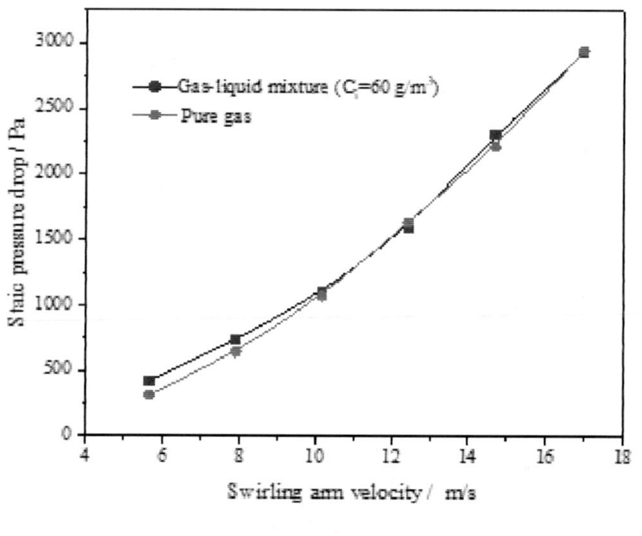

Fig.5 The relationship between pressure drop and velocity of swirling arm

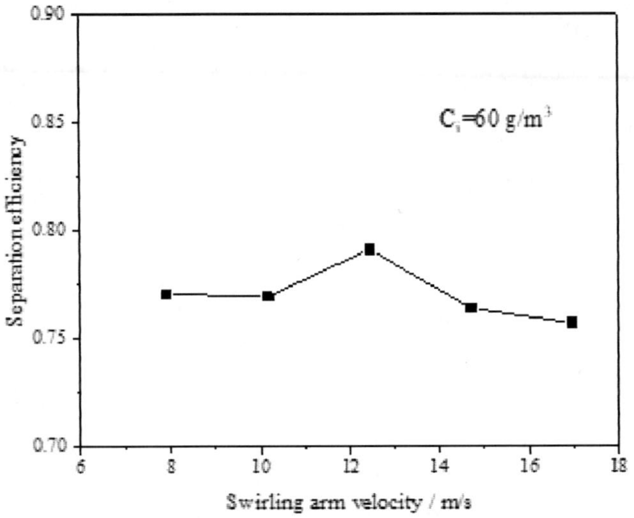

Fig.6 The separation efficiency vs. swirling arm velocity

572636 Magnetophoretic Capture of 5 Nm Sized Superparamagnetic Iron Oxide Nanoparticles Under Different Gradient Field Conditions

Superparamagnetic iron oxide nanoparticles (SPIONs) are widely employed in multiple fields, especially in biology and medicine where they can be used for diagnosis, therapy and disease monitoring due to their outstanding features. Among them, especially important are their high surface to volume ratio, their biocompatibility, their chemical stability and most importantly, superparamagnetism. However, the magnetic separation or manipulation of SPIONs with sizes below of 20 or 30 nm is very challenging as the magnetophoretic motion is hindered by thermal energy and viscous drag.

In this work, we report the magnetophoretic capture of 5 nm SPIONs under different variables and magnetic conditions. Particle suspensions with different concentrations were prepared and placed inside rectangular glass channels, which were then inserted into different separators. More specifically, the particle suspensions were placed inside different quadrupole magnetic sorters (QMS), designed to achieve different magnetic field and gradient conditions, rendering different magnetic forces acting on the SPIONs. The analysis of the SPIONs concentration levels over time and over the length of the glass channel was carried out by image processing.

Multiple variables and parameters were evaluated, including the initial SPIONs concentration, the temperature, the magnetic field and gradient and operation time. Our preliminary results suggest that for the trapping or capture of the SPIONs high magnetic gradients are imperative. Further exploration of the magnetic fields provided accurate quantitative data on the effect of the magnetic field on the SPIONs magnetophoretic motion. Furthermore, the particle separation is enhanced by increasing the initial particle concentration, which suggests that particle-particle interactions (magnetic coupling) is an essential phenomenon involved in the separation. Finally, the capture time has been calculated to be less than 10 minutes, which is a great advantage in comparison to other microfluidic separators that require longer operation times or low flow rates.

564879 Bubbly Flow in Degassing Cyclones and Potential Applications

A degassing cyclone is a separator that can be widely used for the removal of bubbles and dissolved gas from liquids. Gas bubbles move toward the swirling center region in a degassing cyclone due to centrifugal force. The dissolved gas is desorbed due to a decrease in the saturated solubility resulting from the pressure gradient in the degassing cyclone. Additional carrier gas bubbles with lower partial pressure lead to mass transfer at the turbulent bubble interface. The final sate of the dissolved gas and carrier gas bubble in degassing cyclones is an important issue that can be used to develop a theoretical foundation for green separation engineering.

The feasible development of this technology and potential applications are discussed in this work. Gas stripping, as a common method of gas-liquid mass transfer separation in the chemical industry, is widely used in petroleum refining, wastewater treatment, metal smelting, and other production processes. Increasing amounts of sulfur-containing crude oil are used in petroleum refining. The sulfur elements in crude oil are eventually discharged in the form of hydrogen sulfide. Higher requirements have been put forward for hydrogen sulfide stripping technology due to the increased and stricter environmental protection rules. In order to control the content of hydrogen sulfide in produced oil, a hydrogen sulfide stripper usually needs to consume more hot carrier gas. However, the color and stability of the bottom oil may deteriorate due to thermal decomposition. This indicates that the single heating method cannot solve the problem of hydrogen sulfide stripping. Therefore, high-efficiency and low-cost stripping separation of dissolved hydrogen sulfide from oil products has become an important separation method in chemical engineering. The degassing hydrocyclone has proved to be an optimized and feasible solution for this.

568855 Novel Al-Rich Chabazite Compositions Enable Ar PSA

Adsorption 2E

Adsorbent session

Novel Al-rich Chabazite Compositions Enable Ar PSA

Anthony Vallace[1], Grant Kester[1], William Casteel[2], Garret Lau[2],

Roger Whitley[2], and Charles Coe[1]

[1] *Department of Chemical Engineering, Villanova University, Villanova, PA 19085*

[2] *Adsorption Technology Center, Air Products and Chemicals, Inc. 7201 Hamilton Blvd, Allentown, PA 18195*

Carbon molecular sieves enable kinetic based gas separations. It has been shown that single phase chabazite (CHA) in a mixed cation form (Na/KCHA (X_{Kz}~21%)) having an Si/Al (SAR) range of 1.35 to 1.6 has both high oxygen diffusivity (D/r^2) and high kinetic selectivity of O_2 over Ar. Due to the kinetic nature of the separation, the O
of the potassium cation exchange level as well as the SAR of the CHA at low and ambient temperatures. These CHA adsorbents were prepared by interzeolite conversion of Y-type zeolites in both powder and beaded form. Various pretreatments including ion exchange, caustic treatment, calcination, and steaming were performed on beaded and powder forms of NaY. A range of gel composition were studied to map the synthesis space for converting the Y-type zeolite to KCHA (1.44 -2.41). The SAR of the CHA, formed by this interzeolite conversion is controlled by the Si/Al atomic ratio, the total alkalinity, and the water content. It was shown that the SAR and Si/OH in the starting gel strongly affected the CHA product SAR; whereas the water content strongly affected the product phase (higher water content leads to densification). Decreasing the SAR(gel) or Si/OH leads to leaching of more SiO_2 into solution, which in turn leads to a lower SAR product. The consequence of low SAR and Si/OH gels is densification to an undesired P-zeolite (GIS-type) if the gel is too basic. Determining synthetic conditions for controlling the framework SAR is crucial because the mixed cation Na/K CHA is unselective for SAR's>1.7. Activation studies needed for successful adsorption measurements clearly show that the Al content and K^+ exchange level strongly affect the adsorbent's hydrothermal stability. O_2 and Ar isotherms at 175K were obtained for various pure and mixed cation forms of CHA at different SARs prepared from beads and powders of NaY. After it was determined that 3K/unit cell with an SAR of 1.6 was the optimal CHA composition for removing O_2 from Ar, kinetic and equilibrium ion exchange studies were used to optimize the batch ion exchange conditions for both a single and double contact Na^+ exchange to obtain the desired potassium level. The adsorption properties at 175K for the single and double contact Na exchanged chabazites were consistent with those of 3KNaCHA(1.6) prepared from a stoichiometric K^+ back exchange of NaCHA(1.6), validating the batch exchange method. The adsorption properties of different pure and mixed alkali cation forms of CHA(1.4-2.4) at 175K show a significant decrease in Ar adsorption in 3KNaCHA at SAR<1.6. Kinetic adsorption data collected at 30°C also shows an optimal O_2/Ar kinetic selectivity for the 3KNaCHA(1.6) composition. Simulation results predict that these novel Al-rich compositions provide similar Ar recovery and double the Ar productivity, compared to CMS, for an ambient temperature pressure swing adsorption (PSA) process.

566695 A Porous Organic Cage's Structural Evolution through Time

The structural evolution of a Porous Organic Cage, CC3, was studied as a function of synthesis time. The evolution involved three crystal growth stages: rapid crystal growth at short times, crystal fragmentation, and dissolution at intermediate times, and crystal regrowth through Ostwald ripening at long synthesis times. A combination of experimental, and *in silico* calculations of nitrogen isotherms confirm the presence of both CC3a and CC3-amorphous phases present in samples synthesized for various times. Due to this mixture of phases, nitrogen and carbon dioxide adsorption was highly dependent on synthesis time. *in silico* CC3 models and simulations of XRD patterns coincide with experimental patterns indicating CC3a contracts at different synthesis times.

555692 A Technical Overview about the Use of Ultra-Microporous Fluorinated MOF for Separations and Sensing Applications

Youssef Belmabkhout, Mohamed Rachid Tchalala, Amandine Cadiau, Karim Adil, Aleksander Shkurenko, Prashant M. Bhatt, and Mohamed Eddaoudi

Functional Materials Design, Discovery & Development (FMD3), Advanced Membranes and Porous Materials Center, Division of Physical Sciences and Engineering, King Abdullah University of Science and Technology, 4700 King Abdullah University of Science and Technology (KAUST), Thuwal 23955-6900, Kingdom of Saudi Arabia.

In the last few years, the adsorption community witnessed an increasing interest in unveiling ultra-microporous fluorinated MOFs for gas/vapor separation applications. This sub-class of materials mainly limited to structures with **pcu** topology showed a plethora of interesting properties that thousands of discovered MOFs, with a wide variety of topological structures, could not reveal. Further, the structural-adsorptive properties relationships of this unique platform were exploited to uncover their corresponding unique sensing properties. In this paper, an overview about the use of ultra-microporous fluorinated MOFs will be proposed. The authors will describe critically this growing area of research and present their perspectives of the future use of this platform for real applications.

References

[1] Y.Belmabkhout, Z. Zhang, K. Adil, P. M Bhatt, A. Cadiau, V. Solovyeva, H. Xing, M. Eddaoudi. Hydrocarbon recovery using ultra-microporous fluorinated MOF platform with and without uncoordinated metal sites: I- structure properties relationships for C_2H_2/C_2H_4 and CO_2/C_2H_2 separation Chemical Engineering Journal 359, 32-36.

[2] Y. Belmabkhout, P. M Bhatt, Karim A, R. S Pillai, A. Cadiau, A. Shkurenko, G. Maurin, G. Liu, W. J Koros, M Eddaoudi. Natural gas upgrading using a fluorinated MOF with tuned H_2S and CO_2 adsorption selectivity. Nature Energy 3 (12), 1059.

[3] Amandine Cadiau et al. Hydrolytically stable fluorinated metal-organic frameworks for energy-efficient dehydration. *Science* 356 (6339), 731-735.

564861 Study of Lanthanum Carbonate Nanowire for Adsorption of Phosphorus: Material Synthesis and Optimization and Demonstration of Performance

Eutrophication has caused severe damages to water environment, due to excessive release of nutrients (phosphorus and nitrogen). Effective removal of phosphorus is of great importance in prevention of eutrophication. In this study, we report an innovative lanthanum carbonate (LC) nanowire for removal of phosphate. The concentration of urea, temperature and time were first optimized so as to obtain the best LC nanowire for the removal. It was showed that the best adsorption was obtained at pH 3 and pH 5 when pH was not controlled and was controlled, respectively. The LC worked well in pH 3 to 8 for the phosphate removal. The experimental data were fitted well by Langmuir isotherm; the maximum adsorption capacities of 312.5 and 303.03 mg/g were found at pH 5 and 7, respectively. The adsorption equilibrium time was 5 h; the adsorption history was well described by the pseudo-second-order equation. The presence of ionic strength and competitive substances had less influence on the uptake. The spent adsorbent can be regenerated and reused satisfactorily. The mechanism study revealed that the adsorption was associated with ion exchange between carbonate and phosphate ions. Our study demonstrated that the LC nanowire was a potential adsorbent for the phosphate removal.

575845 Development of Oxygen-Selective Sorbents for Low- and High-Temperature Air Separation

Development of Oxygen-selective Sorbents for Low- and High-temperature Air Separation

Qinghe Zheng, Shaojun Zhou, and Marty Lail*

Technology Advancement and Commercialization Division, RTI International, Durham, NC 27709, United States.

*mlail@rti.org

Introduction

We are reporting our recent material development progress on oxygen-selective sorption for air separation. Depending on the gas environment and process conditions, different material development strategies were used based on the material-O_2 affinity. Various types of biomimetic room-temperature (25 °C) and high-temperature (600~1000 °C) oxygen-selective sorbents were developed. Specifically, novel Cobalt (II) salen functionalized task-specific ionic liquids or novel types of Lewis base ligands were synthesized for reversible biomimetic room-temperature oxygen binding. Meanwhile, novel $CaCo_xZr_{1-x}O_{3-d}$ perovskites were developed to allow rapid high-temperature O_2 sorption cycles.

In biological systems, the respiratory pigments (e.g. hemoglobins) are able to reversibly bind dioxygen from the atmosphere without appreciable loss of activity. Transition metals with certain organic coordination environment plays an important role in the oxygen binding in these natural oxygen carriers. In synthetic chemistry, low-molecular biomimetic substances containing a transition metal center and being able to reversibly bind oxygen have attracted considerable interest for air separation application. By far the greatest number of synthetic oxygen complexes are based on cobalt, and one of the most thoroughly investigated systems is a Schiff base complex *N,N′*-Bis(salicylidene)ethylenediaminocobalt(II), i.e. (Co(II)salen), and its ring-substituted derivatives. Unprocessed solid-phase Co(II)salen does not bind oxygen efficiently, and it exhibited degradation after extended sorption cycles. For optimized oxygen binding performance, we focused on modification of Co(II)salen by coordination with delicately-designed ionic liquid functional groups or novel types of Lewis base ligands.[1]

For some applications including glass and steel industries, and combustion processes, it is desirable to supply oxygen at elevated temperature above 300 °C. In these cases, an air separation process including oxygen-selective sorption at high temperature is preferable because a single stream of high-purity (> 99%) O_2 can be obtained by temperature-, pressure-swing, or combined sorption cycles, with O_2-deficient air as the only byproduct in a separate stream. ABO_{3-d} perovskites, where alkaline-earth or lanthanide elements occupy A site and transition metal elements occupy B site, are crystalline ceramics and are ideal for high temperature thermochemical air separation for oxygen production. This is due to the fact that their oxygen nonstoichiometry *d* can be varied in response to changes in temperature and oxygen partial pressure. Previously, we reported outstanding low temperature (< 550 °C) redox property of our patented novel $CaCo_xZr_{1-x}O_{3-d}$ perovskite-type oxygen storage materials (OSMs).[2] In the present study, we further examined the thermochemical oxygen mobility within these materials at higher temperature regime (600 to 1000 °C). These perovskites showed outstanding oxygen sorption performance and they are potential oxygen-selective sorbents for high-temperature air separation.

Experimental

For room-temperature O_2-selective sorbent development, IL ligands composed of different combinations of phosphonium-type cations $[PC_1C_2C_3C_4]^+$ ($C_1 = C_2 = C_3 = 2$, 4, or 6; C4= 5, 8, or 16) and anions including *N*-methylglycinate ([NmGly]-) or/and bis-(trifluoromethanesulfonyl)amide ($[Tf_2N]^-$) were synthesized and complexed with Co(II)salen to yield 11 TSIL samples. Novel types of Lewis-base ligands were also used to coordinate with Co(II)salen. Material characterizations (CHN, ICP-MS, NMR, CHNSO, High resolution MS, FT-IR, DRIFT, UV-vis, and Single-crystal XRD) were used to confirm material composition and structure. The dioxygen sorption capacity and reversibility of the as-synthesized materials were investigated via both gravimetric and volumetric approaches under various gas sorption.

For high-temperature O_2-selective sorbent development, $CaCo_xZr_{1-x}O_{3-d}$ perovskites were synthesized using modified Pechini method. *In situ* thermal XRD was used to study the material structural changes in response to temperature variations in air or inert atmosphere. Temperature programmed reduction was employed to elucidate the relationship between perovskite composition and redox property. O_2 sorption performance was evaluated by isothermal study at various temperature and oxygen partial pressure conditions, as well as long-term absorption-desorption cycle tests.

Results and Discussion

A series of five-coordinate Co(II)salen complex-type O_2 sorbent materials were successfully synthesized, characterized, and screened for room-temperature selective O_2 absorption over N_2. Types and molecular weights of coordinating ligands, and coordination stoichiometry all have significant influences on the gas sorption capacity and kinetics. Among the studied TSIL samples, $[P_{2225}]_2$[NmGly][NTf_2][Co(salen)] showed the highest O_2 absorption capacity and kinetics, with maximum capacity of 4368 uL/g sample and nearly half of the capacity reached (2065 uL/g sample) within 30 min of TOS at 25 °C (Figure 1a). Oxygen sorption isotherms and long-term material stability tests were performed with $CaCo_{0.7}Zr_{0.3}O_{3-d}$ and $CaCo_{0.9}Zr_{0.1}O_{3-d}$, with stable oxygen production of 2.58 wt% and 2.87 wt% respectively shown at 900 °C for 100 cycles (Figure 1b), and 5 minutes for each sorption step. Stable material performance of $CaCo_{0.9}Zr_{0.1}O_{3-d}$ was further demonstrated in extended thermal cycle test with shorter sorption duration (2 minutes for each sorption step), for 500 cycles. An air separation process scheme was proposed to produce > 99% purity O_2 by rapid absorption-desorption redox cycles using the above perovskite-type oxygen-selective sorbents.

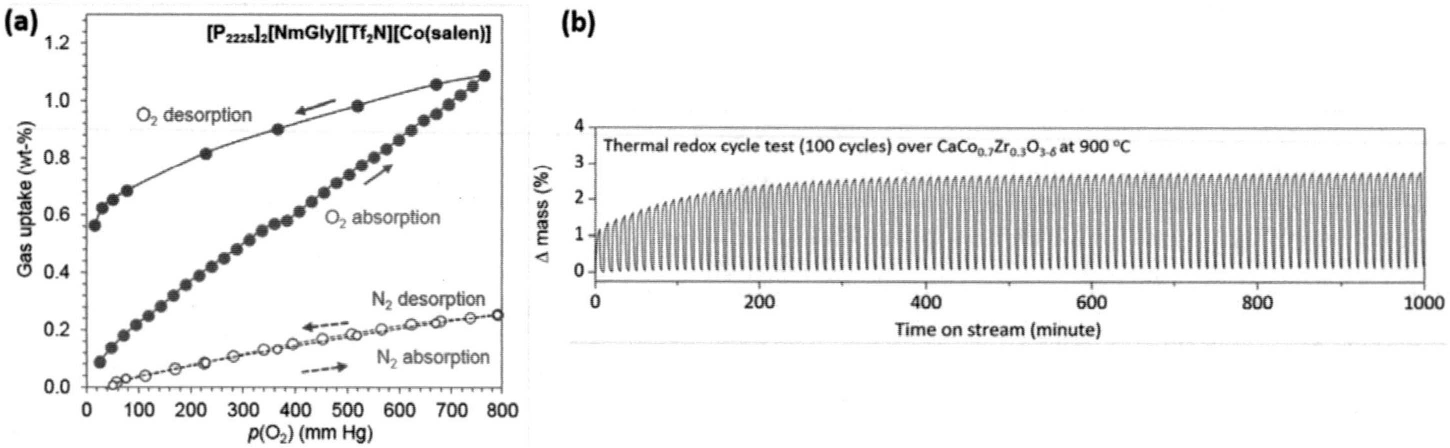

Figure 1. (a) O_2 and N_2 absorption and desorption isotherms of $[P_{2225}]_2$[NmGly][Tf_2N][Co(salen)] measured at 25 °C and between near zero and 760 mmHg. (b) Thermal oxygen sorption cycle tests by novel $CaCo_{0.7}Zr_{0.3}O_{3-d}$ perovskite at 900 °C.

References

[1] Q. Zheng, S. J. Thompson, S. Zhou, M. Lail, K. Amato, A. V. Rayer, J. Mecham, P. Mobley, J. Shen and B. Fletcher, *Ind. Eng. Chem. Res.*, 2019, 58, 334–341.

[2] Q. Zheng, M. Lail, K. Amato and J. T. Ennis, *Catal. Today*, 2019, 320, 30–39.

570355 Fabrication of Metal-Organic Framwork Nanosheets@Cellulose Composites As Adsorbents for Organic Solvent Recovery

Metal-organic frameworks have been widely used in gas-based adsorption, storage and separation due to their large micropore volume. In this work, we demonstrated the utilization of coordinatively unsaturated metal sites of MOFs to adsorb organic molecules that are larger than the pores of MOFs in organic solvents. Firstly, successful growth of MOF nanosheets on cellulose monolith (MOF_NS@CL) was achieved by controlling the heterogeneous nucleation and growth based on proper selections of temperature, solvents and additional base. Four commonly used dyes were used to investigate the adsorption performance of MOF_NS@CL composites. It was found that the obtained composites showed selective adsorption of rose bengal and methyl orange over crystal violet and methylene blue. X-ray diffraction results unveiled the structure change upon dye adsorption, indicating the coordinatively unsaturated copper center played a key role in selective dye adsorption. The mechanism underlying the adsorption of organic molecules over MOFs could be helpful in designing new types of MOF-based materials for sensing, enzyme immobilization and catalysis.

579255 Textural Characterization of Shale Nanostructure

Nanostructure characterization of shales has received tremendous interest in recent years and aids in the understanding of how gases (and in what volume) are stored and permeate or transport through a shale formation. Of particular interest is also how the different components of shale (clay, kerogen, mineral, total organic content, etc) contribute to the nanostructure. The pore size range in shales spans from ultramicropores to macropores, so there is a need to combine several different techniques in order to determine the complete shale pore structure.

Gas sorption, including CO_2 adsorption at 273 K to assess the ultramicropores combined with Ar adsorption at 87 K to assess the mesoporosity, coupled with mercury porosimetry gives a complete micro-, meso-, and macropore size distribution as well as total pore volume. Micro- and mesopore size distributions were obtained from gas adsorption using state-of-the-art density functional theory (DFT) methods. Pore connectivity was explored through a detailed study of the hysteresis loops of the Ar isotherms.

In addition to the study of the shale nanostructure, kerogen was isolated intact from the shale sample and studied independently to compare the porosity of kerogen to the porosity of the shale. Comparison of the micro- and mesoporosity of the shale and its isolated kerogen indicate that kerogen is the main contributor to the porosity in this pore size range.

Finally, transport in shale and the isolated kerogen component was studied via experiments utilizing a zero length column and correlated with the determined structural properties of the materials.

575082 Surface Area Determination of Metal-Organic Frameworks (MOFs) Using the Brunauer-Emmett-Teller (BET) Method: Limitations and Improvements

Surface areas of metal-organic frameworks (MOFs) have been commonly characterized using the Brunauer-Emmett-Teller (BET) method based on adsorption isotherms of non-reactive nitrogen or argon. Recently, however, some discrepancies between surface areas computed from the BET method and those from geometric methods were reported in the literature. In this study, we systematically evaluated the BET and geometric surface areas for hundreds of geometrically diverse MOFs as well as carbon nanotubes (CNTs) with varying pore sizes as model systems to achieve a comprehensive understanding of the limitations of the BET and geometric methods. We compared the BET and geometric surface areas to the true monolayer area which is determined by directly counting the number of molecules included in the monolayer from molecular snapshots. An important concern with the use of the BET method is that the predicted area is highly dependent on the region of the isotherm that is chosen for the computation. The choice of this region is often confusing and ambiguous. Currently, this choice is informed by consistency criteria proposed by Rouquerol et al. While these criteria are useful in general, there are scenarios where their applicability is limited. In this study, we systematically evaluated the implementation of the consistency criteria and their influence on the predicted surface areas. Additionally, in order eliminate the dependence of predicted areas on chosen regions, we also employed the excess sorption work (ESW) method to determine the surface areas. With the ESW method, one simply needs to identify the first minima of a transformation of the isotherm in order to determine the surface area. In this study, we found that the excess sorption work (ESW) method, or a combination of ESW and BET methods, can potentially provide a more accurate estimation of the surface area, particularly in cases where the BET method overestimates the surface area significantly.

570082 Pore Size Estimation from Ultrasonic Measurements during Nitrogen Adsorption Experiment

Pore Size Estimation from Ultrasonic Measurements during Nitrogen Adsorption Experiment

Max A. Maximov[1] and Gennady Gor[1]

[1] *Otto H. York Department of Chemical and Materials Engineering, New Jersey Institute of Technology, 323 Dr Martin Luther King Jr Blvd, Newark, NJ 07029*

Nitrogen adsorption is one of the main techniques for characterization of micro- and mesoporous materials This method can give such information as the BET surface area and pore size distribution. The traditional approaches are based on analysis of the adsorption isotherm obtained by volumetric or gravimetric methods [1]. However, Warner and Beamish showed that the adsorption isotherms can be extracted from measurements of sound wave propagation through the sample during an adsorption experiment [2].

We show that from the analysis of data from the ultrasonic experiments one can extract information about the shear and longitudinal moduli of the fluid adsorbed in the pores. The resulting shear modulus of the confined liquid nitrogen is zero, similarly to a bulk fluid. However, the longitudinal modulus changes abruptly at the point of capillary condensation, and gradually changes after that. To mimic this behavior, we performed grand canonical Monte Carlo simulations of this system and demonstrated a good match with the experimental data [3]. We also performed simulations for the pore sizes from 2 to 8 nm. The model illustrates that the dependence of the elastic modulus at saturation on the inverse pore size is linear so we can estimate the mean pore size. In addition to that, we suggest that the linear dependence of the elastic modulus on the solvation pressure in the pores can be used for the prediction of the pore size distribution.

Overall, the ultrasonic measurements provide a good piece of complementary information to the traditional adsorption methods, namely the shear and longitudinal modulus as a function of vapor pressure, and the data have a nice agreement with simulations. We also demonstrated that our work sets the ground for the pore size distribution prediction from the isotherms obtained using the ultrasonic method.

References:

1. J. Rouquerol, F. Rouquerol, P. Llewellyn, G. Maurin, K. S. Sing, Adsorption by powders and porous solids: principles, methodology and applications, Academic Press, 2013.
2. K. L. Warner, J. R. Beamish, Ultrasonic measurement of the surface area of porous materials, J. Appl. Phys. 63 (9) (1988) p. 4372–4376.
3. M. A. Maximov, G. Y. Gor, Molecular Simulations Shed Light on Potential Uses of Ultrasound in Nitrogen Adsorption Experiments, Langmuir, 34 (51), 2018, p. 15650-15657

576451 A Hydrogen-Bonded Supramolecular Coordinating Complex for Copper Removal from Aqueous Solutions with Unprecedentedly Fast Kinetics

Copper, one of the most widely used heavy metals, is extremely toxic to human health if present in large amounts despite its essential role in human physiology. Albeit a plethora of explored MOFs, relatively few structures have been sought for stable, affordable yet highly effective copper removal from water streams. Few MOFs have been studied either as copper adsorbents or detective sensors. Among them, zeolitic imidazolate framework (ZIF8) was reported to perform an extremely high Cu(II) ions adsorption capacity (up to ~800 mg/g) which makes it one of the current state-of-the-art adsorbents for copper ions removal. However, ZIF8 did not perform well at pH < 3. Adsorbents that are specifically designed to target given ions of interest must be made available and affordable for pragmatic large-scale separation. Compared to frameworks built upon coordination bonds and covalent bonds (e.g. MOFs, covalent organic frameworks – COFs), hydrogen-bonded frameworks have received less attention due to a preconceived notion of a lack of stability due to the non-covalent interacting network. However, recent studies have shown potentials for these material families in selective separation thanks to their flexibility, reversibility and capability for guest-host interaction.

Herein, we demonstrate the synthesis, characterization, and mechanistic understandings of copper adsorption behaviors of a new water-stable copper scavenger with unprecedentedly fast sorption kinetics, relatively high sorption capacity, selective performance at low pH and scalability. In essence, by coupling coordination chemistry and supramolecular assembly, a new crystal material was formed in aqueous medium through the combination of metal-ligand coordination and non-covalent interactions such as inter- and intra-molecular hydrogen bonds. The material structure and behaviors in water environment were studied experimentally and computationally. Performed on our first model ion-of-interest, copper, with immense promise compared to ZIF8, we envision that such a synthetic strategy may pave the way for further interesting functional crystalline structures to be emerged and readily translated to macroscopic platforms.

564852 Shape-Selective Adsorption Separation of C_4 Olefins in Gallate-Based Metal-Organic Frameworks

C_4 olefins including 1,3-butadiene (C_4H_6), 1-butene (*n*-C_4H_8), iso-butene (*iso*-C_4H_8), are widely used in the production of synthetic rubbers and varieties of chemicals. C_4 olefins are generally obtained from C_4 hydrocarbon mixtures produced in petrochemical plants. High-purity individual C_4 olefin is critical for its downstream processes. However, these C_4 hydrocarbons have similar molecular sizes, shapes and physical properties, such as close boiling points and polarizability, making it difficult to separate. Currently, industrial separation of C_4 hydrocarbons is performed by extractive distillations, which are energy-intensive and cost-inefficient. A promising energy-efficient purification method is physisorption with porous materials that can separate gas mixtures by distinguishing the differences in molecular sizes, shapes, polarities, polarizabilities, coordination abilities, and so forth. Metal-organic frameworks (MOFs), emerging as a new class of porous materials, have showed great potential in adsorptive separation of C_4 hydrocarbons.

Herein we report gallate-based metal-organic frameworks, M-gallate (M=Mg, Co, Ni), realized highly efficient separation of C_4 olefins into individual components by shape-size sieving (Scheme 1). Single-component adsorption isotherms of C_4 olefins on Mg-gallate revealed that the uptake of C_4H_6 can reach as high as 1.96 mmol/g (higher than DD3R of 0.83 mmol/g at 303 K and 1 bar), while the uptakes of *n*-C_4H_8 and *iso*-C_4H_8 were as low as 1.46 mmol/g and 0.13 mmol/g, respectively (Figure 1), affording high selectivities of C_4H_6/*n*-C_4H_8, C_4H_6/*iso*-C_4H_8 and *n*- C_4H_8/*iso*-C_4H_8. DFT modelling was performed to gain a better insigh into the adsorption mechanism (Figure 2). Fixed bed breakthrough experiments using a ternary C_4 mixture was conducted to simulate the practical industrial separation processes at 298 K and 1.01 bar on Mg-gallate (Figure 3). *iso*-C_4H_8 broke through the bed of Mg-gallate almost immediately after feeding while *n*-C_4H_8 and C_4H_6 began to breakthrough at approximately 20 min and 38 min. This work demonstrated that M-gallate was one of top-performing MOF materials for adsorption separation of C_4H_6, *n*-C_4H_8 and *iso*-C_4H_8.

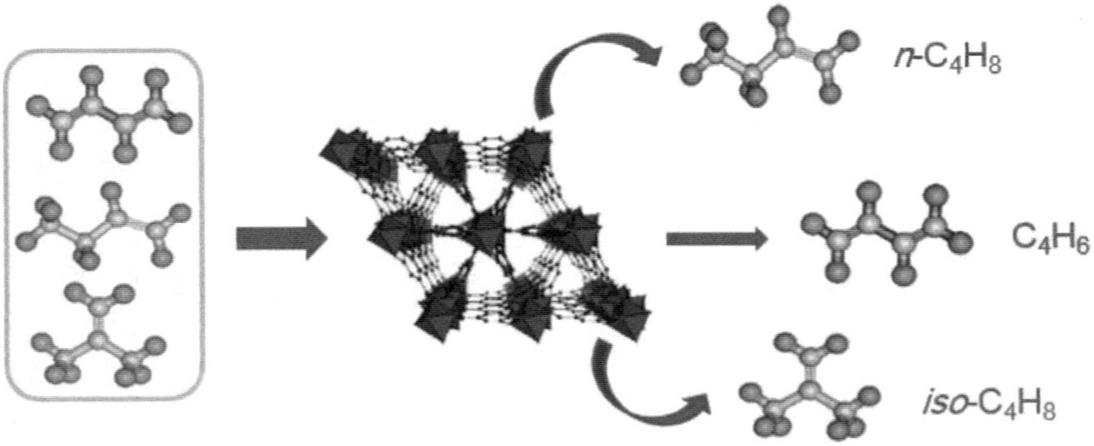

Scheme 1. Schematic illustration of the adsorption separation of C_4 olefins on M-gallate (M=Mg, Co, Ni).

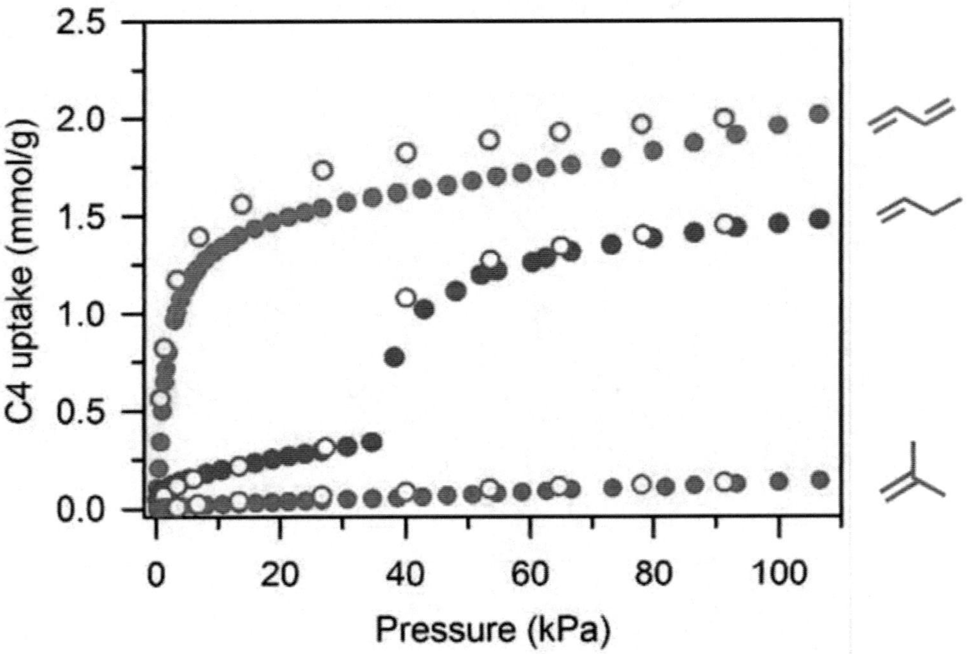

Figure 1. Adsorption isotherms of C_4H_6(red), *n*-C_4H_8(blue), *iso*-C_4H_8(dark green) on Mg-gallate at 298K.

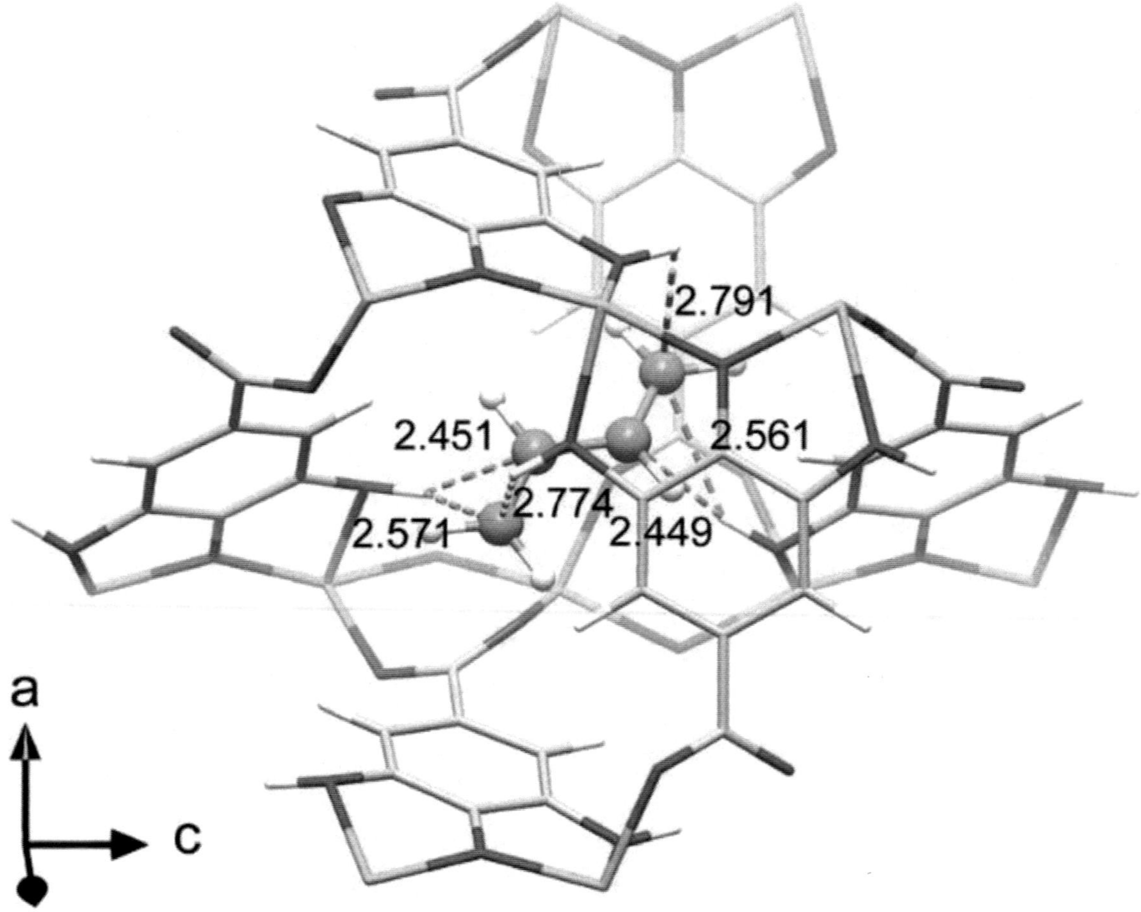

Figure 2. The adsorption binding sites of C_4H_6 in Mg-gallate calculated by DFT method.

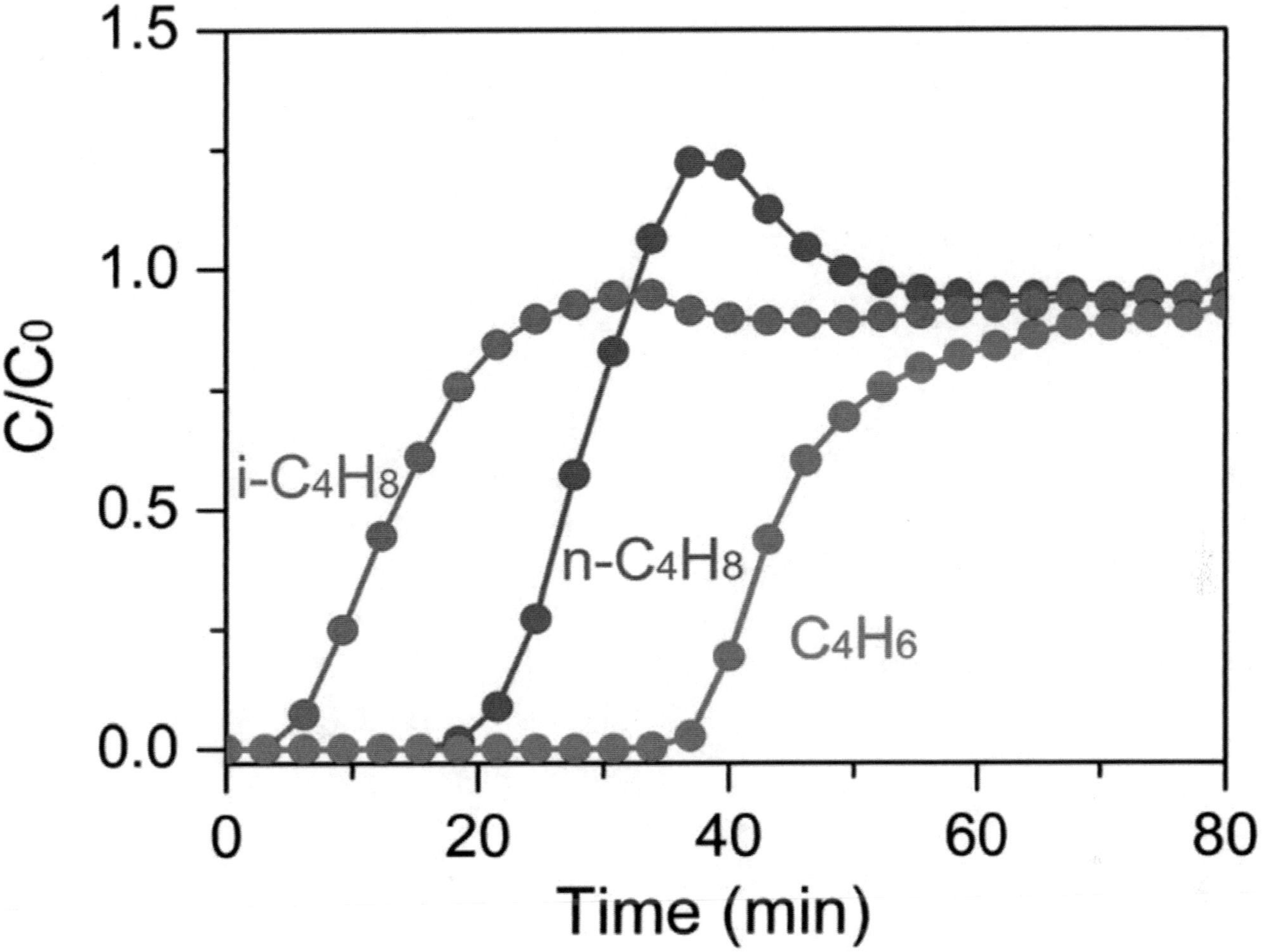

Figure 3. Ternary mixture breakthrough curves on Mg-gallate for C_4H_6/n-C_4H_8/iso-C_4H_8/He (20/10/10/60) at 298 K, with a flow rate of 0.5 mL/min.

Reference

1. P. Q. Liao, N. Y. Huang, W. X. Zhang, J. P. Zhang, X. M. Chen, *Science* **2017**, *356*, 1193-1196.

2. G. Meyer, U. S. Patents, 20030181772, **2003**.

3. Takahashi, R. T. Yang, C. L. Munson, D. Chinn, *Langmuir* **2001**, *17*, 8405-8413.

4. K. Kishida, Y. Okumura, Y. Watanabe, M. Mukoyoshi, S. Bracco, A. Comotti, P. Sozzani, S. Horike, S. Kitagawa, *Angew. Chem. Int. Ed. Engl.* **2016**, *55*, 13784-13788.

5. U. Wagner, H. M. Weitz, *Ind. Eng. Chem.* **1970**, *62*, 43-48.

6. A. Cadiau, K. Adil, P. M. Bhatt, Y. Belmabkhout, M. Eddaoudi, *Science* **2016**, *353*, 137-140.

7. Z. R. Herm, B. M. Wiers, J. A. Mason, J. M. van Baten, M. R. Hudson, P. Zajdel, C. M. Brown, N. Masciocchi, R. Krishna, J. R. Long, *Science* **2013**, *340*, 960-964.

8. J. Pires, M. L. Pinto, V. K. Saini, *ACS applied materials & interfaces* **2014**, *6*, 12093-12099.

9. X. L. Cui, K. J. Chen, B. L. Chen, H. B. Xing, *Science* **2016**, *353*, 141-144.

10. E. D. Bloch, W. L. Queen, R. Krishna, J. M. Zadrozny, C. M. Brown, J. R. Long, *Science* **2012**, *335*, 1606-1610.

559393 Thermal Stability of Imidazolium-Based Ionic Liquids: The Effects of Alkyl Chain Length, Environment, and Heating Conditions

Ionic liquids are room-temperature molten salts that have found increasing use as solvents for polymer and biomaterial processing. This study investigates a series of 1-alkyl-3-methylimidazolium ionic liquids, which have been found to effectively dissolve lignocellulosic biomass. This family of ionic liquids has received significant attention in the literature. Despite this, the thermal stability and degradation behavior of these solvents is not well understood. Furthermore, there have been only a few investigations into the effect of structural variations such as varying the length of their alkyl chains.

In the present study the thermal stabilities of these ionic liquids were investigated with the ethyl, propyl, butyl, hexyl, and decyl alkyl chains, and the chloride and acetate cations. The ionic liquids were studied under both isothermal conditions from 100 to 200 ^{o}C over sixteen hours, and under dynamic heating conditions with heating rates from 5 to 20 ^{o}C/min up to 650 ^{o}C. It was found that the ionic liquids with longer alkyl chains generally displayed decreased thermal stability.

The effect of the environment on the stability of ionic liquids is also significant, and was investigated with 1-ethyl- and 1-butyl-3-methylimidazolium acetate in mixtures with ethanol, methanol, water, and *N,N*-Dimethylformamide (DMF) at elevated temperatures and pressures, up to 150 ^{o}C and 350 bar. A custom high-pressure view cell was used in conjunction with a UV-vis spectrophotometer to capture the degradation of the ionic liquid-solvent mixtures over time. It was found that in protic solvents, including ethanol, methanol, and water degradation was suppressed as reflected by the UV absorption patterns. In contrast DMF, an aprotic solvent, was found to significantly accelerate thermal degradation.

This research is supported by NSF Grant CBET-1509390.

558616 Thermodynamics of the Mixtures of Ammonia and Ionic Liquids

Ionic liquids (ILs) are liquid salts composed entirely of ions, and defined to have a melting point below 373 K. Ammonia (NH_3) is one of the most important and largest volume industrial chemicals produced today. The fundamental thermodynamic knowledge for ILs with NH_3 is essential for exploring the feasibility of ILs in various applications such as gas separation and absorption-refrigeration. This study is divided into three parts. In the first part, the vapor-liquid-equilibria of NH_3 in ILs will be reported. The solubility of NH_3 in ILs was measured using the gravimetric method (XEMIS Microbalance, Hiden Iscohema, UK) at pressures up to 0.7 MPa and at temperatures of 283.15, 298.15, 323.15 and 348.15 K. Experimental vapor-liquid-equilibria data was modeled using Peng-Robinson Equation of State and Non-Random Two Liquid (NRTL) activity coefficient models. In the second part, the direct experimental determination of enthalpies of solution of NH_3 in ILs using high pressure Calvet calorimeter (SETERAM BT 2.15, France) will be reported. Heat of solution is a fundamental thermodynamic property that is related to the nature of interaction between the gas and the IL. The enthalpy of NH_3 absorption in IL can reflect the heat effect due to physical dissolution of NH_3 in the IL and heat effect due to chemical interaction between the two if present. In the third and final part, proton Nuclear Magnetic Resonance (1H NMR) technique will be reported to elucidate the interactions between NH_3 and ILs.

563947 Characterization of 1-Ethyl-3-Methyl-Imidazolium Ionic Liquid Behavior in Non-Aqueous Solvents

We characterize binary systems of 1-Ethyl-3-methyl-imidazolium $[emim]^+$ ionic liquids (ILs) with various molecular solvents including; glycerol, ethylene glycol, acetonitrile, and triethanolamine using precise measurement techniques to determine the conductivity, viscosity, and density of these systems across a wide IL concentration range. We use these three properties in conjunction with the estimated Stoke's radii of the ions calculated from the ubiquitous UNIQUAC and UNIFAC and values, to estimate the ion dissociation based on a method previously described in our previous work.[1,2] Additionally we explore the role of hydrogen bonding in these systems by measuring excess enthalpy across a wide range of IL concentrations. These measurements offer new insights into the driving forces behind ion pair formation in IL/solvent systems. We further consider the solvent effects on ion pair formation, by studying solvents with similar dielectric constants and vary structure (i.e. glycerol, triethanolamine, and ethylene glycol) as well as solvents with similar structure and different dielectric constants (i.e. triethanolamine and glycerol).

(1) Nordness, O.; Simoni, L. D.; Stadtherr, M. A.; Brennecke, J. F. Characterization of Aqueous 1-Ethyl-3-Methylimidazolium Ionic Liquids for Calculation of Ion Dissociation. *J. Phys. Chem. B* **2019**, *123*, 1348–1358. https://doi.org/10.1021/acs.jpcb.8b11892.

(2) Simoni, L. D. Predictive Modeling of Fluid Phase Equilibria for Systems Containing Ionic Liquids, University of Notre Dame, Indiana, 2009.

568983 Investigation of Ion Correlations in Ionic Liquid and Organic Solvent Mixtures through Scalable Screening

Investigation of Ion Correlations in Ionic Liquid and Organic Solvent Mixtures Through Scalable Screening

Ray A. Matsumoto[1,2], Matthew W. Thompson[1,2], and Peter T. Cummings[1,2]

[1]*Multiscale Modeling and Simulation Center, Vanderbilt University, Nashville, TN, 37235-1604, USA*

[2]*Department of Chemical and Biomolecular Engineering, Vanderbilt University, Nashville, TN, 37235-1604, USA*

ABSTRACT

Considerable efforts have been made to improve the performance of supercapacitors, electrochemical devices that exhibit high power density but low energy density. As part of these efforts, room temperature ionic liquids have shown to improve the energy density of supercapacitors due to their wide operational voltage windows in comparison to other electrolytes[1]. Ionic liquids however exhibit high viscosities and slow dynamic properties which can negatively affect the charging rates of such devices. To mitigate slow dynamics, ionic liquids can be solvated in organic solvents without a substantial loss in operational voltage window[2].

We have previously used a screening approach to encompass a larger chemical space and expand our knowledge of interactions between ionic liquids and organic solvents. As a result, we implemented the necessary algorithms and force fields within MoSDeF[3] to conduct screening of $[BMIM^+][Tf2N^-]$ in 22 distinct solvents over a range of 18 different concentrations. Self-diffusivities of ions in these solutions were computed via mean squared displacement (MSD) calculations, while conductivity was estimated via the Nernst-Einstein (NE) equation. From the ability to sample a larger portion of the solvent-ionic liquid design space, we were able to identify several key trends related to ionic liquid diffusivity and conductivity.

Using the previous screening parameter space, we now focus our efforts towards understanding the structure of these ionic liquid and organic solvent mixtures. Ion correlations have been well-characterized for bulk or so-called "neat" ionic liquids, which have shown to impact conductivity through the concept of "ionicity" [4]. On the other hand, ion correlations in ionic liquid and solvent mixtures are still relatively unknown and therefore, we characterize ion correlations within these ionic liquid and organic solvent mixtures by analyzing ion pairing and caging. Furthermore, we compare the Einstein-Helfand (EH) approach[5] to the NE equation for estimating the ionic conductivity of these solutions. Unlike the NE equation, the EH formalism considers ion correlations, which enables us to investigate ionicity through a ratio of the two conductivities. A greater understanding of the relationship between ion correlations and dynamics presented can further guide the selection of solvents and concentration for electrochemical applications of ionic liquids.

References

[1] L. P. Rebelo, J. N. Lopes, J. M. Esperanca, E. Filipe, "On the Critical Temperature, Normal Boiling Point, and Vapor Pressure of Ionic Liquids", Journal of Physical Chemistry B, **2005**, 109, 6040-6043

[2] R. Lin, P. Huang, J. Segalini, C. Largeot, P.L. Taberna, J. Chmiola, Y. Gogotsi, P. Simon, "Solvent Effect on the Ion Adsorption from Ionic Liquid Electrolyte Into Sub-nanometer Carbon Pores", Electrochimica Acta,**2009**, 54, 7025-7032.

[3] "MoSDeF" [Online]. Available: https://github.com/mosdef-hub.

[4] D. R. MacFarlane. *et al.*, "On the Concept of Ionicity in Ionic Liquids", Physical Chemistry Chemical Physics, **2009**, 11, 4962-4967

[5] C. Schroder, M. Haberler, O. Steinhauser, "On the Computation and Contribution of Conductivity in Molecular Ionic Liquids", The Journal of Chemical Physics, **2008**, 128, 134501

573482 *In Situ* Grazing-Incidence Wide Angle X-Ray Scattering (GIWAXS) Investigation of Effects of Silica Nanopore Confinement on Crystallization of [BMIM] Ionic Liquids

Supported ionic liquids (ILs) in nanoporous matrices are being developed to overcome the cost and unknown toxicity of bulk ILs, and to accentuate the microenvironments they offer for separation and catalysis. However, it is not yet known to what extent confinement into nanometer-scale pores alters the surface-induced ordering of ILs. The present work gives insights into how silica mesopore confinement affects the crystallization behavior of two selected ILs based on *in situ* GIWAXS experiments performed at Argonne National Laboratory's Advanced Photon Source.

1-Butyl-3-methylimidazolium hexafluorophosphate ([BMIM]PF_6) and [BMIM]Cl are among the most extensively studied ionic liquids (ILs), which makes bulk crystallization information available for comparison. Silica films with orthogonally oriented 10 nm mesopores are prepared by templating with Pluronic surfactant P123. Some films are modified by covalently tethering 1-(3-trimethoxysilylpropyl)3-methylimidazolium chloride [TMS-MIM]Cl onto the pore wall and XPS is used to verify uniform modification throughout the depth of the films. Both ILs underwent cold crystallization below their glass transition during heating from low temperatures. GIWAXS patterns were collected *in situ* during heating. The crystal phase transition temperatures and corresponding crystal phases were captured and compared for bulk ILs and confined ILs in silica nanoconfinement with and without tethering. Confinement of both ILs in unmodified silica mesopores causes changes in their crystal phases and phase transition temperatures compared with bulk ILs. The melting point, where the long range order completely disappears from the GIWAXS pattern, of [BMIM]PF_6 is about 17.5 °C higher with confinement compared with bulk [BMIM]PF_6 while it is about 190 °C lower for [BMIM]Cl. Confined [BMIM]PF_6 in silica mesopores tethered with [TMS-MIM][Cl] shows a stable and symmetric crystal structure which did not melt at room temperature even though the bulk IL as a film on a Si wafer melts at -11 °C. The melting point depression for [BMIM]Cl is consistent with the Gibbs-Thomson equation with a surface energy of 8.0 mN/m. However, the melting point increases for [BMIM]PF_6 cannot explained by capillary effects. The interaction between ILs and the silica surface (most likely [BMIM] interaction with the negatively charged surface) and molecular rearrangement due to nanoconfinement are believed to be the main reasons for the changes observed. These findings show that when ILs are confined, drastic changes in structure can occur that alter the liquid-state temperature operating range for applications such as ion conduction and supported catalysts.

566210 Impact of Carbon Number of Carboxylic Acid on the Polarity of Hydrophobic Deep Eutectic Solvent

Solvatochromic dyes offer a powerful look into the molecular forces that exist in solutions. Zwitterionic dyes are particularly useful for the investigation of hydrophobic deep eutectic solvents (HDES), which interact via a hydrogen bond. This bond results in a solution with a melting point significantly lower than that of either hydrogen bond acceptor (typically a quaternary ammonium salt) or donator. The hydrogen bond donor, which is usually a carboxylic acid for hydrophobic deep eutectic solvents, is extremely influential on this melting point, with octanoic acid based HDES being liquid at room temperature and dodecanoic acid HDES being a solid until 35-45 °C. By taking advantage of this phenomena, the hydrogen bond acceptor can be changed to impact physical properties like Henry's law constants without dramatically changing polarity or melting point of the solvent. COSMOTherm, a density functional theory based molecular modeling software, is used to explore the impact of carboxylic acid length on the sensitivity of HDES systems to change in N-quaternary ammonium bromide salt characteristics. These results are verified using solvatochromic dyes 4-nitroaniline and 4,4 N-diethyl 4-nitroaniline.

575714 The Effect of Cation Polarity on the Melting Points of Ionic Liquids: An Experimental and Computational Study

A number of molten salts cannot be classified as ionic liquids (ILs) because they are not liquids at or below 100 C. These materials may have interesting or desirable properties however because they remain solids at standard temperatures this is likely to severely restrict their usage in a given process. Lowering the melting points of these materials without significantly affecting their other properties, effectively extending their liquidus range, could facilitate their adoption in industrial processes.

Recently, it was shown that the addition of different substituents to a series of thermally-robust IL cations could have a very substantial effect on the melting point. It was concluded that electron withdrawing effects in these groups create polarity within the cations, which could align, lowering the repulsion in these systems and hence the liquid phase enthalpy, leading to a depression in the melting point. From these results, it remained unclear if this principle was system specific or if it could be reasonably extended to other IL classes.

In this talk we present a systematic study of single fluorine substitutions on the structure and thermodynamics of different homologous series of IL cations. We systematically vary the position of the fluorine substitution in perarylsulfonium, perarylphosphonium, imidazolium, and pyridinium cations paired with bistriflimide, effectively altering the electronic charge and dipole moment of these cations. These ILs were synthesized and their melting points measured along with their enthalpies and entropies of fusion. Furthermore, the crystal structures of the compounds were determined using X-ray crystallography and used to simulate these structures using molecular dynamics. Using two-state simulations, we are able to predict the melting points of these ionic liquids and connect the effect that induced polarity in the IL cation has on the overall melting points.

565090 Change Regulation of Cellulose Depolymerization during Dissolving Process in Ionic Liquids

Abstract

Regenerated cellulose fibers are commercially produced by viscose process using cellulose as raw materials. However, this process exits some drawbacks including high cost, solvent volatility and toxicity. Therefore, it is necessary to develop more efficient and greener technology for cellulose dissolution. Ionic liquids (ILs), as green solvents with attractive properties of non-volatility, thermal stability, and tunable structure, show excellent solubility in dissolving cellulose and spinning, which has been regarded as a promising method for cellulose solution.

However, cellulose degradation is inevitable in the process of cellulose dissolution in ILs, which significantly influences the mechanical properties of regenerated fibers. The mechanical properties of regenerated fiber decreased approximately linearly with cellulose depolymerized. Thus, it's vital to understand the change regulation of cellulose depolymerization in ILs, and hence guides to obtain regenerated fibers with high tensile strength. Usually, the reduction of degree of polymerization (DP) is used to evaluate the degree of cellulose depolymerization.

In this work, microcrystalline cellulose and wood pulp were dissolved respectively in two ILs, i.e., 1-allyl-3-methylimidazolium chloride (AmimCl) and 1-ethyl-3-methylimidazolium diethyl phosphate (EmimDep). The cellulose depolymerization was investigated within 72 hours at 90~130°C. The results indicate that the DP of regenerated cellulose decrease with time, and the decrease in DP of regenerated cellulose became more pronounced when the temperature is above 100°C. And cellulose depolymerization in EmimDep rather than AmimCl has a slight change with the increasing of reaction temperature and time. Therefore, high-quality regenerated cellulose fiber will be acquired when using EmimDep as solvent and no more than 100°C.

Reference

1. Liu, Y. R., K. Thomsen, Y. Nie, S. J. Zhang & A. S. Meyer (2016) Predictive screening of ionic liquids for dissolving cellulose and experimental verification. *Green Chemistry,* 18, 6246-6254.
2. Douglass, E. F., H. Avci, R. Boy, O. J. Rojas & R. Kotek (2018) A Review of Cellulose and Cellulose Blends for Preparation of Bio-derived and Conventional Membranes, Nanostructured Thin Films, and Composites. *Polymer Reviews,* 58, 102-163.
3. Barthel, S. & T. Heinze (2006) Acylation and carbanilation of cellulose in ionic liquids. *Green Chemistry,* 8, 301-306.

553523 Nanoengineered Solvent-Resistant Membranes from Biophenols and Graphene

Separating molecules dissolved in organic solvents is particularly important in pharmaceutical processing, where separations can account for as much as 70% of the total manufacturing cost. Organic solvent nanofiltration (OSN) membranes have the potential to achieve lower energy consumption than conventional separation processes. Despite the increasing interest in the OSN field, it is still an ongoing endeavor to improve separation performance and long-term stability of membranes. Our research provides solutions by designing new membrane nanocomposites for advanced separations in organic media:

- Polybenzimidazole (PBI) was functionalized with hydroxyl groups, followed by covalent grafting of graphene oxide on the polymer matrix, which tripled the solvent permeance (up to 45 L m^{-2} h^{-1} bar^{-1} for acetone) while maintaining high solute rejection with molecular weight cutoff (MWCO) down to 140 g mol^{-1}.[1]
- Polymer of intrinsic microporosity (PIM-1) was blended into a PBI matrix, followed by the reduction of nitrile to amine, and ion-stabilization. The modification simultaneously increased solvent permeance and enabled fine-tuning of the MWCO of the membranes between 190 and 650 g mol^{-1} in various media including polar aprotic solvents.[2]
- A novel membrane fabrication method that realizes the *in situ* polymerization of dopamine during PBI membrane formation was developed. The formation of interpenetrating polymer networks (IPN) endows the membranes with resistance towards polar aprotic solvents. The obtained membranes have MWCO values in the range of 190–850 g mol^{-1} and permeances in the range of 2–19 L m^{-2} h^{-1} bar^{-1}.[3]
- A matrix of membranes comprising of broad combination of biophenol coatings and polymer supports was explored. Nanoengineering of the polymer membrane surface resulted in tunable membrane performance through systematic, versatile, reproducible and easily scalable biophenol coatings.[4]

References

[1] F. Fei, L. Cseri, G. Szekely, C.F. Blanford, Robust Covalently Cross-linked Polybenzimidazole/Graphene Oxide Membranes for High-Flux Organic Solvent Nanofiltration, ACS Applied Materials & Interfaces, 2018, 10, 16140–16147.

[2] G. Ignacz, F. Fei, G. Szekely, Ion-Stabilized Membranes for Demanding Environments Fabricated from Polybenzimidazole and its Blends with Polymers of Intrinsic Microporosity, ACS Applied Nano Materials, 2018, 1, 6349–6356.

[3] D. Zhao, J.F. Kim, G. Ignacz, P. Pogany, Y.M. Lee, G. Szekely. Bio-Inspired Robust Membranes Nanoengineered from Interpenetrating Polymer Networks of Polybenzimidazole/Polydopamine, ACS Nano, 2019, 13, 125–133.

[4] F. Fei, H.A. Le Phuong, C.F. Blanford, G. Szekely, Tailoring the Performance of Organic Solvent Nanofiltration Membranes with Biophenol Coatings, ACS Applied Polymer Materials, just accepted, DOI: 10.1021/acsapm.8b00161

561985 Vapor-Phase Crosslinked Mixed Matrix Membranes with Uio-66-NH_2 for Organic Solvent Nanofiltration

In this work, a series of mixed matrix membranes (MMMs) containing nano-size amine-functionalized UiO-66 particles were vapor-phase crosslinked by tris (3-aminopropyl) amine (TAPA) and investigated for organic solvent nanofiltration (OSN) applications. A thin dense layer was successfully formed on top of MMMs which improved solute selectivity in isopropanol (IPA). By manipulating (1) the size and loading of UiO-66-NH_2 nanoparticles in dope solutions, and (2) the temperature and duration of the TAPA vapor crosslinking reaction, we have fabricated a membrane with a 99.2 % rejection towards negatively charged Rose Bengal (RB, Mw= 1017 $gmol^{-1}$) in IPA and a pure IPA flux of 11.5 $Lm^{-2}h^{-1}$ at 10 bar. Interestingly, the optimal composite membrane developed in this study shows impressive rejections towards positively charged dyes possibly because of electrostatic repulsion (e.g., Donnan exclusion). In addition, the optimized TAPA vapor-phase crosslinked MMMs display stable performance in various organic solvents and the 14-day rejection test. To our best knowledge, this is the first report on organic solvent nanofiltration MMMs made from the vapor-phase crosslinking modification. This modification method may have great potential to develop next-generation OSN membranes for the separation of organic solvents and solutes in pharmaceutical and food industries.

567361 Anti-Tradeoff Graphene Oxide Membrane for Organic Solvent Nanofiltration

Two-dimensional (2D) material membranes are highly attractive in precise separations. However, in most of the 2D material membranes, the permeant transport mostly occurs in single-spacing channels with undesired physical and chemical structures. The permeance-rejection trade-off is always a significant challenge, which has received less focus in 2D membranes. In this presentation, by using graphene oxide (GO) as the building block, we design alternating dual-spacing channel membranes, with locally tailored chemical functionalities, that give high permeance, high rejection and high stability in organic solvent nanofiltration. This unique structure is easily constructed by in situ intercalating and cross-linking scattered sub-5 nm silica nanoparticles in the GO interlayers. The hydrophilic nanoparticles locally widen the interlayer channels to enhance the solvent permeance. In the alternating nanoparticle-free areas, the GO layers simultaneously bend and the p–p interactions retain the narrow and hydrophobic channels, promoting high solute rejection. With a 10-fold increase in water permeance and unaffected rejection, the dual-spacing channel membranes exhibit high permeance especially for high polarity organic solvents. The methanol permeance reaches 290 L /m^2 h bar, with more than 90% rejection of dyes larger than 1.5 nm. This new approach of designing hierarchical channels in 2D materials can be used for a wide spectrum of applications.

Reference:Wang, S.; Mahalingam, D.; Sutisna, B.; Nunes, S*. 2D-Dual-Spacing Channel Membranes for High Performance Organic Solvent Nanofiltration. J. Mater. Chem. A 2019. DOI: 10.1039/C8TA10872B

558690 Torlon® Hollow Fiber Membranes for Organic Solvent Reverse Osmosis Separation of Complex Aromatic Hydrocarbon Mixtures

Abstract

Defect-free polyamide-imide (Torlon®) hollow fiber membranes were fabricated with the aim of investigating the potential for polymer-based organic solvent reverse osmosis (OSRO) separations of complex aromatic hydrocarbon mixtures. The quality of the membranes was assessed by gas permeation, and the membranes used in this study were found to be essentially defect-free (i.e., the permselectivity of the fibers closely matched the permselectivity of dense films). Low molecular weight cut offs (MWCO) of ~180 g mol^{-1} were obtained using a complex mixture of aromatic hydrocarbons relevant to separations downstream of naphtha crackers. Beyond characterizing the membranes, we also demonstrate bulk OSRO-type separations of 80/20 (mol%) mixtures of toluene and 1,3,5-triisopropylbenzene (TIPB, 204 g mol^{-1}) where TIPB was used as a probe molecule of large aromatic hydrocarbons. At an upstream pressure of 70-80 bar (more than 50 bar over the osmotic pressure for this system), we find that the solvent concentration is approximately 98.5-99.0 mol% and that the TIPB rejection coefficient is approximately 90% (separation factor of ~25) in the permeate. These purities and rejections approach 100% as the concentration of the TIPB decreases. Complex mixture of alkyl aromatics were also investigated, and we observe that solvent molecules with molecular weights larger than 185 g mol^{-1} have rejection coefficients in excess of 90%. We observe low solvent (toluene) permeances of approximately 0.01 L/m^2-h-bar, which we find is consistent with the low driving forces in OSRO separations and the general low permeability of Torlon®. However, the ability to reject solutes as small at 185 g mol^{-1} using commercial polymers, commercially-inspired spinning techniques, and without the need of post-fabrication crosslinking suggests that these membranes have the potential to enable challenging solvent-solute or even solvent-solvent separations found in the hydrocarbon processing industry. This work provides a detailed benchmark for OSRO-type separations using polymeric hollow fiber membranes.

571825 Molecular Design of Polymeric Membranes with Tunable Gradient Pores for Organic Solvent Nanofiltration

Organic solvent nanofiltration (OSN) is a robust separation technology for the recovery of organic solvents in the chemical and pharmaceutical industries. A handful of polymeric membranes have been tested for OSN; the pore size and pore size distribution are found to play a key role in membrane performance. Recently, there is increasing interest to develop membranes with gradient pores for separation, which however has not been realized for OSN membranes. Based on molecular simulations, we propose a strategy to computationally design polymer membranes with gradient pores by merging a series of layers of polymers with different chain lengths, and examine their performance for OSN. To illustrate, the polybenzimidazole (PBI) and polymer of intrinsic microporosity (PIM-1) membranes are constructed. For each membrane, the density, pore size distribution, swelling behavior are examined. Compared with the layer composed of shorter chains, the layer of longer chains possesses a lower density, larger pores and a higher swelling degree, thus offering more free volume for permeation. This computational study provides a proof-of-concept study to assess the effects of gradient pores on membrane structures and properties, and suggests that the pore sizes can be tuned via adjusting chain length to develop high-performance OSN membranes.

576069 Novel Metal Organic Brush Membranes (MOBs) for Organic Solvent Nanofiltration

2019 AICHE Annual Meeting, Orlando, Florida (November 10th -15th, 2019)

Session: Separations: 02D12 Membrane-Based Organic Solvent Separations - Oral Session

Chairs: Ryan Lively, Andrew Livingston and Guanghui Zhu

Novel metal organic brush membranes (MOBs) for organic solvent nanofiltration

Pranav Ramesh, Weiwei Xu, Miao Yu and Georges Belfort

Howard P. Isermann Department of Chemical and Biological Engineering and the Center of Biotechnology and Interdisciplinary Studies, Rensselaer Polytechnic Institute, Troy, NY 12180-3590

Abstract

Polymer membranes suffer from selectivity limitations for liquid applications that result from relatively wide pore size distributions and polymer flexibilities. Although these membranes have been widely successful in many industries and dominate membrane applications in general, their use has been limited to large differences in solute size. The long-term objective of this project is to narrow the pore size distribution and stiffen the polymer structure by incorporating metal organic framework (MOF) entities and cross-linking special designed brush membranes. Below, we report on preparing the brushes and then incorporating the MOFs into the brushes and on their characterization and filtration performance.

Previously, we and others have reported on the use of Activators Regenerated by Electron Transfer - Atom Transfer Radical Polymerization (ARGET-ATRP) for grafting polymer brush membranes on a porous support. In this work, we focus on Single Electron Transfer–Living Radical Polymerization (SET-LRP) due to the 'green' advantages of this method such as the ability to carry out the reaction at room temperature and utilizing copper metal as opposed to dissolving copper powder in solution. Several vinyl monomers were evaluated based on carbon chain length, functional end-groups and polarity. More importantly, this work further investigates the role of incorporated MOFs in the selective brush layer. Metal Organic Framework (MOF) were formed *in-situ* within the brush upper layer. Characterization of MOB membranes as well as the underlying brush membranes were carried out using a combination of ATR-FTIR and EDX to understand the chemistry followed by SEM and AFM for morphological information. The membranes were also tested for their ability to remove dyes from organic solvents by filtration.

The use of MOFs in membranes allows for greater structural rigidity as well as to control pore size. MOFs may be altered to obtain a series of membranes which may have the potential to replace energy chugging distillation columns. Due to their stability, tunability and ease of scaleup, we expect MOBs to play a critical role in replacement of thermal systems for chemical and petrochemical industries.

556377 Two-Dimensional Molybdenum Disulphide Membranes for Organic Solvent Nanofiltration: Stability, Structural Manipulation and Separation Performances

Two-dimensional (2D) materials such as molybdenum disulfide (MoS_2) present tremendous opportunities in membrane-based molecular separation. Compared to graphene oxide, MXene and other 2D inorganic materials, MoS_2 show exceptional stability in organic solvents. A few studies have reported the preparation of MoS_2 membranes for separation in aqueous solutions. However, there have been no investigations on the application of MoS_2 membranes for organic solvent separation. In addition, multi-layer and mono-layer MoS_2 nanosheets may be prepared by different approaches and have both been used for membrane preparation. Though studies have revealed that monolayer MoS_2 structure is unstable in the ambient conditions, there have been no reports on the stability of MoS_2 membranes.

This presentation covers our recent progress in 2D MoS_2 membranes for organic solvent separation. In the first part, we prepared both multi-layer and mono-layer MoS_2 nanosheets using different methods, and compared their structure, stability and performances after made into membranes. It's found that multi-layer MoS_2 membranes are looser in structure, and hence result in higher permeability to isopropanol and lower rejection to dyes. On the other hand, though mono-layer MoS_2 membranes show better rejection to dyes in the short tests (within 3 hours), the rejection keeps decreasing over a 5-day tests, which is attributed to the limited stability of such nanosheets that lead to aggregation and breakage. To achieve high rejection and good stability, a 'bridging' agent is employed to regulate the interlayer spacing and nanosheet alignment of multi-layer MoS_2 membranes, leading to improved rejection and stable performances over one week.

In the second part, mixed matrix membranes containing MoS_2 nanosheets based on a 3-step layer-by-layer (LbL) method have been prepared, which provides a more cost-effective and scalable way to fabricate 2D membranes. Very interestingly, while the original LbL polyelectrolyte layer is impermeable to organic solvents, mixed matric membranes containing MoS_2 nanosheets open the pathways to solvents while rejecting solutes whose molecular weight is around 300 Dalton. The phemonena is attributed to the lower of surface energy by MoS_2 and its bridging effects. With a thin layer of < 50 nm, the mixed matrix membrane can achieve ethanol permeance –selectivity that surpasses the hypothetical upper bound, e.g., 92.9 % rejection for Victoria blue B (VBB, MW 506.08) with an ethanol flux of 11.4 L m^{-2} h^{-1} bar^{-1}.

562350 Toward Equant Shaped Crystal Habit through a 3-Stage Cyclic Process: Design and Applications

Crystallization is a fundamental unit operation in many chemical processes where a product in solid form is obtained, and several parameters are of interest when assessing the quality of a crystalline powder. The size and the shape of the crystals are certainly among their most important physical attributes, and contribute to determining properties such as the flowability, filterability, and, in the case of an active pharmaceutical ingredient (API), its bioavailability.

Multiple ways to modify the crystal size and shape have been investigated and are applied at the industrial scale, e.g. temperature cycles, dry milling, wet milling, micronization [1-2]. A 3-stage process [3], consisting of cycles of crystallization, milling, and dissolution, has been recently proposed to efficiently reduce the aspect ratio of needle-like crystals, so as to facilitate downstream processing, while avoiding common problems such as amorphization and formation of sub-micron particles.

In this contribution, we elucidate the link between compound-dependent, yet general features of the crystals, namely the ratios of growth and dissolution rates along their width and length, and the attainable regions through the 3-stage process in terms of average sizes. The process variables spanning the attainable regions are the number of temperature cycles performed and the rotor speed of the wet milling device. Depending on the aforementioned ratios, for the same combinations of process variables, the amplitude of the regions was observed to vary significantly, thus strongly affecting the effectiveness of the process. [4]

Further attention was dedicated to the dissolution step, which aims at reducing the number of fines, and is commonly applied to dissolve a given mass ratio of the solids in suspension. In the frame of the 3-stage process, we show alternative procedures of performing the dissolution step, which aim at removing the fine particles while extending the attainable regions.

The presented results combine experimental work in a laboratory-scale setup and modeling based on two-dimensional PBEs for needle-like particles, and are given in the perspective of performing a robust and quick process design.

[1] Lovette, M. A.; Muratore, M.; Doherty M. F. *AIChE J.*, **2012**, *58*, 1465–1474.

[2] Rasenack, N.; Mueller, B. W. *Pharm. Dev. Technol.*, **2004**, *9*, 1–13.

[3] Salvatori, F.; Mazzotti, M. *Ind. Eng. Chem. Res.*, **2017**, *56*, 9188–9201.

[4] Salvatori, F.; Binel, P; Mazzotti, M. *Chem. Eng. Sci. X*, **2019**, *1*, 100004.

554406 A Quick Design of Spherical Agglomeration of Cefotaxime Sodium Based on Adhesion Free Energy

Spherical crystallization could transform the microcrystalline drug itself into spherical agglomerates improving flowability, compressibility, mean size and compactibility of the drug. Solvent system for spherical crystallization is significant. However, much of the work concerning on the design of wetting liquid was conducted by trials and errors, wasting time and energy. Therefore, this work reports a method to select the proper wetting liquid for spherical agglomeration of cefotaxime sodium by combining theoretical prediction and experimental validation. The theoretical prediction is achieved by calculating the adhesion free energy between the solid surfaces and the liquid (solvents). Various solvents have been screened and candidates with a negative adhesion free energy such as water and chloroform are selected for spherical agglomeration experiments. Experimental results also validate that water and chloroform can be acted as the wetting liquid to form spherical agglomerates of cefotaxime sodium, but mixtures of them can produce high-quality spherical agglomerates. The size distribution, morphology, cohesion and compactibility of the agglomerates have been characterized in terms of process variables such as agitation speed and temperature. An agitation speed of 200 rpm and a temperature of 297.15 K are found to give the optimal result. Finally, the spherical agglomeration process is visualized by PVM and a possible agglomeration mechanism is proposed.

575417 Impact of Polyvinylpyrrolidone (PVP) on the Polymorphism of Succinic Acid Formed Via Monodisperse Droplet Evaporation

Particle structure, whether it is an amorphous form or one of multiple crystalline polymorphs, directly impacts the performance of crystalline products with applications in the food product and pharmaceutical industry. Based on the structure, the final product can have different behavior ranging from varied solubility and bioavailability to surface functionality. Additionally, morphology can impact surface properties or downstream processing ability. The formation of amorphous structures is often achieved by forming particles using the addition of a polymer excipients to the system. Sometimes, however, these polymer additives can impact the resulting polymorphism of the small molecule of interest, when the amount of polymer present is below the amount needed to form a fully amorphous particle. Previous work has shown that monodisperse droplet evaporation can be a valuable experimental method to study polymorphism of small molecule systems at room temperature, whereas spray drying typically takes place at notably elevated temperature.

In this work, we highlight our recent work on the quantified formation of the alpha polymorph of succinic acid, which is not stable at room temperature. The particles are generated through monodispersed droplet evaporation using a vibrating orifice aerosol generator. The resulting polymorphs are determined using powder X-ray diffraction (pXRD), and are quantified through comparison to standardized pXRD peak intensities. Through evaporation from isopropanol or ethanol the metastable alpha polymorph of succinic acid is not present; however, it is formed with the addition of low loadings of polyvinylpyrrolidone (PVP). The amount of alpha polymorph formed with a small PVP loading (< 5%) is greater for ethanol solutions compared to isopropanol solutions. However, both solutions reach an equivalent maximum amount of alpha present at levels of PVP > 10 wt%. We explain the polymorphic behavior based on PVP's inhibition of solute nucleation, allowing the droplet to achieve higher levels of supersaturation before the crystals form. This inhibition reaches a maximum due to thermodynamic limitations. Further, through evaporation from methanol solutions, the alpha polymorph is dominant, and increasing the amount of PVP in the solution slowly decreases the amount of alpha present. This behavior is explained based on the hydrogen bonding donating capability of methanol relative to ethanol and isopropanol. Further, the addition of PVP reduces the impact of that hydrogen bond donating ability due to it being a hydrogen bond acceptor. Finally, we will discuss the resulting particle morphologies as well as the potential of this work to extend to more general methods to understand polymorph formation of similar systems.

567529 Enhanced Nucleation at Liquid-Liquid Interfaces Due to Local Supersaturation

Enhanced Nucleation at Liquid-Liquid Interfaces due to Local Supersaturation

David McKechnie [1,2], Saraf Zahid [1], Paul A. Mulheran [1], Jan Sefcik [1,3], Karen Johnston [1]

[1] Department of Chemical and Process Engineering, University of Strathclyde, 75 Montrose Street, Glasgow, UK

[2] Doctoral Training Centre in Continuous Manufacturing and Advanced Crystallisation, University of Strathclyde, Glasgow, UK

[3] EPSRC Future Manufacturing Hub in Continuous Manufacturing and Advanced Crystallisation, University of Strathclyde, Glasgow, UK

Crystallisation is an important separation process used across a wide range of industries including pharmaceuticals, fine chemicals, and food. Crystallisation studies are often performed in small-scale setups that allow for large numbers of experiments to be performed simultaneously. However, these differ from the typical macroscale crystallisations carried out in industrial settings in a number of ways. One key difference is the presence of interfaces at high surface area to volume ratios. Microfluidic [1] and microwell [2] set-ups are commonly used for such experiments, investigating small volumes of solution that are partially or fully surrounded by immiscible liquids. Heterogeneous nucleation is known to occur at a much greater rate than homogeneous nucleation, and so the presence of these interfaces at such high surface area to volume ratios could allow heterogeneous nucleation to dominate the behaviour of the system. The aim of this work is to study heterogeneous nucleation using a model system of aqueous glycine solution in contact with tridecane.

Glycine is commonly selected for nucleation studies as it is the simplest amino acid, is easy to handle, is polymorphic [3], and has been observed to form mesoscale clusters [4]. Tridecane was used in previous microwell studies [2] due to its immiscibility with water and low vapour pressure. We will use a combined experimental and modelling approach to understand the impact of a liquid-liquid interface on the crystallisation of glycine compared to crystallisation from aqueous solution.

Experiment

A large number of vials of aqueous glycine solutions were prepared at concentrations ranging between 275 – 450 g/kg$_{solvent}$ and monitored by webcam to determine the isothermal induction times. The nucleation rate in unagitated samples was found to be negligible at the range of concentrations investigated, with no crystallisation occurring at any concentration below 450 g/kg$_{solvent}$. At 450 g/kg$_{solvent}$ only 5% of vials crystallised within a two week period (see Table 1).

Vials with a layer of tridecane covering the solution were prepared and observed alongside experiments at concentrations ranging from 275 – 333 g/kg$_{solvent}$. A diagram of the vial setups is shown in Figure 1. The presence of the liquid-liquid interface resulted in a major increase in the nucleation rate of glycine. At the lowest concentration investigated, 63% of vials with an oil interface nucleated within three days, and as the concentration increased the nucleation rate increased, as expected. This significant increase in nucleation rate demonstrates the profound effect the interface has on nucleation behaviour. The percentage of vials that crystallised for each experiment is presented in Table 1.

Figure 1: Diagram of prepared vials with and without

the oil - solution interface.

Table 1: Percentage of vials that crystallised at each concentration with and without an oil interface.

Concentration ($g/kg_{solvent}$)	Vials Crystallised without Oil (%)	Vials Crystallised with Oil (%)
275	0	63
307	0	83
333	0	95
400	0	-
450	5	-

Simulation

Classical molecular dynamics (MD) simulations were performed to provide insight to the behaviour of the system at a molecular level that is not currently accessible by experimental techniques. Simulations of a glycine solution in contact with a layer of tridecane were performed (see Figure 2 top). Simulations were performed in the NPT ensemble at 298 K and 1 atm. The simulation box was of the dimensions 3.45 x 3.45 x 20 nm and allowed to adjust in the z direction. 240 glycine, 4000 water and 256 tridecane molecules were simulated, giving a concentration of 250 $g/kg_{solvent}$ corresponding to the experimental saturation point for 298 K.

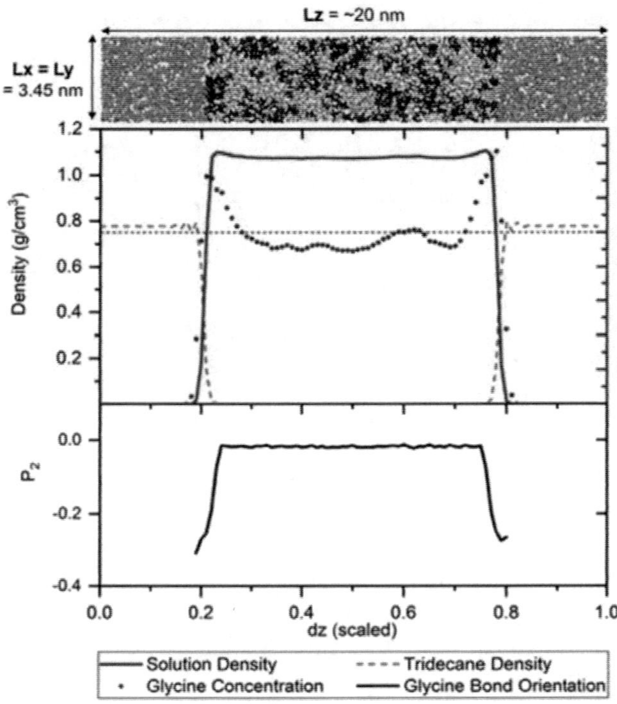

Figure 2: Top: Snapshot of the simulated system: glycine, water and tridecane molecules are blue, red, and green, respectively. Middle: Solution and tridecane density (primary axis) and glycine concentration (secondary axis). The blue dashed line shows the average concentration of the glycine solution (250 g/kg$_{solvent}$). Bottom: bond orientation profile of the glycine molecules.

Structural properties were analysed with respect to distance from the liquid-liquid interface, where dz = z/Lz. Concentration profiles show an increased concentration of glycine near the interface as shown in Figure 2 (middle). A concentration of 367 g/kg$_{solvent}$ is obtained at the interface which is approximately 1.5 times higher than the average concentration of 250g/kg$_{solvent}$. This gives a local supersaturation at the interface that would drive crystal nucleation.

The bond orientation, P_2, of the carbon-carbon bond in glycine molecules with the interface was analysed, where P_2 is defined as:

$$P_2 = \frac{3}{2}\langle\cos^2\theta\rangle - \frac{1}{2}$$

where θ is the angle of the bond vector and the z-axis, where z is normal to the interface. The values of P_2 range from -0.5 to 1. A P_2 of -0.5 indicates the bonds are oriented parallel to the interface, a value of 1 corresponds to perpendicular orientation, and a value of 0 represents random orientation of the bonds. The bond orientation profile is presented in Figure 2 (bottom). The orientation of the bonds is random in the centre of the film (dz = 0.35 – 0.65), and as the glycine molecules approach the liquid-liquid interface they order with the carbon-carbon bond parallel to the interface.

Conclusions

Our MD simulations clearly demonstrate that oil-solution interfaces have an increased concentration and ordering of glycine molecules at the interface. This high concentration at the interface provides the driving force for the increased crystallisation rate observed in experiments. Future work will explore the effect of the interface on the solution dynamics. This work highlights the need to account for, or control, interface effects during the design of experiments.

Acknowledgements

The authors would like to thank EPSRC and the Future Continuous Manufacturing and Advanced Crystallisation Research Hub (Grant Ref: EP/P006965/1) for funding this work. We would also like to acknowledge that this work was carried out in the CMAC National Facility supported by UKRPIF (UK Research Partnership Fund) award from the Higher Education Funding Council for England (HEFCE) (Grant ref HH13054). Computational results were obtained using the ARCHIE-WeSt High Performance Computer (www.archie-west.ac.uk) based at the University of Strathclyde.

References

[1] Ildefonso, M. et al., Organic Process Research & Development 16, 556-560 (2012); Teychene, S and Biscans, B, Chemical Engineering Science 77, 242-248 (2012)

[2] Little, L. J., et al. Crystal Growth & Design 15, 5345-5354 (2015); Little, L. J., et al. Journal of Chemical Physics 147, 144505 (2017)

[3] Boldyreva, E. V., et al. Journal of thermal analysis and calorimetry, 73, 409-418 (2003)

[4] Jawor-Baczynska, A., et al. Crystal Growth & Design 13, 470-478 (2013)

577974 Emergence of Faceted Spiral Patterns during Eutectic Crystallization

Engineered self-organization, including nonlinear pattern formation in crystallization, presents a bottom-up approach to design new multi-phase materials with potentially exotic morphologies and superior properties. Particularly appealing are eutectic systems – mixtures of two or more distinct solid phases that emerge from a parent liquid phase – which exhibit outstanding electromechanical properties because their microstructures act as natural composite materials. Eutectics comprise a wide range of pattern-forming systems, dictated by a number of factors including material properties and growth conditions. In only a few documented cases, the solid phases arrange in intricate spiraling patterns, somewhat akin to a DNA helix. The intrinsic chirality of spiral eutectics offers a new strategy for rapid fabrication – or templates for additive manufacturing – of large-area photonic crystals. Unfortunately, our progress is hindered largely by the lack of mechanistic understanding of spiral eutectic crystallization.

Herein, we demonstrate the two-step formation pathway of faceted spiral eutectics upon directional solidification in the Zn-Mg alloy system. These two-phase Zn-$MgZn_2$ microstructures are periodic, metastable, and intrinsically chiral. We trace the emergence of such structures from the parent liquid through a correlative and multiscale investigation encompassing 3D measurements – namely X-ray nano-tomography aided by machine learning and electron backscatter diffraction – along with *in situ* synchrotron X-ray diffraction and atomic-resolution electron microscopy. Altogether, the results reveal the morphological and heteroepitaxial relationships between the eutectic phases and the origin of spiral growth. The results also identify a thus-far neglected nucleation mechanism involving crystallographic defects within 'hidden' polytetrahedral phases, which provide a template for spiral eutectic crystallization. Our findings provide the necessary benchmark data for simulations of complex crystallization patterns, thus expanding the horizon for the bottom-up design of next-generation alloys.

567138 Characterization and Formation Mechanism of Crystal Defects of Drug a

In the pharmaceutical industry, the internal defects of organic crystals seriously affects the quality and purity of the drug products. However, the mechanism of forming crystal defects has not been systematically reported so far. In this work, drug A, applied for curing cerebrovascular diseases, was selected as model compound. Firstly, it is found that there are three types of defects in drug A such as pores, cracks and inclusions. Then the effects of crystallization conditions such as solvent, temperature, supersaturation and stirring rate on the forming crystal defects were investigated. The Miller index of each crystal plane was determined by combining the results of X-ray diffraction and computer simulation. The interaction energy between each crystal face and the solvent molecules was calculated. The on-line SEM and AFM analysis of the crystal surface determined the local growth inhibition and the crystal growth mechanism of the specific crystal surface was investigated by the molecular arrangement characteristics, which is one of the important reasons for the formation of the crystal defects.

561425 Centrifugal Reverse Osmosis (CRO) - a Novel Process for Achieving Desalination at the Thermodynamic Restriction

For a specified feed concentration and flowrate, and membrane rejection, the water recovery for an equilibrium reverse osmosis (RO) stage is uniquely related to the osmotic pressure differential across the membrane. This relationship defines the thermodynamic restriction for RO desalination. Fig. 1 shows a plot of the pressure in bar versus the fractional water recovery for which the concave-upward line shows the thermodynamic restriction. The area under this line (i.e., the PdV work) is the minimum energy required to achieve a desired water recovery; e.g., a 0.65 recovery requires a transmembrane pressure (TMP) of 80 bar and a minimum energy of 2.46 kWh/m^3, assuming a 100% pump efficiency; this required energy can be reduced if an Energy-Recovery Device (ERD) is used to recover the pressure energy of the retentate brine. However, considerably more energy is required in practice when the feed is pressurized to the maximum pressure required for the desired water recovery; e.g., a 0.65 recovery requires 3.43 kWh/m^3 corresponding to the shaded red and green areas. This required energy increase is caused by pumping all the feed to 80 bar. The required energy can be reduced by increasing the pressure in steps via successive stages. The green-shaded area in Fig. 1 shows the energy saved by increasing the pressure in steps from 0 to 40, 40 to 60 and 60 to 80 bar. However, this requires interstage pumping that increases the capital and maintenance costs as well as the process complexity. RO at the minimum energy dictated by the thermodynamic restriction would require stepping the pressure in differential increments from ambient to the maximum using an infinite number of stages. The Closed-Circuit-Desalination (CCD) process approaches RO at the thermodynamic restriction by recirculating the feed while gradually increasing its pressure in a closed system. However, CCD is a batch process that is not amenable for large-scale desalination or for using an ERD on the retentate brine.

The novel Centrifugal Reverse Osmosis (CRO) process recently patented by the Singapore Membrane Technology Center approximates using an infinite number of stages to increase the pressure, thereby reducing the pumping costs to the minimum defined by the thermodynamic restriction curve. CRO is advantageous relative to CCD since it is a continuous process amenable to using an ERD. The CRO process involves rotating a 'sandwich' of parallel circular flat sheet RO membranes and associated spacers about the axis-of-symmetry of the CRO module. The feed enters through ports in a rotating hollow tube, concentric with the axis-of-symmetry, and flows out radially through the feed channels. Since the back-pressure at the downstream ends of the feed channels is controlled, the centrifugal acceleration causes the pressure on feed side to increase with increasing radial distance from the axis-of-symmetry. Since the permeate channels vent to the ambient, the rotation does not increase the pressure within them, thereby allowing water permeation from the high-pressure feed channels. Using rotation about an axis to create the pressure for RO has been advanced in prior studies. However, these prior studies used the centrifugal force only to create the maximum pressure required for the desired water recovery and did not capitalize on the differential increase in centrifugal force that permits RO at TMPs only slightly higher than the minimum defined by the thermodynamic restriction. The retentate brine in CRO is discharged through a series of jets directed at stationary vanes at the bottom of the module, thereby serving as an ERD to recover its pressure energy.

A mathematical model has been developed for the CRO process based on design at the thermodynamic limit and assuming 100% pump and ERD efficiencies. Fig. 2 compares the predictions for the gross (without an ERD) and net (with an ERD) specific energy consumption (SEC) as a function of the fractional water recovery for conventional single-stage reverse osmosis (SSRO) and the novel CRO process. Table 1 compares the gross and net SEC for 50% and 65% recoveries for conventional SSRO and the CRO process. The required TMP for 50% recovery is 56 bar and that for 65% recovery is 80 bar, the maximum sustainable pressure for currently available commercial RO membranes. The CRO process reduces the SEC_{net} relative to SSRO by 31% and 43% for 50% and 65% water recoveries, respectively. The CRO process not only significantly reduces the energy required for the RO separation, but also reduces the total cost of the water production by permitting economic operation at higher water recoveries, thereby decreasing the normalized pretreatment and brine-disposal costs.

Table 1. Comparison of specific energy consumption for Single-Stage Reverse Osmosis (SSRO) and Centrifugal Reverse Osmosis (CRO)

	50% Water Recovery at 56 bar		65% Water Recovery at 80 bar	
	SEC_{gross} (kWh/m^3)	SEC_{net} (kWh/m^3)	SEC_{gross} (kWh/m^3)	SEC_{net} (kWh/m^3)
SSRO	3.12	1.56	3.43	2.23
CRO	2.56	1.08	2.46	1.26

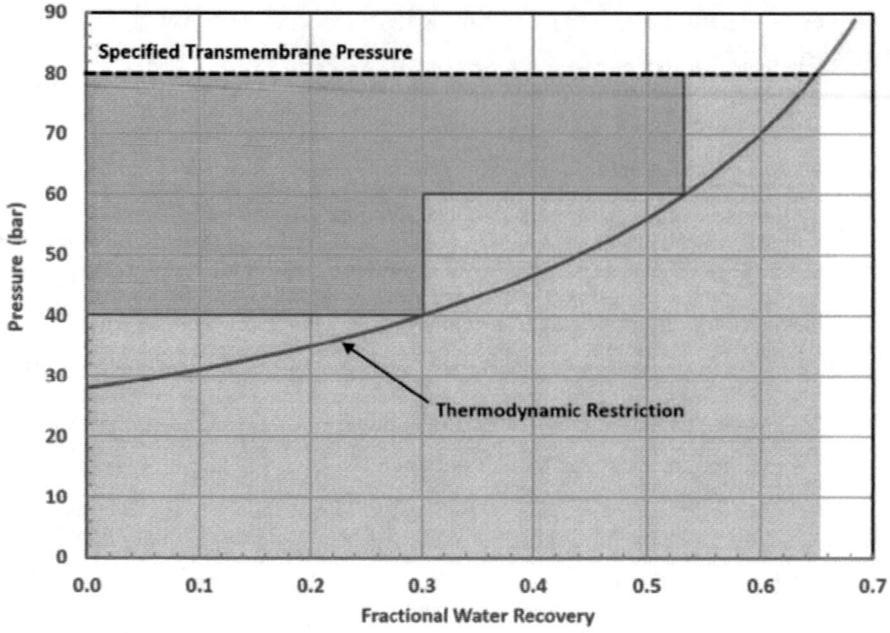

Figure 1. Pressure versus fractional water recovery showing the thermodynamic restriction line, the area below which is the minimum energy required for RO desalination; the energy saved by increasing the transmembrane pressure in three steps rather than applying the maximum pressure to the feed is shown by the green shading.

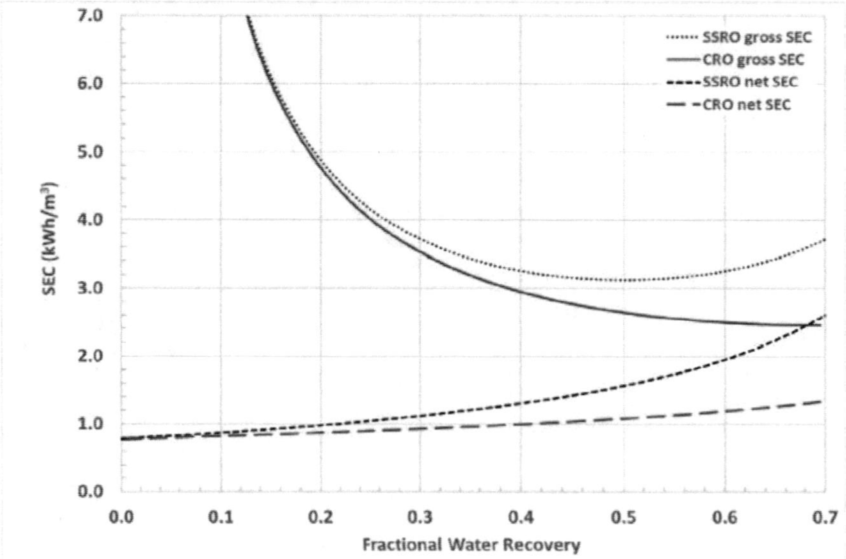

Figure 2. Gross and net specific energy consumption versus fractional water recovery for single-stage reverse osmosis (SSRO) and centrifugal reverse osmosis (CRO); the net specific energy consumption employs an energy-recovery device (ERD) to recover pressure energy from the retentate.

569059 Functionalized Cellulose Nanocrystal Nanocomposite Membranes with Controlled Interfacial Transport for Improved Reverse Osmosis Performance

Thin film nanocomposite membranes (TFNs) are a recent class of materials that use nanoparticles to provide improvements over traditional thin film composite (TFC) reverse osmosis membranes by addressing design challenges such as low flux for brackish water sources, biofouling, etc. In this study, TFNs were produced using as-received cellulose nanocrystals (CNCs) and TEMPO-oxidized cellulose nanocrystals (TOCNs) as nanoparticle additives. CNCs are broadly interesting due to their high aspect ratios, low cost, sustainability, and potential for surface modification. Two methods of membrane fabrication were used in order to study the effects of nanoparticle dispersion on membrane flux and salt rejection: a vacuum filtration method and a monomer dispersion method. In both cases, various quantities of CNCs and TOCNs were incorporated into a polyamide TFC membrane via in-situ interfacial polymerization. The flux and rejection performance of the resulting membranes was evaluated, and the membranes were characterized via ATR-FTIR, TEM, and AFM. The vacuum filtration method resulted in inconsistent TFN formation with poor nanocrystal dispersion in the polymer. In contrast, the dispersion method resulted in more consistent TFN formation with improvements in both water flux and salt rejection observed. The best improvement was obtained via the monomer dispersion method at 0.5 wt% TOCN loading resulting in a 260% increase in water flux and an increase in salt rejection to 98.98 ± 0.41 % compared to 97.53 ± 0.31 % for the plain polyamide membrane. The increased flux is attributed to the formation of nanochannels at the interface between the high aspect ratio nanocrystals and the polyamide matrix. These nanochannels serve as rapid transport pathways through the membrane. Further studies with varying CNC surface moieties are warranted to explore and control these nanochannels to further enhance membrane performance.

577959 Evaluating Pretreatment Technologies to Mitigate Fouling/Scaling in Membrane Distillation

Membrane distillation (MD) is a thermally-driven membrane-based technology that has attracted attention for treating high-salinity wastewater, such as produced water streams resulted from unconventional oil and gas explorations. Lower operating temperature compared to other conventional distillation-based technologies, the potential to harvest low-grade heat and integrate with renewable resources, and modularity of membrane units are some of the attractive features that can make MD a cost-effective technology for treating high-salinity water streams in distributed locations. Currently, there are two major challenges preventing MD from commercialization: 1) low energy efficiency and 2) fouling/scaling. Energy efficiency can be improved by intensifying MD and by designing better heat integration technologies. Fouling is a known to be a more challenging issue. Without addressing fouling/scaling issue properly, not only the maintenance cost will increase but also the whole treatment may fail in a short time.

Fouling could be mitigated by designing better membranes (engineering the materials and the surface) or by designing proper pretreatment technologies. The former is extensively studied, while the latter is rarely studied for MD applications. There is a need to better understand the fouling mechanism, foulants, and scaling materials in MD. Additionally, designing cost-effective pretreatment approaches to treat high-salinity water is critical. In this presentation, we discuss our efforts to alleviate the fouling and scaling in MD. To enhance fouling, we decided to choose vacuum MD (VMD); high salinity produced water from Permian Basin was utilized as the wastewater. Different pretreatment methods, including flocculation, electrocoagulation, filtration, acidification, accelerated precipitation, as well as combined treatments were investigated. Long-term VMD tests were carried out and membranes were characterized after filtration tests to characterize foulants. We will provide a preliminary techno-economic analysis of the pretreatment technologies studied in our efforts.

565776 Selective Separation of Industrial Relevant Metal Ions from High Tds Water Using Wafer-Enhanced Electrodeionization (WE-EDI)

There is a need for an efficient and economical process of ion-selective separation. In hydraulic fracturing, for instance, several million gallons of water are used in a site, most of which are disposed of by deep well injection taking the water from the ecosystem and causing earthquakes. However, due to the high amount of ions in wells at fracking sites, pressure driven processes for water recovery are impossible to use. This project will demonstrate an emerging environmentally friendly Wafer Enhanced Electrodeionization (WE-EDI) technology which attempts to eliminate the shortcomings that come with normal EDI and separates effluent streams into concentrated ions and clean water. In WE-EDI, ion-exchange resins are replaced by wafers that consist of cation and anion exchange beads and a polymer binding agent. In this study,selective separation of these ion exchange wafers and the effect of electro-osmotic drag on the membrane structure have been performed for various anions and cations (Na^+, K^+, Ca^{2+}, Cl^-) and improved selectivity and higher water recoveries have been observed while significant modeling has taken place to predict the selective separation as a strong function of the equilibrium constants between the ions and the beads.

574225 Development of Robust, Ion-Selective Polymeric Ionic Liquid Composites for Reduced Energy Consumption within Water Purification Via Electrodialytic Separations

The onset of climate change and rising global population has increased water stresses around the globe. This heightened demand for water drives researchers to develop improvements to water treatment and water recovery technologies. Among these technologies, membrane-based separations processes play a vital role for the future of water treatment as desalination and industrial wastewater recovery become increasingly important to ensure global water stability. Electrodialysis (ED) and Electrodeionization (EDI) are mature technologies for water treatment; however, its application has been limited by scaling, lack of ion selectivity, and high energy consumption for high salinity feeds. This study investigated the potential of polymeric ionic liquid-based composites within electrodialytic separation systems for improvements in both ion selectivity and energy consumption with ED and EDI. Materials and membranes were characterized through NMR, FTIR, SEM, AFM, contact angle measurement, EIS, and separation performance. Results suggest that while imidazolium coatings were incompatible with ED due to rapid degradation during water treatment, phosphonium-based polymer ionic liquid composites exhibited improvements to ion selectivity and specific energy consumption. This work demonstrates that modification of ion-exchange membranes with thin-film PIL coatings can exhibit enhanced ion selectivity with comparable separation efficiency to uncoated membrane, resulting in improved productivity for water treatment in various applications, such as industrial wastewater reuse, agriculture, and hydraulic fracturing.

578016 Electrochemical Stripping for Selective Nitrogen Recovery from Wastewaters

Electrochemical stripping was shown to selectively recover nitrogen as ammonium sulfate from wastewaters of varying compositions based on the charge and volatility of ammonia. Electrochemical water treatment replaces chemical inputs with electricity, allowing for fine-tuning of unit processes through control of applied current and potential. These modular systems are flexible to varying degrees of centralization and robust to handle intermittent flows and shock loadings. Recovering nitrogen from wastewater can potentially reduce costs and energy of wastewater nitrogen removal and fertilizer production. Through benchtop experiments with a novel three-chamber, two-membrane system, we demonstrate the recovery of nitrogen as ammonium sulfate using electrochemical stripping, a combination of electrodialysis and membrane stripping. Electrochemical stripping combines electromigration of cations with a cation exchange membrane and separation of volatile species with a polypropylene, gas-permeable membraneNitrogen was selectively recovered with 93% efficiency in batch experiments with real wastewater and required 30.6 MJ kg N^{-1} in continuous-flow experiments (slightly less than conventional ammonia stripping). The effects of solution chemistry on nitrogen flux, electrolytic reactions, and reactions with electro-generated oxidants were evaluated using synthetic wastewater solutions. Fates of wastewater-relevant trace organic contaminants, including electrochemical oxidation and reaction with electro-generated chlorine, were investigated with a suite of common pharmaceuticals. Trace organics (<0.1 mg L^{-1}) and elements (<30 mg L^{-1}) were not detected at appreciable levels in the ammonium sulfate fertilizer product. We have recently compared bulk kinetic parameters for each nitrogen transformation and transport step. This novel approach holds promise for selective recovery of nitrogen from concentrated liquid waste streams such as source-separated urine, anaerobic digester effluent, hydrothermal liquefaction effluent, and nitrogen-rich food waste.

574798 CO_2/H_2O Sorption Behaviour of Nickel Nanowires Immobilized in Silica Aerogels for the CO_2 Hydration Reaction

The rise of carbon dioxide (CO_2) in atmosphere is a major factor responsible for increasing global surface temperature. A possible solution to mitigate climate change is CO_2 capture, storage and utilization (CCSU). CCSU could be useful for controlling and moderating anthropogenic emissions of greenhouse gases.

Several technologies have been considered to remove CO_2 from flue gas streams including absorption, adsorption, membrane separations and cryogenic distillations. Another promising CCSU technology is removing CO_2 by using the CO_2 hydration reaction (CHR) which works well in the presence of catalyst such as carbonic anhydrase (CA). Other promising catalysts are nickel nanoparticles (NiNPs) developed by Siller et al. (17-19) which have catalytic activity to accelerate CHR. They are also stable in water based environments. Their catalytic activity is affected by surface area, surface structure of different dimensionalities of nanomaterials. In this work we study immobilized nickel nanowires (NiNWs) as 1D nanomaterials in SiO_2 aerogels as a heterogeneous catalyst for the CHR and compared them with NiNPs as 0D nanomaterials. Data for the differences in catalytic activities for CHR in a vapor phase when two different forms of nickel nanomaterials (NiNPs and NiNWs) are immobilized in silica aerogels will be presented. CO_2/H_2O co-sorption measurements were employed to study the catalytic activity. Vacuum Dynamic Vapor Sorption Analyzer (DVS Vacuum) was employed to study CO_2/H_2O adsorption-desorption kinetics and isotherms.

References:

1. Betts, R. A.; Jones, C. D.; Knight, J. R.; Keeling, R. F.; Kennedy, 411
2. $_2$ rise. Nat. Clim. Change 2016, 6, 806-810
3. Minju, N.; Abhilash, P.; Nair, B. N.; Mohamed, A. P.; 421 Ananthakumar, S. Amine impregnated porous silica gel sorbents synthesized from water-glass precursors for CO_2 capturing. Chem. Eng. J. 2015, 269, 335−342
4. Kong, Y.; Shen, X.; Fan, M.; Yang, M.; Cui, S. Dynamic capture of low-concentration CO_2 on amine hybrid silsesquioxane aerogel. Chem.Eng. J. 2016, 283, 1059−1068
5. Zhang, W.; Liu, H.; Sun, C.; Drage, T. C.; Snape, C. E. Capturing CO2 from ambient air using a polyethyleneimine-silica adsorbent in fluidized beds. Chem. Eng. J. 2014, 116, 306−316
6.
7.
   mineralization carbon capture and storage. Catal. Sci. Technol. 2013, 3, 1234−1239
8.
   Nanomater. 2015, 2015, 1−13.
9. Khalil T. Hassan, Jiabin Wang, Xiao Han, Jon J. Sharp, Gaurav A. Bhaduri, Vladimir Martis, and Lidija Šiller Catalytic Performance of Nickel Nanowires Immobilized in Silica Aerogels for the CO_2 Hydration Reaction, ACS Omega 2019, 4, 1, 1824-1830

576173 Effects of Acid Gases in a Chemically Stable Metal-Organic Framework

Metal-organic frameworks (MOFs) are promising materials for a variety of applications, including adsorption separation processes. Some of these potential gas separation applications involve exposure to acid gases (e.g. sulfur dioxide and nitrogen oxides), which may cause framework degradation or poisoning of adsorption sites, limiting the applicability of MOFs in adsorption processes or making their repeated use more difficult. In this work, MIL-101(Cr)—a chemically stable MOF predicted in literature to experience poisoning by acid gases and derivatives—is exposed to SO_2 and NO_2 at varying concentrations under both dry and humid conditions. Post-exposure nitrogen physisorption measurements demonstrate a loss of adsorption capacity—as measured by BET surface area—caused by exposure to SO_2 and NO_2, but without structural degradation (as indicated by powder x-ray diffraction). Repeated single-component breakthrough experiments (with sample regeneration conducted *in situ*) done with both SO_2 and NO_2 on MIL-101(Cr) show that this loss in adsorption capacity applies to multiple adsorbates and cannot be completely reversed by normal activation conditions. X-ray photoelectron and infrared spectroscopy conducted post-exposure and after regeneration conditions show retention of acid gas species and their derivatives in the MIL-101(Cr) samples, suggesting that the cause of the reduced post-exposure capacity includes acid gas species strongly bound to the framework structure. These results experimentally confirm a difficulty that acid gases may present in MOF adsorbents in addition to well-known issues of chemical stability. Hence, the results draw attention to a detrimental effect of acid gases to be accounted for in future work on chemically stable MOFs.

578310 Biomimetic, Molecularly-Designed Adsorbents for Selective Nitrogen Separations in Wastewater

The presence of nitrogen compounds, such as ammonium, in wastewater is concerning due to their contribution to eutrophication, which has a wide range of negative environmental effects. Removal of ammonium in wastewater streams is typically conducted through microbial nitrification-denitrification pathways that convert ammonium into dinitrogen gas. Unfortunately, this reverses the Haber-Bosch process, which consumes roughly 1% of global energy in order to convert dinitrogen gas into ammonia. A more sustainable and economically advantageous approach for wastewater treatment plants would be to implement technologies that efficiently remove ammonium from wastewater directly, then use it for value-added products, such as ammonium sulfate fertilizer. Ion exchange is a well-established technology for removing ionic impurities from water, but the resulting eluent stream currently requires additional cost-ineffective steps to further separate ions before value-added products can be made.

Therefore, we have fabricated an acrylamide-based, ammonium-selective adsorbent material, inspired by the KS-AMT5 ammonium transporter protein found in ammonium-oxidizing bacteria, for use in wastewater ion-exchange processes. Through the combined effect of multiple functional groups in close proximity, this material selectively binds to ions (like ammonium) that can engage in hydrogen bonding in addition to ionic attraction and cation-pi interaction. Ubiquitous monoatomic wastewater cations like sodium and potassium can only engage in the latter two. Batch and continuous adsorption tests reveal that our adsorbent exhibits higher ammonium adsorption capacity compared to commercial adsorbents. Ion competition experiments exhibit negligible sodium, potassium, and lithium adsorption compared to ammonium (high nitrogen selectivity). Additionally, the adsorbent is resistant to chemical degradation by the regenerant, 0.1 M sulfuric acid, which both removes ammonium and forms ammonium sulfate concurrently. By using this material, wastewater treatment plants will be able to directly remove ammonium and close the industrial nitrogen loop while preserving economic viability.

562428 Metal-Catecholate Incorporated UiO Type MOFs for Natural Gas Upgrading and Toxic Gas Capture

Metal-organic frameworks (MOFs) have been a very captivating topic for researchers in various fields including gas separation, catalysis, sensing, drug delivery etc. owing to their ordered, chemically diverse, tailorable structures. There have been many studies illustrating the effectiveness of pristine MOFs for various gas separation tasks such as CO_2/N_2, CH_4/H_2 etc. where MOFs show highly favorable interaction for a specific adsorbate while much less affinity for the other adsorbate.

More recently, it has been demonstrated that metal catecholate incorporation into MOFs can further alter the affinity of MOFs towards a particular sorbate that could impart a more selective gas separation capability.[1] In this work, we investigate the potential of metal-catecholate doped UiO type MOFs (UiO-66, UiO-67, UiO-68) for natural gas upgrading and toxic gas removal at DFT level using both cluster and periodic models. Considering the binding energy differences of various sorbates, we identify the structures that could be useful for those two objectives. Moreover, we describe the correlations between the calculated binding energies in cluster and periodic models for each sorbate studied.

Reference

(1) Kim, K. C.; Moghadam, P. Z.; Fairen-Jimenez, D.; Snurr, R. Q. Computational Screening of Metal Catecholates for Ammonia Capture in Metal–Organic Frameworks. *Ind. Eng. Chem. Res.* **2015**, *54*, 3257–3267.

568536 Shape-Controlled Cuprous Chloride Crystals for Selective CO Adsorption

CO-containing emissions are environmentally harmful, and for PEM fuel cells a trace amount of CO leads to precious metal catalyst poisoning; therefore, separation/removal of CO is highly demanded. Cuprous chloride, due to the presence of Cu (I) ion, is a promising CO adsorption medium. Many attempts have been made to synthesize CuCl-decorated adsorbent to achieve highly selective adsorption of CO from gas mixture stream; however, less attention has been paid to the effect of CuCl crystal morphology. We conducted CO adsorption under a dynamic condition to understand how different crystallographic planes of CuCl affect selectivity and capacity of CO adsorption. The solid copper substrates were reacted with dissolved $CuCl_2$ in a variety of solvents for shape engineering. The adsorption behavior of the prepared samples was investigated using CO pulse injection and breakthrough curve analysis. In addition, the cation dopant effect was studied for CO adsorption. It was found that the dopants in the reacting solution also change the growth mode of the CuCl crystals.

570804 Selective Removal of Trace Propyne from Propylene Using Ion-Containing Materials

Selective removal of trace propyne from propylene using ion-containing materials

Tian Ke,[1] Jin Shen,[1] Xin He,[2] Qingju Wang,[1] Zongbi Bao,[1] Zhiguo Zhang,[1] Huabin Xing,[1] Qiwei Yang,[1*] Qilong Ren[1]

[1]College of Chemical and Biological Engineering, Zhejiang University, Hangzhou 310027, China.

[2]Department of Chemistry,Massachusetts Institute of Technology, Cambridge, MA 02139, USA

E-mail: yangqw@zju.edu.cn

Selective removal of trace propyne from propylene remains one of the most challenging processes in petrochemical industry. Herein we report several new room temperature molten salts and ion-containing porous solid materials as alternative absorbents/adsorbents for propyne/propylene separation. The Henry's constants of propyne and propylene and separation selectivity were measured for the room temperature molten salts, and both showed superiority to those of reference absorbents. The absorption-desorption cycles could be easily achieved by evaporation at moderate temperature because of the ultralow volatility of molten salts. For the ion-containing porous solid materials, through the optimization of pore chemistry and size, both the single-component adsorption isotherms and the breakthrough experiments showed high adsorption capacity of trace propyne along with excellent propyne/propylene selectivity. Meanwhile, cycling tests showed that the separation performance could remain unchanged for several cycles even for water-containing propyne/propylene mixtures. Density functional theory calculations were performed to simulate the host-guest interactions in both room temperature molten salts and porous solid materials, and the hydrogen bonds between anion and propyne showed a key role in the selective removal of trace propyne from propylene using these ion-containing materials.

Acknowledgement: This research was supported by the National Natural Science Foundation of China (No. 21725603), Zhejiang Provincial Natural Science Foundation of China (No. LZ18B060001), and the National Program for support of Top-notch Yong Professionals (H. X.)

References:

1. L. F. Yang, X. L. Cui, Z. Q. Zhang, Q. W. Yang, Z. B. Bao, Q. L. Ren, and H. B. Xing*. Angew. Chem. Int. Ed. 2018, 57, 13145 –13149.
2. L. F. Yang, X. L. Cui, Q. W. Yang, S. H. Qian, H. Wu, Z. B. Bao, Z. G. Zhang, Q. L. Ren, W. Zhou, B. L. Chen, and H. B. Xing*. Adv. Mater. 2018, 30, 1705374.

578576 Trace Removal of CO_2 from LNG Using Advanced Materials

The proposed research project will focus upon utilizing advanced physisorbent materials to efficiently remove trace levels of carbon dioxide (CO2) from methane (CH4). This application is especially relevant to natural gas purification (sweetening) prior to liquefaction, which requires CO2 levels to be less than 50 ppm. Liquid amines are used to capture CO2 in the conventional process for removing CO2 and they require a high-energy cost to be regenerated. Physisorbents are treated as an important alternative because of lower regeneration energy. Whereas existing classes of physisorbent materials, such as MOFs and zeolites, are generally selective towards CO2 over CH4, their selectivity is not high enough to remove trace levels (e.g. 1%) of CO2 from CH4. Whereas selective capture of CO2 from gas mixtures in which CH4 or N2 are the major components has already been demonstrated, selective capture of trace levels of CO2 from CH4 relevant to natural gas sweetening has not yet been achieved by physisorbents. This project will determine the feasibility of using Different solid sorbents such as silicates, activated carbon, metal organic frameworks, and zeolites have been studied for CO2 capture. Zeolites are crystalline, microporous alum inosilicates based upon tetrahedral building units. These frameworks are anions and so they allow for ion exchange and their robust nature enables reversible dehydration with most common zeolites type X and A. Two types of zeolites including ZE1A and ZE1B has been characterized and tested magnetic suspension balance (MSB) equipment for CO2 adsorption at pressure range of 0-10 bar. Thermo-gravimetric analysis (TGA) shows that ZSM-5 material has very strong thermal stability while there is a rapid decrease in ZE1B weight. XRD characteristic patterns were observed for ZE1A and ZE1B peaks at 8.90° and 7.74°, respectively. The CO2 rubotherm balance results indicates that ZE1B has higher CO2 uptake capacity and can capture up to almost 9.6 mmol/g of CO2 due to the larger pore volume.

Acknowledgment: This research was made possible by a grant from the Qatar National Research Fund under its National Priorities Research Program award number NPRP 10-0107-170119. Its contents are solely the responsibility of the authors and do not necessarily represent the official views of the Qatar National Research Fund.

562408 High-Throughput Screening of Metal-Organic Frameworks to Identify Best Materials for Biogas Upgrading

With the increasing industrialization and world population, the energy need of societies has grown substantially. Currently, a big portion of the worldwide energy demand is still met by fossil fuels despite the environmental and health problems that their consumption can result in. Biogas is an alternative, renewable energy resource which mainly involves methane whose combustion is cleaner than those of fossil fuels. As the biogas mixtures may involve various contaminants (CO_2, N_2, NH_3, H_2S etc.) in considerable amounts, the biogas upgrading is crucial before its use.

Metal-organic frameworks (MOFs) have emerged as promising candidates for many gas separation challenges as they possess high surface area, chemical versatility, and ease of functionalization. As their name implies, they are constructed by the combination of two moieties, a metal node and an organic linker. Since there are numerous metal node and organic linker types, the number of combinations (thus structures) that can be generated is enormous. Relatively recently, a collection of synthesized MOFs is computationally cleaned and made ready for computational exploration which is named Computation-Ready Experimental (CoRE) MOF database[1].

In this computational study, CoRE MOFs are investigated for their efficiency in biogas upgrading using a hierarchical screening procedure and the best performing MOFs are identified on the basis of the performance metrics of selectivity, working capacity, and regenerability.

Reference

(1) Chung, Y. G.; Camp, J.; Haranczyk, M.; Sikora, B. J.; Bury, W.; Krungleviciute, V.; Yildirim, T.; Farha, O. K.; Sholl, D. S.; Snurr, R. Q. Computation-Ready, Experimental Metal–Organic Frameworks: A Tool To Enable High-Throughput Screening of Nanoporous Crystals. *Chem. Mater.* **2014**, *26*, 6185–6192.

574979 Topological Effects on Separation of Alkane Isomers in Metal-Organic Frameworks

The separation of linear and branched alkanes from mixtures is an important industrial process, particularly for enriching the octane rating of gasoline. These separations are commonly done using distillation; however, adsorption offers a less energy-intensive alternative. Metal-organic frameworks (MOFs) are nanoporous, crystalline materials made from inorganic nodes connected via organic linkers. Owing to the large number of nodes, linkers, and functional groups that can be combined to make a unique MOF, there are a virtually unlimited number of possible MOFs. Computational screening can be used to choose materials tailored for a specific application, such as alkane separations.

Nodes and linkers can also be connected in different topologies, or connection networks. In some cases the same set of nodes and linkers may be able to form multiple MOFs with different topologies. These polymorphs are made from the same chemical building blocks but may have radically different pore structures that can affect the adsorption of guest molecules. In this work, we examine the adsorption of C1-C6 hydrocarbons in a large set of topologically-diverse MOFs including many families of polymorphs. We will describe structure-performance relationships in this set of MOFs as they relate to the selective adsorption of linear and branched alkane isomers. We will discuss the influence of the topology on the adsorption capacity and selectivity for separating branched isomers of these alkanes based on the shape and size of the MOF pores and suggest some novel materials that exhibit good selectivity for branched alkane separations.

561418 Microporous MOFs with Specific Binding Sites for Trace Gas Contaminants Removal

Microporous MOFs with Specific Binding Sites for Trace Gas Contaminants Removal

WANG Jun [1], *DENG Shuguang* [1,2]

[1] School of Resource, Environmental and Chemical Engineering, Nanchang University, Nanchang 330031, Jiangxi, China

[2] School for Engineering of Matter, Transport and Energy, Arizona State University, 551 E. Tyler Mall, Tempe, AZ 85287, USA

Abstract: Excessive energy consumption caused by ever-growing population and economy, which compels the massive utilization of low-grade coal and fuels. Sulfur dioxide (SO_2) emissions induced by the combustion of these fossil fuels are serious threat to human health and detriment on environment. Meanwhile, aromatic hydrocarbons contains various volatile organic compounds (VOCs) are direct haze to human health and induce the formation of PM 2.5. However, it is very challenging to achieve efficient and deep removal of gas contaminants, especially, in the cases of extremely low contaminant concentrations. Constructing MOF materials with specific binding sites could significantly enhance the capture ability toward these trace gas contaminants. Thus, we have applied ELM-12 with polar $CF_3SO_3^-$ groups, and the specific recognition sites enable the highly efficient SO_2 removal from gas mixtures. Meanwhile, the novel microporous MOF of NCU-1 have showed superior adsorption capacities of various VOCs, for instance, it displayed a high benzene uptake of 2.9 mmol g^{-1} at 0.1 mbar and 303 K.

570337 Hierarchical MOF Nanoparticles for Natural Gas Liquids Adsorption

Metal-organic frameworks (MOFs) are three-dimensionally ordered porous materials consisting of metal nodes and organic linkers. MOFs (such as MOF-5, and MOF-74) have been shown to adsorb gases in large amounts, many of which are of practical importance, such as H_2, CO_2, C_2 and C_3 hydrocarbons. This gives MOFs great potential in gas storage. In addition to gas adsorption and storage, MOF materials are also used for gas separation by adsorption. Controlled chemical structures usually give MOFs selectivities in gas separations (such as open metal sites in MOF-74). Although gas separation with MOFs with tunable chemical structures have been well studied, there has been no exploration of hierarchical porosity in MOFs for gas separation. Pore size control among MOFs have been shown in isoreticular synthesis. However, these syntheses involve custom-synthesized linkers, which usually require dedicated organic synthesis labs. There have been very few reports on expanding pore size distribution by crystal engineering only. In this work, we show a general 2-step procedure to synthesize hierarchical MOFs with mesoporosity by controlling synthesis conditions, without the use of linkers of various lengths. The method reported in this work maintained the structural integrity of the MOFs during intergrowth, where the seed particle structures are preserved intact. These hierarchical features were shown to increase the adsorption of hydrocarbons. Adsorption of methane, ethane, ethene, propane, and propene pure gases were performed on the hierarchical MOFs as well as MOF seed particles at pressures ranging from 10^{-5} Pa – 1 MPa (absolute). Results from adsorption have shown that the hierarchical MOFs not only adsorbed large amount of gas than the seed particles, they have also shown selectivity of ~50 in propane/CH_4 adsorption (from 1 atm 20% CH4 + 80% propane mixture, according to IAST).

571863 Metal-Organic Frameworks As Food Spoilage Indicators

Food spoilage is associated with the release of Volatile Organic Compounds (VOCs), which usually accumulate into the food package headspace and provide useful information about the quality of the food product itself. Smart packages that integrate miniaturized optical sensors targeted towards food spoilage indicators may be used to monitor the products quality in real-time. Affinity layers represent a core component ofchemical sensors for the selective and sensitive detection of analytes. In this work, Metal-Organic Frameworks (MOFs) were evaluated as affinity layer for food spoilage indicator applications. A novel innovative experimental methodologywas developed to determine the enrichment of individual spoilage indicators from a mixture at ppm level in the nanopores of the MOFs.

Multicomponent breakthrough curves of a mixtureof ethanol, ethyl acetate, 2-butanone and 2,3-butanedione)were measured on two highly stable MOFs, namely ZIF-8 and UiO-66, using Selected Ions Flow Tube Mass Spectrometry (SIFT-MS) for monitoring. Both MOFs share preferentiality towards the adsorption of hydrophobic molecules. Very large enrichment factors (between 100 and 5000) show that these materials can increase many-fold the adsorbed phase over gas phase concentration, even when present at ppm levels, which is crucial for the good operation of affinity layer based sensors. Furthermore,VOCs adsorption on UiO-66 unveils intricate competitive mechanisms, leading to multiple steps of ethanol displacement while simultaneous adsorption of ethyl acetate, 2-butanon and 2,3-butanedione occurs.

574072 Gas Adsorption Properties of NH_2-MIL-53(Al) Metal Organic Framework: Experimental and Modeling Insights

Gas adsorption properties of NH_2-MIL-53(Al) metal organic framework: Experimental and modeling insights

Mudit Baurai, Prudhviraj Medikonda, Prashant Mishra, Satyannarayana Edubilli, **Sasidhar Gumma**[*]

Department of Chemical Engineering, Indian Institute of Technology Guwahati

Guwahati - 781039, Assam, India.

Corresponding author's email: s.gumma@iitg.ac.in (S. Gumma)

During the past two decades, Metal Organic frameworks (MOFs) has gained much research attention in the field of gas adsorption/separation due to their fascinating characteristics like high and selective adsorption capacity, tunable pore structure and ease of post synthetic modification towards specific applications.

MIL (Material Institut Lavoisier) series of MOFs were extensively studied for gas separation applications. Among these materials, MIL-53(Al), which exhibits structural transformation form **lp** (large pore) domain to **np** (narrow pore) domain and vice versa (so-called breathing phenomenon) upon adsorption of certain guest molecules (like CO_2, CH_4 etc), by changing temperature or by applying mechanical pressure was interesting and studied by many researchers. The amine functionalized form of this MOF i.e. NH_2-MIL-53(Al) also shows a similar behavior. However, there are certain differences in the nature such as after activation at room temperature MIL-53(Al) exhibits large pore (**lp**) form, whereas NH_2-MIL-53(Al) show a very narrow pore (**vnp**) form due to the hydrogen bonding between NH_2 group, oxygen atoms and OH group of the framework. Interestingly, higher value of CO_2 capacity and selectivity are reported on NH_2-MIL-53(Al) for sub-atmospheric pressure. However, to the best of our knowledge, a detailed adsorption study including several adsorbates (covering a wide range of polarity and polarizability), high pressure and multiple temperatures seems to be unavailable in the literature. Hence, in this work, we measured pure gas adsorption properties of CO_2, CH_4, N_2 and CO on NH_2-MIL-53(Al) at three different temperatures *viz.* 293, 318 and 354 K over a pressure ranging 0 - 100 bar. In addition, attempts are made to relate the physical properties of gases with the structural transformation of NH_2-MIL-53(Al).

Key words

Adsorption, Metal organic frameworks (MOFs), flexible MOFs, Gas separation

573744 Ultrahigh CO_2/CH_4 and CO_2/N_2 Adsorption Selectivities on a Cost-Effectively L-Aspartic Acid Based Metal-Organic Framework

sing porous materials to selectively adsorb CO_2 from flue gas or natural gas is a promising method for purifying methane and mitigating CO_2 emission. Here, we synthesized an L-aspartic acid based microporous metal-organic framework (MIP-202) and studied its adsorptive separation performance for CO_2/CH_4 and CO_2/N_2 mixtures by measuring isotherms and breakthrough curves. Results show that MIP-202 had ultrahigh CO_2/CH_4 (72.9 and 241.5 for CO_2/CH_4 = 50/50 and 10/90, respectively) and CO_2/N_2 (1950000 and 2129.1 for CO_2/N_2 = 50/50 and 15/85, respectively) IAST selectivities at 298 K and 100 kPa. The high selectivity was verified by the calculations of Henery's law selectivity and breakthrough experiments. Metropolis Monte Carlo simulation calculations show that CO_2 with greater polarizability and quadruple moment tends to concentrate in the large cages (containing high-density amino groups), while the less polar CH_4 or N_2 majorly being adsorbed in the small cages, resulting in the ultrahigh CO_2/CH_4 and CO_2/N_2 separation selectivities. MIP-202 exhibits low adsorption enthalpy (17.2-30.7 kJ/mol), superior persistent reusability after five cyclic adsorption experiments and easy desorption performance at 298 K. The not bad water and moisture stability of MIP-202 was demonstrated by immersing the material in water for 1 and 3 days and exposing the material to 80% RH for 20 days. MIP-202 has low ligand cost , and it can be shaped into MIP-202 extrudates by an extrusion molding method using HPC as the binder. This work shows that MIP-202 is an industrially promising material for the separation of CO_2/CH_4 and CO_2/N_2 mixtures.

556011 Custom Formulation of Ultra-High Permeability Mixed-Matrix Membranes for Post-Combustion Carbon Capture

Mixed matrix membranes (MMMs) have great potential for separation applications including post-combustion carbon capture. To reduce the cost of capture and make them economically viable, new formulations with ultra-high permeability and moderate selectivity must be identified. Here, we present a bottom-up design of MMMs wherein the constituent materials have been designed for high gas separation performance and compatibility. MOF nanoparticles and polymers are decorated by desired functional groups that can chemically interact with each other to form defect free MMMs. The custom formulation of MMMs is important in scale-up production for practical applications. Successful formulation requires a blend of art and science. Here, two, three or four-components, were chosen for their compatibility, where each component has a different function and all are formulated with specific quantities to fine-tune the physicochemical properties for defect free custom-formulated mixed-matrix membranes (CF-MMMs). PIM-1 and MEEP80 were combined with various MOF nanoparticles with different surface areas, pore sizes, and binding sites to conduct a parametric analysis of ten different membranes (CF-MMM-1 to CF-MMMs-10). The chemical interaction between the MOF nanoparticles and the polymer was the key facet for optimizing the interfacial compatibility between the polymer matrices and MOF fillers. Systematic variation of the compositions of these components and the type of MOF nanoparticles were performed to optimize the membrane separation performance affording enhanced CO2 permeability (4200-25670 barrer) and CO2/N2 selectivity (17-34) compared with the neat polymers. The permeability/selectivity values surpass Robeson upper bound and suggest the potential of these membranes for practical CO2 separations. The scaling-up of these membranes were performed by preparing their components in a kilogram-scale, while their precise formulation allows technical feasibility of CF-MMMs fabrication.

561527 Highly Propylene-Selective Mixed-Matrix Membranes By *in-Situ* MOF Formation Using a Polymer-Modification Strategy

Despite the potentials of C_3H_6/C_3H_8 separation, there have been no industrial applications of ZIF-8 mixed-matrix membranes (MMMs) due to the moderate separation performances and several challenging processing issues. Herein, we present a new paradigm of MMM fabrication, named polymer-modification-enabled *in-situ* metal-organic framework formation (PMMOF), enabling *in-situ* formation of ZIF-8 nanocrystals inside 6FDA-DAM polymer. PMMOF consists of four steps including hydrolysis of polymer, ion-exchange, ligand treatment, and imidization. Each step was thoroughly analyzed and important processing parameters were identified, enabling the structural control of MMMs by PMMOF. The binary C_3H_6/C_3H_8 separation performance of the MMMs showed much higher separation factors than conventionally-prepared MMMs at similar filler loadings, satisfying the commercial C_3H_6/C_3H_8 separation performance criteria. PMMOF was successfully applied for other MOFs, demonstrating the process could be general. Finally, as a proof-of-concept, asymmetric mixed-matrix hollow fiber membranes with ultrathin selective skin layers were prepared by PMMOF, showing promising C_3H_6/C_3H_8 separation performances.

562593 Natural Gas Sweetening By MOF-Polymer Hybrid Membranes with Tailored Formulation

Natural Gas Sweetening by MOF-Polymer Hybrid Membranes with Tailored Formulation

Yang Liu[1], *Zhijie Chen*[2], *Gongping Liu*[1], *Youssef Belmabkhout*[2], *Mohamed Eddaoudi*[2,*], *William J. Koros*[1,*]

[1] School of Chemical & Biomolecular Engineering, Georgia Institute of Technology, 311 Ferst Dr. NW, Atlanta, GA 30332, USA

[2] Functional Materials Design, Discovery and Development research group (FMD^3), Advanced Membranes & Porous Materials Center, Division of Physical Sciences and Engineering, King Abdullah University of Science and Technology, Thuwal 23955-6900, KSA

Worldwide energy consumption has grown steadily with continuing increases in world population and economic growth of developing countries. Natural gas is particularly attractive to meet these energy needs, and consumption of natural gas will continue to rise over the next several decades. Although CH_4 is the main component of natural gas, acidic CO_2 and H_2S usually must be removed before delivery to the pipeline to avoid corrosion of processing and transporting equipment. Pipeline specifications for H_2S (< 4 ppm in U.S.) are much stricter than for CO_2 (< 2% in U.S.) because of the additional extremely toxic nature of H_2S. Currently, roughly 30% of proven natural gas resources in the U.S. require treatment due to excessive CO_2 and H_2S concentration. In addition, 40% of worldwide natural gas resources are estimated to be contaminated with high concentration (up to 20 mol.%) of H_2S. Fortunately, membrane processes offer potential compact, energy-saving approaches to address such challenges, as compared to traditional thermally driven gas amine absorption processes. Unfortunately, organic polymer membranes face a trade-off limit between permeability and selectivity, while inorganic molecular sieving membranes face a challenge of reproducibility and economical scale-up. A practical approach to overcome these drawbacks is to create hybrid mixed matrix materials (MMM) by dispersing a molecular-sieve phase (filler) in a continuous polymer (matrix). Such hybrids combine the advantages of both inorganic membranes, e.g. high intrinsic permeability and selectivity, and organic polymer membranes, e.g. robust processing and mechanical properties. Here, we report promising features of MMMs based on metal organic frameworks (MOFs) with tailored internal electrostatic environments for simultaneous removal of H_2S and CO_2 from natural gas. Using M-**fcu**-MOFs and 6FDA-based polyimides, MOF-polyimide hybrid membranes show extraordinary performance on H_2S/CH_4 separation under various H_2S concentrations, *well above the trade-off limit of polymer membranes*. Moreover, the membranes maintain high CO_2/CH_4 separation efficiencies. Among the M-**fcu**-MOFs studied, Zr-*fum*-**fcu**-MOF is considered as the best performing filler with the highest H_2S and CO_2 removal efficiencies. Our interpretation of both sorption and diffusion behaviors for these hybrids will outline key factors to optimize membrane performance for such advanced membranes for H_2S and CO_2 removal for natural gas sweetening.

565581 Enhanced Gas Separation Performance and Plasticization Resistance through Mixed Matrix Membrane Approach Incorporating Multi-Dimensional Metal-Organic Frameworks

Metal-organic frameworks (MOFs) are an emerging class of crystalline materials that are formed by coordination reactions between metal cations and organic ligands. MOFs exhibit high internal surface area, tunable pore size, and tunable chemical functionality, thus making these materials interesting for gas storage and separation applications. In regard to gas separations, mixed-matrix membranes (MMMs) are formed by combining MOFs into polymers to achieve high separation performance while maintaining processability. This study reports a new type of multi-dimensional MOF structure that has 1D branches that are interconnected into 3D hierarchical structures. The multi-dimensional MOFs were incorporated into a polyimide matrix to form MMMs with uniform MOF dispersion. At relatively low MOF loadings, the high effective aspect ratio of interconnected branches creates a percolation network, which results in an increase in gas permeability far above the theoretical permeability predicted by the Maxwell Model. Additionally, the hierarchical architecture leads to a suppression of polymer chain relaxation, thereby reducing effects from plasticization. This study focuses on two multi-dimensional MOFs with distinct mechanisms for molecular transport performance: HKUST-1, which has open-metal sites, has beneficial sorption selectivity, and ZIF-8, which has an appropriately sized molecular-sieving window, has beneficial size-sieving selectivity. Both structures are formed into multi-dimensional MOFs, and a comparison is made between their separation performance and plasticization resistance relative to their bulk counterparts in MMMs. The multi-dimensional structure can improve separation performance for certain separations and simultaneously improve resistance to plasticization.

568419 Ultra-Permeable Mixed Matrix Membranes Containing ZIF Nanoparticles for CO_2 Capture

Membrane technology is an attractive approach for CO_2 capture from flue gas or syngas derived from coal-fired plants, due to its inherent advantages such as high energy-efficiency, small footprint, and potentially low cost. The state-of-the-art membranes are based on polar poly(ethylene oxide) (PEO), which exhibit high CO_2 permeability and high CO_2/N_2 and CO_2/H_2 selectivity. In this work, these PEO containing materials were doped with zeolitic imidazolate framework (ZIF) nanoparticles to improve CO_2 permeability. Specifically, ZIF-8 with uniform particle sizes of ~100 nm were synthesized and incorporated into polymers prepared from poly(ethylene glycol) diacrylate (PEGDA). Increasing the ZIF-8 loading dramatically increased CO_2 permeability and slightly decreased the CO_2/N_2 selectivity. For example, adding 10 wt% ZIF-8 increased the CO_2 permeability from 130 Barrers in a polymer prepared from PEGDA to 320 Barrers without changing the CO_2/N_2 selectivity. At a loading of 50 wt%, the mixed matrix membranes (MMMs) exhibited a CO_2 permeability of 1300 Barrers and CO_2/N_2 selectivity of 33 at 35 ^{o}C, which was one of the best separation properties reported in the literature. Furthermore, an open-cocoon-like ZIF with particle sizes of ~500 nm were synthesized and incorporated into the polymer. Adding 10 wt% open-cocoon ZIF nanoparticles in the polymer increased CO_2 permeability to 370 Barrers with a 15% increase in CO_2/H_2 selectivity and a 10% decrease in CO_2/N_2 selectivity. While the CO_2/light gas selectivity in these MMMs can be satisfactorily described using the Lewis-Nielsen model, the model underestimates the gas permeability. Despite a body of studies of MMMs comprising ZIFs, this work represents one of the few on the MMMs based on amorphous rubbery polymers elucidating the effect of ZIF loading on the polymer properties and demonstrating their impact in achieving superior CO_2/H_2 and CO_2/N_2 separation properties.

574160 Adsorptive and Destructive Mixed Matrix Membranes for Chemical Protection

Flexible semi-permeable membrane filters which selectively allow moisture transport and block or destroy toxic gases/vapors are of significant interest for protective garments. Adsorptive and reactive nanostructured materials such as metal organic frameworks (MOFs) and metal oxides/hydroxides when incorporated in membranes can create a mixed matrix membrane (MMM) having those types of capabilities. We will report the results of our efforts to incorporate MOFs such as UiO-66-NH_2 in a polymeric membrane. We will also report results of barrier performance of such MMMs vis-a-vis toxic gases e.g., ammonia, chlorine from an environment along with our results of moisture vapor transmission rate through such a protection medium.

575975 Ultrathin MOF-Based Mixed Matrix Hollow Fiber Membranes for Highly Efficient Carbon Capture

Ultrathin MOF-based Mixed Matrix Hollow Fiber Membranes for Highly Efficient Carbon Capture

Shouliang Yi[1,2*], Patrick Muldoon[3], Ali Sekizkardes[1,2], Fangming Xiang[1], Victor Kusuma[1,2], Lingxiang Zhu[1], Nathaniel Rosi[3], Kevin.Resnik[1,2], David Hopkinson[1]

[1]U.S. Department of Energy – National Energy Technology Laboratory, Pittsburgh, PA, 15236

[2]Leidos Research Support Team, Pittsburgh, PA, 15236

[3]Department of Chemistry, University of Pittsburgh, Pittsburgh, PA 15260, USA

The National Energy Technology Laboratory (NETL) is developing the next generation of advanced CO_2 capture concepts. Post-combustion carbon capture from flue gas stream produced by conventional pulverized coal power plants after fuel combustion in air involves the separation of CO_2 (~10–16%) from nitrogen, the primary constituent of flue gas. Membrane-based processes have been considered as one of the most promising technologies for post-combustion carbon capture due to numerous advantages including lower energy consumption, smaller environmental footprint, and the potential to be installed in an existing power plant as a true bolt-on technology. Polymeric membranes are currently considered the most promising candidate for industrial application due to their additional benefits of being inexpensive and easy to manufacture. However, the performance of polymer membranes are limited by a trade-off upper bound between permeability and selectivity. To overcome the limitations of polymer membranes, mixed matrix membranes (MMMs) have been developed by incorporating inorganic fillers into the polymer matrix. Recently, metal-organic frameworks (MOFs), consisting of metal ions/clusters connected by organic linkers, have received increasing interest as promising candidates for MMMs owing to their diverse functionalities and tunable pore structures. Although a significant improvement in gas separation performance has been observed with the mixed matrix dense/flat sheet membranes, thin-film composite hollow fiber membranes are superior for large-scale industrial application due to their higher packing density, higher gas permeation rate, and lower consumption of selective layer materials.

In this work, we match the transport properties and interphase compatibility between MOF and polymer materials. A series of MOF-based mixed matrix hollow fiber composite membranes have been developed by incorporating different types of MOF crystals (e.g., surface functionalized UiO-66 and ZIF-8) into a variety of polymers (e.g., polymers of intrinsic microporosity, and poly(ether-block-amide) copolymers) for highly efficient CO_2 capture. Preliminary results showed that the CO_2 permeance of these MOF-based ultrathin film composite mixed matrix hollow fiber membranes was higher than 1000 GPU with a CO_2/N_2 ideal selectivity of >20. Testing these membranes using real flue gas with humidity and minor contaminants at the National Carbon Capture Center (NCCC) will also be presented. We demonstrated that MOF-based mixed matrix hollow fiber membrane is a very promising platform to provide high-performance and highly scalable gas separation membranes for the application of post-combustion CO_2 capture to minimize the cost of an integrated carbon capture system.

563161 Fouling-Resistant Block Copolymer Membranes By Simultaneous Selective Swelling and Zwitterization

We report on a new strategy to produce zwitterization-enabled fouling-resistant membranes, in which zwitterization and formation of membrane pores take place simultaneously. This strategy is enabled by selective swelling of block copolymers (BCPs) in the presence of zwitterization agents. We synthesized the BCPs of poly(2-dimethylaminoethyl methacrylate) and polystyrene (PDMAEMA-*b*-PS) via reversible addition fragmentation chain transfer (RAFT) polymerization. The BCP was coated onto macroporous supports to produce composite structures, which were then soaked in hot ethanol to induce the selective swelling of the PDMAEMA microdomains. 1, 3-propane sultone (1, 3-PS) was dosed into ethanol to simultaneously zwitterizate the tertiary amine groups in the PDMAEMA chains. During the swelling process, the PDMAEMA chains spontaneously migrated onto the interfaces (including membrane surface and pore walls), and the surface-enriched PDMAEMA chains were transformed into zwitterionic poly(sulfobetaine methacrylate) (PSBMA) by the quaternization with 1,3-PS. Upon evaporation of ethanol after removing the membranes from the swelling bath, well-defined nanoporous structures in the BCP layers were formed with PSBMA chains uniformly along the pore walls. Nanoporous BCP films served as the separation layers of thus-obtained composite membranes. On account of the zwitterionic polymers, the composite membranes exhibit excellent perm-selectivity and fouling-resistance. Moreover, their separation properties can be readily regulated by swelling conditions as well as concentrations of 1, 3-PS.

576973 Membranes with Non-Stick Surface to Mitigate Fouling for Water Purification

The conventional approach to mitigate fouling is to increase membrane surface hydrophilicity or surface energy to be as close to water (72 mJ/m^2) as possible. On the other hand, surfaces with low energy (≈20 mJ/m^2) such as Teflon and silicon rubber have demonstrated excellent antifouling properties due to their non-stick nature. In this work, we demonstrate two facile methods to reduce the surface energy of UF membranes by (1) direct coating of "non-stick" Teflon AF1600 (an amorphous glassy perfluoropolymer), and (2) two-step coating of polydopamine (PDA) followed by an amine-functionalized tetramethyldisiloxane ($Si\text{-}NH_2$). The surface modification significantly decreases the surface energy and adhesion of a model foulant of bovine serum albumin (BSA). For example, as the Teflon AF1600 coating layer thickness increased from 0 to 7.0 nm, the surface energy decrease from 42 to 15 mJ/m^2. Consistently, the BSA adsorption decreased from 14 $\mu g/cm^2$ to 2.5 $\mu g/cm^2$. The effect of the coating on the membrane MWCO, pure water permeance, and antifouling properties was systematically investigated. In dead-end filtration systems, the UF membranes coated with ≈5 nm Teflon AF1600 exhibit a 35% reduction in water permeance, less than the unmodified one (50% reduction) for obtaining the same amount of permeated water. Using a constant-flux crossflow system, the Teflon AF1600-coated membranes exhibit a much lower rate of transmembrane pressure (TMP) increase (i.e., lower fouling rate) than the unmodified ones. For instance, the membrane coated with ≈5 nm Teflon AF1600 showed a dTMP/dt value of 0.42 bar/h at a water flux of 70 LMH, much lower than that of the pristine membrane (1.0 bar/h). Similar behaviors have also been observed for the membranes modified with $Si\text{-}NH_2$. Specifically, the Si-modified membranes exhibit dTMP/dt values of ≈20% lower than the pristine ones. In contrast to the conventional approach of increasing surface hydrophilicity, this work demonstrates a new facile method of achieving non-stick surface to enhance membrane antifouling properties for water purification.

578040 Surface Modification of PVDF Membranes for Treating Produced Waters By Direct Contact Membrane Distillation

Direct contact membrane distillation is a promising unit operation for treating hydraulic fracturing flow back and produced water. However, while a hydrophobic membrane is essential to prevent passage of water from the feed to the permeate side, fouling by dissolved organic species can compromise membrane performance and result in wetting of the membrane pores. Here 4 monomers, hydroxyethylmethacrylate, acrylic acid, 1-vinyl-3-allylimidazolium bromide, and 1-vinyl-3-hexylimidazolium bromide have been grafted from the surface of a PVDF membrane. The modified and base membranes were tested in a direct contact membrane distillation system. All membranes were challenged with real produced water. In addition, base membranes and membranes modified by grafting 1-vinyl-3-allylimidazolium bromide were challenged with produced water that was pretreated by electrocoagulation. These membranes were also challenged with a synthetic wastewater made by adding to DI water the major inorganic compounds present in the produced water. The highest fluxes were obtained for the membrane grafted with 1-vinyl-3-allylimidazolium bromide chains. The membrane surface after membrane distillation, was analyzed by scanning electron microscopy and energy-dispersive X-ray (EDX) spectroscopy. For all membranes the interaction between adsorbed organic and inorganic species determines the degree of fouling and hence the loss in flux and membrane stability. Polyionic liquid chains that contain a repeating charged species and hydrophobic segments minimized fouling by organic species and improved the flux and membrane stability. The results suggest that by carefully tuning the properties of the monomer units in the polymer chains, membrane stability and performance can be improved.

565098 Crosslinked Mixed Matrix Polyimide Membranes Surface-Modified By Amine Vapor for Organic Solvent Nanofiltration

The modification of a cross-linked mixed matrix polyimide membrane substrate by tris (3-aminopropyl) amine (TAPA) vapor is performed in this work to modify its surface properties for organic solvent nanofiltration (OSN). The TAPA modification is found to increase the surface hydrophilicity and narrow the pore size of the membrane to make it suitable for OSN applications. The effects of the temperature and time of TAPA vapor treatment are investigated. The newly developed TAPA-modified OSN membranes demonstrate good stability in various solvents (water, methanol, ethanol, isopropanol, dimethylformamide, tetrahydrofuran and hexane) and the permeances of these solvents are found to be proportional to the products of their solubility parameters, molar volumes and viscosities ($d_{p,s} MV_s^{-1} \eta_s^{-1}$). The study illustrates that the optimal composite membrane retains more than 90% for dyes with a molecular weight higher than 828 $gmol^{-1}$ in isopropanol. Moreover, compared with neutral and negatively charge solutes, TAPA-modified membranes show impressive higher rejections towards positively charged dyes possibly because of electrostatic repulsion (e.g., Donnan exclusion). 14-day tests are also conducted to separate Rose Bengal (Mw= 1017 $gmol^{-1}$) from isopropanol/dye solutions, both the fluxes and rejections of membranes do not fluctuate significantly. These findings indicate that amine vapor modification is an advantageous technique to develop next-generation OSN membranes for in food and pharmaceutical industries applications.

568181 Ultrathin MOF/Zeolite Membrane with Enhanced Separation Performance

It is always a great challenge to prepare ultrathin, molecular sieving membranes to realize high permeance and selectivity in the separation field. Recently, two-dimensional materials, with nanometer-thick and high length/thick ratio, provide an effective pathway to maximize the membrane throughput and break the traditional trade-off limitation of membrane-based separation. Herein, we present an ultrathin MOF/zeolite membrane with a thickness of less than 200 nm, resulting in high gas permeation flux and excellent selectivity. What's more, organic-inorganic MOF exhibits an important role in the improvement of the membrane separation performance.

578302 Janus Membranes Via Physicochemical Area Selective Initiated Chemical Vapor Deposition

Development of scalable methods for fabrication of flexible, porous, and chemically heterogeneous thin films are of high interest, especially in membrane-based separation and purification technologies. Here, we introduce a simple method to fabricate thin film composite membranes through area-selective initiated chemical vapor deposition (iCVD). Using iCVD, we gained exceptional control over thickness, pore size, and porosity of the produced flexible films. Our method uses physicochemical templating of the surface to allow for area-selective deposition of polymers. We studied different conditions to identify the effective time constant during which the film growth can be suppressed. We used both free radical scavenger and monomer diffusivity in the templated domain, as the controlling parameters, and inhibited radical chain polymerization. We achieved control over spacing between the pores, pore diameter, and thickness of the films at the submicron resolution and on the planar surfaces (e.g., Si wafer). The chemical and physical properties of the porous films were analyzed through scanning electron microscopy, atomic force microscopy (AFM), X-ray diffraction (XRD), and X-ray photoelectron spectroscopy (XPS). Furthermore, we extended this approach to the 3-Dimensional (3D) substrates and created composite membranes by selective deposition of fluoropolymers on porous hydrophilic supports. We evaluated the performance of these membranes in seawater desalination through direct contact membrane distillation and recorded a stable permeate flux of 24 ± 3 kg m^{-2} h^{-1} and a 100% rejection for the solute.

570970 Thin Film Nanocomposite Hollow Fiber Membranes Comprising Na^+-Functionalized Carbon Quantum Dots for Brackish Water Desalination

We have incorporated Na^+-functionalized carbon quantum dots (Na-CQDs) into the polyamide layer via interfacial polymerization reaction and developed novel thin film nanocomposite (TFN) hollow fiber membranes for brackish water desalination. Comparing with the conventional thin film composite (TFC) membranes, the TFN membranes comprising Na-CQDs have a larger effective surface area, thinner polyamide layer and more hydrophilic oxygen-containing groups in the polyamide layer. Besides, the interstitial space among the polyamide chains becomes larger due to the presence of Na-CQDs. As a result, the incorporation of 1 wt% Na-CQDs into the polyamide layer could improve the pure water permeability (PWP) of the membranes from 1.74 LMH/bar to 2.56 LMH/bar by 47.1% without compromising their NaCl rejection of 97.7%. Interestingly, stabilization of the TFN hollow fiber membranes containing 1wt% Na-CQDs at 23 bar could further promote the PWP to 4.27 LMH/bar and the salt rejection to 98.6% under the same testing conditions due to the deformation of the membranes under a high hydraulic pressure. When using a 2000 ppm NaCl aqueous solution as the feed, the optimal water flux and rejection of the newly developed TFN membranes at 15 bar are 57.65 ± 3.26 LMH and $98.6\% \pm 0.35\%$ respectively. The Na-CQDs incorporated TFN hollow fiber membranes show promising applications in the field of brackish water desalination.

562620 The Influence of Temperature Gradients on Selective Charge Transport in Electrodialysis Systems

Electrodialysis (ED) is an established technique for the desalination of brackish water or the concentration of industrial streams. An electric field is applied over a stack of alternating cation and anion exchange membranes to selectively transport ions, resulting in a concentrated and a diluted product stream. The number of ions transported through the system, indicated by the measured current, increases linearly (following Ohm's law) with increasing driving potential difference, until there is a mismatch between the supply and removal rates of ions at the membrane interface. Concentration polarization, the development of a depleted boundary layer at one side of the membrane and a concentrated boundary layer on the opposite site occurs at a critical potential. Transport through the depleted boundary layer, which is dictated by diffusion and electromigration limits the overall transport through the system. As a result, a system specific so-called ``limiting current density" is observed for all electrodialysis systems. Increasing the transport through the boundary layer can enhance the total transport through the system and increase the efficiency of ED at higher potentials.

In this research [1-2], we experimentally investigate the influence of temperature and temperature gradients on the transport of various ions in electrodialysis systems. Temperature has a large influence on the physical properties of the ionic solutions in the system, as the viscosity decreases with increasing temperature and the diffusivity and mobility of the ions increases with increasing temperature. Increasing the temperature of the fluid on the depleted side of the membrane interface can thus enhance the transport of ions through the depleted boundary layer. After initial theoretical work [3] indicating the possible enhancement of selective transport through a charge selective interface in the presence of a temperature gradient, experiments were conducted using a commercially available, lab-scale electrodialysis stack. The influence of temperature and temperature gradients on the transport of different ions was measured in both the Ohmic and limiting current regimes.

For measurements in the Ohmic regime using a 1;1 electrolyte (NaCl), a clear increase in total ion transport was measured when increasing the temperature of either the depleted or enriched stream by 20°C when compared to having both streams at a lower temperature. The energy required for the ED process was reduced by 9% when heating one of the feed streams. Increasing the temperature in both streams enhanced the total transport even further and reduced the required energy input for the ED process by 15%. However, in contrast to our expectations, no significant change in selective transport was measured for any of the different temperature configurations in the Ohmic regime. The reduced power input is mainly attributed to the increasing diffusivity of the ions at elevated temperatures, and the viscosity of the solutions was found to be of less influence.

In the limiting current regime, where the depletion boundary layer is dictating the overall transport through the stack, the influence of a temperature gradient is more pronounced. We find that the direction of the temperature gradient influences the total transport of ions and the selectivity towards transport. A significant improvement of the total transport is found when the depleted stream is heated by 20 °C, when compared to heating the concentrated stream. Additionally, for systems containing a mixture of monovalent (Na^{+}) and divalent (Mg^{2+}) ions, a change in selectivity was found when applying a temperature gradient. The selectivity favored the separation of divalent ions when the dilute stream was heated when compared to isothermal systems. This is attributed to competitive transport of the mono- and divalent ions that is influenced by their different response on temperature.

The results obtained in these experiments indicate that the application of temperature gradients has a positive influence on the required power for electrodialysis systems and can be of use in enhancing the selective transport between mono- and divalent ions. This is of potential use in industrial systems where heavy metals have to be selectively removed and in water softening technologies exchanging divalent for monovalent ions. The utilization of industrial waste heat, available from various processes and at various locations, for the application of a temperature gradient is a promising method for the improvement of overall process efficiency.

[1] Benneker, Anne M., et al. "Influence of temperature gradients on mono-and divalent ion transport in electrodialysis at limiting currents." Desalination 443 (2018): 62-69.

[2] Benneker, Anne M., et al. "Effect of temperature gradients in (reverse) electrodialysis in the Ohmic regime." Journal of membrane science 548 (2018): 421-428.

[3] Wood, Jeffery A., Anne M. Benneker, and Rob G.H. Lammertink. "Temperature effects on the electrohydrodynamic and electrokinetic behaviour of ion-selective nanochannels." Journal of physics: Condensed matter 28.11 (2016): 114002.

576695 Thin Film Nanocomposite (TFN) Membranes Containing Functionalized Zeolite Nanoparticles for Desalination with High Boron Rejection

Inadequacy of natural water resources has become a worldwide concern and production of potable water from brackish or seawater via reverse osmosis (RO) stands out as the leading technology. Mixed matrix membrane (MMM) concept, which combines desirable properties of a filler and a polymer, have been utilized extensively for gas separation applications and more recently extended to water separation applications. Several studies on nanoparticle (zeolites, carbon nanotubes, graphene, metal organic frameworks, metal oxides, etc.) incorporation into the thin film polyamide structure during interfacial polymerization process reported improved RO membrane performance, i.e. improved salt rejection, water flux and anti-fouling properties. These mixed matrix RO (also called TFN) membranes offer considerable promise for achieving desalination while concurrently reducing the boron concentration to the level recommended by the World Health Organization (< 2.4 ppm for potable water and < 0.5 ppm for irrigation water). In this work, we explore the possibility of tailoring the desalination performance to add new functionality of high boron rejection to RO membranes by incorporating functionalized zeolite 4A nanoparticles into the polyamide layer. The 3-D structure of zeolite A offer the advantages of easy embedding in membrane structure and lack of the need for orientation over 1-D and 2-D structures such as carbon nanotubes and graphene. In-house synthesized zeolite 4A nanoparticles were surface functionalized with hexadecyltrimethylammonium (HDTMA) to hinder the passage of boric acid molecules. Zeta potential measurements were used for characterization of functionalized zeolites. TFN polyamide layer was prepared on commercial ultrafiltration support membranes by interfacial polymerization using m-phenylene diamine (MPD) and tri-mesoyl chloride (TMC) as monomers and zeolites were incorporated in the TMC solution. TFN membranes were characterized by XPS, SEM, AFM, contact angle measurements and crossflow filtration tests for salt and boron removal. TFN membranes had improved boron rejection and higher pure water flux than TFC membranes without compromising salt rejection performance at neutral pH.

(This work is supported by the Scientific and Technological Research Council of Turkey (TUBITAK) through Project No. 114Y165.)

571654 High Concentration Ternary Saline Water Treatment Via Superhydrophobic and Fabricated Antifouling Membrane Via Vacuum Membrane Distillation

High concentration ternary saline water treatment via superhydrophobic and fabricated antifouling membrane via vacuum membrane distillation

Yushan Shao, Yingqi Wang, Gaohong He , Xiaobin Jiang*
State Key Laboratory of Fine Chemicals, School of Chemical Engineering, Dalian University of Technology, Dalian, Liaoning 116024, China

E-mail: xbjiang@dlut.edu.cn

Abstract

Vacuum membrane distillation (VMD) is a promising technology for high salinity wastewater treatment[1-3]. One of the key issues in VMD research is improving the antifouling performance of the hydrophobic porous membrane under proper interface modification theory and approach[4-6]. In this work, the superhydrophobic polypropylene (PP) composite membrane with modified surface were successfully fabricated by loading SiO_2 nanoparticles and attaching low surface energy 1H,1H,2H,2H-perfluorodecyltriethoxysilane. The model concerning the deposited nanoparticles size, surface contact angle and roughness was constructed to predict the influence of interface geometric parameters on membrane surface induced nucleation barrier. The Classical heterogeneous nucleation theory was introduced to illuminate the improved performance of developed membrane with targeted surface morphology on inhibiting the initial interfacial nucleation process[7] (**Figure 1**). The improved superhydrophobic surface with water contact angle of 158. 5° enhanced the anti-wetting ability of the membrane for VMD purpose as shown in **Figure 2**.

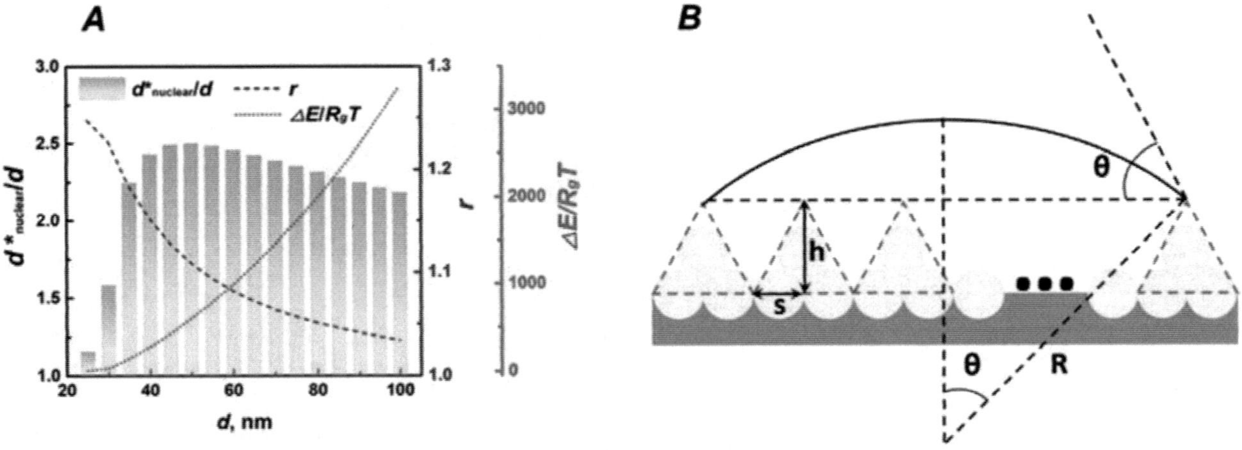

Figure 1. (A) Simulation results of d^*_{nuclear}/d, surface roughness r and heterogeneous nucleation energy ΔE^*_{het} with various particle size d and the same loading capacity (1 wt % particles). (B)Variables in membrane based nucleation model defined by geometric method.

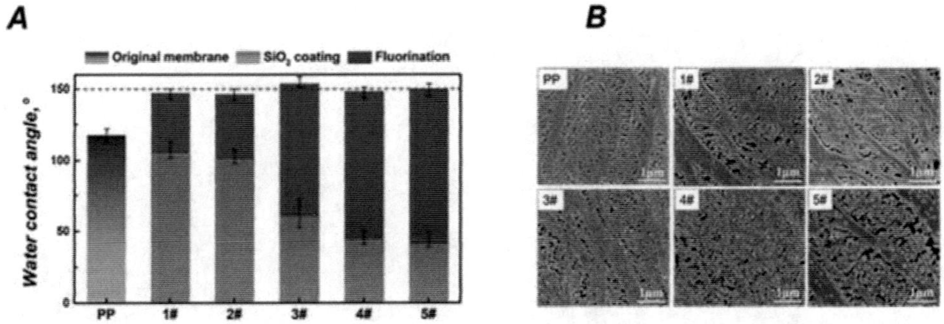

C

TEOS:Ethanol:NH_3·H_2O, in mole ratio	–	1:50:2.4	1:50:2.8	1:50:3.2	1:50:3.6	1:50:4.0
SiO_2, wt%	–	0.4	0.7	1.0	1.6	2.4
$\Delta E_{ha}/RT$	198.3	26.56	100.45	650.01	892.75	1630.66
d, nm	–	26	32	51	57	73
r	–	1.364	1.268	1.062	1.058	1.028
d*, nm	265.0	135.69	211.30	393.50	437.36	534.52
d*/d	–	5.22	6.60	7.72	7.67	7.32

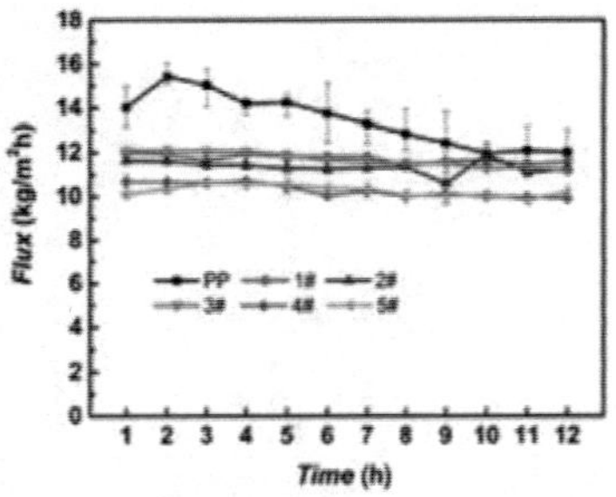

Figure 2. Membrane properties with different fabrication conditions (different TEOS: Ethanol: NH_3•H_2O mole ratio). (A) Water contact angle of the membrane surface. (B) SEM images of modified membranes. (C) Membrane surface characterization results (mass fraction of loading SiO_2, average diameter of SiO_2 particles of the membrane surface). (D) Comparison of permeate flux of different membranes.

The fabricated superhydrophobic PP membrane exhibited unique anti-fouling and anti-wetting abilities, as well as extremely stable permeating flux even under low feed rate and highly concentrated NaCl (15 wt%)/$MgCl_2$ (from 3 to 9 wt%) saline solution system during long-term continuous and batch operation in VMD (**Figure 3 and Figure 4**). The fouling rate of the fabricated membrane was decreased to 20% of the one of the original membrane under 15 wt% high salinity feed condition. In addition, it was also indicated that sodium ion and magnesium ion with distinct features in the structure of water-ion solution system had the potential impact on the permeate flux decline and concentration polarization effect, which was helpful for the further assessing and improving the stability and durability of this series modified membrane for the harsh treatment solution system.

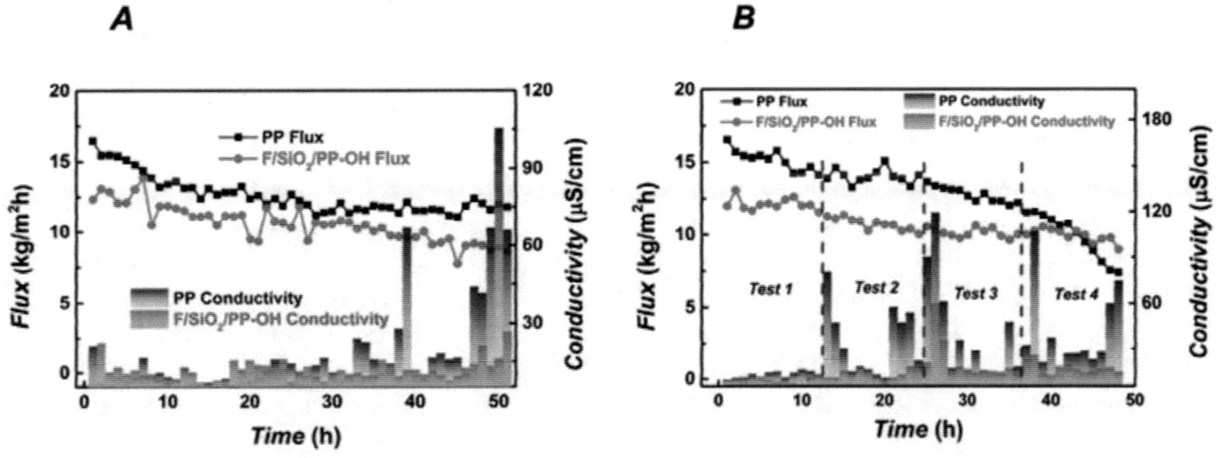

Figure 3. Permeate flux and conductivity of the original PP membrane and the F/SiO_2/PP-OH membrane in 3.5 wt% NaCl solution as feed for (A)long-term continuous operation and (B) long-term batch operation (each batch is 12 hours, the retention time is not shown in the figure).

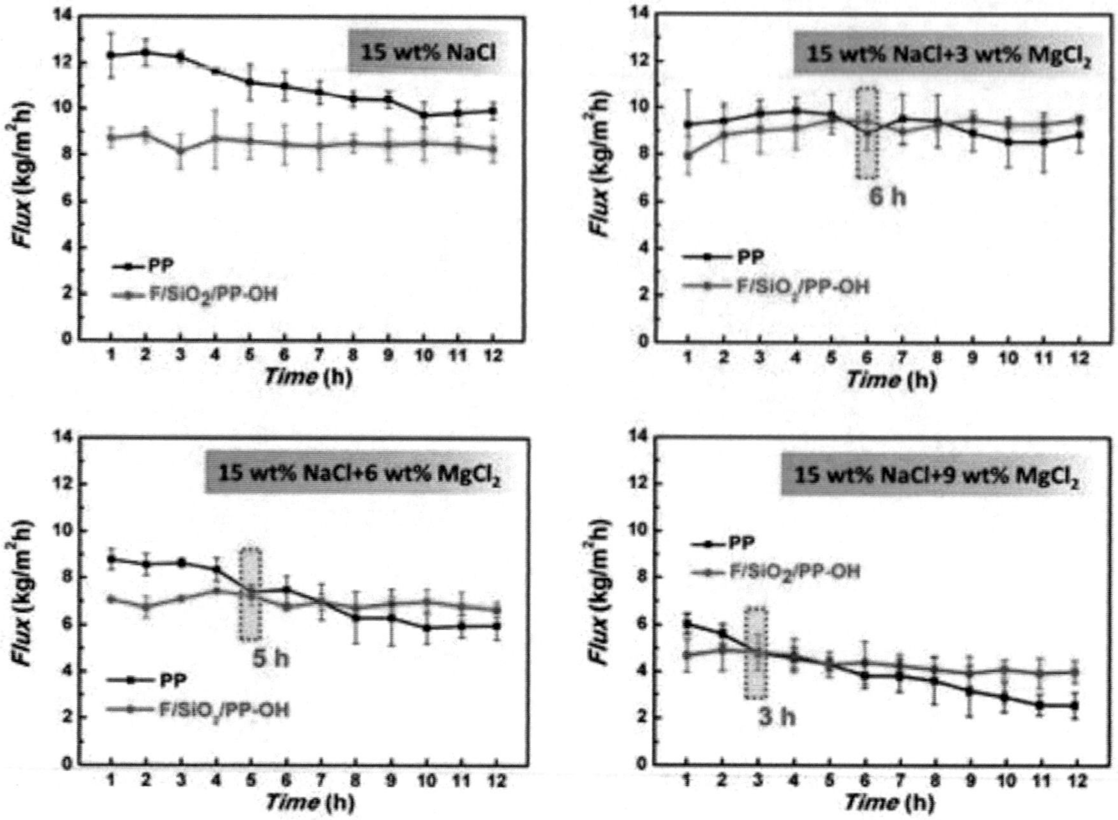

Figure 4. Effect of different concentrations on the flux of PP membrane and F/ SiO_2 / PP-OH membrane. (feed flow rate: 90 mL/min; ¡÷*P*: 0.096 MPa; feed solution temperature: 80 °C)

Acknowledgment

We acknowledge the financial contribution from National Natural Science Foundation of China (Grant No. 21527812, 21676043, 21606035), the Fundamental Research Funds for the Central Universities (DUT16TD19, DUT17ZD203) and MOST innovation team in key area (No. 2016RA4053).

References

1. Deshmukh, A.; Boo, C.; Karanikola, V.; Lin, S.; Straub, A. P.; Tong, T.; Warsinger, D. M.; Elimelech, M., Membrane distillation at the water-energy nexus: limits, opportunities, and challenges. *Energy & Environmental Science* **2018,** *11* (5), 1177-1196.

2. Eykens, L.; De Sitter, K.; Dotremont, C.; Pinoy, L.; Van der Bruggen, B., How To Optimize the Membrane Properties for Membrane Distillation: A Review. *Industrial & Engineering Chemistry Research* **2016,** *55* (35), 9333-9343.

3. Wang, Q.; Gao, X.; Zhang, Y.; He, Z.; Ji, Z.; Wang, X.; Gao, C., Hybrid RED/ED system: Simultaneous osmotic energy recovery and desalination of high-salinity wastewater. *Desalination* **2017,** *405*, 59-67.

4. Laqbaqbi, M.; Sanmartino, J.; Khayet, M.; Garc¨ªa-Payo, C.; Chaouch, M., Fouling in Membrane Distillation, Osmotic Distillation and Osmotic Membrane Distillation. *Applied Sciences* **2017,** *7* (4).

5. Tijing, L. D.; Woo, Y. C.; Choi, J.-S.; Lee, S.; Kim, S.-H.; Shon, H. K., Fouling and its control in membrane distillation¡ªA review. *Journal of Membrane Science* **2015,** *475*, 215-244.

6. Warsinger, D. M.; Swaminathan, J.; Guillen-Burrieza, E.; Arafat, H. A.; Lienhard V, J. H., Scaling and fouling in membrane distillation for desalination applications: A review. *Desalination* **2015,** *356*, 294-313.

7. Curcio, E.; Curcio, V.; Profio, G. D.; Fontananova, E.; Drioli, E., Energetics of Protein Nucleation on Rough Polymeric Surfaces. *J. Phys. Chem. B* **2010,** *114*, 13650-13655.

563501 Membrane Degradation of Forward Osmosis (FO) in Treating MED Brine: Comparison between Cellulose Triacetate (CTA) and Thin-Film Composite (TFC) Polyamide Membranes

Membrane degradation of forward osmosis (FO) in treating multi-effect distillation (MED) brine: comparison between cellulose triacetate (CTA) and thin-film composite (TFC) polyamide membranes

Ye Yang (speaker) , Yuzhu Sun*, Jianguo Yu*

State Key Laboratory of Chemical Engineering, College of Chemical Engineering, East China University of Science and Technology, Shanghai 200237, China

Abstract

The discharge of reject brine from thermal desalination processes is a threat to marine ecosystems because of its high salinity, high temperature, and chemical dosages. In this research, the forward osmosis (FO) was used to treat multi-effect distillation (MED) brine. The changes in surface chemical properties of two commercial forward osmosis (FO) membranes resulting from exposure to MED brine for 30 days was characterized via X-ray photoelectron spectroscopy (XPS), Fourier-transform infrared spectroscopy (FTIR), Contact angle and zeta potential. The permeability (P), forward solute flux (FSF) and specific reverse solute flux (SRSF) were measured to evaluate the membrane performance degradation. Besides, the extended Derjaguin¨CLandau¨CVerwey¨COverbeek (XDLVO) approach was used to predict the changes in the membrane fouling propensity. The results showed that the cellulose triacetate (CTA) membrane suffered deacetylation and chain cleavage due to the alkaline condition of the MED brine, while the polyamide thin-film composite (PA-TFC) membrane experienced degrafting of poly(sulfobetaine methacrylate) (pSBMA) brushes due to the high temperature. The selectivity reductions of both membranes were observed according to the increased FSF and SRSF. Moreover, the results indicated that CTA membrane fouling of sodium alginate (SA) was ameliorated as the attractive XDLVO interaction energies decreased, which was attributed to the increased hydrophilicity caused by the CTA hydrolysis. On the contrary, the fouling resistance of the PA-TFC membrane was severely reduced as the attractive interaction energies increased, which was resulted from the exfoliation of pSBMA brushes. This study reveals the necessity for the further design and fabrication of the FO membrane with long-term stability against MED brine.

Keywords: forward osmosis, degradation, brine

Graphic Abstract

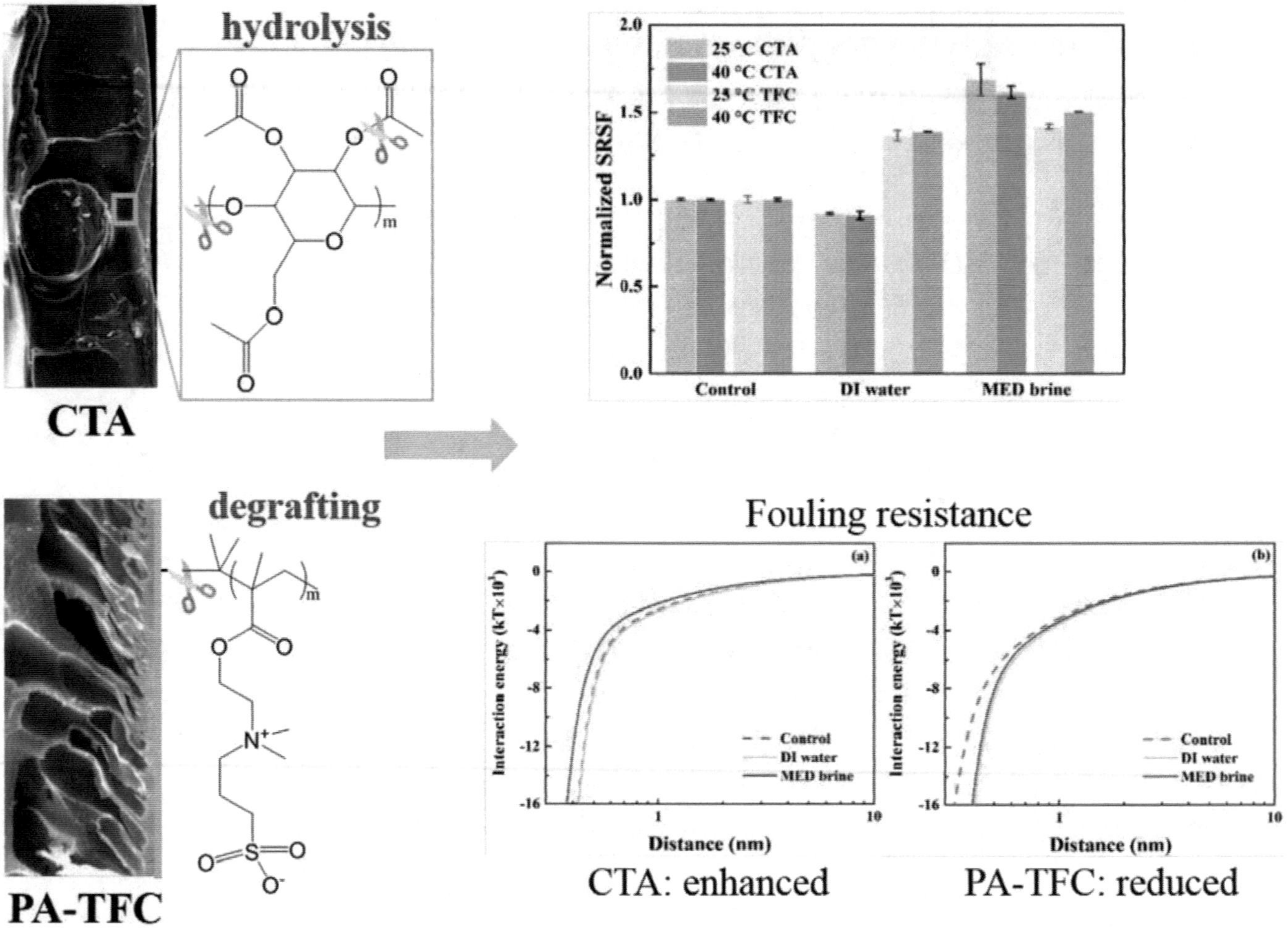
hydrolysis
CTA
degrafting
PA-TFC
25 °C CTA
40 °C CTA
25 °C TFC
40 °C TFC
Normalized SRSF
Control
DI water
MED brine
Fouling resistance
Interaction energy (kT×10³)
Distance (nm)
(a)
(b)
Control
DI water
MED brine
CTA: enhanced
PA-TFC: reduced

553501 Highly Permeable Dual-Layer MOF Composite Membranes for Post-Combustion CO_2 Separation

Highly Permeable Dual-Layer MOF Composite Membranes for Post-combustion CO_2 Separation

Sameh K. Elsaidi,* Surendar R.Venna, Mona H. Mohamed and David P. Hopkinson

ABSTRACT: We present a novel concept for the fabrication of custom-designed dual-layer metal organic framework (MOF) membranes that show great promise in improving permeability and selectivity performance for CO_2/N_2 separation as compared with the individual constituent MOF materials. A highly selective thin MOF layer with narrow pore size (shell layer) was grown on the surface of a highly permeable MOF layer with large pore size (core layer). Most importantly, the interaction between the two layers was optimized through the chemical binding between the two MOF materials affording continuous, defect-free membranes. The two MOF@MOF membranes: SIFISIX-3-Ni@SIFISIX-1-Cu and SIFISIX-3-Ni@HKUST-1, represent the first examples of inorganic membranes constructed by dual-layer MOF composites. The CO_2/N_2 selectivities of the HKUST-1 and SIFSIX-1-Cu membranes were improved from 5.2 to 13 and 3.4 to 7.8, respectively, after the growth of a selective layer of SIFSIX-3-Ni. This work emphasizes the importance of the custom-design of the membrane material rather than the screening approach of dozens of materials.

565940 Facile and Scalable Fabrication of High-Performance CHA Zeolite Hollow Fiber Membranes and Modules for Gas Separations

Efficient separation of gases such as CO_2 from other molecules such as CH_4, N_2, and hydrocarbons is a key problem in many energy production and greenhouse gas reduction schemes. Zeolite materials have demonstrated attractive gas separation properties, but their industrial application as membranes in gas separations has been elusive because of high cost and low reproducibility of conventional fabrication techniques. We demonstrate a facile, one-step method for fabrication of high-silica CHA zeolite membranes on low-cost alumina hollow fiber substrates along with detailed characterization of the CHA hollow fiber membranes. To demonstrate the scalability of the method, 20 hollow fiber membranes were synthesized in the same batch and assembled into a module. The permeation characteristics of these membranes will be discussed in detail. For example, under dry conditions, the membranes showed high CO_2 permeances up to 3500 GPU (1 Gas Permeation Unit = 3.348×10^{-10} mol m^{-2} s^{-1} Pa^{-1}) and excellent CO_2 separation performance (e.g., CO_2/CH_4 separation factors up to 210 in the 1-7 bar feed pressure range). Additionally, the membranes exhibited considerable increases in CO_2/N_2 selectivity under humid (saturated water vapor) conditions with moderate permeance drops. This work opens a promising new path for accelerated development of low-cost zeolitic membranes for gas separation technology.

566170 Sub-200-Nm Thick Carbon Molecular Sieve Membranes and Modulation of Pore-Size-Distribution By Room Temperature Oxygen Functionalization

Carbon molecular sieves (CMS), hosting slit-like nanopores formed by a disordered packing of aromatic carbon strands driven by the volume exclusion effects, have emerged as a highly promising membrane material.[1–3] CMS can be synthesized with a tunable pore-size-distribution, yielding a high molecular selectivity with a sub-angstrom resolution in molecular differentiation. However, the permeance of the CMS membranes remains low due to the micron-thick selective layer, typically afforded in such films. It is challenging to reduce the thickness below 1 μm without incorporating pinhole defects which compromises the molecular selectivity. To the best of our knowledge, CMS films with thickness below 300 nm, yielding attractive gas separation performance have not been reported. Moreover, currently, there is no room temperature pore modification route to tune the molecular selectivity from the CMS membranes.

In this presentation, I will discuss two membrane fabrication routes, namely transfer and masking techniques,[4] leading to 100 – 200 nm CMS film by preventing the infiltration of the CMS precursor in the pores of the membrane support. The 100-nm-thick CMS film yielded attractive gas-sieving performances with H_2 permeance reaching up to 3060 gas permeation units (GPU) with corresponding H_2/CH_4 selectivity of 18 at 150 °C. Furthermore, a rapid (of the order of 15 sec) and highly-tunable post-synthetic modification method based on room temperature ozone treatment will be discussed. Gas selectivity could be improved by several folds by shrinking the CMS micropores by a fraction of an angstrom. The optimized membranes yielded H_2 permeance of 507 GPU and H_2/CH_4 selectivity of 50.7. Other membrane yielded H_2 permeance of 453 GPU and H_2/CH_4 selectivity of 106.

References

(1) Rungta, M.; Wenz, G. B.; Zhang, C.; Xu, L.; Qiu, W.; Adams, J. S.; Koros, W. J. Carbon Molecular Sieve Structure Development and Membrane Performance Relationships. *Carbon N. Y.* **2017**, *115*, 237–248.

(2) Adams, J. S.; Itta, A. K.; Zhang, C.; Wenz, G. B.; Sanyal, O.; Koros, W. J. New Insights into Structural Evolution in Carbon Molecular Sieve Membranes during Pyrolysis. *Carbon N. Y.* **2019**, *141*, 238–246.

(3) Koh, D.-Y.; McCool, B. A.; Harry W., D.; Ryan P., L. Reverse Osmosis Molecular Differentiation of Organic Liquids Using Carbon Molecular Sieve Membranes. *Science.* **2016**, *353*.

(4) Huang, S.; Villalobos, L.F.; Babu, D. J.; He, G.; Li, M.; Züttel. A.; Agrawal, K. V. Ultrathin Carbon Molecular Sieve Films and Room-temperature Oxygen Functionalization for Gas-sieving. Submitted, under review.

576080 Tuning Carbon Molecular Sieve Hollow Fiber Membranes for Gas Separation through Hyperaging

We developed carbon molecular sieve (CMS) membranes using polyimide precursors, and used the CMS to recover hydrogen from mixtures of hydrogen with short-chain (C1-C3) hydrocarbons. A post-treatment method called “hyperaging” was invented to create highly selective CMS hollow fiber membranes. It was found hyperaging benefited small molecular such as hydrogen much more than larger molecular such as ethylene. Through the utilization of hyperaging method, H_2/C_2H_4 selectivity was significantly enhanced while H_2 permeance decreased moderately. Hyperaging conditions on CMS fiber separation properties were investigated, and hyperaging mechanism was discussed. Hyperaging temperature was found to be a primary factor to influence separation performance. The workable hyperaging temperature is beyond 90 ^{o}C but lower than 250 ^{o}C. Too high temperature causes an unrecoverable damage to fiber separation properties. The post-treatment using hyperaging can also stabilize CMS materials, and maintain separation performance for the hydrogen separation application over extended periods.

576522 Microporous Inorganic Membranes for Ammonia Separation

Ammonia (NH_3) is synthesized by the Haber-Bosch process where nitrogen (N_2) and hydrogen (H_2) gases are reacted at high pressure (150-200 bar) and temperature (400-500 ^oC). The process has two drawbacks: i) low NH_3 yield (~10-15%) due to the tradeoff between reaction kinetics and thermodynamics; ii) energy intensive processes are required for NH_3 removal and recycling of unreacted gases. Considering the harsh separation conditions for NH_3 recovery, microporous inorganic membranes, such as polycrystalline zeolite membranes, composed of zeolites with uniform molecular-sized pores, are a promising candidate due to their tunable surface properties and excellent thermal and chemical stabilities. The goal of this project is to develop a zeolite membrane that can separate NH_3 from a mixture gas containing NH_3, H_2 and N_2 at high pressure and elevated temperature to drastically improve the energy efficiency of NH_3 synthesis. Considering the differences in size and polarity of NH_3 (kinetic diameter: 0.260 nm), H_2 (kinetic diameter: 0.289 nm) and N_2 (kinetic diameter: 0.364 nm), membrane separation can be accomplished by two mechanisms: i) adsorption-based separation at relatively low temperature and ii) molecular-sieving separation at relatively high temperature. Preliminary data at low temperature and low pressure shows that the developed microporous membrane has great potential for NH_3 separation. At 50 psig, NH_3/H_2 selectivity of >250 and NH_3 permeance of >400 GPU were obtained at > 200 ^oC. High pressure experiments at 200 ^oC resulted in minimum NH_3/H_2 selectivities of 201, 213 and 209 and NH_3 permeances of 173 GPU, 163 GPU and 151 GPU for 100, 300 and 500 psig, respectively. The long-term testing results also indicated good membrane stability. Our results suggest the potential use of the developed membrane for NH_3 separation.

577884 Development of Evaluation Method for the Ideal Separation Property of Microporous Membrane: A Case Study for Ag-X Membrane for Propylene/Propane Separation

Since zeolite membrane has polycrystalline structures, there are two kinds of pathways across a membrane, intracrystalline (micropore of zeolite) and intercrystalline (defects among crystals) pathways. Separation property of zeolite membrane is always influenced to varying degrees by defects present in a membrane. Generally, a few intercrystalline pathways readily spoil separation performance of zeolite membrane, and thus reducing the intercrystalline pathways would help zeolite membranes to improve permselectivity.

It is important to develop a method to evaluate the ideal selectivity when molecules permeat only through the intercrystalline pathways, excluding the influence of permeation through the intercrystalline voids. We recently reported Ag-X membrane for propylene/propane separation based on selective adsorption.[1] In this study, we developed a novel method to evaluate the ideal separation property of Ag-X membrane for propylene/propane.

Na-X membrane was synthesized by a secondary growth method on a porous a-alumina tubular support (outer diameter, 10 mm; inner diameter, 7 mm; length, 30 mm, Noritake Co. Ltd.). The mean pore diameter of support was ca. 150 nm. The support was seeded by means of a dip coating method with a USY seed slurry. A seeded support was immersed in a synthesis solution for hydrothermal treatment. Hydrothermal treatment was carried out at 343 K for 24 h. After the hydrothermal treatment, Na-X membrane obtained was rinsed with distilled water and dried at 343 K overnight.

Ag-X membranes were prepared by an ion exchange method for Na-X membranes obtained. Na-X membrane was immersed into the nitrate salt aqueous solution (10 mM) and kept while stirring for 1 h. Ion exchanged X membranes were washed with distilled water and dried at 343 K overnight prior to use.

Separation test to evaluate the ideal selectivity was carried out as follows. A propylene/propane equimolar mixture was fed to Ag-X membrane at 313 K and atmospheric pressure. The permeate side was swept with flowing helium. To eliminate the effect of permeation through the intercrystalline pathway, the permeation side was pressurized from 0.1 to 0.4 MPa by using a back pressure valve.

When the permeation side was pressurized, we observed that the flux of propane drastically decreased while that of propylene was not deceased so much, and thus the separation factor increased with increasing pressure. For example, the propane fluxes at 0.1 and 0.4 MPa of pressure in the permeation side were 3.19 x 10^{-2} and 0.704 x 10^{-2} mmol m^{-2} s^{-1}, respectively, resulting in that separation factor improved from 29 to 102. These results of permeation test suggest that propane permeation through the intercrystalline voids would be inhibited by helium counter-flowed from the permeate side.

This evaluation method could reduce the influence of permeation through intercrystalline voids and be effective to estimate an ideal selectivity of porous crystalline membrane. In addition, we can conclude that there is still room for improvement of the separation property for propylene/propane by reducing defects existing in Ag-X membrane.

(1) Sakai, M.; Sasaki, Y.; Tomono, T.; Seshimo, M.; Matsukata, M. *Olefin Selective Ag-Exchanged X-Type Zeolite Membrane for Propylene/Propane and Ethylene/Ethane Separation. ACS Appl. Mater. Interfaces 2019*, 11, 4145-4151.

Acknowledgment

This work was partially supported by JST CREST (Japan Science and Technology agency, Create REvolutionary technological seeds for Science and Technology innovation program), Grant Number JPMJCR1324, Japan.

I&EC research
Industrial & Engineering Chemistry Research

Article

Cite This: Ind. Eng. Chem. Res. XXXX, XXX, XXX−XXX

pubs.acs.org/IECR

Fabrication of Hydrogen-Selective Silica Membranes via Pyrolysis of Vapor Deposited Polymer Films

Bryan Nguyen, Sasan Dabir, Theodore Tsotsis, and Malancha Gupta*

Mork Family Department of Chemical Engineering and Materials Science, University of Southern California, 925 Bloom Walk, Los Angeles, California 90089, United States

ABSTRACT: Efficient separation of hydrogen under steam reforming conditions is important for the development of clean energy sources. Although high-temperature and steam-stable membranes with high fluxes and large separation factors would be valuable for such an application, their fabrication remains a challenge. Silicon-based ceramic membranes are particularly promising due to their high temperature resistance and excellent chemical stability. In this study, we propose a new synthetic route for fabricating nanoporous, asymmetric membranes via the pyrolysis of silicon-containing polymer films deposited by initiated chemical vapor deposition (iCVD) on macroporous silicon carbide supports. Specifically, we systematically investigated the change in the chemical structure of poly(2,4,6,8-tetravinyl-2,4,6,8-tetramethyl cyclotetrasiloxane) films at different pyrolysis temperatures and found that the complete transition to a silica membrane occurred at ~1100 °C. Three different supports composed of silicon carbide powders of varying sizes were tested for membrane preparation. It was found that membranes formed with our process were microporous with separation factors several times above the corresponding Knudsen factors. Our synthetic route, therefore, offers a scalable and solventless method for producing silicon-based ceramic membranes for high-temperature separation and sensor applications.

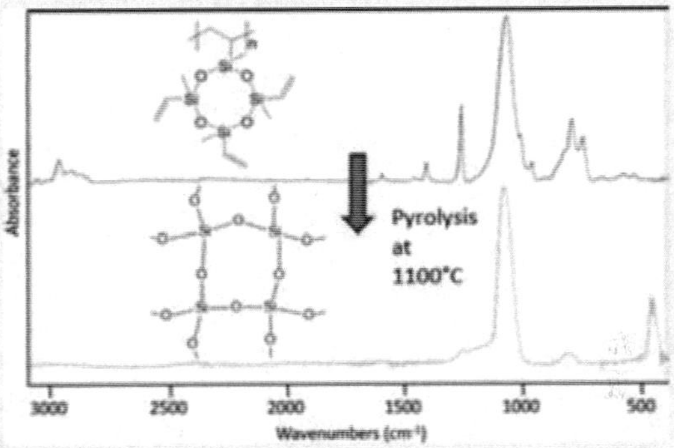

■ INTRODUCTION

As the global energy demand continues to rapidly grow, hydrogen has been increasingly discussed as a viable alternative energy source for applications such as in hydrogen fuel cells. Approximately 50% of commercial hydrogen is currently produced via steam methane reforming (SMR), where methane reacts with high-pressure steam to form hydrogen and carbon monoxide, and the hydrogen is then extracted using adsorption[1] or membrane separation.[2,3] In order to reduce the cost of hydrogen for practical applications, it is important to optimize its production through SMR.[2,4] This highly endothermic reaction requires a large input of energy, with process temperatures as high as 850 °C typically employed.[5,6] In order to improve the efficiency of the reaction, membrane reactors have been studied as a promising alternative to the more conventional packed-bed reactor processes.[7]

One challenge in using such reactors is that the membranes, in addition to having good hydrogen permselectivity, must also be able to function in these extreme environments.[4−6] Inorganic membranes have several inherent advantages over their polymeric counterparts due to their superior chemical, thermal, and structural stability.[7−9] Current materials used as membranes for hydrogen separation include palladium,[9−11] zeolites,[4,9,12] and carbon molecular sieves.[13−15] Palladium and palladium-alloy membranes are currently the most commonly employed in SMR membrane reactor studies. However, these membranes are expensive and are susceptible to coking and hydrogen embrittlement which impart weaker thermal and mechanical properties causing the membranes to develop cracks under stress.[4,16] In addition, poisoning from sulfur-containing compounds negatively impacts separation performance. Zeolite membranes, while made of common materials, can have poor hydrogen selectivity under SMR conditions due to the presence of intercrystalline voids, and they also lack long-term stability in the presence of steam and sulfur-containing compounds.[4,9] Carbon-based membranes are highly selective toward hydrogen,[4] and we are presently field-testing them in power generation applications.[13] However, these membranes have an upper temperature limit of operation of ~325 °C, and therefore, they cannot be used in the SMR process which employs significantly higher temperatures.

Silicon-based membranes, such as those with chemical structures ranging from amorphous silica to pure SiC, have been shown to possess excellent hydrogen separation properties and do not have issues with hydrogen embrittlement and sulfur poisoning.[4,8,17] Current methods for producing these ceramic membrane films include solution-based techniques such as spin-coating[18] and dip-coating.[19−21] However, films fabricated with these solution-based methods have disadvantages such as substrate compatibility issues and difficulties with

Received: May 28, 2019
Revised: July 23, 2019
Accepted: July 25, 2019

ACS Publications

disposal of solvents.[22] Silicon-based membranes have also been produced using chemical vapor deposition (CVD)[17,23,24] and plasma-enhanced chemical vapor deposition (PECVD)[25,26] techniques. These deposition techniques often require high energy inputs that can damage the chemical functionality of the precursors, which hinders the ability to control the structure of the final ceramic produced.[22] In addition, PECVD reactors often employ processing conditions that induce a competition between deposition and etching of the film, which presents a challenge for controlling the thickness of the deposited coating.[22]

In this study, we introduce a new synthetic method to create robust asymmetric, nanoporous silica-based membranes using initiated chemical vapor deposition (iCVD) and subsequent pyrolysis. The iCVD process is a one-step, solventless process that utilizes a thermally cleavable compound, such as *tert*-butyl peroxide, to initiate a free radical polymerization process.[22,27] Compared to other CVD processes, the iCVD process requires very low energy and mild reactor conditions and can employ a broad range of monomers to generate a variety of polymer films.[22] It can be used to produce thin, conformal films over a wide variety of planar,[28] curved,[29] and porous surfaces.[30] In addition, the iCVD process can be scaled-up for roll-to-roll operation.[31] In this paper, we deposited poly(2,4,6,8-tetravinyl-2,4,6,8-tetramethyl cyclotetrasiloxane) (PV4D4) on a variety of silicon carbide supports fabricated from powders of various sizes. The pyrolysis of silicon-based polymers has been shown previously to form a variety of ceramics, ranging from silica to silicon oxycarbide and silicon carbide, for use in different applications including membranes.[32−36]

The macroporous support used during membrane preparation needs to be mechanically strong to withstand the high temperatures typically utilized in SMR and the high trans-membrane pressure gradients imposed across during membrane reactor operation; it must also be resistant to the corrosive SMR atmosphere and be able to withstand thermal and pressure cycling.[4,6] We, therefore, selected in our study silicon carbide as our support material due to its superior chemical and thermal stability.[8] We deposited our PV4D4 film onto these supports via iCVD and systematically studied the conversion of the polymer into a ceramic membrane at different pyrolysis temperatures using infrared spectroscopy (IR) and X-ray photoelectron spectroscopy (XPS). To our knowledge, we are the first group to report the fabrication of a microporous ceramic membrane from the pyrolysis of polymer films deposited via iCVD.

■ EXPERIMENTAL SECTION

Support Preparation. We prepared supports by cold-pressing and sintering SiC powders into flat disks. We employed two different materials: a pure 0.6 μm β-SiC powder (Superior Graphite Co.) and a <80 nm β-SiC powder (US Research Nanomaterials). Three different supports were prepared: one composed solely of the 0.6 μm powder, one composed solely of the <80 nm powder, and one composed of a 50/50 blend of the 0.6 μm and <80 nm powders. Prior to being pressed into a flat disk pellet, the SiC powder was mixed with boron carbide (0.1 wt %) and phenolic resin (4 wt %) serving as sintering aids, employing acetone as the dispersing medium. For thorough mixing of the SiC powder and the sintering aids, the resulting slurry was ultrasonicated for 20 min, then manually mixed with a spatula, and then ultrasonicated again for another 20 min. The whole procedure was repeated before the mixture was placed in a fume-hood to dry for 24 h. The resulting material was then ground into a powder using a pestle, and oleic acid and toluene as pressing aids were then gradually added to the powder and mixed manually using a pestle and then allowed to dry for 30 min. One gram of powder was then loaded into a disk-like mold and pressed into a pellet using a hydraulic press for 2 min at a pressure of 86 MPa. The pellets were then placed in a high-temperature graphite furnace (Thermal Technology, Inc., Model 1000-3060-FP20) where they were heated (3 °C/min) in flowing Ar to a sintering temperature of 1900 °C, where they were kept for 3 h and were then cooled down (6 °C/min) to room temperature. Prior to deposition, they were sonicated in acetone for 20 min and then dried.[37,38]

A dense iCVD layer is necessary for separation of the molecules, and therefore, an intermediate layer was first formed on the macroporous support. Without an intermediate layer, the iCVD film would uniformly coat the pores of the macroporous support instead of forming a dense layer since the pore size of the support is similar to the thickness of the iCVD layer. The intermediate layer solution was prepared by mixing 10 wt % of allyl-hydridopolycarbosilane (AHPCS, Starfire Systems), 1.25 wt % polystyrene (2500 MW, Sigma-Aldrich), and 5 wt % fine SiC powder (100−200 nm) in hexane. The fine SiC powder was prepared from the original 0.6 μm SiC powder using a precipitation method described elsewhere.[39] The support was dip-coated by immersing it in this solution for 12 s before raising it out of the solution at a rate of 3 cm/min. The disk was then heated to 250 °C (at a rate of 3 °C/min) in flowing Ar in a tube furnace (Lindberg/Blue, Model STF55433C), where it was held for 1 h. It was then heated to 400 °C, where it was held for another hour, and then finally heated to 750 °C, where it was held for 2 h, before being cooled down to room temperature at a rate of 3 °C/min.

iCVD Deposition and Subsequent Pyrolysis. The iCVD method was used to deposit poly(2,4,6,8-tetravinyl-2,4,6,8-tetramethyl cyclotetrasiloxane) (PV4D4) onto the SiC support disks, as shown in Figure 1. The custom-designed

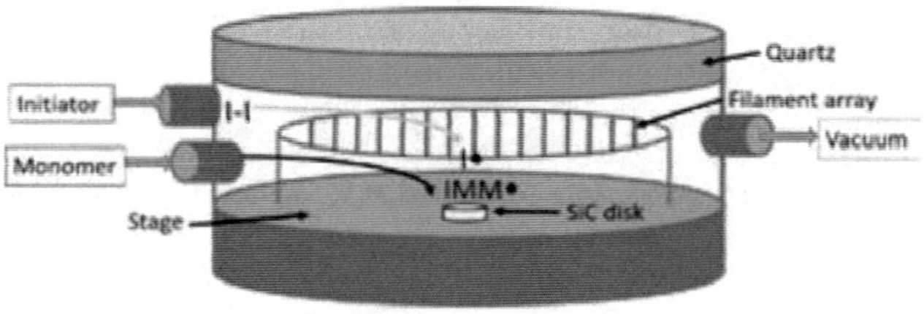

Figure 1. Schematic of the iCVD reactor.

vacuum reactor (250 mm in diameter, 48 mm in height, GVD Corporation) has a quartz top that allows for in situ thickness measurements on a reference silicon wafer via a helium−neon laser interferometer (Industrial Fiber Optics). The support disks were placed on the reactor stage which was kept at a constant temperature of 40 °C using a recirculating chiller (Thermo Scientific Haake A25). A piece of aluminum foil was tightly wrapped and taped around the support to ensure that deposition only occurred on the top surface. A rotary-vane vacuum pump (Edwards E2M40) was used to hold the reactor under vacuum, and the pressure was regulated at 190 mTorr by a throttle valve (MKS 153D) with active feedback control from a capacitance manometer (MKS 622C01TDE Baratron). The initiator and monomer liquids were loaded into stainless steel

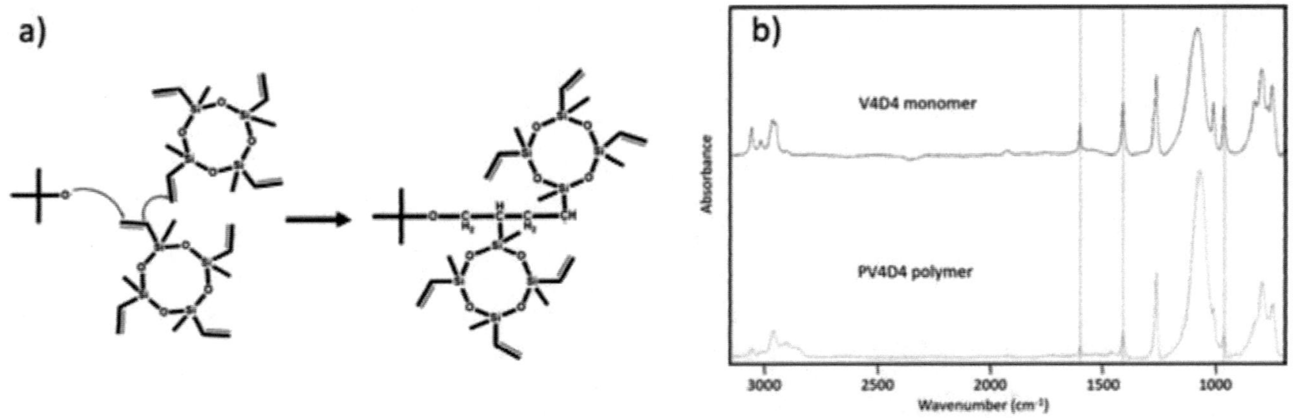

Figure 2. (a) Schematic of the iCVD polymerization of V4D4 to PV4D4. (b) FTIR spectra of the monomer and the resulting polymer. The dashed lines indicate the vinyl bond modes, which have a noticeable decrease in the IR spectra of the polymer.

jars and introduced into the reactor through heated lines. The monomer 2,4,6,8-tetravinyl-2,4,6,8-tetramethyl cyclotetrasiloxane (V4D4) (Gelest, Inc.) was heated to 50 °C, and the flow rate was maintained at 0.7 sccm. The initiator *tert*-butyl peroxide (Aldrich, 98%) was kept at room temperature, and the flow rate was maintained at 1 sccm. A nichrome filament array (Omega Engineering) was placed inside the reactor and heated to 250 °C to cleave the initiator. Approximately 1 μm of PV4D4 was deposited on each support as measured via interferometry on a reference silicon wafer.

After the deposition of the polymer, the disk was placed in a tube furnace (Lindberg/Blue, Model STF55433C) and pyrolyzed at the desired temperature over flowing argon. Prior to this pyrolysis step, flowing argon was introduced into the furnace chamber for 30 min to purge the chamber from ambient air. The sample disk was heated at a rate of 3 °C/min to 250 °C, where it was held for 1 h, and then to 400 °C, where it was held for another hour, and finally to the desired pyrolysis temperature, where it was held for two more hours, before being cooled down to room temperature at a rate of 3 °C/min.

Chemical Characterization. To determine the optimal film pyrolysis conditions, diffuse reflectance infrared fourier transform spectroscopy (DRIFTS) (COLLECTOR II, Thermo Scientific) was utilized. DRIFTS monitors the change in the IR spectra of a powder substrate in situ as a function of the temperature of the sample.[40] Due to SiC having a high IR radiation absorption in the mid-IR region, DX-type α-Al_2O_3 powder was chosen in this study for the DRIFTS experiments. To coat the alumina power with PV4D4, it was first spread onto a silicon wafer using a razor blade and 1 μm of PV4D4 was deposited on the alumina powder as measured on a reference wafer by interferometry. After the deposition, the PV4D4-coated powder was mixed manually with barium fluoride (BaF_2), a nonabsorbing matrix, at a ratio of 1:9 powder to BaF_2. The powder was then placed into the sample cup of the DRIFTS cell. Flowing Ar was introduced into the cell at a pressure of 170 kPa to ensure an inert atmosphere. The temperature of the DRIFTS chamber was raised in increments of 10 °C, and the IR absorbance spectrum was measured every 100 °C. The DRIFTS absorbance spectra of the uncoated alumina powder were used as the background spectra.

We also measured the IR absorbance spectra of PV4D4 films ex situ by deposition onto clean silicon wafers and subsequent pyrolysis at preset temperatures. For the pyrolysis step, the PV4D4-coated wafers were placed in a tube furnace which was then purged with Ar for 30 min to ensure that no oxygen remains in the furnace. Throughout the pyrolysis process, Ar was allowed to flow through the furnace in order to ensure an inert atmosphere. The wafers were first heated to 200 °C at a rate of 2 °C/min and then kept at this temperature for 1 h. They were then heated to 400 °C at the same rate, where they were again kept for 1 h, and then finally brought up to the desired pyrolysis temperature, where the they were kept for two additional hours. The wafers were then cooled back to room temperature at a rate of 3 °C/min. The slow heating and cooling rates are known to prevent cracking in the film.[41] After the wafers were cooled down to room temperature, the IR spectra were recorded using a FTIR spectrophotometer (Nicolet iS10, Thermo Scientific). The DRIFTS and FTIR spectra were both analyzed using the freeware SpectraGryph 1.2, and the deconvolution was performed using Origin Peak Analyzer.

X-ray photoelectron spectroscopy (XPS) analysis was used to determine the elemental composition and bonding environments of the films before and after pyrolysis at various temperatures. The Kratos Axis Ultra DLD XPS spectrometer equipped with a magnetic immersion lens and charge neutralization system and a monochromator Al X-ray source was used to collect the spectra. Survey scans were performed at a pass energy of 160 eV, while high-resolution scans were captured at a pass energy of 40 eV. A charge correction value was applied to the C 1s environment so that the peak value was shifted to 284.8 eV. The same correction factor was applied to the Si 2p and O 1s orbitals, which resulted with the O 1s peak value being shifted to ~532 eV. Since there were signals generated from both the Si $2p_{1/2}$ and Si $2p_{3/2}$ electron spin states, it was necessary to deconvolute each silicon environment into two separate peaks. The peak area for the Si $2p_{1/2}$ was set equal to half of the peak area for the Si $2p_{3/2}$, and the full-width half-maximum (fwhm) values of each spin state were set to be equal. In addition, the energy difference between the two peaks was kept constant at 0.65 eV. The number of environments we fitted per peak was based on the data obtained from our FTIR analysis. The XPS data were analyzed using the software CasaXPS. A scanning electron microscope (SEM) (Nova NanoSEM 450 Field Emission Scanning Electron Microscope) at a 15 kV accelerating voltage was also used to image the membranes.

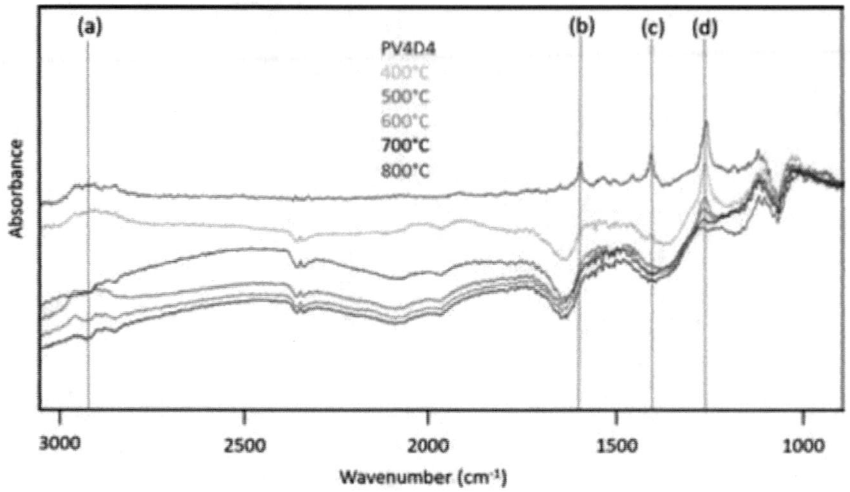

Figure 3. Real-time DRIFTS data showing the IR absorbance spectra for PV4D4 films at different pyrolysis temperatures. The dashed lines represent: (a) symmetric $-CH_2$ stretching, (b) the stretching of the C=C bond, (c) the wagging mode of the vinyl bond, and (d) the symmetric Si$-CH_3$ stretching.

Permeation Testing. The membranes were tested by measuring the single-gas permeance (mol/m^2·s·Pa) of helium, hydrogen, and argon. The ideal binary separation factors were calculated by dividing the permeances of the two gases. Prior to testing, the flat disks were glued to metal fender washer supports using an epoxy (J-B Weld) and allowed to cure for 24 h. To measure the permeance, the metal supports were placed in between the two half-cells of the permeation system and sealed using silicone O-rings to ensure no leaks. Compressed air was used to seal the chambers via a hydraulic mechanism. The gas to be measured was introduced to the bottom half-cell at a set pressure and allowed to permeate through the membrane and to exit in the top cell. The pressure of the permeate side was kept at atmospheric conditions and measured using a pressure gauge (OMEGA Engineering, DPG1000B-15G). The pressure gradient in between the top and bottom cells was also kept constant and measured using a differential pressure transducer. The pressure on the feed side was controlled using the regulator attached to the gas cylinder and a needle valve installed on the reject-side tube. The flow rate of the permeated gas was measured with bubble-flow meters of different volumes (1 mL, 10 mL, 100 mL). The gas was allowed to fill the half-cells and permeate through the membrane for at least 30 min before data were taken to ensure that steady state conditions prevail. For the permeation properties reported here, the temperature of the cell was kept at 200 °C and the transmembrane pressure difference at 241 kPa (for further details about the permeation setup, see refs 38 and 39).

■ RESULTS AND DISCUSSION

Figure 2a shows the iCVD polymerization of V4D4 to PV4D4. The monomer consists of a Si−O−Si cyclical structure with four vinyl bonds. Figure 2b displays the IR spectra for both the monomer and the polymer formed via iCVD. The peaks at 960, 1400, and 1600 cm^{-1} in the monomer spectra represent the vinyl bonds. The reduction in the size of these peaks in the polymer spectra along with the increase of the symmetric stretching of CH_2 at 2920 cm^{-1} confirms that there is polymerization taking place through the vinyl bonds. In addition, there is a shift of the broad Si−O−Si bond from 1080 cm^{-1} to 1065 cm^{-1}, which is associated with the polymer.[42,43] The resulting PV4D4 polymer is highly cross-linked, which has been shown to be beneficial to retain silicon groups during pyrolysis. For example, Carlsson et al.[44] found that cross-linking PDMS films, which were cast from toluene onto silicon wafers, minimized chain-scission reactions during pyrolysis which in turn prevented loss of silane groups.

To study the transition of the polymer film into a ceramic membrane at elevated temperatures, the PV4D4 film was deposited onto alumina particles and pyrolyzed in flowing Ar. The changes in the IR absorbance spectra were recorded in situ in a DRIFTS cell as its temperature was raised up to the maximum rated temperature of the cell (~900 °C). As shown in Figure 3, several of the distinct peaks of the PV4D4 film prior to pyrolysis were detected such as the symmetric $-CH_2$ stretching related to the polyethylene backbone centered at 2920 cm^{-1} (3a), the stretching of the C=C bond at 1600 cm^{-1} (3b), the wagging mode of the vinyl bond at 1415 cm^{-1} (3c), and the symmetric Si-CH_3 stretching at ~1260 cm^{-1} (3d). There was no significant change in the spectra when heating to 300 °C, indicating that the chemical structure of the polymer film stays relatively intact in this range of temperatures. There were significant changes at 400 °C such as the disappearance of the peaks corresponding to the stretching of the C=C bond and the wagging mode of the vinyl CH_2, indicating that there were no vinyl bonds left in the material. At 600 °C, there was a significant reduction in the size of the peaks corresponding to the polyethylene bonds and the symmetric Si−CH_3 bond at 1260 cm^{-1}, indicating that these bonds were likely cleaved. The polyethylene peak continued to be present in small quantities until ~800 °C. The results of the DRIFTS experiments demonstrate that 600 °C is a critical temperature during which most of the hydrocarbon bonds are removed, signaling the transition to a pure silica film. Our results agree with those of Narisawa,[45] who reported that the decomposition of cross-linked polysiloxane resins with methyl side groups begins first with the removal of methyl groups at ~600 °C, and the formation of silica begins beyond 1000 °C even in an oxidizing environment.

Since PV4D4 does not completely convert into the ceramic until 1000 °C and the maximum operating temperature of the

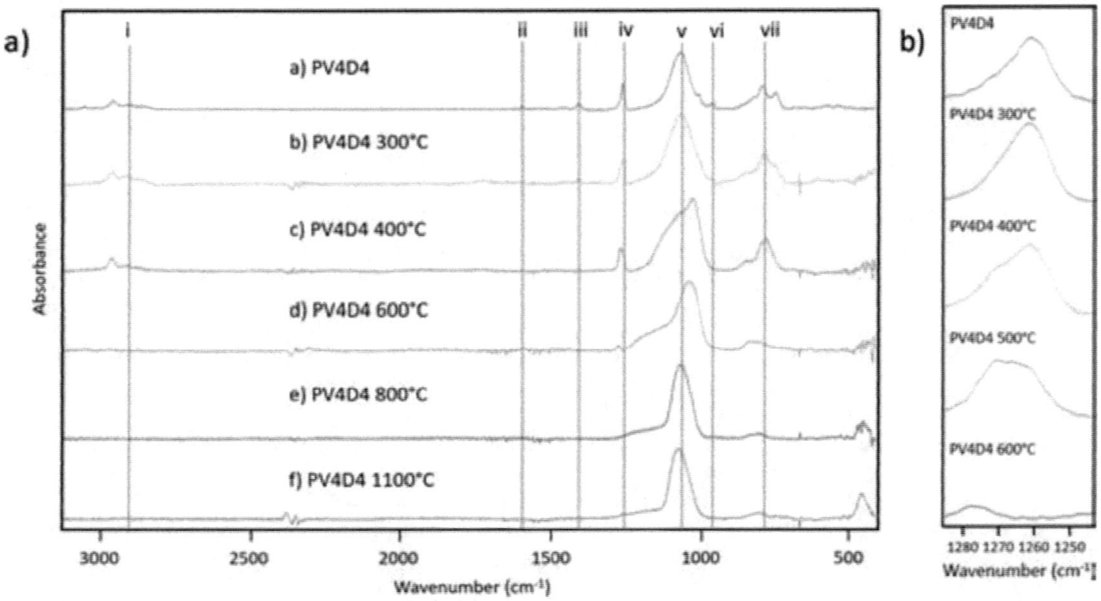

Figure 4. (a) IR absorbance spectra showing the evolution of the PV4D4 film into a silica film. The absorbance bands labeled above correspond to (i) the hydrocarbon bonds from the polyethylene bond, (ii, iii, vi) vibration modes corresponding to vinyl bonds, (iv) stretching of methyl bonds, (v) asymmetric stretching of Si−O−Si bonds, and (vii) asymmetric rocking in the methyl group. (b) Expanded view of the IR region from 1250 to 1280 cm^{-1} showing the methyl stretching band.

DRIFTS cell is ∼900 °C, we also carried out complementary ex situ FTIR measurements to study pyrolysis at higher temperatures. For these experiments, PV4D4 was deposited onto silicon wafers using the iCVD process, pyrolyzed in a tube furnace in flowing Ar under select pyrolysis conditions, and then the IR spectra of the resulting films were measured (Figure 4). We monitored the change in the oxidation state of the film by monitoring the following four silicon environments: 'M' representing Si bonded to a single oxygen atom (SiO_1R_3), 'D' representing Si bonded to two oxygen atoms such as in the network polymer film (SiO_2R_2), 'T' representing Si bonded to three oxygen atoms such as in the silsesquioxane cage structure (SiO_3R), and 'Q' representing Si bonded to four oxygen atoms such as in a pure silica environment (SiO_4). There is no significant change between the polymer before pyrolysis (4a) and the polymer after pyrolysis at 300 °C (4b), which is consistent with our DRIFTS experiment. For the sample pyrolyzed at 400 °C (Figure 4c), the absorbance band at 2850−2950 cm^{-1} corresponding to the polyethylene carbon chain decreased slightly, indicating the cleavage of some of the Si−C polyethylene bonds. In addition, the bands at 960 cm^{-1}, 1415 cm^{-1}, and 1600 cm^{-1} corresponding to the vinyl bonds completely disappeared, in agreement with our results from the DRIFTS experiment. Previously, Trujillo et al.[42] studied the annealing of PV4D4 films in air up to a temperature of 410 °C, and they found that the films oxidized to form silsesquioxane-type structures. Our FTIR results show that although there are bands at 1028 cm^{-1}, corresponding to a small bond angle Si−O−Si, and at 1120 cm^{-1}, corresponding to a silsesquioxane cage ('T'), the band for the intact network at 1065 cm^{-1} ('D') still dominates the structure. This is likely because we pyrolyzed our samples under inert conditions and, therefore, there were fewer oxygen atoms readily available to allow the network rings to form the silsesquioxane cages. The $Si-CH_3$ stretching band shifted from 1260 cm^{-1} to a double peak at 1262 cm^{-1} and 1275 cm^{-1}, and the $Si-CH_3$ asymmetric rocking band at 800 cm^{-1} transitioned to a double peak at 780 cm^{-1} and 800 cm^{-1}, which provides further evidence of the oxidation of some of the network groups.[42,46] For the sample pyrolyzed at 600 °C (Figure 4d), there was an overall shift in the Si−O−Si peak from 1028 cm^{-1} to 1047 cm^{-1}. The polyethylene band centered at 2920 cm^{-1} and the bands at 1270 cm^{-1} and 2965 cm^{-1} mostly disappeared at this pyrolysis temperature. These observations are consistent with the findings of Mantz et al.,[47] who reported that cylic dimethylsiloxane oligomers started decomposing at ∼400 °C and that the silsesquioxane ring lost its structure in the temperature range from 450 °C to 650 °C due to methyl abstraction. The deconvolution of the IR region between 1250 cm^{-1} and 1270 cm^{-1} can be used to determine the ratio of M, D, and T groups in the bulk film at each pyrolysis temperature.[48,49] Figure 4b shows an expanded view of this region. It can be seen that 'D' groups dominated the deposited polymer film prior to pyrolysis. At pyrolysis temperatures of 300 and 400 °C, 'D' groups still dominate the structure, accounting for ∼81% and ∼73%, respectively. However, at a pyrolysis temperature of 500 °C, there was a noticeable transition to a silsesquioxane structure, with 'T' groups accounting for 48% of the structure. At a pyrolysis temperature of 600 °C, the methyl peak decreased and 'T' groups accounted for 94% of the remaining methyl bonds. Beyond 600 °C, the methyl stretching band was enveloped in the larger Si−O−Si asymmetric stretching band and it was not possible to deconvolute it. At a pyrolysis temperature of 800 °C (Figure 4e), the Si−O−Si band continued transitioning to a silica structure with a peak shift to 1065 cm^{-1} and the silsesquioxane peak, the methyl bond, and the polyethylene absorption bands were greatly reduced.[42,45] Figure 4f shows that pyrolysis at 1100 °C led to the formation of silica thin films as indicated by the presence of the following characteristic peaks of Si−O−Si: a rocking mode at ∼450 cm^{-1}, a weak bending mode near 800 cm^{-1}, and a tall stretching vibration at ∼1080 cm^{-1}.[50−52] The shifts to these peaks, signaling the change to a pure silica 'Q' environment, were not apparent in our spectra at pyrolysis temperatures lower than 1100 °C. We also pyrolyzed PV4D4 films at 1200 °C and 1300 °C, and the FTIR spectra showed no further change in the structure of the ceramic. Burns et al.[33] and Babonneau et al.[53] reported that the pyrolysis of siloxanes resins beyond 1300−1500 °C resulted in the scission of Si−O bonds, with oxygen loss of up to 80% at 1600 °C. Therefore, we did not test pyrolysis at those temperatures. The consistency among the FTIR results, for which the PV4D4 film was deposited on a silicon wafer, and the DRIFTS results, for which the PV4D4 film was deposited on alumina powder, indicates that the processes that take place during the conversion of the polymer into the ceramic are substrate independent.

The pyrolyzed PV4D4 films on silicon wafers were also analyzed by XPS to determine their chemical structure and composition. Survey scans were performed on each of the pyrolyzed films to determine the atomic fraction of carbon, oxygen, and silicon present pre- and post-pyrolysis, and the results are shown in Table 1. In agreement with the IR data,

Table 1. XPS Elemental Analysis of the PV4D4 Films Pre- and Post-pyrolysis

Element	% Silicon	% Oxygen	% Carbon
PV4D4, theoretical	20	20	60
PV4D4, experimental	23	28	49
PV4D4, pyrolyzed at 400 °C	22	25	53
PV4D4, pyrolyzed at 600 °C	37	58	5
PV4D4, pyrolyzed at 800 °C	37	59	4
PV4D4, pyrolyzed at 1100 °C	38	60	2

these results indicate that organic groups are removed from the PV4D4 film in increasing amounts as the temperature of pyrolysis increases. The deposited PV4D4 films had atomic compositions that were comparable to previously reported PV4D4 films deposited via iCVD.[42] There was no noteworthy shift in the elemental composition at a pyrolysis temperature of 400 °C, but there was a significant decrease in carbon content when the temperature was increased to 600 °C. This observation is consistent with the IR studies that show the disappearance of the polyethylene backbone and methyl groups, thus causing the composition of the film to shift toward the pure silica structure. At a pyrolysis temperature of 1100 °C, the silicon fraction was higher than expected from a film composed purely of 'Q' groups which would have a composition of 33.33% silicon and 66.66% oxygen.[54] The likely reason for this is the penetration of the X-ray photons through the pyrolyzed film to the underlying silicon substrate, since these films are porous due to the void space left by the organic groups that cleave during pyrolysis. The small amount of carbon is likely indicative of surface contamination or the presence of trace amounts of carbon embedded within the structure.

In order to confirm the bonding environments in the film during the pyrolysis process, high resolution XPS scans of the silicon environment were collected. Alexander et al.[55] have shown previously that in siloxanes, the oxidation state is the biggest contributing factor to the shifts in the binding energy of silicon, with reported binding energies of $Si(\text{-}O_1)$ = 101.5 eV, $Si(\text{-}O_2)$ = 102.1 eV, $Si(\text{-}O_3)$ = 102.8 eV, and $Si(\text{-}O_4)$ = 103.4 eV. O'Hare et al.[56] identified similar binding energies for these components of a siloxane film: 'M' = 101.63 eV, 'D' = 101.99 eV, 'T' = 102.67 eV, and 'Q' = 103.47 eV. The high resolution scans of the PV4D4 films pyrolyzed at different temperatures are shown in Figure 5. The areas corresponding to each oxidation state are shown in Table 2. As was previously seen in the IR data, there was a small shift from the network structure and small bond angle Si−O−Si 'D' groups to silsesquioxane 'T' groups at a pyrolysis temperature of 400 °C. At 600 °C, there was a dramatic shift toward a mixed oxidation state, with the 'Q' groups dominating. This observation is consistent with the IR data that show that the removal of the polyethylene backbone and the methyl groups also occurs primarily between 400 °C and 600 °C. However, as with the FTIR data, there is evidence in the XPS data of residual hydrocarbon bonds, with a small number of the 'D' groups remaining. Overall, we found that there were similarities in the ratios of the silicon environments in the bulk and the surface as determined by FTIR and XPS, respectively, up to a pyrolysis temperature of 400 °C. At a pyrolysis temperature of 600 °C, the bulk showed a greater number of 'T' groups than 'D' groups, whereas the surface showed a greater number of 'D' groups than 'T' groups. This is consistent with the observations of Corriu et al.,[57] who investigated the pyrolysis of siloxane gels and found that after a pyrolysis temperature of 600 °C, 'T' siloxane materials transitioned to both 'D' and 'Q' structures due to restructuring of Si−C and Si−O bonds. Since methyl groups are cleaved and diffuse out of the film, there is likely a higher level of recombination at the surface compared to the bulk. At ~800 °C, these groups mostly transitioned to 'Q' groups; however, residual 'D' groups were still present. The peak shifted to 103.35 eV at a pyrolysis temperature of 1100 °C, indicating the transformation of the film into silica. In summary, based on the FTIR/DRIFTS and XPS analysis results, the formation of silica film occurs at a pyrolysis temperature of 1100 °C.

The asymmetric membranes are composed of the iCVD polymer layer deposited on SiC disks consisting of a macroporous support and an intermediate layer prepared as described in the Experimental Section. For the preparation of the membranes, the composite structure is then pyrolyzed following the procedure previously described with the final pyrolysis temperature being 1100 °C. Figure 6 shows SEM images of the resulting asymmetric structure with the polymer film deposited on the top (prior to its pyrolysis) confirming that the polymer forms a distinct layer on the surface of the SiC support disk.

Table 3 shows the permeance data for three distinct asymmetric membranes made with different supports: SiC supports composed of 50% of particles with an average diameter less than 80 nm and 50% of particles with average diameter of 0.6 μm, supports made of pure 0.6 μm SiC powders, and supports made with pure <80 nm particles. Table 3 shows the average measured permeances of He, H_2, and their ideal separation factors (defined here as the ratio of permeances) over Ar for each different type of membrane. It is important to note that the permeances and ideal separation factors for each membrane were calculated from three distinct samples prepared under the same conditions. As can be seen in the table, all silica films are microporous with ideal separation factors which are several times higher than the corresponding Knudsen separation factor $\alpha_{i/j} = \sqrt{\frac{m_j}{m_i}}$, where m_j is the molecular weight of Ar and m_i the molecular weight of He or H_2. The mixed powder supports (0.6 μm/80 nm) resulted

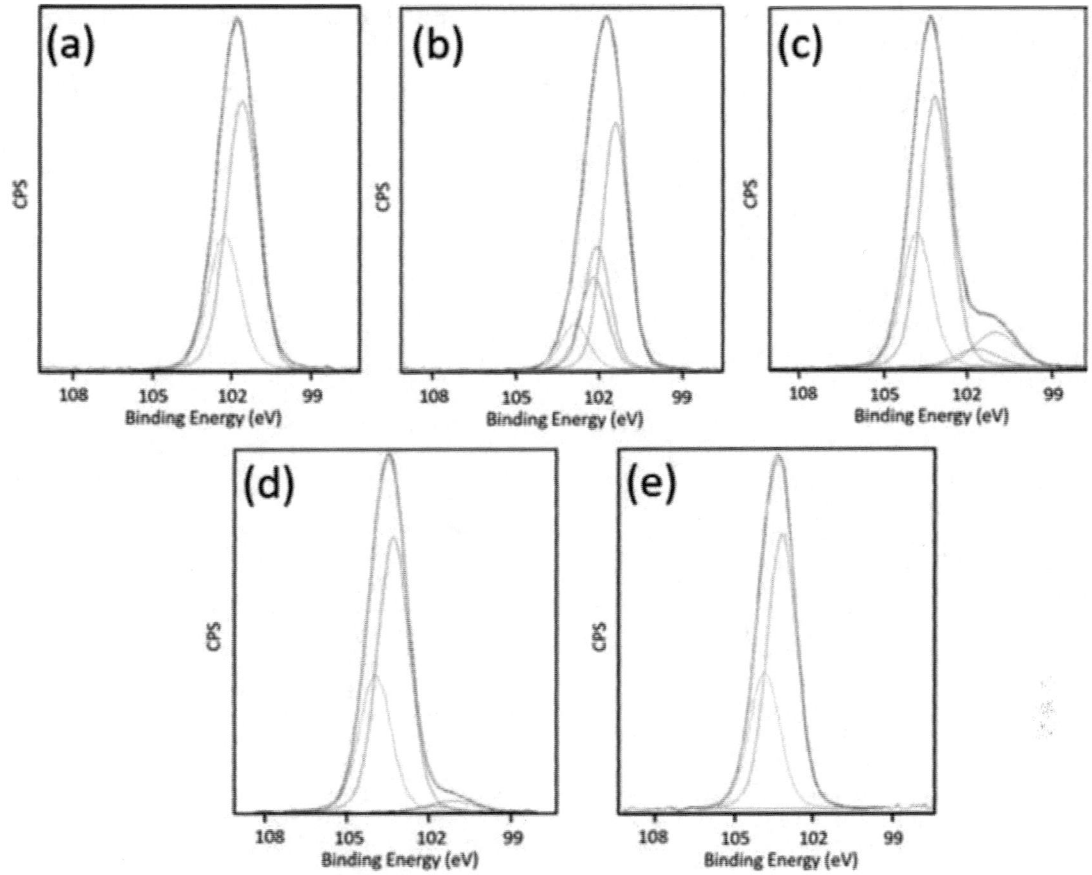

Figure 5. High resolution scans of Si 2p for the (a) PV4D4 polymer and for the polymer after pyrolysis at (b) 400 °C, (c) 600 °C, (d) 800 °C, and (e) 1100 °C.

Table 2. Peak Areas of the Silicon Oxidation State from High Resolution XPS Analysis

Oxidation levels	SiO_1	SiO_2	SiO_3	SiO_4
Binding energy	-	101.8	102.6	103.4
PV4D4	0	100	0	0
400 °C	0	72	28	0
600 °C	0	15	0	85
800 °C	0	6	0	94
1100 °C	0	0	0	100

in an asymmetric membrane with a higher separation factor than the pure powder supports (either 0.6 μm or 80 nm). This is consistent with prior observations by our group,[38] which show that utilizing larger particle sizes for supports results in bigger pores and, therefore, higher permeances and lower separation factors. However, using a mixture of powders with distinct particles sizes allows for better packing and, therefore, better separation factors with only a small sacrifice in permeance. It is important to note that there have been other silica membranes produced by both solution-phase[8,14]

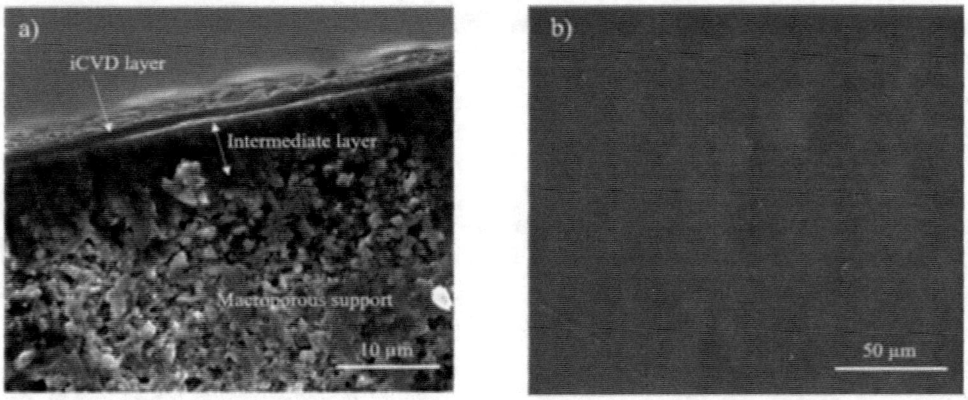

Figure 6. (a) Cross-sectional view of the asymmetric composite structure consisting of the iCVD polymer layer, and the intermediate and the macroporous SiC layers. (b) Top-down image of the dense iCVD layer on top of the SiC support.

Table 3. He and H_2 Permeances and Ideal Separation Factors of He and H_2 over Ar for Membranes Resulting from the Pyrolysis at 1100 °C of PV4D4 Films Deposited on SiC Supports[a]

Composition of the support	He permeance (mol/m²·Pa·s)	H_2 permeance (mol/m²·Pa·s)	He/Ar separation factor	H_2/Ar separation factor
100% <80 nm	4.19×10^{-8}	2.25×10^{-8}	20.7	11.0
100% 0.6 μm	7.45×10^{-7}	4.81×10^{-7}	16.4	10.6
50% (<80 nm)/50% (0.6 μm)	2.74×10^{-7}	1.89×10^{-7}	25.1	17.4

[a]The testing pressure was 342 kPa, and the testing temperature was 473 K.

and conventional CVD[17,23] methods with separation factors several times larger than ours. Our future work will involve optimizing the thickness, composition, and density of the iCVD polymer film prior to its pyrolysis to improve the separation factor and permeance of the resulting membranes.

■ CONCLUSION

In this study, we fabricated microporous silica membrane films capable of selectively separating lighter gases (e.g., H_2 and He) from bulkier gases (e.g., Ar) through the deposition and subsequent pyrolysis of PV4D4 films. We used in situ DRIFTS and complementary ex situ FTIR studies to investigate the chemical structure of PV4D4 at different pyrolysis temperatures and found that the network and silsesquioxane-dominated bonding environment shifted to a largely silica 'Q' bonding environment at 600 °C. It was not until the pyrolysis temperature reached 1100 °C, however, that we saw a complete transformation into a silica structure. In addition to the DRIFTS/FTIR studies, we also used XPS to analyze the films after pyrolysis. The XPS results confirm that organic groups were removed from the film structure as the temperature of pyrolysis increased. The high resolution XPS analysis allowed us to identify the oxidation state of each film and to establish that 1100 °C was the pyrolysis temperature at which a pure silica film was formed. Therefore, we used this pyrolysis temperature to prepare silica membrane films on SiC support disks to measure their permeation properties. The resulting membrane films were shown to be microporous and capable of separating light gases. The nature of the supports also has an influence on the properties of the resulting membrane films with mixed-powder supports exhibiting higher separation factors.

■ AUTHOR INFORMATION

Corresponding Author

*E-mail: malanchg@usc.edu.

ORCID

Sasan Dabir: 0000-0001-8671-9873

Malancha Gupta: 0000-0002-6828-7445

Notes

The authors declare no competing financial interest.

■ ACKNOWLEDGMENTS

The authors acknowledge the support of the ACS Petroleum Research Fund (PRF) and the National Science Foundation (Award# CBET-1705180). XPS data and SEM images were acquired at the Core Center of Excellence in Nano Imaging at the University of Southern California.

■ REFERENCES

(1) Song, C.; Liu, Q.; Ji, N.; Kansha, Y.; Tsutsumi, A. Optimization of Steam Methane Reforming Coupled with Pressure Swing Adsorption Hydrogen Production Process by Heat Integration. *Appl. Energy* **2015**, *154*, 392–401.

(2) da Silva Veras, T.; Mozer, T. S.; da Costa Rubim Messeder dos Santos, D.; da Silva César, A. Hydrogen: Trends, Production and Characterization of the Main Process Worldwide. *Int. J. Hydrogen Energy* **2017**, *42* (4), 2018–2033.

(3) Tsuru, T.; Morita, T.; Shintani, H.; Yoshioka, T.; Asaeda, M. Membrane Reactor Performance of Steam Reforming of Methane Using Hydrogen-Permselective Catalytic SiO_2 membranes. *J. Membr. Sci.* **2008**, *316* (1–2), 53–62.

(4) Ockwig, N. W.; Nenoff, T. M. Membranes for Hydrogen Separation. *Chem. Rev.* **2007**, *107*, 4078–4110.

(5) Beurden, P. va. On The Catalytic Aspects Of Steam Reforming Methane - A Literature Survey. *ECN* **2004**, 1–27.

(6) Gallucci, F.; Comite, A.; Capannelli, G.; Basile, A. Steam Reforming of Methane in a Membrane Reactor: An Industrial Case Study. *Ind. Eng. Chem. Res.* **2006**, *45* (9), 2994–3000.

(7) Baker, R. W. *Membrane Technology and Applications, Third*; John Wiley & Sons: West Sussex, 2012.

(8) Elyassi, B.; Sahimi, M.; Tsotsis, T. Inorganic Membranes. *Encyclopedia of Chemical Processing* **2009**, 1–16.

(9) Gallucci, F.; Fernandez, E.; Corengia, P.; van Sint Annaland, M. Recent Advances on Membranes and Membrane Reactors for Hydrogen Production. *Chem. Eng. Sci.* **2013**, *92*, 40–66.

(10) Uemiya, S. Brief Review of Steam Reforming Using a Metal Membrane Reactor. *Top. Catal.* **2004**, *29* (1/2), 79–84.

(11) Fernandes, F. A. N.; Soares, A. B. Modeling of Methane Steam Reforming in a Palladium Membrane Reactor. *Lat. Am. Appl. Res.* **2006**, *36* (3), 155–161.

(12) Xu, X.; Yang, W.; Song, C.; Liu, J.; Lin, L. Hydrogen Separation by Zeolite Membranes. *ACS Div. Fuel Chem. Prepr.* **2003**, *48* (1), 284–285.

(13) Parsley, D.; Ciora, R. J.; Flowers, D. L.; Laukaitaus, J.; Chen, A.; Liu, P. K. T.; Yu, J.; Sahimi, M.; Bonsu, A.; Tsotsis, T. T. Field Evaluation of Carbon Molecular Sieve Membranes for the Separation and Purification of Hydrogen from Coal- and Biomass-Derived Syngas. *J. Membr. Sci.* **2014**, *450*, 81–92.

(14) Lin, Y. S.; Kumakiri, I.; Nair, B. N.; Alsyouri, H. Microporous Inorganic Membranes. *Sep. Purif. Methods* **2002**, *31* (2), 229–379.

(15) Shelekhin, A. B.; Dixon, A. G.; Ma, Y. H. Theory of Gas Diffusion and Permeation in Inorganic Molecular Sieve Membranes. *AIChE J.* **1995**, *41* (1), 58–67.

(16) Jimenez, G.; Dillon, E.; Dahlmeyer, J.; Garrison, T.; Garrison, T.; Darkey, S.; Wald, K.; Kubik, J.; Paciulli, D.; Talukder, M.; et al. A Comparative Assessment of Hydrogen Embrittlement: Palladium and Palladium-Silver Subjected to Hydrogen Absorption/Desorption Cycling. *Adv. Chem. Eng. Sci.* **2016**, *6* (3), 246–261.

(17) Khatib, S. J.; Oyama, S. T.; de Souza, K. R.; Noronha, F. B. Review of Silica Membranes for Hydrogen Separation Prepared by Chemical Vapor Deposition. *In Membrane Science and Technology* **2011**, *14*, 25–60.

(18) Satvekar, R.; Phadatare, M. R.; Karande, V.; Patil, D.; Tiwale, B. M.; Pawar, S. H. Influence of Silane Content on the Optical Properties of Sol Gel Derived Spin Coated Silica Thin Films. *Int. J. Basic Appl. Sci.* **2012**, *1* (4), 468–476.

(19) Saputra, R. E.; Astuti, Y.; Darmawan, A. Hydrophobicity of Silica Thin Films: The Deconvolution and Interpretation by Fourier-Transform Infrared Spectroscopy. *Spectrochim. Acta, Part A* **2018**, *199*, 12–20.

(20) Kahlib, N. A. Z. Fabrication of Silica Ceramic Membrane via Sol-Gel Dip-Coating Method at Different Nitric Acid Amount. *IOP Conf. Ser. Mater. Sci. Eng.* **2018**, *290*, 012042

(21) Boffa, V.; Blank, D. H. A.; Elshof, J. E. Hydrothermal Stability of Microporous Silica and Niobia - Silica Membranes. *J. Membr. Sci.* **2008**, *319* (1−2), 256−263.

(22) Seidel, S.; Riche, C.; Gupta, M. Chemical Vapor Deposition of Polymer Films. *Encycl. Polym. Sci. Technol.* **2011**, DOI: 10.1002/0471440264.pst467

(23) Iarikov, D. D.; Hacarlioglu, P.; Oyama, S. T. Amorphous Silica Membranes for H_2 Separation Prepared by Chemical Vapor Deposition on Hollow Fiber Supports. In *Membrane Science and Technology*; 2011, Vol. *14*, pp. 61−77.

(24) Khatib, S. J.; Oyama, S. T. Silica Membranes for Hydrogen Separation Prepared by Chemical Vapor Deposition (CVD). *Sep. Purif. Technol.* **2013**, *111*, 20−42.

(25) Nagasawa, H.; Minamizawa, T.; Kanezashi, M.; Yoshioka, T.; Tsuru, T. High-Temperature Stability of PECVD-Derived Organosilica Membranes Deposited on TiO_2 and SiO_2-ZrO_2 Intermediate Layers Using HMDSO/Ar Plasma. *Sep. Purif. Technol.* **2014**, *121*, 13−19.

(26) Nagasawa, H.; Minamizawa, T.; Kanezashi, M.; Yoshioka, T.; Tsuru, T. Microporous Organosilica Membranes for Gas Separation Prepared via PECVD Using Different O/Si Ratio Precursors. *J. Membr. Sci.* **2015**, *489*, 11−19.

(27) Baxamusa, S. H.; Im, S. G.; Gleason, K. K. Initiated and Oxidative Chemical Vapor Deposition: A Scalable Method for Conformal and Functional Polymer Films on Real Substrates. *Phys. Chem. Chem. Phys.* **2009**, *11* (26), 5227−5240.

(28) Gupta, M.; Gleason, K. K. Initiated Chemical Vapor Deposition of Poly(1H,1H,2H,2H-Perfluorodecyl Acrylate) Thin Films. *Langmuir* **2006**, *22* (24), 10047−10052.

(29) Cheng, C.; Gupta, M. Surface Functionalization of 3D-Printed Plastics via Initiated Chemical Vapor Deposition. *Beilstein J. Nanotechnol.* **2017**, *8* (1), 1629−1636.

(30) Haller, P. D.; Flowers, C. A.; Gupta, M. Three-Dimensional Patterning of Porous Materials Using Vapor Phase Polymerization. *Soft Matter* **2011**, *7*, 2428−2432.

(31) Cheng, C.; Gupta, M. Roll-to-Roll Surface Modification of Cellulose Paper via Initiated Chemical Vapor Deposition. *Ind. Eng. Chem. Res.* **2018**, *57* (34), 11675−11680.

(32) Wilson, A. M.; Zank, G.; Eguchi, K.; Xing, W.; Yates, B.; Dahn, J. R. Polysiloxane Pyrolysis. *Chem. Mater.* **1997**, *9* (7), 1601−1606.

(33) Burns, G. T.; Taylor, R. B.; Xu, Y.; Zangvil, A.; Zank, G. A. High-Temperature Chemistry of the Conversion of Siloxanes to Silicon Carbide. *Chem. Mater.* **1992**, *4* (6), 1313−1323.

(34) Naserifar, S.; Goddard, W. A.; Liu, L.; Tsotsis, T. T.; Sahimi, M. Toward a Process-Based Molecular Model of SiC Membranes. 2. Reactive Dynamics Simulation of the Pyrolysis of Polymer Precursor to Form Amorphous SiC. *J. Phys. Chem. C* **2013**, *117* (7), 3320−3329.

(35) Gao, H.; Wang, H.; Zhao, Z.; Niu, M.; Su, L.; Wei, Y. Reactive Dynamics Simulation Study on the Pyrolysis of Polymer Precursors to Generate Amorphous Silicon Oxycarbide Structures. *J. Phys. Chem. C* **2018**, *122* (10), 5767−5773.

(36) Sarkar, S.; Gan, Z.; An, L.; Zhai, L. Structural Evolution of Polymer-Derived Amorphous SiBCN Ceramics at High Temperature. *J. Phys. Chem. C* **2011**, *115* (50), 24993−25000.

(37) Elyassi, B.; Deng, W.; Sahimi, M.; Tsotsis, T. T. On the Use of Porous and Nonporous Fillers in the Fabrication of Silicon Carbide Membranes. *Ind. Eng. Chem. Res.* **2013**, *52* (30), 10269−10275.

(38) Deng, W.; Yu, X.; Sahimi, M.; Tsotsis, T. T. Highly Permeable Porous Silicon Carbide Support Tubes for the Preparation of Nanoporous Inorganic Membranes. *J. Membr. Sci.* **2014**, *451*, 192−204.

(39) Dabir, S.; Deng, W.; Sahimi, M.; Tsotsis, T. Fabrication of Silicon Carbide Membranes on Highly Permeable Supports. *J. Membr. Sci.* **2017**, *537*, 239−247.

(40) ThermoSpectraTech. *COLLECTOR ™ II User's Manual.*.

(41) Raj, R.; Pederiva, L.; Narisawal, M.; Soraru, G. D. On the Onset of Fracture as a Silicon-Based Polymer Converts into the Ceramic Phase. *J. Am. Ceram. Soc.* **2018**, *102* (3), 924−929.

(42) Trujillo, N. J.; Wu, Q.; Gleason, K. K. Ultralow Dielectric Constant Tetravinyltetramethylcyclotetrasiloxane Films Deposited by Initiated Chemical Vapor Deposition (iCVD). *Adv. Funct. Mater.* **2010**, *20* (4), 607−616.

(43) Lubguban, J.; Rajagopalan, T.; Mehta, N.; Lahlouh, B.; Simon, S. L.; Gangopadhyay, S. Low-k Organosilicate Films Prepared by Tetravinyltetramethylcyclotetrasiloxane. *Appl. Phys.* **2002**, 92 (2), 1033−1038.

(44) Carlsson, D. J.; Cooney, J. D.; Gauthier, S.; Worsfold, D. J. Pyrolysis of Silicon-Backbone Polymers to Silicon Carbide. *J. Am. Ceram. Soc.* **1990**, *73* (2), 237−241.

(45) Narisawa, M. Silicone Resin Applications for Ceramic Precursors and Composites. *Materials* **2010**, *3* (6), 3518−3536.

(46) Ryan, E. T.; Gates, S. M.; Cohen, S. A.; Ostrovski, Y.; Adams, E.; Virwani, K.; Grill, A. Effect of Low-Frequency Radio Frequency on Plasma-Enhanced Chemical Vapor Deposited Ultra Low-κ Dielectric Films for Very Large-Scale Integrated. *Appl. Phys.* **2014**, *115*, 144107.

(47) Mantz, R. A.; Jones, P. F.; Chaffee, K. P.; Lichtenhan, J. D.; Gilman, J. W.; Ismail, I. M. K.; Burmeister, M. J. Thermolysis of Polyhedral Oligomeric Silsesquioxane (POSS) Macromers and POSS - Siloxane Copolymers. *Chem. Mater.* **1996**, *8* (6), 1250−1259.

(48) Burkey, D. D.; Gleason, K. K. Temperature-Resolved Fourier Transform Infrared Study of Condensation Reactions and Porogen Decomposition in Hybrid Organosilicon-Porogen Films. *J. Vac. Sci. Technol., A* **2004**, *22* (1), 61−70.

(49) Burkey, D. D.; Gleason, K. K. Structure and Mechanical Properties of Thin Films Deposited from 1,3,5-Trimethyl-1,3,5-Trivinylcyclotrisiloxane and Water. *J. Appl. Phys.* **2003**, 93 (9), 5143−5150.

(50) Samitier, J.; Marco, S.; Morante, J. R. Analysis by FT-IR Spectroscopy of SiO_2-Polycrystalline Structures Used in Micromechanics: Stress Measurements. *Sens. Actuators* **1992**, *32*, 347−353.

(51) Sanaeishoar, H.; Sabbaghan, M.; Mohave, F. Synthesis and Characterization of Micro-Mesoporous MCM-41 Using Various Ionic Liquids as Co-Templates. *Microporous Mesoporous Mater.* **2015**, *217*, 219−224.

(52) Shokri, B.; Firouzjah, M. A.; Hosseini, S. I. FTIR Analysis of Silicon Dioxide Thin Film Deposited by Metal Organic-Based PECVD. In *19th international symposium on plasma chemistry society, Bochum, Germany*; 2009; pp 1−4.

(53) Babonneau, F.; Thorne, K.; Mackenzie, J. D. Dimethyldiethoxysilane/Tetraethoxysilane Copolymers: Precursors for the Silicon-Carbon-Oxygen System. *Chem. Mater.* **1989**, *1* (5), 554−558.

(54) Mabboux, P.; Gleason, K. K. Chemical Bonding Structure of Low Dielectric Constant Si: O: C: H Films Characterized by Solid-State NMR. *J. Electrochem. Soc.* **2005**, *152* (1), F7−F13.

(55) Alexander, M. R.; Short, R. D.; Jones, F. R. A Study of HMDSO/O2 Plasma Deposits Using a High-Sensitivity and -Energy Resolution XPS Instrument: Curve Fitting of the Si 2p Core Level. *Appl. Surf. Sci.* **1999**, *137* (1−4), 179−183.

(56) O'Hare, L.; Parbhoo, B.; Leadley, S. R. Development of a Methodology for XPS Curve-Fitting of the Si 2p Core Level of Siloxane Materials. *Surf. Interface Anal.* **2004**, *36* (10), 1427−1434.

(57) Belot, V.; Corriu, R.; Leclercq, D.; Mutin, P. H.; Vioux, A. Thermal Reactivity of Hydrogenosilsesquioxane Gels. *Chem. Mater.* **1991**, 3 (1), 127−131.

572041 Membranes for Continuous, on-Demand Dehydration of Anhydrous Solvents in Flow Chemistry Manufacturing Processes

Solvents can be critical in the development and production of new pharmaceutical molecules, affecting yield, product quality and process efficiency to name just a few key elements. One of the ways to minimize the environmental impact of solvents is to recycle and reuse them. However, it is not always possible to recycle solvents if their use must be at anhydrous levels and use in a process renders them less pure than is ideal for the process. On-demand production of anhydrous solvents is not always trivial given the temperature, quality and process requirements of APIs, particularly in flow chemistry systems. Membrane dehydration and purification can provide an avenue for efficient and cost-effective, on-demand production of these anhydrous solvents. Compact Membrane Systems has installed several fluoropolymer membrane pervaporation systems into flow chemistry manufacturing processes to allow inline dehydration of solvents. CMS membranes have been specifically developed to be thermally and chemically resistant to allow for their use in harsh environments, including organic solvents. One example of this work was published in OPR&D along with authors from Pfizer and Asymchem and provides a new route to manage solvents and allow recycling and on-demand production of anhydrous solvents in flow chemistry processes. This talk will discuss solvent purification and management in flow chemistry processes using pervaporation membranes and will also cover specific operational issues such as how to manage low flow rates or small hold up volumes.

557376 Influence of Molecular Interactions, Membrane Swelling and Plasticization on Pure and Mixed Fluid Transport in OSN Membranes

The vast majority of industrial chemical synthesis occurs in organic solution. Solute concentration and solvent recovery consume approximately 50% of the energy required to produce chemicals and pose problems that are as relevant as the synthesis process itself. Novel, energy-efficient technologies based on polymer membranes are emerging as a viable alternative to distillation. Despite organic solvent nanofiltration (OSN) could revolutionize the chemical industry, its development is still in its infancy for two reasons: *i)* the instability of traditional polymer materials in chemically challenging environments, and *ii)* the lack of fundamental knowledge of elemental transport phenomena in OSN membranes. Most of available transport data refer to composite membranes, where the presence of a fabric backing makes it difficult to provide a fundamental description of solvent and solute sorption and transport in the active layer. The lack of fundamental information has hampered the development of rational methods to design better materials for OSN application.

In this study, the solubility of several pure and mixed organic species in Celazole®, a commercial polybenzimidazole, has been investigated. Polymer volumetric dilation upon liquid sorption has been investigated as well, using the optical method. The role of polymer-penetrant interactions, membrane degree of swelling, and penetrant clustering on small molecule sorption and transport in Celazole® has been discussed and several structure-property correlations were identified. Methanol was selected as a model penetrant to run sorption and diffusion experiments in the activity range 0-1. Methanol and other polar lower alcohols cause severe matrix plasticization. We have hypothesized thc possible mechanism of Celazole® plasticization. Specifically, methanol (as well as other polar penetrants) likely break the inter-chain hydrogen bonds in favor of polymer-penetrant hydrogen bonds, according to a mechanism that we could define as *competitive hydrogen bonding*. The breaking of the original polymeric network would increase, in turn, the distance between polymer chains, thus enhancing their mobility. This picture helps to rationalize mixed methanol-PEG400 sorption data. In contrast, Celazole® is very stable in non-polar aliphatic and aromatic hydrocarbons.

Finally, the polymer mechanical properties were measured before and after soaking in liquid water and methanol. The results provide further proof that polar liquids severely plasticize Celazole®.

573144 Surface-Modified Uio-66-NH_2 Mixed Matrix Membranes for Xylene Isomer Separation

Mixed matrix membranes (MMMs) have been shown to greatly improve gas separation performance over purely polymeric membranes. However, their potential for organic solvent separations remains less explored, particularly in organic solvent reverse osmosis (OSRO) applications. Here, we fabricate MMMs based on the metal-organic framework (MOF) UiO-66-NH_2 and the polyimide Matrimid® and investigate their ability to separate para and ortho-xylene in both vapor permeation and hydraulic modes of permeation. In addition, we present a new approach to overcome interfacial defects for the UiO-66-NH_2/Matrimid® system. Pyromellitic diimide is used to homogenize the interface between filler and polymer and improved polymer-MOF adhesion is observed as a result. This improvement is tentatively attributed to charge-transfer complexes between the diimide, MOF ligand, and polyimide. Adhesion and stability was further improved through diamine crosslinking. The effects of the diimide compatibilizer on the interface are examined through FTIR and DSC measurements. MMM performance is measured on custom-built Wicke-Kallenbach and OSRO cross-flow cell systems. Pure component diffusion coefficients are estimated from isotherm measurements and the solution-diffusion model. The xylene isomer separation performance is shown to be significantly enhanced over Matrimid® and unmodified MMMs. At 10 weight percent MOF loading, surface modified MMMs show para-xylene permeability of 350 Barrer and a selectivity of 5.8 from a 50/50 feed mixture at unit activity compared to 3.2 Barrer and 1.05 for native Matrimid®.

562008 Separation of Vegetable Oil Compounds and Solvent Recovery Using Commercial Organic Solvent Nanofiltration Membranes

Solvent recovery and separation of oil compounds by means of thermal evaporation processes consume a huge amount of energy. Organic solvent nanofiltration (OSN) is an energy-efficient separation technique that can be potentially used for solvent recovery and separation of oil compounds in the vegetable oil industry. However, there are only a few studies on the separation of oil compounds using OSN. Almost no studies have been conducted on membrane fouling and long-term stability using highly concentrated oil/solvent feeds. This study explores the separation of oil compounds such as triglycerides and free fatty acids as well as solvent recovery using a series of Evonik commercial membranes. Under a static testing condition, the permeance of Duramem 500 reduces severely from 1.02 to 0.06 LMH/bar with an increase in oil concentration while its rejection to glyceryl trilinoleate decreases slightly from 86 to 84.8% when the feed concentration increases from 5 to 50 wt%. The severe permeance decline arises from the high viscosity of the permeate and oil layers accumulated on the membrane surface. In contrast, the membrane is able to separate glyceryl trilinoleate from linoleic acid with permeances approximately 3-6 times of those obtained from the static testing mode under 1-week cross-flow tests using a 20 wt% oil/acetone feed.

564752 Design of a Pervaporation Assembly Using Static Mixers

Chemically inert membrane-based separation assemblies have been widely employed in numerous industries during processes of phase separation, purification, extraction, and many more. Pervaporation is a powerful and efficient membrane-based process to separate the minor components from a liquid mixture solution by partial vaporization and permeation. In addition, pervaporation can be implemented in a continuous synthesis process that allows for smaller dead volumes, intensified heat and mass transfer, better process control, safer operation, and the potential handling of reactions that are too dangerous or difficult to perform in batch.

In general, pervaporation process suffers from mass transfer limitations due to poor mass transport in the feed side (liquid side). A common model to describe the mass transfer mechanism in a pervaporation device is the resistance-in-series model, in which the overall mass transfer coefficient (k_{all}) is linked with each of k_{liquid}, $k_{membrane}$, and k_{gas} ($1/k_{all} = 1/k_{liquid} + 1/k_{membrane} + 1/k_{gas}$, and $1/k_{gas}$ can be safely ignored in most scenarios). In our design, we employ static mixers to create complex flow dynamics thus enhance mass transfer in the feed side (increasing k_{liquid}). The static mixers can be either commercial ones or self-designed and fabricated using 3D printing technologies.

In this work, we introduce a combination of simulations and experiments approach for pervaporation module design with static mixers as key components in continuous flow process. We design the pervaporation module with static mixers using computer-aided design (CAD) programs, and evaluate its performance by coupling and simulating the hydrodynamics and mass transfer performance. Moreover, the device is tested with ethanol – water and diethyl ether – water systems experimentally. From both approaches, we confirm that static mixers are beneficial to higher separation rates. In addition, we are able to fit the membrane mass transfer coefficient ($k_{membrane}$) to show that our simulations are capable of reproducing the experimental data. Then we can use simulations to show whether the devices could fulfill the desired functionality. Therefore, fundamental understanding is gained using the proposed methodology and applied to otherwise more challenging and more complicated membrane processes.

561619 Fabrication of Graphene Oxide (GO) Composite Membranes for Organic Solvent Nanofiltration (OSN)

An ultrathin graphene oxide (GO) laminar composite membrane has been fabricated in-house via pressure-assisted filtration. The membrane substrate is a lab-fabricated Matrimid® membrane crosslinked with 1,6-hexanediamine (HDA) to improve its chemical stability. GO dispersions are centrifuged and diluted before filtration to minimize the defects of the resultant GO selective layer. The morphologies, pore sizes as well as the microstructures of the prepared membranes are characterized by field emission scanning electron microscopy (FESEM), filtration test and position annihilation spectroscopy (PAS), respectively. When being treated with various solvents, the GO laminar composite membrane exhibits good stability in both polar and nonpolar solvents. Experimental results suggest that the Donnan exclusion is less effective in organic solvents than in aqueous system, while the size exclusion and solute-membrane affinity turn out to be the dominant factors in determining separation performance. Some active pharmaceutical ingredients (APIs) and food additives, e.g. tetracycline, roxithromycin, vitamin B12 and lecithin, are used to demonstrate the separation capability of the newly developed membrane in real-life applications. Three 7-day tests of vitamin B12 in isopropanol and ethanol as well as lecithin in hexane also confirm the membrane integrity and separation performance for continuous uses.

557443 Understanding the Salt Transport Properties of Graphene Oxide Membranes: Permeation Measurements and Electrokinetic Modeling

Graphene oxide (GO)-based membranes have gained tremendous interest in nanofiltration (NF) and reverse osmosis (RO) applications. A major challenge in GO membranes is to improve their salt rejection in NF and RO processes. Previous researchers proved that the interlayer spacing between the GO layers as well as the charge density of the GO layers can strongly influence salt rejections from ideal/dilute solutions. However, many practical desalination and dewatering processes operate at much higher concentrations and extreme pH conditions. In this work, our goal is twofold: to systematically measure the effect of salts concentration (0.01 – 0.5 M) and pH (3-13) on GO membrane properties particularly in extreme conditions, and to understand and predict these properties by electrokinetic modeling. The correlations between feed concentration/pH, interlayer spacing, and charge density of these GO membranes were determined by tracer experiments and ion adsorption models. The modified Donnan Steric Pore Model was then used to model, fit, and predict the salts rejections at different conditions. We investigate in detail the contributions of convection, diffusion, and electro-migration as a function of feed conditions. Finally, the combined effects of initial charge density and interlayer distances on salts rejection are evaluated through the plotting of 2D "heatmaps". Our study indicates for neutral pH, under low salt concentrations (~ 0.01 M), modifying the charge density of GO will be more helpful for increasing salt rejections. On the other hand, under high concentrations (above 0.1 M), decreasing the interlayer distance of the GO membranes (i.e., reduction of effective pore size) would be a better strategy. Finally, we examine the possibility of accurate prediction of multicomponent salt behavior using single-component data. Overall, this study provides mechanistic insight into salt transport and rejection in GO membranes, and useful guidance for modification of GO membranes in a number of desalination processes.

570312 Large-Scale Computational Study of Zeolite Nanosheets As Pervaporation Membranes for Ethanol Extraction

Large-Scale Computational Study of Zeolite Nanosheets as Pervaporation Membranes for Ethanol Extraction.

Changlong Zou[1] and Li-Chiang Lin[1]*

[1]William G. Lowrie Department of Chemical and Biomolecular Engineering, The Ohio State University, Columbus, OH 43210, USA

Membranes based pervaporation separation is one of the promising energy-efficient and cost-effective approaches for anhydrous ethanol extraction from the dilute biomass fermentation broth. By employing molecular dynamics simulations, we have carried out a large-scale computational study to investigate zeolite nanosheets with diverse structural features as potential candidates for this application. Our results show that zeolite nanosheets can offer better separation factors (i.e., highly selective toward ethanol over water) and orders of magnitudes higher fluxes as compared to other membranes reported in the literature. The roles of the bulk part (i.e., non-surface region) and the surface of nanosheets are also explored at an atomic level in their overall separation performance. We find that the adsorption selectivity of the bulk zeolites and the ethanol concentration at the nanosheets membrane surfaces can significantly impact the overall separation factor. Furthermore, our results also suggest that the surface ethanol concentration, particularly the ethanol concentration at the channel entrance, has a strong correlation with the number of silanol groups around the entrance and the diameter of the entrance. Strategically selecting zeolites with a high adsorption selectivity and surface topologies that maximize the ethanol concentration at the entrance region of pore/channels are the key to the design of highly selective zeolite nanosheets membranes. On the order hand, the overall permeation flux is found to strongly correlate with the channel density of zeolites along the permeation direction and their pore limiting diameters. Overall, this work has demonstrated the enormous potential of zeolite nanosheets as pervaporation membranes. Results obtained here from a large-scale computational study also provide atomic-level guidelines into the future design and discovery of the zeolite nanosheet membranes for ethanol/water pervaporation separation and potentially for other separation applications.

578058 The Effects of Additives on PES Membrane Formation Vis Nips from Dissipative Particle Dynamics Simulations

Polymeric membranes produced by Non-solvent Induced Phase Separation (NIPS) have been extensively used across a wide array of microfiltration and ultrafiltration processes. However, despite their broad applications, current fabrication processes are largely empirical, determined by trial and error. Better insights into the factors affecting membrane morphology would be of great use. To that end, Dissipative Particle Dynamics (DPD), a mesoscale simulation tool, was used to simulate membrane formation process involving polyethersulfone (PES) in methyl-2-pyrrolidinone (NMP) with H_2O as a nonsolvent. The DPD simulations were performed for a total of several million beads and over several microseconds. The interactions between these beads are determined according to Flory-Huggins thermodynamic theory and continuum transport equation. In addition, the effects of additive polyethylene glycol (PEG) and its concentration on the thermodynamics and kinetics of PES membrane formation were investigated. Membrane morphology was found to depend strongly on the amount of additive included.

569235 Molecular Dynamics Simulations of Polyamide Water Purification Membranes

Nanofiltration and reverse osmosis membranes are widely used for water treatment and purification. Here we explore the molecular level properties of the membrane nanostructure and charge as well as nanoscale transport of water and contaminants using molecular dynamics simulations. Both meta-phenylenediamine (MPD) based polyamide membranes (typical of reverse osmosis membranes) and piperazine based polyamide membranes (typical of nanofiltration membranes) are constructed using a virtual polymerization technique. In both cases, accounting for charged carboxylate end groups is a challenging computational issue. The distribution of open volume within the molecular structure and the polymer density are compared between the MPD and piperazine membranes. Non-equilibrium simulations are performed in which water, modeled using the TIP3P model, is forced at high pressure through both types of membranes. We also consider uncharged organic solutes to assess the transport of neutral species. Finally, we consider ion complexation with the charged carboxylate end groups to examine the impact of solute-membrane charge interaction. This approach of non-equilibrium molecular dynamics simulations to examine water and solute transport in polyamide membranes at the molecular level holds great promise for designing the next generation of membranes for desalination and other applications. Funded by the Northwestern University Center for Water Research.

563160 Fast Desalination By Multilayered Covalent Organic Framework (COF) Nanosheets

Covalent organic frameworks (COFs) are penetrated with uniform and ordered nanopores, implying their great potential in molecular/ion separations. As an imine-linked, stable COF, TpPa-1 is receiving tremendous interest for molecular sieving membranes. Theoretically, atomically thin TpPa-1 monolayers exhibit extremely high water permeance but unfortunately no rejection to ions because of its large pore sizes (~1.58 nm). COF monolayers tend to stack to form laminated multilayers, but how this stacking influence water transport and ion rejections remain unknown. Herein, we investigate the transport behavior of water and salt ions through multilayered TpPa-1 COFs by nonequilibrium molecular dynamics simulations. By analyzing both the interfacial and interior resistance for water transport we reveal that with rising stacking numbers COF multilayers exhibit increasing ion rejections at the expense of water permeance. More importantly, stacking in the offset eclipsed fashion significantly reduces the equivalent pore size of COF multilayers to 0.89 nm and ion rejection is correspondingly increased. Remarkably, 25 COF monolayers stacked in this fashion give 100% $MgCl_2$ rejection while the water permeance remain 1 to 2 orders of magnitude higher than that of commercial nanofiltration membranes. This work demonstrates the rational design of fast membranes for desalination by tailoring stacking number and fashion of COF monolayers.

571885 A Molecular Simulation Protocol for Membrane Swelling and Organic Solvent Nanofiltration

Polymer membranes have received considerable interest for organic solvent nanofiltration (OSN). The OSN performance is governed by many complex factors and quantitative understanding is experimentally challenging. In addition, it is a prerequisite to characterize membrane swelling in solvents prior to nanofiltration. A molecular simulation protocol is developed herein to investigate the swelling of three polymer membranes namely polybenzimidazole (PBI), polyimide (PI) and polymer of intrinsic microporosity (PIM-1) in four organic solvents (methanol, ethanol, acetonitrile and acetone), and then examine the OSN through the swollen membranes. The simulation protocol can highly efficiently simulate membrane swelling, which is approximately one order of magnitude faster than normal simulation method. The predicted swelling degrees of the three polymer membranes follow experimentally measured trend, exhibit good relationships with the polymer-solvent solubility differences and the mean pore sizes of swollen membranes. For the OSN of a dye (methylene blue) in the four organic solvents, 100% rejection is found by the swollen PBI, PI and PIM-1 membranes. The predicted solvent permeabilities through PIM-1 and PI membranes match fairly well with experimental data, and their dependences on solvent viscosity, pore size and polymer-solvent interaction are examined. From this simulation study, the microscopic insights into the swelling of polymer membranes and the permeation of organic solvents are provided, which are useful for the development of new membranes for high-performance OSN.

564959 Promoted Metallization and Mineralization By Zwitterion Grafting: A Potential Modification of TFC Membranes for Maximized Anti-Fouling Properties Process

A comparative study is conducted to investigate the influence of zwitterions on membrane metallization and mineralization processes. Metallized and mineralized membranes are developed via *in-situ* formation of silver nanoparticles (AgNPs) and deposition of silver chloride (AgCl) particles on thin-film composite (TFC) membranes pretreated by grafting with two different zwitterions (SO_3^--based and COO^--based), respectively. Both metallized and mineralized membranes with zwitterion pretreatment exhibit much higher AgNPs and AgCl loading amount compared to that without zwitterion pretreatment, resulting in significantly improved anti-microbial and anti-fouling properties without compromising the separation performance. Moreover, it is found for the first time that COO^--based zwitterion achieves better results in promoting the metallization and mineralization degrees than SO_3^--based zwitterion. The underlying mechanisms are further revealed by the Gibbs free energy of the binding between zwitterions and Ag^+ calculated via Density Functional Theory (DFT).

560822 Novel Thin Film Composite Hollow Fiber Membranes Incorporated with Carbon Quantum Dots for Osmotic Power Generation

By means of carbon quantum dots incorporation, we have developed novel thin film composite (TFC) membranes for osmotic power generation. The newly developed TFC membrane exhibits a peak power density as high as 34.20 W/m^2 at 23 bar using 1.0 M NaCl and deionized water as the feed pair. To our best knowledge, this is the highest ever power density reported in the literature. The carbon quantum dots (CQDs) are a new class of carbon nanomaterials with advantages of excellent hydrophilicity, low toxicity, environmental friendliness, easy synthesis and low cost. The CQDs are incorporated into the polyamide selective layers via the conventional interfacial polymerization reaction. The effects of incorporating different CQDs and their loadings on membrane morphology, properties and PRO performance have been examined. It is found that the addition of Na^+–functionalized CQDs not only increases the existence of hydrophilic oxygen-containing groups and surface area of the polyamide layer, but also changes the morphology with a looser and thinner polyamide network. The TFC membrane comprising 1 wt% Na–CQD-9 has the optimal performance. Compared with the control, the water flux and power density at 23 bar increase from 44.52 to 53.54 LMH and 28.44 to 34.20 W/m^2 respectively, while the reverse salt flux remains unchanged.

569114 Functionalized High Aspect Ratio PCN-222 Nanoparticles for Tunable Reverse Osmosis Performance

Nanorods of PCN-222, a large-pore, zirconium-based porphyrinic metal-organic framework (MOF), have been prepared through coordination modulation for incorporation into reverse osmosis thin film nanocomposite (TFN) membranes. Myristic acid can be loaded into the MOF pores via solvent-assisted ligand incorporation to alter channel dimensions and pore size distribution. TFNs are a class of materials that rely on functional nanoparticles to provide improvements over traditional thin film composite (TFC) reverse osmosis membranes. TFN membranes containing PCN-222 nanoparticles modified with varying levels of myristic acid were fabricated via dispersion of nanoparticles in the aqueous phase of an interfacial polymerization, and the resulting flux and rejection capabilities of each membrane were evaluated. The membranes were characterized using scanning electron microscopy (SEM), transmission electron microscopy (TEM), and contact angle goniometry. The best result was observed at 0.01 wt% PCN-222 loading with a 10:1 myristic acid to TCPP linker modification, where membrane flux increased by 190% and salt rejection decreased marginally compared to the plain polyamide control membrane ($96.0 \pm 2.4\%$ vs. $97.5 \pm 0.3\%$). This flux increase is attributed to water transport through the nanoparticle pore structure, and thus further studies are necessary to determine the optimal PCN-222 loading and pore size modifications, leading to the fabrication of TFN membranes with tunable properties.

Contents lists available at ScienceDirect

Chemical Engineering Journal

journal homepage: www.elsevier.com/locate/cej

Arsenate removal by reactive mixed matrix PVDF hollow fiber membranes with UIO-66 metal organic frameworks

Peng Wan[a,b,1], Mengxi Yuan[a,1], Xiaolong Yu[d], Zheng Zhang[e], Baolin Deng[c,*]

[a] Department of Chemical Engineering, University of Missouri, Columbia, MO 65211, USA
[b] Shenzhen Institutes of Advanced Technology, Chinese Academy of Sciences, Shenzhen 518055, China
[c] Department of Civil & Environmental Engineering, University of Missouri, Columbia, MO 65211, USA
[d] School of Environmental Science and Engineering, Southern University of Science and Technology, Shenzhen, China
[e] China Shanghai Architectural Design & Research Institute Co., Ltd, Shanghai 200333, China

HIGHLIGHTS

- Reactive MOF PVDF HFM was successfully fabricated.
- DFT calculations and experimental characterizations were applied.
- Adsorption and membrane filtration were combined for arsenate treatment.

GRAPHICAL ABSTRACT

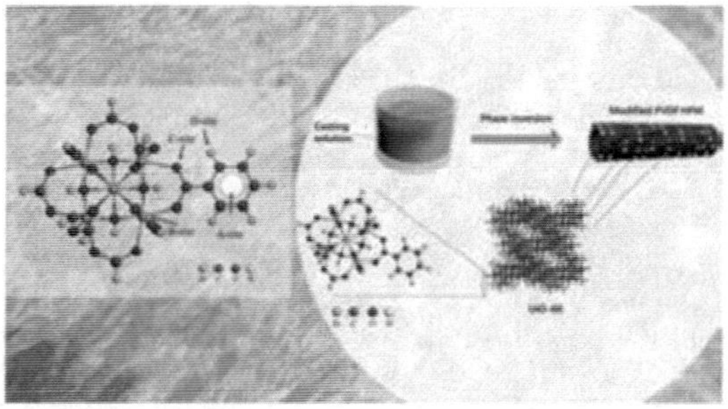

ARTICLE INFO

Keywords:
Hollow fiber membrane
Adsorption
Metal organic frameworks
Arsenate removal
PVDF

ABSTRACT

High arsenic concentration in drinking water is a great threat to human health. In this study, reactive mixed matrix polyvinylidene fluoride (PVDF) hollow fiber membrane (HFM) was fabricated by embedding UiO-66 particles, a water stable metal organic frameworks (MOFs), for the removal of arsenate from aqueous solutions. Adsorption and membrane filtration were combined for arsenate treatment and the results showed that arsenate could be removed efficiently by this strategy. Adsorption of arsenate onto UiO-66 particles was first investigated using batch experiments for the determination of adsorption isotherm, pH edge, and adsorption kinetics. The maximum arsenic adsorption by UiO-66 was found to reach 267 mg/g at pH 4.7, which was ranked among the top of all arsenic sorbents reported in the literature. UiO-66 particles were then mixed in the dope of PVDF to produce the mixed matrix HFM for arsenate removal from water. The removal efficiency was studied for 6 types of HFMs with an increasing amount of UiO-66 particles. The pristine PVDF HFM without UiO-66 particles could not remove arsenate from water as expected; the pores of ultrafiltration membranes are too large to reject arsenate. With UiO-66 particles embedded into the membrane matrix, the adsorption capacity of membrane was increased significantly, resulting in arsenate removal primarily by sorption during the membrane filtration. The UiO-66 particles-embedded HFM was characterized by morphology, hydrophilicity, arsenate rejection, and ATR-FTIR. This study demonstrated that the reactive ultrafiltration membrane had a great premise to treat drinking water with arsenic.

* Corresponding author at: Department of Civil & Environmental Engineering, University of Missouri, Columbia, MO 65211, USA.
E-mail address: dengb@missouri.edu (B. Deng).
[1] P.W. and M.Y. contributed equally.

https://doi.org/10.1016/j.cej.2019.122921
Received 26 July 2019; Received in revised form 16 September 2019; Accepted 20 September 2019
1385-8947/

1. Introduction

Arsenic present in drinking water supplies in many parts of the world has been a major concern because of its known carcinogenic and toxic effects. The high arsenic level is mostly derived from natural weathering processes, although it can also be caused by the discharge of untreated industrial wastewater and mining and smelting activities [1]. Canada, United States, Chile, Bolivia, El Salvador, Nicaragua, Slovenia, Mexico, Bangladesh and India are among the countries facing the greatest threat of arsenic contamination [2]. Arsenic disturbs enzymatic activities and transcriptional events in cells that lead to health complications such as cancers of the skin, kidney, lung, and bladder, as well as diseases of skin, neurological, and cardiovascular system. The World Health Organization (WHO) and US Environmental Protection Agency (USEPA) has set the maximum contaminant level of arsenic in drinking water at 10 ppb [3], a standard set based not only on the health effect but also treatment cost consideration. To remove arsenic efficiently in a cost-effective way, many approaches have been explored including coagulation/precipitation [4,5], lime softening [6], ion exchange [7], adsorption [8,9], and membrane filtration processes [10–15].

Adsorption is generally considered a promising method for removing a low concentration of arsenic. Based on the adsorption capacity, conventional adsorbents like activated alumina [16], commercial TiO_2, and activated carbon [17] are not competitive to nano-adsorbents such as zero-valent iron nanoparticles. When the size and structure of an adsorbent reach the nanometer domain, there will be two very favorable characteristics: large surface area and likely fast mass transfer rate for the sorption process. Metal-organic frameworks (MOFs), a kind of microporous material with large surface area and fast mass transfer rate, have a great potential for industrious applications. Applications of MOFs in arsenate adsorption have been relatively new but several materials including ZIF-8 [18], MIL-53 [19], and UiO-66 [20,21] MOFs had been tested for their applications. Zirconium-based MOF, UiO-66, shows an extraordinarily high arsenate adsorption capacity of 303 mg/g, ranked among the highest reported in the literature [22]. The high stability of UiO-66 in the presence of a strong acid (HCl) and a strong base (NaOH) makes it suitable for applications within a wide pH range of 1 to 10 [23]. It should be noted that UiO-66 has high structural stability resisting to high temperature as well, up to 375°C [23]. Besides, a rapid adsorption rate of arsenate also makes the UiO-66 adsorbent unique [22].

Despite of excellent performances of UiO-66 for arsenate sorption, the size of UiO-66, existing as nanoscale powders, is pretty small, and cannot be readily separated and recycled after the sorption process [24]. For practical industrial applications, the powder materials could be supported by other matrices. Attaching powder materials to support materials have been explored for applications in membranes, filters, or columns. Tunable, cost-efficient, and chemically inert materials can be considered for supporting materials [25].

Reverse osmosis (RO) and nano-filtration (NF) are applicable for arsenate treatment [13,14]. NF has an efficiency of 90% while it can be 91%–98% by RO [2]. However, RO and NF require high pressure as the driving force, which results in high energy consumption. On the other hand, the low-pressure micro and ultrafiltration techniques can only reject particulate and colloidal materials, so they are not suitable for soluble arsenate removal. A combined system of adsorption and microfiltration/ ultrafiltration can increase the efficiency of the filtration process for arsenate treatment. Between flat-sheet and hollow fiber membranes, HFMs have advantages of high packing density and self-supporting structure, so it's suitable for developing such a combined system.

In this study, PVDF, a semi-crystalline polymer containing a crystalline phase and an amorphous and/or rubbery phase, was used to fabricate the mixed matrix HFMs containing an increasing amounts of UiO-66 nanoparticles for arsenic removal. PVDF has high thermal stability and chemical resistance [26,27] in comparison with other polymers such as polysulfone and polyethersulfone [28]. The functional MOFs HFMs were fabricated on a custom-designed apparatus in our laboratory [29]. The molecular binding mechanisms of arsenate species with the MOFs were investigated using density functional theory (DFT) calculations and experimental characterizations. The prepared HFMs were also characterized in terms of the membrane structure, water contact angle, permeability, rejection and adsorption performance. This study provided new insights into the adsorption mechanisms of arsenate to MOFs, and facilitated the development of MOF-based membrane technology for arsenate removal.

2. Materials and methods

2.1. Materials

Chemicals for UiO-66 synthesis including zirconium (IV) chloride ($ZrCl_4$, 98%, anhydrous), 1,4-benzenedicarboxylic acid (BDC, 98%), dimethylformamide (extra pure DMF, 99.9%), and acetone (99.8%) were purchased from Acros-Organic. Chemicals for fabricating the polyvinylidene difluoride PVDF HFM including PVDF, N-methyl-2-pyrrolidone (NMP, 99.5%), and polyvinylpyrrolidone (PVP) were purchased from Sigma-Aldrich. Others were obtained either from Sigma-Aldrich (sodium arsenate dibasic heptahydrate ($Na_2HAsO_4{\cdot}7H_2O$, 98%) and ethanol (99.9%)) or Fisher-Scientific (sodium hydroxide, nitric acid (70%), and acetic acid (99.7%)).

Stock arsenic solution of 1000 mg/L was prepared by dissolving $Na_2HAsO_4{\cdot}7H_2O$ in deionized water, and was diluted to obtain solutions with lower arsenate amounts. Nitric acid and sodium hydroxide were used for pH adjustments. Acetate buffer was used to control pH 4.7, which was obtained by adding sodium hydroxide (NaOH) to acetic acid. Sodium nitrate at 50 mM was used to control the ionic strength of all arsenate working solutions.

2.2. Synthesis of UiO-66 particles

UiO-66 particles were prepared by a reported procedure in the literature [31], in which $ZrCl_4$ (2.378 g), BDC (1.695 g), DI water (0.180 g), and DMF (365.480 g) were mixed in a molar ratio of Zr^{4+}/BDC/H_2O/DMF = 1:1:1:500. After stirring for 60 min, the mixture was transferred into a 500 mL Teflon-lined stainless steel autoclave and heated to 120 °C for 48 h. Upon cooling to room temperature, the white precipitates produced were separated from the solvent via centrifugation at 4000 rpm for 6 min and then washed for 3 times with acetone. The obtained white precipitates were dried at 120 °C for 24 h and then ground to fine powders prior to further experiments.

2.3. Preparation of UiO-66 embedded PVDF HFMs

Membranes were produced via the phase inversion method as reported previously [30]. The dope solution of PVDF, PVP and NMP was spiked with different amounts of UiO-66 particles and extruded to make six types of hollow fiber membranes (HFMs). In the dope solution, the weight ratio of UiO-66 particles to PVDF in the membranes were 0, 5, 10, 20, 30, and 40%, named M0, M5, M10, M20, M30, and M40, respectively. The respective amounts of UiO-66 particles used were 0, 0.24, 0.48, 0.96, 1.44 and 1.92 g, all dissolved in NMP (23.7 g). Each suspension was mixed under the ultrasound for 1 h to obtain homogeneous dispersion. Afterward, PVP (1.5 g) and PVDF (4.8 g) were added, and stirred at 60 °C for 8 h on a combined hot-plate magnetic-stirrer device. The dope solution was left to cool for at least 10 h for eliminating bubbles.

A custom-designed apparatus was used for HFM fabrication as illustrated by Fig. 1. A hollow fiber was formed when the casting solution was pressurized by nitrogen cylinder through a spinneret, with DI water being pumped by a syringe pump as the bore solution simultaneously. The nascent fiber was immersed in a coagulation bath of tap water. The

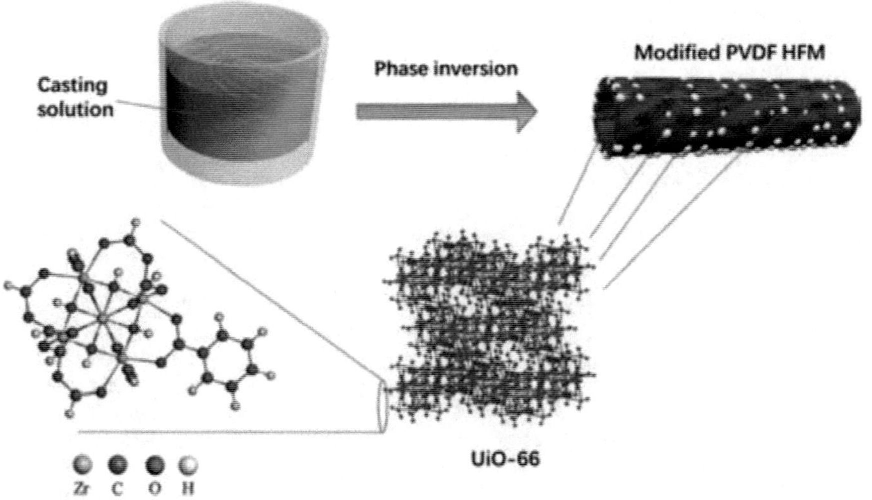

Fig. 1. A schematic representation for fabricating HFMs.

details of the fabrication parameters are given in Table S2. HFM modules assembled in the lab for filtration test were shown in Fig. S3. Briefly, ten pieces of hollow fibers with 45 cm in length were used to fabricate each module by gluing two ends of the fibers to a 1-milliliter pipette tip.

2.4. *Material characterization for UiO-66 and its ability for arsenate sorption*

UiO-66 particles were analyzed for the Brunauer–Emmett–Teller (BET) surface area by Coulter SA 3100 Surface Area and Pore Size Analyzer. The sample was outgassed at 100 °C for 2 h to remove adsorbed moisture and other gasses. The Fourier transform infrared spectra of the material were collected from 400 cm^{-1} to 4000 cm^{-1} with 32 scans at resolution of 2.0 cm^{-1} by using a Cary 660 FTIR (Agilent Technologies, Santa Clara, CA, USA).

Adsorption isotherms of arsenate were measured in vials with 10.00 mL suspensions containing 10.0 mg UiO-66 particles but various initial arsenate concentrations (50, 100, 200, 300, 500, 700, 800, 900, 1000 and 1100 mg/L). The pH values were controlled by 5 mM acetate buffer (pH 4.7) and the ionic strength was controlled by 50 mM sodium nitrate. The sealed vials were mixed on a vertical rotational shaker (60 rpm) for 24 h prior to filtration through 4.5 μm cellulose filter to collect filtrate that was analyzed for arsenate concentration by the atomic absorption spectroscopy (AAS, 210 VGP, Buck Scientific, USA). The detection limit for arsenic was 0.2 μg/L.

The effects of pH on arsenate adsorption were examined in a pH range of 1.5–10. Each vial with 10.0 mL arsenate solution (100 mg/L) was spiked with 10.0 mg UiO-66 particles, and then mixed on a rotational mixer for 24 h to reach equilibrium. The solution pH was adjusted to a pre-determined value by adding HNO_3 or NaOH solution every 8 h. All tests contained 50 mM $NaNO_3$ for the control of near constant ionic strength.

Arsenate adsorption kinetics were studied at pH 4.7 in solution with 1000 ppm of arsenate, 5 mM acetic buffer and 50 mM sodium nitrate. Vials containing 10 mL arsenate solution were each spiked with 10 mg UiO-66 particles, then after mixing for different times (1, 2, 5, 10, 20, 30, 40, 50, 60, and 120 min), the liquid samples were collected by filtration and analyzed for arsenate concentrations.

2.5. *Membrane characterization and performance assessment*

Functional groups of membrane surfaces were evaluated by attenuated total reflection-Fourier transform infrared spectroscopy (ATR-FTIR) on the Nicolet 4700 equipped with a multi-reflection Smart Performer® ATR accessory. The samples were dried in the air for 24 h before measurement. All spectra were captured from 64 scans at a resolution of 2 cm^{-1}. The HFM surface and cross-section were imaged by scanning electron microscopy (SEM) on a Quanta FEG 600 (FEI, Hillsboro, OR). The HFM samples were dried at 25 °C and the cross-section was prepared by freeze-fracturing in liquid nitrogen. The samples were sputtered with platinum to increase the sample conductivity. SEM images were analyzed using ImageJ to determine average pore diameters [31]. Emitech K575x Sputter Coater was used for coating a layer of silver on the membranes. The membranes were imaged by EI Quanta 600F ESEM, and their chemical compositions were analyzed by energy dispersive spectroscopy (EDS) using Bruker Quantax 200 Silicon Drift Detector under high vacuum mode (10 kV), a working distance of 10 mm, a spot size of 3 and an aperture 3. The topography of the membrane surface was imaged by Everhart-Thornley Detector under high vacuum mode, 10 kV with a working distance of 10 mm, a spot size of 3 and an aperture of 6. The membrane contact angle was measured by a video contact angle system (VCA-2500XE, AST products, Billerica, MA) using the sessile drop method.

The solution flux and solute rejection of membranes were evaluated by a cross-flow filtration system, and calculated by Eq. (1) and Eq. (2), respectively.

$$J = \frac{V_p}{A \cdot t} \quad (1)$$

$$R = \left(1 - \frac{C_p}{C_f}\right) \times 100\% \quad (2)$$

Here J (L/m^{2}h) is the flux of permeate solution, V_p (L) the permeate volume, A (m^2) the membrane area, t (h) the filtration time, R the rejection ratio, C_p the concentration of permeate, and C_f the concentrations of feed solution. The humic acid (HA) concentration was determined by UV–visible spectrophotometry at 254 nm and arsenic concentration by the atomic absorption spectroscopy.

2.6. *DFT computation*

First principles calculation based on density functional theory (DFT) was performed to help understand the interactions between UiO-66 adsorption sites and arsenate species. $DMOL^3$ code within Accelrys software package Materials Studio (v7.0) was adopted [32]. The UiO-66 crystal structure was obtained from the Cambridge Crystallographic Data Centre. In order to save computational resources, one cluster model with 12-coordinated Zr node $[Zr_6O_4(OH)_4(-CO_2)_{12}]$ and one linker molecular were modified to represent the unit-cell structure of UiO-66. The feasibility of using this kind of cluster model has been verified in other researches [33–35]. For DFT calculation process, spin-polarized generalized gradient approximation with exchange-correlation function of Perdew-Wang 1991 [36] was employed. The convergence thresholds for self-consistent field tolerance was set as 1.0×10^{-5} hartree (Ha). The DFT semi-core pseudopots [37] was selected for the core treatment. The double numerical plus basis set and a real-space orbital cutoff of 4.4 Å basis file were applied. Since the adsorption reaction occurred at the water/solid interface, the conductor-like screening model [38] was used to monitor solvents effects of water with its dielectric constant of 78.54. The above settings were consistently applied for all the computation processes discussed in this paper.

The structure of arsenate molecules and the cluster model were geometrically optimized before energy calculation. The thresholds of convergence for geometry optimization were arranged as medium quality with 1.0×10^{-5} Ha, 2.0×10^{-3} Ha/Å gradient and 5.0×10^{-3} Å displacement. After optimization, the arsenate molecules (i.e., H_3AsO_4, $H_2AsO_4^-$, $HAsO_4^{2-}$ and AsO_4^{3-}) were placed over four spatial locations of cluster model to learn the complexes (Fig. 7). The four sites respectively modelled the interaction between arsenic species and linker benzene rings (A), Zr (B), oxygen in Zr–O–C (C) and C–H groups on benzene rings (D). For each complexation, the binding energy (E_b, kJ/mol) was defined as the energy difference between the total energy of adsorption complexes ($E_{UiO\text{-}66\text{-}As}$, kJ/mol) in equilibrium state and the independent total energy of cluster model of UiO-66 ($E_{UiO\text{-}66}$, kJ/mol) and arsenate molecules (E_{As}, kJ/mol) before adsorption. The Eq. (3) represents the calculation process. A negative E_b value indicates that the adsorption process of arsenate molecules towards UiO-66 sites is an exothermic reaction. Moreover, a higher absolute value of E_b represents a more easily-occurred adsorption process.

$$E_b = E_{UiO-66-As} - (E_{UiO-66} + E_{As}) \qquad (3)$$

3. Results and discussion

3.1. Characteristics of UiO-66 particles

The Brunauer–Emmett–Teller (BET) specific surface area of UiO-66 was 918 m^2/g. In comparison with the specific surface areas of several arsenate adsorbents (Table 1), the surface area of UiO-66 is significantly higher, which could facilitate arsenate adsorption.

On the FTIR spectra of UiO-66 particles (Fig. S9), peaks at 748 and 667 cm^{-1} could be attributed to Zr-(μ3)O, and the peak at 1018 cm^{-1} was related to the bending vibrations of hydroxyl groups on metal oxide clusters (Zr–OH) [44]. The peak at 1400 cm^{-1} was in accordance with C–O bond and C–OH group of carboxylic acid [45]. The carboxylate group was confirmed by characteristics peaks at 1656 and 1700 cm^{-1}. The peaks at 1509 cm^{-1} and 1579 cm^{-1} belonged to C=C in an aromatic compound of organic linker, which was related to the BDC linker on UiO-66. The functional group information from the FTIR spectra was consistent with the chemical formula of UiO-66 ($Zr_6O_4(OH)_4(CO_2C_6H_4CO_2)_6$) [44], indicating successful synthesis of UiO-66.

Table 1
Comparison of arsenate adsorption among prevalent adsorbents and UiO-66.

Arsenate adsorbents	Surface area m^2/g
Iron-impregnated tablet ceramic [39]	45.6
Copper (II) oxide nanoparticles [40]	52.11
Superparamagnetic high-surface-area Fe_3O_4 nanoparticles [41]	179
Self-assembled mesoporous γ-Al_2O_3 spherical nanoparticles [42]	231–497
Magnetic mesoporous Fe/carbon aerogel structures [43]	487
UiO-66	918

3.2. Adsorption properties of UiO-66 towards arsenate

Fig. 2 shows the plot of arsenate adsorption data at pH 4.7 using Langmuir isotherm and Freundlich isotherm [46,47]. When the initial As (V) concentration was 1100 mg/L, the arsenate adsorption on UiO-66 reached 267 mg/g. Langmuir isotherm model fitted the data with a correlation coefficient (R^2) of 0.940. The values for Q_{max} and K_l were 363 mg/g and 0.0030 L/mg, respectively. When the data were fitted to Freundlich isotherm, the obtained R^2 was 0.982, K_f was 11.4 (mg/g)/(mg/L)^n and n was 0.468. Higher correlation value of R^2 obtained from Freundlich isotherm suggested that the adsorption of arsenate onto UiO-66 was a heterogeneous surface adsorption process.

Adsorption kinetics of As (V) on UiO-66 at pH 4.7 is shown in Fig. 3, which was evaluated under an initial concentration of As (V) at 1000 mg/L. A sharp decrease in arsenate concentration, from 1000 mg/L to 842 mg/L, occurred in the first 1 min, followed by a more graduate process reaching the adsorption equilibrium within 120 min. Fig. 4 (a) shows the plot for pseudo-first-order kinetic model. The correlation coefficient R^2 was 0.975, k_1 was 0.0265 (min^{-1}) and Q_e was 264.0 (mg/g). Fig. 4 (b) was the plot based on the pseudo-second-order model, showing that Q_e was 262.5 mg/g and k_2 was 0.0008 ($g{\cdot}mg^{-1}{\cdot}min^{-1}$) with an R^2 of 0.991. The pseudo-second-order model fitted the kinetics data better, which suggested a chemical adsorption. In addition, Q_e predicted from pseudo-second-order model was in a slightly better agreement with the adsorption capacity from the experiment.

Fig. 5 presents the effect of pH on the arsenate adsorption by the UiO-66 particles, tested at an initial arsenic concentration of 100 mg/L and solid loading of 1.00 g/L. A significant arsenate sorption occurred in a wide pH range from 1 to 10. The sorbed amount was the highest at pH 2.5, which was consistent with a published result [44]. The UiO-66 adsorbent performed well at acidic conditions, the amounts sorbed were all more than 70 mg/g at pH 2 to 3.5. Arsenate species in water varied with pH condition, H_3AsO_4, $H_2AsO_4^-$, $HAsO_4^{2-}$ and AsO_4^{3-} existed in water with pH values from low to high, respectively [48]. According to the literature [49], the pH_{zpc} (pH at the surface with a net zero charge) of UiO-66 was around 4. The positively charged surface of UiO-66 under acid condition (pH < 4) might benefit the adsorption of negatively charged arsenate species via electrostatic attraction [22]. As a result, arsenate adsorption amount was higher at the low pH condition. As pH value increased, the adsorbed amount was somewhat decreased but was always higher than 66 mg/g at pH < 10. This was significant because the pH values of ground water with arsenic concerns were mostly lower than 10, so UiO-66 could be applied to remove groundwater arsenic without pH adjustment. When pH was above 4.0, the surface charge of UiO-66 turned to negative. It was therefore, the other driving forces instead of electrostatic attraction were accounted for arsenate adsorption. For instance, published research [44] documented that the exchange between BDC ligands and arsenate species as well as formation of arsenic complexes by establishing Zr-O-As coordination bonds were responsible for the adsorption. The adsorption capacity, however, clearly dropped at pH 11 to approximately 46 mg/g, which could be caused by the competition between AsO_4^{3-} and OH^- in alkaline condition on the adsorption sites [33].

3.3. DFT computation results

The binding energies of various arsenate species with four possible binding sites, obtained from the DFT computation, were tabulated in Table 2. The E_b values were all negative, indicating that the adsorption of arsenate species to UiO-66 was a favorable process that could occur with arsenate species at all protonation levels and on four possible surface binding sites. The negative values were the biggest for C site,

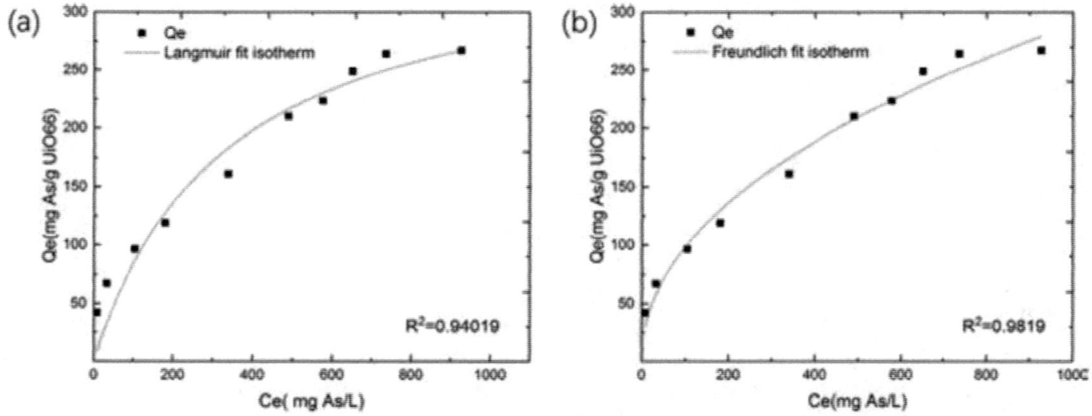

Fig. 2. (a) Langmuir isotherm and (b) Freundlich isotherm plot for the adsorption of arsenate onto UiO-66.

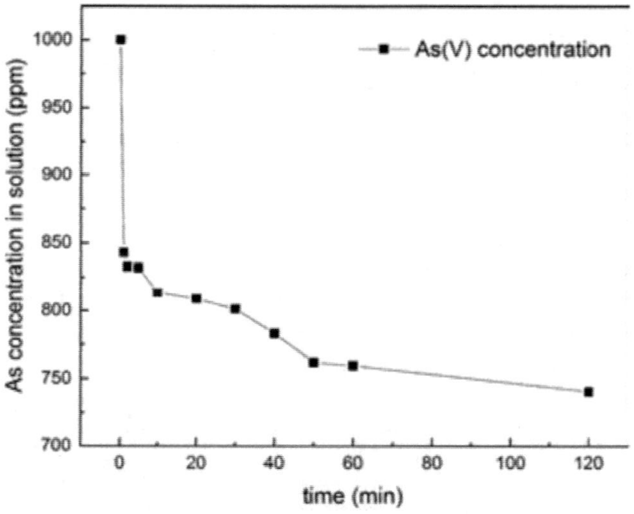

Fig. 3. Sorption kinetics of arsenate onto UiO-66.

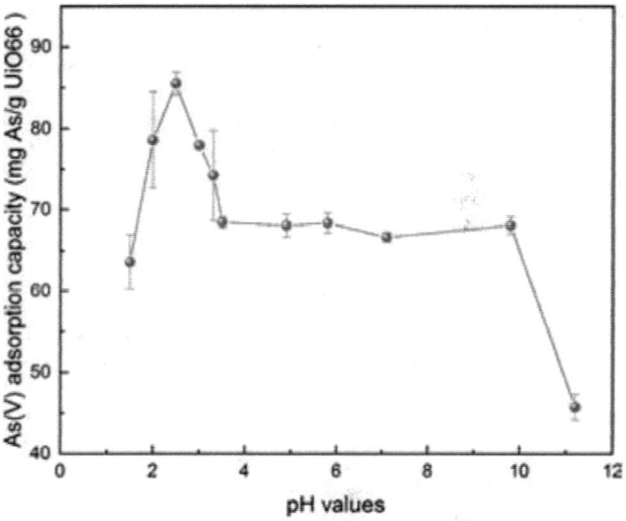

Fig. 5. Effect of pH on arsenate adsorption by UiO-66.

i.e., the oxygen of Zr-O-C, suggesting that this site was the most favored one for arsenic uptake. The fully-protonated arsenate (H_3AsO_4, $pK_{a1} = 2.20$) dominated at highly acidic conditions, its comparatively low E_b for each adsorption site implied a relatively low adsorption affinity toward UiO-66 than other arsenate species. As pH increases, $H_2AsO_4^-$ became the dominant species under the acidic to near neutral pHs from 2.2 to 6.97, and $HAsO_4^{2-}$ the dominant species at pHs from 6.97 to11.53. The binding energies of $H_2AsO_4^-$ with the surface sites were found to be significantly more negative than those of H_3AsO_4. These computational results were consistent with the experimental observation (Fig. 6), which showed that the sorbed amount at pH 1.8 was significantly lower than at pH 2.5. The surface of UiO66 was positively charged below pH 4, so the electrostatic interactions could have

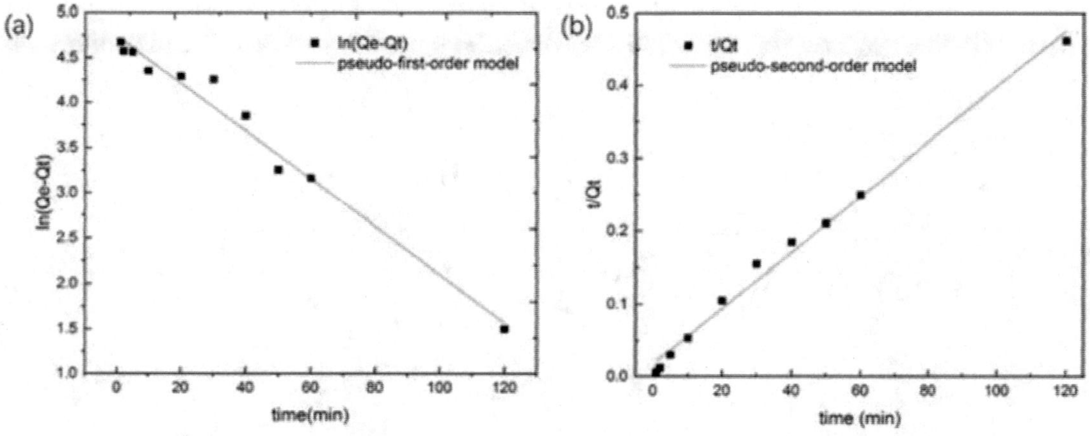

Fig. 4. Sorption kinetics plot using (a) Pseudo-first-order model and (b) Pseudo-second-order model.

Table 2
Binding energy (KJ/mol) between arsenate species and supposed adsorption sites on UiO-66.

	H_3AsO_4	$H_2AsO_4^{-1}$	$HAsO_4^{2-}$	AsO_4^{3-}
A	−17.97	−59.32	−78.89	−143.85
B	−3.50	−61.78	−69.63	−150.27
C	−48.05	−128.01	−112.37	−153.80
D	−26.49	−77.23	−65.44	−143.71

contributed to the highest adsorption around pH 2.5 as well. Once pH was higher than around 4, the surface of UiO-66 became negatively charged based on its pH_{zpc} [49], which could negatively impact sorption of $H_2AsO_4^-$ species. The amount of arsenic sorption observed at pH 3.5 and higher was indeed lower, likely resulted from the unfavorable electrostatic interactions. The sorbed amount decreased further when pH was higher than 10, this could be explained by the stronger repulsive interactions between an increasingly more negative UiO-66 surface and negative charged $HAsO_4^{2-}$.

Based on the DFT calculation, the biding energies of AsO_4^{3-} with the surface sites had the most negative values when compared with other arsenic species. However, AsO_4^{3-} was not the major species until pH was above11.53, and under this condition the surface would be even more negatively charged that repelled its sorption, so we believed its contribution to the overall arsenic sorption would likely be minimal.

3.4. Analysis of UiO-66 embedded PVDF HFMs

ATR-FTIR spectra were collected to confirm whether the UiO-66 particles were successfully embedded in the membrane (Fig. S11). Functional group Zr-(μ_3) O was indicated by the peak at 667 cm^{-1}. The peak at 667 cm^{-1} can be observed for M5, M10, M 20, M30 and M40. C=C in an aromatic compound of the organic linker was displayed by the peak at 1509 cm^{-1} and the peak at 1579 cm^{-1} can be obviously observed on the spectra for M30 and M40. The observation of these three peaks confirmed the existence of UiO-66 particles [44].

Fig. 6 depicts the surface and cross-section morphologies of the HFMs. The load of UiO-66 particles on the HFM influenced both porosity and pore size. With more UiO-66 particles embedded in the HFMs, the pore size of the membrane surface was increased. The surface micrographs indicated that more pores were observed on the membrane surfaces with UiO-66 (M30, M40) than the unmodified HFM (M0). More pores could be created on the membrane surface with a higher content of UiO-66. Similar phenomena were found by other researchers reporting that the load of zirconia to PVDF flat sheet membrane could result in more membrane pores [50]. The largest pore size of M40 was 600 nm, while the pore size of M0 was 30 nm (measured by ImageJ software). During the phase inversion process, the hydrophilic UiO-66 enhanced the demixing rate by facilitating the transport of water as a coagulant and the membrane pore size was also affected by the viscosity of dope solution as reported in the literature [27]. In addition, the porous structure of the modified HFM enabled arsenate solution to be in contact with the UiO-66 particles, which were embedded in the membrane matrix.

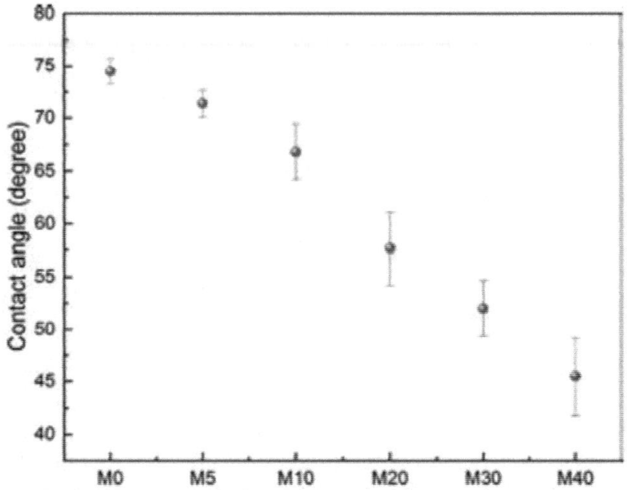

Fig. 7. Pure water contact angles of the pristine PVDF and various modified PVDF HFMs.

Hydrophilicity of membrane was obtained by measuring contact angle. As shown in Fig. 7, the contact angle of the HFM was significantly reduced from 75° to 46° as more UiO-66 particles were embedded into the mixed matrix membranes. The UiO-66 in the

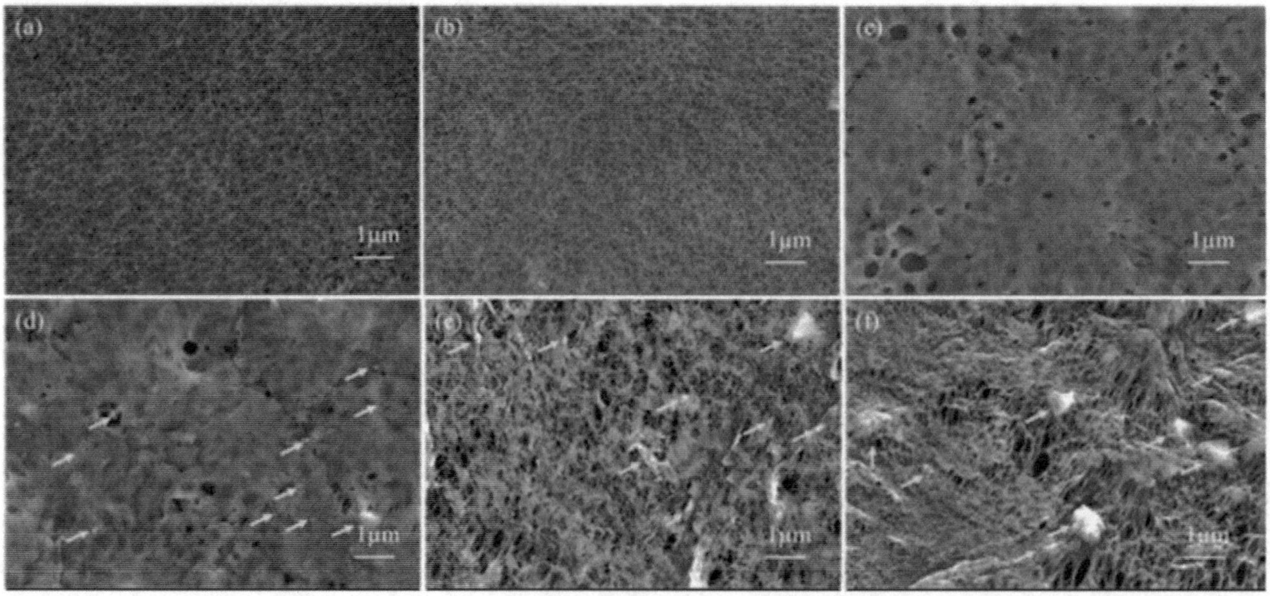

Fig. 6. Representative SEM images of surface morphologies of HFMs: (a) M0, (b) M5, (c) M10, (d) M20, (e) M30, (f) M40. (The scale bar is 1 μm in each surface image.)

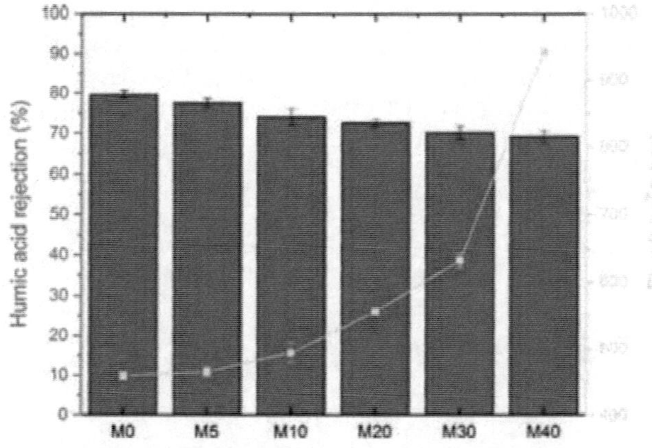

Fig. 8. Humic acid rejections and pure water fluxes of the pristine PVDF and various modified PVDF HFMs.

membrane had hydrophilic carboxylic acid and hydroxyl groups, which were hydrophilic and known to increase the membrane hydrophilicity [51].

The HFMs in this study has a pore size range for ultrafiltration and thus are capable of removing suspended solids. Humic acid is a natural organic matter in water resources including upland streams, dystrophic lakes, and ocean water, which may cause fouling of membranes. The removal efficiency of humic acid was therefore tested in this work. The membrane rejection for humic acid and pure water flux are showed in Fig. 8. The humic acid rejections were all above 69%. As the amount of UiO-66 particles in the membrane was increased, the rejection decreased accordingly. Pristine PVDF HFM (M0) had a rejection value of 79%, while M40 had only 69%. UiO-66 particles reduced the rejection of humic acid due to the enhancement of the pore size of the membrane, which was consistent with the results in Fig. 6. The membrane pure water flux increased with the addition of UiO-66 in the membrane matrix due to the improvement of membrane hydrophilicity and the increase of the membrane pore size.

Fig. 9 (a) shows the filtration results of arsenic solution for both pristine PVDF HFM (M0) and modified HFMs (M5, M10). The influent arsenate concentration was 100 ppb and the filtration rate was controlled at 0.395 L/h. At initial stage, the effluent concentration of PVDF HFM (M0) was higher than 10 ppb of arsenic. The filtration curve of M0 leveled off with only 0.2 L passing through the membrane, by this time the arsenate concentration of the influent and the effluent became the same. This was as expected because the membrane pore size (20 nm) was much larger than arsenate anions, so the pristine PVDF HFM could not reject arsenate. M5 and M10 containing UiO-66 could treat about

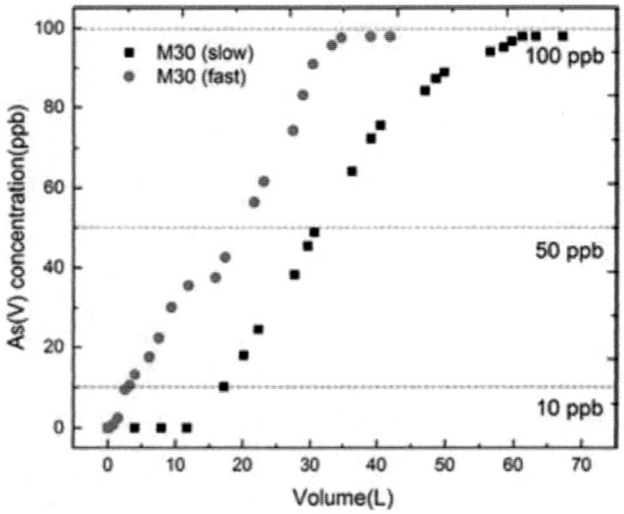

Fig. 10. Filtration curves of M30 under slow flow rate (0.395 L/h) and fast flow rate (1.444 L/h).

Table 3
Regeneration results of M40 by using 4 types of solution.

	NaOH	Na_2CO_3	NaClO	$NaHCO_3$
Concentration	0.1 (mol/L)	0.1 (mol/L)	0.1 (mol/L)	0.1 (mol/L)
Results	Damaged	Damaged	7.11 L	26.4 L
Recovery rate	0%	0%	20%	76%

0.1 L arsenic solution (100 ppb), keeping the effluent concentration below the arsenic drinking water standard of 10 ppb. The membrane's arsenate removal efficiency was primarily determined by the UiO-66 particles. While M5 and M10 could treated more arsenic solution than M0, their performance was not so good because of low amounts of UiO-66. Fig. 9 (b) presents the filtration results of M20, M30 and M40 with the same filtration rate of 0.395 L/h. As more solution passed through the membranes, arsenate concentration of the effluent increased. Prior to the breakthrough of the effluent exceeding 10 ppb, M20, M30 and M40 could treat 8 L, 17 L and 34 L, respectively. The arsenate removal capacity of the membrane was gradually increased as the UiO-66 content in the membrane increased. As the UiO-66 particles gradually became saturated with arsenic, the ability of the membranes to remove arsenate was lost and effluent arsenic concentrations reached the influent concentration. This result was consistent with the reported trend in the literature [28].

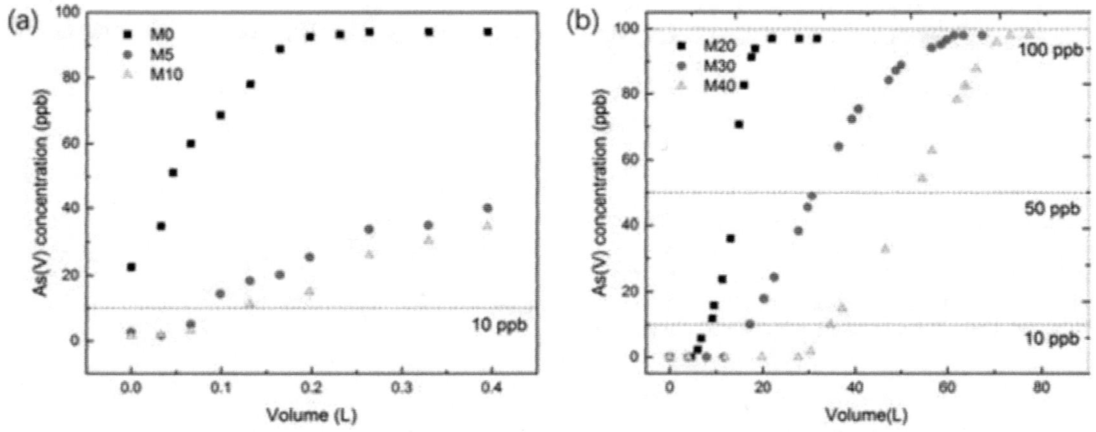

Fig. 9. Filtration curves of the pristine PVDF and various modified PVDF HFMs: (a) M0, M5 and M10; (b) M20, M30 and M40.

In this research, the highest weight ratio of UiO-66 particles to PVDF was controlled at 0.4, under this condition, we could easily control the pore size of the mixed matrix PVDF membranes with UiO-66. No UiO-66 was detected in the filtrate, which indicated the excellent long-term stability of the HFMs under the continuous filtration conditions. For drinking purpose, it is recommended that a person should take eight 8-ounce glasses of water per day, which equals about 2 L [52]. The amount of water treated by M20 could satisfy one person's drinking need for one week, and the water produced by M40 could support one person's need for more than two weeks.

In addition to the amounts of UiO-66 in the membrane, the flow rate also affected the ability of membrane for arsenate removal. Fig. 10 shows the arsenic removal by M30 under two different flow rates of membrane modules tested at 100 ppb original arsenic concentration. At the fast flow rate of 1.44 L/h (membrane flux 102 L/m^2.h), M30 could only treat about 3 L of water with arsenate to meet the 10 ppb standard. While at the slow flow rate of 0.395 L/h (membrane flux 27.9 L/m^2.h), the membrane could treat 17 L water prior for the effluent to reaching 10 ppb. UiO-66 in the membrane had longer contact time with arsenate solution with a lower filtration rate, resulting in a higher capacity for arsenate removal. For some developing countries that might use drinking water arsenic standard higher than 10 ppb [53], the mixed matrix membranes could be used for a longer duration.

As with any sorption-based process, the ability of mixed matrix membrane for arsenate removal is lost when the reactive sites on the UiO-66 particles are saturated; therefore, regeneration of the used membrane, if possible, would make the practical application more economical. Table 3 presents the regeneration results of the PVDF HFM M40 by washing with 4 types of solution. It was found that the membranes could be damaged by NaOH and Na_2CO_3. To illustrate, regeneration by NaOH washing resulted in membrane damage and transformed the membranes from light white color to dark brown (Fig. S7 (a)). When the used membranes were treated by NaClO and $NaHCO_3$, no apparent change was observed, and the color remained the same as the original membrane (Fig. S7 (b)).

Following regeneration, the removal capacity of membrane could be partially recovered for arsenate removal as demonstrated by Fig. S8. When the exhausted M40 membrane was regenerated by $NaHCO_3$, the amount of water that could be treated (prior to 10 μg/L in the effluent) was 26.4 L, which represented a 76% recovery in comparison with 34 L for the pristine membrane. When NaClO was used, only 7.11 L of clean water with lower than 10 ppb arsenic could be produced, indicating a recovery of 20%.

4. Conclusions

The surface area of UiO-66 was observed to be 918 m^2/g via BET analysis. The batch adsorption experiment of UiO-66 particles revealed that arsenate adsorption was effective in a wide pH range of 1 to 10. The adsorption equilibrium could be reached within 120 min at the initial concentrations of 100 ppb, and pseudo-second-order model could represent the adsorption kinetics well. At pH 4.7, the maximum arsenate adsorption capacity was 267 mg/g, the adsorption isotherm could be fitted by the Freundlich model. DFT simulation suggested that the sorption could occur on different sites on UiO-66, with oxygen in Zr-O-C being the most favorable site. UiO-66 embedded PVDF HFMs were synthesized via a phase inversion method. Since the pristine PVDF HMF could not adsorb arsenate, the removal of arsenic is by sorption onto the UiO-66 surfaces. The volume of clean water that could be treated increased with the content of UiO-66 in the membrane. The mixed matrix membrane could be used for efficient adsorption removal of arsenate in a wide pH range.

Acknowledgements

We would like to thank Prof. Pingle Liu and Dr. Shaofeng Xiong at College of Chemical Engineering, Xiangtan University for their co-operation on DFT calculation. This work was supported by the United States Geological Survey through Missouri Water Resources Research Center and the Program for Guangdong Introducing Innovative and Entrepreneurial Teams (2017ZT07Z479).

Appendix A. Supplementary data

Supplementary data to this article can be found online at https://doi.org/10.1016/j.cej.2019.122921.

References

[1] P. Song, Z. Yang, G. Zeng, X. Yang, H. Xu, L. Wang, R. Xu, W. Xiong, K. Ahmad, Electrocoagulation treatment of arsenic in wastewaters: a comprehensive review, Chem. Eng. J. 317 (2017) 707–725.
[2] T.I. Aimen Sohail, Ramisha Ashraf, Isra Shahid, Ghazala Yaqub, Latest technologies for arsenic removal in drinking water, Pharm. Chem. J. 3 (2016) 11–16.
[3] F. Di Natale, A. Erto, A. Lancia, D. Musmarra, Experimental and modelling analysis of As(V) ions adsorption on granular activated carbon, Water Res. 42 (2008) 2007–2016.
[4] K.N. Scott, J.F. Green, H.D. Do, S.J. McLean, Arsenic removal by coagulation, J.-Am. Water Works Assoc. 87 (1995) 114–126.
[5] J.G. Hering, P.-Y. Chen, J.A. Wilkie, M. Elimelech, S. Liang, Arsenic removal by ferric chloride, J.-Am. Water Works Assoc. 88 (1996) 155–167.
[6] L.S. McNeill, M. Edwards, Arsenic removal during precipitative softening, J. Environ. Eng. 123 (1997) 453–460.
[7] **D. Clifford, Ghurye, G., Tripp, A., and Tong, Jian, Field studies on arsenic removal in Albuquerque, New Mexico using the University of Houston/EPA Mobile Drinking Water Treatment Research Facility, Final Project Report for the City of Albuquerque, (1997) 126.**
[8] W. Driehaus, M. Jekel, U. Hildebrandt, Granular ferric hydroxide—a new adsorbent for the removal of arsenic from natural water, J. Water Supply: Res. Technol.-Aqua 47 (1998) 30–35.
[9] Z. Gu, J. Fang, B. Deng, Preparation and evaluation of GAC-based iron-containing adsorbents for arsenic removal, Environ. Sci. Technol. 39 (2005) 3833–3843.
[10] S.D. Chang, Hector Ruiz, William D. Bellamy, Carl W. Spangenberg, Debra L. Clark, Removal of arsenic by enhanced coagulation and membrane technology. No. CONF-940789-, Am. Soc. Civ. Eng., New York, NY (United States) (1994).
[11] P. Brandhuber, G. Amy, Alternative methods for membrane filtration of arsenic from drinking water, Desalination 117 (1998) 1–10.
[12] M. Kang, M. Kawasaki, S. Tamada, T. Kamei, Y. Magara, Effect of pH on the removal of arsenic and antimony using reverse osmosis membranes, Desalination 131 (2000) 293–298.
[13] J. Fang, B. Deng, Rejection and modeling of arsenate by nanofiltration: Contributions of convection, diffusion and electromigration to arsenic transport, J. Membr. Sci. 453 (2014) 42–51.
[14] J. Fang, B. Deng, Arsenic rejection by nanofiltration membranes: Effect of operating parameters and model analysis, Environ. Eng. Sci. 31 (2014) 496–506.
[15] A. Seidel, J.J. Waypa, M. Elimelech, Role of charge (donnan) exclusion in removal of arsenic from water by a negatively charged porous nanofiltration membrane, Environ. Eng. Sci. 18 (2001) 105–113.
[16] S.S. Tripathy, A.M. Raichur, Enhanced adsorption capacity of activated alumina by impregnation with alum for removal of As(V) from water, Chem. Eng. J. 138 (2008) 179–186.
[17] M.A. Al-Ghouti, I. Yousef, R. Ahmad, A.M. Ghrair, A.A. Al-Maaitah, Characterization of diethyl ether adsorption on activated carbon using a novel adsorption refrigerator, Chem. Eng. J. 162 (2010) 234–241.
[18] M. Jian, B. Liu, G. Zhang, R. Liu, X. Zhang, Adsorptive removal of arsenic from aqueous solution by zeolitic imidazolate framework-8 (ZIF-8) nanoparticles, Colloids Surf., A 465 (2015) 67–76.
[19] T.A. Vu, G.H. Le, C.D. Dao, L.Q. Dang, K.T. Nguyen, Q.K. Nguyen, P.T. Dang, H.T.K. Tran, Q.T. Duong, T.V. Nguyen, G.D. Lee, Arsenic removal from aqueous solutions by adsorption using novel MIL-53(Fe) as a highly efficient adsorbent, RSC Adv. 5 (2015) 5261–5268.
[20] Y. Khabzina, J. Dhainaut, M. Ahlhelm, H.J. Richter, H. Reinsch, N. Stock, D. Farrusseng, Synthesis and shaping scale-up study of functionalized UiO-66 MOF for ammonia air purification filters, Ind. Eng. Chem. Res. 57 (2018) 8200–8208.
[21] Z.J. Lin, H.Q. Zheng, Y.N. Zeng, Y.L. Wang, J. Chen, G.J. Cao, J.F. Gu, B. Chen, Effective and selective adsorption of organoarsenic acids from water over a Zr-based metal-organic framework, Chem. Eng. J. 378 (2019).
[22] C. Wang, M. Lee, X. Liu, B. Wang, J. Paul Chen, K. Li, A metal-organic framework/alpha-alumina composite with a novel geometry for enhanced adsorptive separation, Chem. Commun. (Camb.) 52 (2016) 8869–8872.
[23] L. Valenzano, B. Civalleri, S. Chavan, S. Bordiga, M.H. Nilsen, S. Jakobsen, K.P. Lillerud, C. Lamberti, Disclosing the complex structure of UiO-66 metal organic framework: a synergic combination of experiment and theory, Chem. Mater. 23 (2011) 1700–1718.
[24] T.Y. Liu, H.G. Yuan, Y.Y. Liu, D. Ren, Y.C. Su, X. Wang, Metal-organic framework nanocomposite thin films with interfacial bindings and self-standing robustness for high water flux and enhanced ion selectivity, ACS Nano 12 (2018) 9253–9265.
[25] D. Wisser, F.M. Wisser, S. Raschke, N. Klein, M. Leistner, J. Grothe, E. Brunner,

S. Kaskel, Biological chitin-MOF composites with hierarchical pore systems for air-filtration applications, Angew. Chem. Int. Ed. Engl. 54 (2015) 12588–12591.
[26] D. Wang, K. Li, W.K. Teo, Preparation and characterization of polyvinylidene fluoride (PVDF) hollow fiber membranes, J. Membr. Sci. 163 (1999) 211–220.
[27] P. Wan, J. Yin, B. Deng, Seven-bore hollow fiber membrane (HFM) for ultrafiltration (UF), Chem. Eng. Res. Des. 128 (2017) 240–247.
[28] J. He, T. Matsuura, J.P. Chen, A novel Zr-based nanoparticle-embedded PSF blend hollow fiber membrane for treatment of arsenate contaminated water: material development, adsorption and filtration studies, and characterization, J. Membr. Sci. 452 (2014) 433–445.
[29] J. Yin, G. Zhu, B. Deng, Multi-walled carbon nanotubes (MWNTs)/polysulfone (PSU) mixed matrix hollow fiber membranes for enhanced water treatment, J. Membr. Sci. 437 (2013) 237–248.
[30] P. Wan, Z. Zhang, B. Deng, Photocatalytic polysulfone hollow fiber membrane with self-cleaning and antifouling property for water treatment, Ind. Eng. Chem. Res. 58 (2019) 3339–3348.
[31] J. Yin, E.S. Kim, J. Yang, B. Deng, Fabrication of a novel thin-film nanocomposite (TFN) membrane containing MCM-41 silica nanoparticles (NPs) for water purification, J. Membr. Sci. 423–424 (2012) 238–246.
[32] B. Delley, An all-electron numerical method for solving the local density functional for polyatomic molecules, J. Chem. Phys. 92 (1990) 508–517.
[33] J. Wei, W. Zhang, W. Pan, C. Li, W. Sun, Experimental and theoretical investigations on Se(iv) and Se(vi) adsorption to UiO-66-based metal-organic frameworks, Environ. Sci. Nano 5 (2018) 1441–1453.
[34] J. Li, Q. Wu, X. Wang, Z. Chai, W. Shi, J. Hou, T. Hayat, A. Alsaedi, X. Wang, Heteroaggregation behavior of graphene oxide on Zr-based metal-organic frameworks in aqueous solutions: a combined experimental and theoretical study, J. Mater. Chem. A 5 (2017) 20398–20406.
[35] K. Liu, S. Zhang, X. Hu, K. Zhang, A. Roy, G. Yu, Understanding the adsorption of PFOA on MIL-101(Cr)-based anionic-exchange metal-organic frameworks: comparing DFT calculations with aqueous sorption experiments, Environ. Sci. Technol. 49 (2015) 8657–8665.
[36] J.P. Perdew, Y. Wang, Accurate and simple analytic representation of the electron-gas correlation energy, Phys. Rev. B 45 (1992) 13244–13249.
[37] B. Delley, Hardness conserving semilocal pseudopotentials, Phys. Rev. B Condens. Matter Mater. Phys. 66 (2002) 1551251–1551259.
[38] A. Klamt, G. Schüürmann, COSMO: a new approach to dielectric screening in solvents with explicit expressions for the screening energy and its gradient, J. Chem. Soc., Perkin Trans. 2 (1993) 799–805.
[39] R. Chen, Z. Zhang, Z. Lei, N. Sugiura, Preparation of iron-impregnated tablet ceramic adsorbent for arsenate removal from aqueous solutions, Desalination 286 (2012) 56–62.
[40] A. Goswami, P. Raul, M. Purkait, Arsenic adsorption using copper (II) oxide nanoparticles, Chem. Eng. Res. Des. 90 (2012) 1387–1396.
[41] L. Feng, M. Cao, X. Ma, Y. Zhu, C. Hu, Superparamagnetic high-surface-area Fe 3 O 4 nanoparticles as adsorbents for arsenic removal, J. Hazard. Mater. 217 (2012) 439–446.
[42] A.K. Patra, A. Dutta, A. Bhaumik, Self-assembled mesoporous γ-Al_2O_3 spherical nanoparticles and their efficiency for the removal of arsenic from water, J. Hazard. Mater. 201 (2012) 170–177.
[43] Y.-F. Lin, J.-L. Chen, Magnetic mesoporous Fe/carbon aerogel structures with enhanced arsenic removal efficiency, J. Colloid Interface Sci. 420 (2014) 74–79.
[44] C. Wang, X. Liu, J.P. Chen, K. Li, Superior removal of arsenic from water with zirconium metal-organic framework UiO-66, Sci. Rep. 5 (2015) 16613.
[45] H.R. Abid, H. Tian, H.-M. Ang, M.O. Tade, C.E. Buckley, S. Wang, Nanosize Zr-metal organic framework (UiO-66) for hydrogen and carbon dioxide storage, Chem. Eng. J. 187 (2012) 415–420.
[46] I. Langmuir, The adsorption of gases on plane surfaces of glass, mica and platinum, J. Am. Chem. Soc. 40 (1918) 1361–1403.
[47] H. Freundlich, Über die Adsorption in Lösungen, Z. Phys. Chem. (1907) 385.
[48] S.A. Baig, T. Sheng, Y. Hu, J. Xu, X. Xu, Arsenic removal from natural water using low cost granulated adsorbents: a review, CLEAN–Soil Air, Water 43 (2015) 13–26.
[49] C. Chen, D. Chen, S. Xie, H. Quan, X. Luo, L. Guo, Adsorption behaviors of organic micropollutants on zirconium metal-organic framework UiO-66: analysis of surface interactions, ACS Appl. Mater. Interfaces 9 (2017) 41043–41054.
[50] Y.-M. Zheng, S.-W. Zou, K.N. Nanayakkara, T. Matsuura, J.P. Chen, Adsorptive removal of arsenic from aqueous solution by a PVDF/zirconia blend flat sheet membrane, J. Membr. Sci. 374 (2011) 1–11.
[51] S. Kasemset, L. Wang, Z. He, D.J. Miller, A. Kirschner, B.D. Freeman, M.M. Sharma, Influence of polydopamine deposition conditions on hydraulic permeability, sieving coefficients, pore size and pore size distribution for a polysulfone ultrafiltration membrane, J. Membr. Sci. 522 (2017) 100–115.
[52] H. Valtin, "Drink at least eight glasses of water a day." Really? Is there scientific evidence for "8 × 8"? Am. J. Physiol. Regul. Integr. Comp. Physiol. 283 (2002) R993–R1004.
[53] K. Henke, Arsenic: Environmental Chemistry, Health Threats and Waste Treatment, John Wiley & Sons, 2009.

565995 Functionalized Copper-Metal Organic Framework Ultrafiltration Membrane with Improved Dye Rejection and Flux for Water Treatment

Dyes are known as one of the serious major water pollutants. Traditional methods cannot decolorize dyes. Dyes are current threats to human and environmental health and pose a risk to the safety of groundwater and surface water. In this study, we developed and compared two surface functionalization methods including the dip-coating and in-situ filtration processes for a polyethersulfone (PES) ultrafiltration (UF) membrane by coating a polydopamine layer incorporated with the copper-based metal-organic framework (MOF). The membranes and fabricated Cu-MOF nanoparticle structure characterized by transmission electron microscopy (TEM), scanning electron microscopy (SEM), X-ray powder diffraction (XRD), X-ray photoelectron spectroscopy (XPS), Atomic Force Microscope (AFM), Fourier-transform infrared spectroscopy (FTIR), Ultraviolet-visible spectroscopy (UV-Vis), and Zeta potential. A cross-flow filtration unit employed for antifouling, dye removal, and permeation examination of the modified membranes. The addition of Cu-MOF nanoparticles through dip-coating approach increased the water flux as well as dye rejection up to 90% compared to pristine PES UF membrane.

561984 Mitigation of Inorganic Fouling on Pressure Retarded Osmosis (PRO) Membranes By Coagulation Pretreatment of the Wastewater Concentrate Feed

Utilization of wastewater or wastewater concentrate as the feed solution in pressure retarded osmosis (PRO) via thin film composite (TFC) membranes causes rapid and significant fouling within the porous substrate and onto the polyamide selective layer facing the substrate. Scaling of calcium phosphate salts has been identified as the predominant cause for the severe PRO membrane fouling. In this study, an acidic coagulant, $AlCl_3$, and a caustic coagulant, $NaAlO_2$, are employed to remove the phosphate from the wastewater concentrate. Their optimal dosages are investigated in order to most effectively remove phosphate and maintain a mild pH environment. 1000 ml of $AlCl_3$ and $NaAlO_2$ pretreated and untreated wastewater concentrates are then used as the feed solutions in the PRO fouling studies. The untreated wastewater concentrate that is saturated with calcium phosphate causes a dramatic flux reduction of 45.2% during the first 360 min of the tests and further reduces to 30.7% after a 25% recovery of the wastewater concentrate. Results show that both $AlCl_3$ and $NaAlO_2$ pretreatments can increase the initial water flux to 25.5 LMH and 24.8 LMH at 20 bar respectively, and increase the normalized water flux to 66% and 64% respectively upon a total permeation volume of 250 ml. However, the $AlCl_3$ pretreatment shows a lower water flux recovery after a hydraulic backwash and a faster flux reduction in the repeated fouling tests. Elemental analyses reveal that $AlCl_3$ is less effective in silica removal from the wastewater concentrate, which causes silica condensation underneath the selective layer of the TFC membrane. $NaAlO_2$ appears to be the more effective coagulant in both phosphate and silica removal for a prolonged PRO operation. Moreover, a critical water flux framework is developed to analyze the PRO calcium phosphate fouling, in order to determine the stable water flux at a given phosphate concentration or the maximum allowable phosphate concentration to achieve the target PRO flux and power density. This pretreatment will be implemented on a 250 m^3/day PRO pilot with reverse osmosis seawater brine and wastewater concentrate as the draw and feed pair, to demonstrate osmotic power generation in Singapore.

572410 Polyamide/CNT Thin Film Nanocomposite (TFN) RO Membranes with Improved Boron Rejection

Thin film nanocomposite (TFC) RO membranes have remarkably improved in terms of salt removal; however, removal of boron from seawater is an emerging problem. According to World Health Organization (WHO), boron concentration should not exceed 2.4 ppm for potable water and 0.5 ppm for irrigation water used for crops such as citrus fruits and nuts. Since boron is present in seawater as boric acid and its effective diameter is close to hydrogen-bonded cluster of water molecules, conventional TFC membranes fail to achieve effective boron removal. Incorporating nanomaterials into TFC membranes has potential for improving selectivity as well as water flux and anti-fouling properties. However, this requires optimizing the conditions for loading, size, type, and functional groups of the nanomaterials as well as the interfacial polymerization and post-treatment protocols. In this study, we investigate the influence of embedding carbon nanotubes functionalized with different functional groups into polyamide RO membranes on their boron rejection performance. Deposition and alignment of SWCNTs on a commercial polysulfone UF membrane was carried out by vacuum filtration. The polyamide layer was prepared by interfacial polymerization using m-phenylene diamine (MPD) and tri-mesoyl chloride (TMC) as monomers. Membranes were characterized by SEM, FTIR, XPS, polarized Raman spectroscopy, and crossflow filtration tests. Results showed that concentration of CNT significantly affects dispersion and alignment of CNTs inside the polymeric matrix as well as homogeneity, thickness and crosslinking ratio of polyamide layer. The salt and boron rejection performance of the TFN membranes depends on the functional group and CNT loading; some membranes exhibited increased performance relative to neat polyamide membranes.

(This work is supported by the Scientific and Technological Research Council of Turkey (TUBITAK) through Project No. 114Y165.)

577102 Graphene Oxide-Titanium Dioxide Nanohybrids for Designing Nanofiltration Membranes with Enhanced Permeability and Rejection in Dye Desalination

This project aims to hybridize TiO_2 onto graphene oxide (GO) nanosheets and use for preparation of loose nanofiltration membrane for dye desalination application. The effect of the functionalization time on the water flux, dye and salt rejections, and physical and chemical properties of the obtained membranes were studied. Results indicated that the TiO_2-functionalized GO membranes (F-GO) could provide very high water flux, about 4-6 times of the GO nanofiltration membrane, while keeping high dye rejection (~99% for Congo Red, Direct Red, and Evan Blue). Furthermore, F-GO membrane showed lower rejection towards Na_2SO_4 salts indicating the successful engineering of the channel size achieved by the functionalization to provide a membrane with high water flux, high dye rejection, and low salt rejection. The differences in water flux and dye rejection has been attributed to the channel size alterations by spacing created by the TiO_2 nanomaterials between layers of GO membrane.

566500 Facilitated Transport Membranes with Tunable Amine-CO_2 Chemistry for Highly Selective CO_2/H_2 Separation

CO_2-selective, amine-containing facilitated transport membranes are of great interest for syngas purification since high-pressure H_2 can be retained upon CO_2 removal. A variety of amine-containing polymers have shown decent chemical and thermal stability at aggressive conditions, but their CO_2/H_2 separation properties are largely limited by the hydrostatic compaction and severe carrier saturation associated with the high syngas pressure. Herein, we report a new approach to enhance the CO_2 permeability by manipulating the steric hindrance of the amine carriers. A series of a-aminoacids with different alkyl or hydroxyethyl substituents are deprotonated by 2-(1-piperazinyl)ethylamine, leading to nonvolatile amine carriers with different degrees of steric hindrance. For hosting the low MW amine carriers, a highly crosslinked poly(vinyl acetal) is synthesized as a water-swellable polymer network. In order to avoid membrane compaction, perforated graphene oxide mono-sheets are dispersed as reinforcement fillers. In the presence of moisture, a bulkier alkyl substituent to the amino site (increasing steric hindrance) destabilizes the carbamate adduct and thus drastically increases the chemisorption of CO_2, while the incorporation of ethylene oxide groups provides additional physisorption of CO_2. The enhanced CO_2 solubility significantly mitigates the carrier saturation behavior, and an unprecedented CO_2/H_2 selectivity greater than 100 is demonstrated at 107 °C and 12.5 atm of CO_2 partial pressure. As the CO_2 partial pressure reduces to 0.4 atm, a less hindered amine yields a higher reactive diffusivity of CO_2, resulting in a CO_2 permeance of 435 GPU with a selectivity greater than 500. These reaction-mediated polymeric membranes are well above the theoretical upper bound, and they are of great interest for designing a highly-selective membrane process for syngas purification.

557302 Polymer Genome: An Open Tool for the Polymer Membrane Community

Predicting polymer permeabilities *a priori* is a long-standing challenge within the membrane research community that has important applications for membrane process design and ultimately widespread adoption of membrane technology. From early attempts based on free volume and cohesive energy to more recent group contribution methods, the ability to predict membrane permeability has improved in terms of accuracy. However, these models usually stay "within the paper", i.e., limited model details are provided to the wider community such that adoption of these predictive platforms is limited. In this work, we combined an advanced structure fingerprinting method with a large experimental database of gas permeabilities to give unprecedented prediction ability over a large range of polymer classes. No prior knowledge of the polymer is needed for the prediction other than the repeating unit chemical formula. In addition, we have incorporated this model into the existing Polymer Genome project to make it open to the membrane research community. In this interactive presentation, we will also demonstrate how membrane researchers can benefit from Polymer Genome to design advanced polymer for gas separation as well as accompanying processing parameter predictions.

569342 The Perfluoropolymer Upper Bound

Perfluoropolymers have fundamentally distinct thermodynamic partitioning properties compared to those of their hydrocarbon counterparts. However, current upper bound theory assumes hydrocarbon solubility behavior for all polymers, regardless of their chemical composition. In this work, the effect of variable solubility behavior on upper bound position is critically assessed for perfluoropolymers and hydrocarbon-based polymers. Based off distinct solubility relationships, theoretical perfluoropolymer upper bounds are established, showing a positive shift of the upper bound front for certain gas pairs, including N_2/CH_4, He/H_2, He/N_2, He/CH_4, and He/CO_2, as a result of beneficial solubility selectivity. The perfluoropolymer upper bounds are compared to currently available experimental data. Within the framework of the solution-diffusion model, the approaches of achieving solubility selectivity through the use of perfluoropolymers and improving diffusion selectivity through the use of rigid hydrocarbon-based polymers are compared for their impact on the upper bound position. This comparative analysis demonstrates the significant benefit that can be achieved by considering both the chemical and morphological features of solid-state macromolecules when designing membrane materials. Finally, promising future applications for perfluoropolymers are discussed and preliminary data exploring the effect of fluorination on sorption and diffusion is presented.

555881 Polymers with Intrinsic Micro Porosity (PIM-1) Based Compatible Blend Membranes for CO_2 Separation Applications

Polymers with intrinsic micro porosity (PIM-1) based compatible blend membranes for CO_2 separation applications.

Ali K. Sekizkardes[1,2], Samir Budhathoki,[1,2] Surendar Venna,[1,2] Victor Kusuma,[1,2] David Hopkinson [1]

[1] U.S. Department of Energy National Energy and Technology Laboratory,

Pittsburgh, Pennsylvania 15236, USA.

[2] Leidos Research Support Team, Pittsburgh, Pennsylvania 15236, USA.

Microporous polymers such as polymers with intrinsic microporosity (PIMs) have shown great potential in gas separation membrane applications due to their facile synthesis, easy processability and high gas permeability. However, the polymer structure of PIM-1 consists of non-polar and large micro pores (>1nm) resulting in lower CO_2 selectivity and weak mechanical properties when processed in membrane films. We address these drawbacks of PIM-1 by developing blend membranes fabricated with PIM-1 and polyphosphazenes. Characterization of these membranes showed an improvement in the brittle film properties of PIM-1 by greatly altering the strain to failure of PIM-1 from 1% to as high as 10.5%, which is indicative of a very flexible and durable film formation. In addition, very high CO_2 permeability (~2000 Barrer) and selectivity (~30) were recorded for the membranes. The membranes lie on the Robeson upper bound which is an empirical limit for the gas separation performance of polymer membranes. Here we also studied the structure property relationship between these two blending polymers, both experimentally and computationally, by altering the chemical structure of the polymers. Membrane fabrication, characterization and gas transport testing will be presented.

Reference

Sekizkardes, A. K.; Kusuma, V. A.; McNally, J. S.; Gidley, D. W.; Resnik, K.; Venna, S. V.; Hopkinson, D. "Microporous polymeric composite membranes with advanced film properties: pore intercalation yields excellent CO_2 separation performance" *Journal of Materials Chemistry A* 2018, 6, 22472-22477.

567466 Facile Infusion of Size-Sieving Macrocyclic 3D Cavities into Polymers of Intrinsic Microporosity for Ultra-Efficient Gas Separation and Enhanced Stability Against Aging

Efficient membrane-based gas separation typically requires strong molecular-sieving characteristic in the constituting materials, but achieving excellent size-sieving effect in polymeric membranes without sacrificing their favorable processability, scalability or mechanical stability has remained a challenge. As a new and facile approach to enhance the molecular-sieving nature of polymeric membranes for gas separation, 4-sulfocalix[4]arene (SCA4), an organic macrocycle which possesses uniquely sized 3-dimensional (3D) open cavities, is for the first time infused into amidoxime-functionalized polymers of intrinsic micro-porosity (AOPIM) to create an interconnected porous network coupled with additional 3D channels that execute ultra-effective size discrimination against gas molecules. Owing to its susceptibility for hydrogen-bonding and electrostatic interactions, SCA4 is able to be steadily and evenly distributed within the AOPIM-1 polymers via a simple and scalable solvent exchange process. In addition, the organic nature of SCA4 allows it to possess naturally strong affinity for organic polymers, which enables the as-fabricated AOPIM-SCA4 membranes to completely preserve or even improve on their mechanical strength and stability. The unique size-sieving 3D cavities of SCA4 also contribute to a drastic enhancement on both the H_2/N_2 and H_2/CH_4 selectivity of the AOPIM-SCA4 membranes, while the H_2 permeability is only minimally compromised. After being infused with only 3 mol.% of SCA4, the AOPIM membrane demonstrates 4.6 and 10.6 times higher H_2/N_2 and H_2/CH_4 selectivity respectively than the pristine sample, with merely a 15.6% decrease in H_2 permeability. As a result, the overall H_2/N_2 and H_2/CH_4 separation performance of the AOPIM-SCA4 membranes can approach or even surpass the proposed 2015 upper bounds, and is comparable to or even better than many of the state-of-the-art PIM membranes and carbon molecular sieve (CMS) membranes. The separation performances of H_2/CO_2, O_2/N_2 and CO_2/CH_4 are also greatly improved after SCA4 infusion, and are all able to surpass the 2008 upper bounds. Meanwhile, the size-sieving effect of the infused SCA4 cavities becomes increasingly prominent as a function of time because the collapse of excess free volume during physical aging reduces the amount of less selective regions in the membrane. For an AOPIM membrane infused with 5 mol.% of SCA4 that has been aged for two months, there is a 76% and 39% increase in H_2/N_2 and H_2/CH_4 selectivity respectively, whereas the H_2 permeability drops only around 18.4%. As such, comparing with the fresh samples, the aged AOPIM-SCA4 membranes exhibit completely uncompromised or even further improved overall H_2/N_2 and H_2/CH_4 separation efficiency, demonstrating good performance stability with time. This simple yet effective approach offers an exciting new platform for enhancing the molecular-sieving nature of polymeric membranes, especially towards hydrogen purification applications, and to also draw more research attention to the utilization of organic macrocycles with gas-favorable cavity size in membrane-based gas separation.

568111 Super-Rigid, Intrinsically Microporous Ladder Polymers for Enhanced Gas Separation Performance and Physical Aging Resistance

High-free-volume glassy polymers such as PIM-1 have shown exceptional promise for advancing current gas separation membrane technology. However, high-free-volume polymers are often limited by severe declines in permeability over time due to reductions in their largely conformation-based free volume from the densification that occurs during physical aging. Here, we present an innovative macromolecular design of ladder-like copolymers that exploit the unique molecular architecture of pentiptycene as building blocks for non-collapsible, yet tunable, configuration-based free volume microporosity. The structural hierarchy of the pentiptycene unit imbues a more permanent, intrinsic free volume into the polymer for increased size sieving while providing a versatile framework that allows tunable ultra-microporosity. Monomers utilizing different shapes of the pentiptycene framework (i.e., S- or C-shaped) were synthesized with tailorable substituent groups and incorporated into copolymers with PIM-1 to study the effect of configurational free volume on gas transport. Achieved copolymers have shown remarkable potential as gas separation membranes, surpassing the 2008 upper bound for several gas pairs. For example, incorporation of 33 mol % of the S-shaped pentiptycene unit provided exceptional combinations of permeability and selectivity for O_2/N_2, with an O_2 permeability of 2370 Barrer and an O_2/N_2 selectivity of 4.6. Additionally, a thin film integrated with only 17 mol % of the C-shaped pentiptycene unit had a hydrogen permeability of 5900 Barrer with upper bound selectivities for typical hydrogen-based separations (i.e., $(H_2/CH_4) = 4.9$; $(H_2/N_2) = 7.7$). A preliminary aging study showed that the 17 mol % C-shape polymer had 30% increases in permeability across all gases after aging 120 days, while maintaining selectivity. This is in sharp contrast to existing ladder polymers like PIM-1 which show significant decreases in permeability over short periods of time due to physical aging. Permeation testing of CO_2/CH_4 in a 50:50 feed ratio was performed within the polymer series and showed surprising mixed-gas separation performance, with the 33 mol % S-shaped copolymer displaying a CO_2 permeability of 2700 Barrer with a CO_2/CH_4 selectivity of 16.6. In this talk, synthesis and characterization of these new ladder copolymers will be presented. Discussions will focus on elucidating the fundamental structure-property relationships within the polymer series.

562217 Microporosity Characterization in Ultra-High Free Volume Glassy Polymers

The development of innovative materials for gas separation membranes application is mainly focused on high free volume glassy polymers, due to their peculiar features that provide excellent permeability and separation performances [1]. Among the others, the new class of polymers with intrinsic microporosity (PIMs) has been extensively studied for a variety of molecular separations, counting on the large excess of free volume produced by the poor packing of the macromolecules, purposely designed as rigid and contorted. For this reason, PIMs show ultra-high gas permeabilities and interesting selectivity values for several gas pairs [2].

In view of their very high free volume, such systems are often treated as porous media, as if permanent pores are present in the glassy matrix, and their characterization typically include nitrogen or argon sorption at cryogenic temperatures (77 or 87 K, respectively). The data obtained are treated by the BET model, in the same way as for rigid solid adsorbents, aiming to obtain material information about the alleged pore size and distribution. However, this method has been questioned for the characterization of microporous materials, as the central assumptions of the BET theory are fail in the case of narrow pore size, approaching the nanometer scale. Several works revealed the inappropriateness of the BET method for the characterization of microporous media, such as MOFs, zeolites, and polymer networks [3].

It is also questionable if the same physical interpretation is applicable also for glassy polymers, even in the case they are endowed with a large excess of free volume or, in other words, if a porous structure with continuous solid wall can be envisaged in these materials. Interestingly, indeed, the experimental N_2 sorption at 77 K in several high free volume polymers (including PIMs) is characterized by the same features observed for conventional dense glassy polymers that represent the typical footprints of non-equilibrium systems. Large sorption/desorption hysteresis are obtained for most of the materials, as well as apparent effects given by sample previous history.

A different approach has been recently proposed for the description of cryogenic sorption data in glassy polymers [4], both conventional commodities (e.g. PS, PMMA or PVC) as well as high free volume systems (e.g. PTMSP, PMP or AF2400). To such aim, the non-equilibrium thermodynamics for glassy polymers (NET-GP) [5], suitable to evaluate the penetrant solubility in polymer phases, treated as macroscopically uniform and homogeneous phases out of their equilibrium conditions, is applied, with no need for any information about the size and the distribution of the existing "free volume elements".

The NET-GP approach provides a description of the thermodynamic behavior of penetrant/polymer mixtures below T_g, accounting for the polymer density as an additional state variable, to describe the non-equilibrium state of the glass. The model thus represents a proper extension of conventional equilibrium equation of state approaches to the non-equilibrium glassy states. In particular, the lattice fluid theory by Sanchez and Lacombe is considered, thus leading to the non-equilibrium (NELF) model.

The sorption of different penetrants, namely N_2, Ar, and CO_2 in cryogenic conditions in a series of high free volume glassy polymers, and in PIMs in particular, has been analyzed and described by the NELF approach. The case of PIM-1 has been inspected more in detail, together with a series of novel PIMs recently developed (e.g. PIM-EA-TP, PIm-6FDA-OH, or PIM-Trip-TB). The model provides a very effective and accurate representation of the experimental isotherms, in wide ranges of temperature and penetrant activity, identifying a significant volume swelling of the polymer matrix associated to gas sorption, very relevant in case of large penetrant uptakes [6]. The model allows the evaluation of the gas solubility in the same glassy polymer up to room temperature or above, with no need of any additional parameters and in a purely predictive fashion. Therefore, the analysis suggests that the same mechanism is driving the sorption process in the entire broad temperature range, from cryogenic to above room temperature.

Interestingly, two main regions can also be identified in the cryogenic isotherm, the first one, in the low activity range, is associated to an iso-volumic sorption in the excess of free volume, while the second, located in the central portion of the isotherm, indicates that the physical dissolution is accompanied by a significant swelling of the polymer. These peculiar features can be associated to specific material properties of the glassy polymers examined, such as the fractional free volume and the tendency to plasticization of the membrane.

Based on these findings, an alternative use of the sorption isotherms in cryogenic conditions is proposed, which may serve as a more appropriate metric for comparison between different glassy polymers.

References

[1] N.B. McKeown, P.M. Budd, K.J. Msayib, B.S. Ghanem, H.J. Kingston, C.E. Tattershall, S. Makhseed, K.J. Reynolds, D. Fritsch, Polymers of intrinsic microporosity (PIMs): bridging the void between microporous and polymeric materials, Chemistry 11 (2005) 2610–2020.

[2] Y. Wang, X. Ma, B.S. Ghanem, F. Alghunaimi, I. Pinnau, Y. Han, Polymers of intrinsic microporosity for energy-intensive membrane-based gas separations, Materials Today Nano 3 (2018) 69-95.

[3] M.F. de Lange, L.-C. Lin, J. Gascon, T.J.H. Vlugt, F. Kapteijn, Assessing the surface area of porous solids: limitations, probe molecules, and methods, Langmuir 32 (2016) 12664–12675.

[4] M. Minelli, D.R. Paul, G.C. Sarti, On the interpretation of cryogenic sorption isotherms in glassy polymers, J. Membr. Sci. 540 (2017) 229–242.

[5] M.G. De Angelis, G.C. Sarti, Solubility of gases and liquids in glassy polymers, Annu. Rev. Chem. Biomol. Eng. 2 (2011) 97–120.

[6] M. Minelli, B.R. Pimentel, M.L. Jue, R.P. Lively, G.C. Sarti, Analysis and utilization of cryogenic sorption isotherms for high free volume glassy polymers, Polymer 170 (2019) 157–167.

564135 Modeling of Organic Fouling Based on Interaction between Membrane and Foulants for Reverse Osmosis Membrane

Reverse osmosis (RO) technology has progressed steadily over the last few decades and is now the leading technology in the desalination industry. Those gains were achieved through improvements in both RO membrane element performance and energy recovery technologies. For further energy saving and RO plant cost reduction, continuous improvement of membrane element performance and process efficiency are needed. A major problem in RO plant operation is fouling on membranes or spacers that increase specific energy consumption and operating cost. Therefore, to evaluate accurately the effect of developing a new membrane element or a new process, RO plant cost must be estimated with consideration of chronological changes caused by fouling.

Several fouling models have been proposed for membrane separation so far, but most of them rely on parameters that were obtained from filtration experiments or field plant operations, as opposed to fouling mechanisms. These models with experimental or empirical parameters are useful to estimate the chronological changes of water permeate flux or resistance of foulant layers when the parameters of fouling model are known for the specific combination of membrane, element, feed water composition, and other operating conditions. However, for the estimation of gains from new membranes, new elements, or new processes, additional filtration experiments or plant tests are needed. Therefore, it is desirable for the estimation of gains from new membranes, new elements, or new processes to have a fouling model that is less dependent upon experimental or empirical parameters.

This study is focused on the modeling of organic fouling, which is the main cause of RO fouling. In this study, the organic fouling rate was assumed to be sum of the attachment rate of foulant on the membrane surface, the detachment rate from membrane surface, and the deposition rate of foulant due to the presence of obstacles or insufficient cross-flow velocity. For the estimation of the attachment rate, the attachment probability is determined from interactions between the membrane and foulants or among the foulants. The detachment rate is modeled to caused by rolling of attached foulants by cross-flow velocity, and the rolling rate and the detachment probability were estimated. The deposition rate was estimated from the residence time of foulants that were not attached to the membrane surface. The chronological flux transition estimated from a fouling model proposed in this study was compared with the experimental results which use sodium alginate and bovine serum albumin (BSA) as model foulants for polysaccharide and protein, respectively.

566363 Trade-Off between Selectivities and Permeabilities in Planar Anisotropic Membranes for Gas Separations

To date the design of membranes for gas separations has mainly relied on the use of isotropic materials. In isotropic membranes separation occurs as a result of differences in flux magnitude of the species involved. Recently, we have introduced an alternative approach for the design of membrane materials by considering engineered anisotropic membranes. In anisotropic membranes separation occurs as a result of controlling both the magnitude and the direction of flux. A fundamental physical consequence of controlling the flux direction is the development of spatially dependent permeabilities and selectivities. In this talk we discuss the capabilities and performance advantages of anisotropic membranes in a simple planar configuration. We show how anisotropic materials can be used to improve the separation ability of current membranes, pushing the limits of existing technologies. We discuss the possibility of increasing the selectivity of membrane separations via flux directional control and also the existence of a trade-off between selectivity, permeability and the collected fraction of the permeate. Our findings on anisotropic planar membranes are applied to the separation of binary mixtures of O_2/N_2 and H_2/CH_4.

566109 Rigorous Process Simulation Using Reverse Osmosis (RO) Membrane Unit Operation for Water Desalination

OLI is the recognized leading provider of simulation software for electrolyte chemical systems, as well as for the prediction of corrosion. This USA-based company has served the chemical process and oilfield industries for over 48 years. OLI technology specializes in the rigorous prediction of multi-phase, multi-component chemical processes. Its temperature and pressure ranges are extreme (-60°C and 4000 bar), and the breadth of chemistry covered include 80+ elements, aqueous and non-aqueous organics, and their behavior in water, vapor, organic, solid, and interfacial environments. The combination of fundamental thermodynamics, property models, numerical algorithms, and databanks enable this unique capability. OLI has multiple thermodynamic models to cover this chemistry and phenomena.

This paper will present a novel scientific approach for process engineers to use while simulating and predicting the behavior of reverse osmosis membranes under various conditions. This approach is based on first-principles models to predict membrane permeability using simultaneous electrolyte thermodynamic models, species-transport models, chemical speciation, water coordination and ion hydration (and their respective radii). The paper will consist of three sections. In the first section, we will discuss the electrolyte modeling capabilities in process simulation. In the second section, we will discuss the theory behind RO membrane unit operation block. The third section will consist of a case study using the said RO membrane unit operation block.

This paper will illustrate the OLI Mixed Solvent Electrolyte thermodynamic model. Sample chemical systems will be potash, sulfuric acid and lithium double salts.

To model the transport mechanism inside the RO membrane, a solution-diffusion approach has been incorporated. In the solution-diffusion model, it is proposed that transfer of ions and water through polymeric membranes occurs via a solution diffusion mechanism because of dissolution of permeates in the membrane materials.

In the example case study, the simulation of a desalination plant located in Egypt will be considered with the goal of producing potable water with a TDS (total dissolved solids) less than 100 ppm reduced from a brackish feed water of over 10,000 ppm TDS.

573622 A Novel Methodology to Correctly Compare the Energy Efficiency between Membranes and Distillation

Among the separation technologies portfolio, distillation continues to be the preferable method for majority of the separations in the Chemical and Petrochemical plants. Just in the U.S. alone, it is estimated that about 90-95 % of all separations in chemical plants and refineries are performed by means of distillation columns [1]. Although distillation is widely used, it is frequently perceived to have a low energy efficiency, whereas in contrast, membrane separations are classified as a highly energy efficient processes which could potentially replace distillations on a wide scale [2], [3] [4].

However, this conclusion is not well supported since it has been mainly derived from comparisons made only with respect to distillation operated with heat, ignoring other operation modes that could be more energy efficient. Apparently, there is a widespread confusion in considering distillation as an absolute heat driven device, which sometimes requires a high amount of thermal energy input at the reboiler. The comparison between work driven membranes and heat driven distillations often leads to inaccurate conclusions. In reality, it is feasible to operate distillation solely with work input without requiring any external heat [5], and for many cases, this operating mode drives into an important improvement of the energy efficiency.

The current research work centers on the development of a consistent procedure to compare the energy performance between membranes and distillation. In this methodology, membranes and distillation are compared in terms of the same energy basis (electrical work). In addition, both separation technologies are optimized as an attempt to evaluate them at their best operating conditions. The proposed methodology is applied to the analysis of two important separation cases; p-xylene/o-xylene and propylene/propane.

Bibliography

[1] U.S. Department of Energy, "Materials Research for Separation Technologies: Energy and Emission Reduction Opportunities," 2005.
[2] D. S. Sholl and R. P. Lively, "Seven chemical separations to change the world," Nature, vol. 532, pp. 435-437, 2016.
[3] J. W. Koros, "Evolving Beyond the Thermal Age of Separation Processes: Membranes Can Lead the Way," AIChE Journal, vol. 50, no. 10, pp. 2326-2334, 2004.
[4] R. P. Lively and D. S. Sholl, "From water to organics in membrane separations," Nature Materials, vol. 16, pp. 276-279, 2017.
[5] A. A. Shenvi, D. M. Herron and R. Agrawal, "Energy Efficiency Limitations of the Conventional Heat Integrated Distillation Column (HIDiC) Configuration for Binary Distillation," Ind. Eng. Chem. Res., vol. 50, pp. 119-130, 2011.

572728 Chemically Dependent Transport of Small Polar Solutes in a Cross-Linked Hii Phase Lyotropic Liquid Crystal Membrane

The pores that constitute cross-linked lyotropic liquid crystal (LLC) membranes are chemically complex and uniform in size which makes them promising for selective separations. We have developed an atomistic molecular model of an LLC membrane that is maximally consistent with experimental X-ray diffraction for the purpose of studying small molecule transport within it.

We have characterized transport of water, sodium ions and 20 small polar solutes within the pores of our model. We found that the transport rate of a species is dependent not only on molecular size, but on chemical functionality as well, leading to anomalous diffusive behavior. Electrostatic interactions between the membrane and solutes provide an interesting diversity of mechanistic behaviors. In general, all solutes perform intermittent hops between lengthy periods of entrapment. Three different trapping mechanisms are responsible for this behavior. First, solutes that drift out of the pore can become entangled among the dense monomer tails. Second, solutes can donate hydrogen bonds to the monomer head groups. Third, solutes can coordinate with sodium counter ions. The degree to which a solute is affected by each mechanism is dependent on the chemical functionality of the solute.

We have used our detailed mechanistic understanding in order to construct a stochastic model which we can use to project the long term behavior of the solutes and ultimately predict selectivity. Using the insights developed in this study, we can begin to think about how to redesign existing LLC membranes in order to perform solute-specific separations.

578386 Process Simulation-Based Comparative Study on Selecting Membrane Distillation Configuration

Membrane distillation is an emerging thermally-driven membrane technology proposed for the separation of salts and other non-volatile compounds from wastewater streams. The process offers a unique solution for the treatment of water streams with high total dissolved solids. There are four configurations proposed for MD applications: Direct Contact MD, Vacuum MD, Sweeping Gas MD, and Air Gap MD. Each of these configurations has its own advantages and disadvantages, and none of them has been commercialized, yet. Studying some of these MD configurations (e.g. DCMD) have been limited to laboratory-scale, while some (e.g. SGMD) have been investigated for pilot scale applications. Comparative analysis of these configurations has been hindered by lack of infrastructure and technology limitations. Here we present a comprehensive analysis of MD configurations based on process-level Aspen simulations. Four MD configurations has been simulated using Aspen Custom Modeler. Several performance metrics such as energy efficiency, capital and operating costs, flux delivery, and potential heat integration technologies are investigated, and results will be presented. We will report the energy requirements of each configuration to treat one cubic meter of water ($\$/m^3$).

AUTHOR INDEX

AUTHOR INDEX

AUTHOR INDEX

AUTHOR INDEX

AUTHOR INDEX

AUTHOR INDEX

AUTHOR INDEX

AUTHOR INDEX

AUTHOR INDEX

AUTHOR INDEX

AUTHOR INDEX

AUTHOR INDEX

AUTHOR INDEX